APPLICATIONS OF RADIO-FREQUENCY POWER TO PLASMAS

SEVENTH TOPICAL CONFERENCE

SC

AIP CONFERENCE PROCEEDINGS 159

RITA G. LERNER
SERIES EDITOR

APPLICATIONS OF RADIO-FREQUENCY POWER TO PLASMAS

SEVENTH TOPICAL CONFERENCE

KISSIMMEE, FL 1987

EDITORS:
S. BERNABEI & R. W. MOTLEY
PLASMA PHYSICS LABORATORY
PRINCETON UNIVERSITY

AMERICAN INSTITUTE OF PHYSICS NEW YORK 1987

L.C. Catalog Card No. 87-71812
ISBN 0-88318-359-5
DOE CONF-870570

Printed in the United States of America

FOREWORD

The Seventh APS Topical Conference on Applications of Radio-Frequency Power to Plasmas was held in Kissimmee, Florida on May 4–6, 1987 under the auspices of the Princeton Plasma Physics Laboratory and was sponsored by the American Physical Society and the U.S. Department of Energy. The Conference was formerly called RF plasma heating. It seemed appropriate to change the name since the Conference was not only extended to applications other than heating—such as current drive, MHD stabilization, impurity control, etc.—but was also opened to fields other than fusion—i.e., ionospheric plasmas, material plasma etching, etc. Another change was in the nomination of an international rather than national Program Committee. This was partly due to recognition of the increased emphasis that foreign programs place on RF methods, and partly to the desire to have the broadest possible discussion of RF applications.

The Program Committee, chaired by R. W. Motley, consisted of D. Batchelor, F. DeMarco, J. Jacquinot, F. Leuterer, S. Luckhardt, R. Prater, and T. Watari. Of the 129 participants, 104 were from the U.S. and Canada, 18 from the European community, and 7 from Japan; there were 110 papers presented, including 9 invited papers.

The Conference spanned two and one half days and was followed by a "Science Court on ICRH Models of Toroidal Plasmas" organized by L. Hively and W. Sadowski of the Department of Energy.

I would like to thank Mrs. Carol Phillips, who provided counseling. I would also like to especially thank and give credit to Mrs. Phyllis Schwarz, who handled all the arrangements and mailings and prepared the proceedings.

Stefano Bernabei
Chairman

CONTENTS

CHAPTER 1
ELECTRON CYCLOTRON RANGE OF FREQUENCIES

CHAPTER 2
LOWER HYBRID RANGE OF FREQUENCIES

CHAPTER 3
ION CYCLOTRON RANGE OF FREQUENCIES

Chapter 4
ALFVÉN WAVES

Chapter 5
GENERAL THEORY OF RF WAVES AND NON-FUSION APPLICATIONS OF RF POWER

A PROGRESS REPORT ON ECR HEATING AND CURRENT DRIVE IN TOKAMAKS

A C Riviere

Culham Laboratory, Abingdon, Oxon, OX14 3DB, UK
(Euratom/UKAEA Fusion Association)

ABSTRACT

Plasma heating at the electron cyclotron resonance and its harmonics has provided a range of important results in tokamaks, stellarators and mirror machines although only tokamaks are considered here. The ability to heat the electrons locally has enabled T_e profile control, m=1 and m=2 mode stabilisation and heat transport to be studied. At high powers the energy confinement degrades in an L-mode like manner in some cases. At very low and at high densities however confinement improves with heating. Current drive has been observed in the WT-2, WT-3 and CLEO tokamaks. In this paper the present state of ECRH, the advances since the last review [1] and the current drive results from CLEO are reported.

INTRODUCTION

Electron cyclotron resonance heating (ECRH) has been demonstrated at power levels up to 2 MW [2] leading to central electron temperatures $T_e(o) > 4$ keV. At very high power densities β_p values close to theoretical limits have been achieved [3,4]. A number of other interesting effects have been observed including mode stabilisation and ECR current drive. However, there are several questions which remain unanswered.

ECRH PHYSICS

Cyclotron damping occurs where the resonance condition

$$\ell\omega_c = \gamma\, \omega_{rf}\, \left(1 - n(v_{\parallel}/c) \cos\theta\right) \qquad (1)$$

is satisfied. Here $\omega_c = eB/m$, ℓ is the harmonic number, $\gamma = (1-v^2/c^2)^{-\frac{1}{2}}$, n is the refractive index and θ is the angle between the wave vector k and the magnetic field B. For major radii R > 1m, electron temperatures $T_e > 1$ keV and $\ell = 1$ and 2 the plasma is essentially a black body for a wide range of $v_{\parallel}/c$, θ and electron density, n_e. The relativistic and Doppler shifts in (1) allow the possibility of absorption at values of B far from the resonance value given by $B = m\,\omega_{rf}/e\ell$. Strong absorption at substantially down-shifted frequencies is predicted [5] in plasmas with T_e of a few keV.

WIDTH OF DEPOSITION PROFILE

For the O-mode at $\omega_{rf} = \omega_c$, perpendicular propagation and optical depth $\tau > 3$ the width of the absorption layer for a narrow beam is independent of tokamak size or electron temperature. At the optimum density

$$\Delta X \approx 2.2\ B^{-1}(T)\ \text{cm} \qquad (2)$$

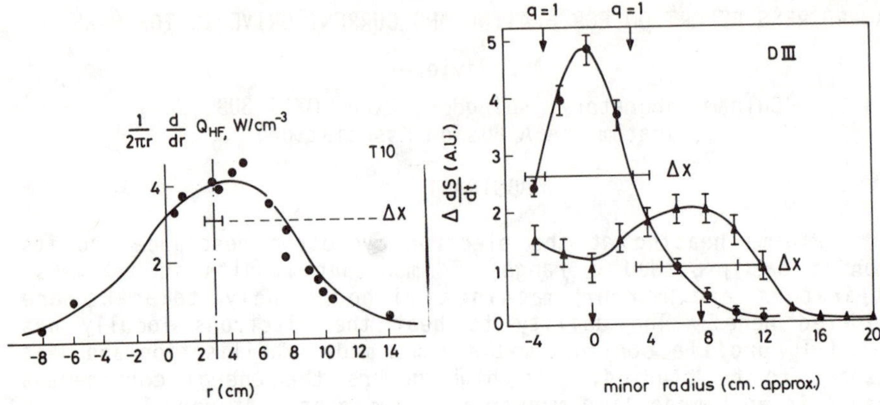

Fig.1 Power absorption profile in T10 [6] *deduced from ECE signal. (P=0.74 MW, I=0.24 MA, $\bar{n}_e$= 1.6x10^{19}m^{-3}, a=0.34 m, LFS 90°).*

Fig.2 Power absorption profiles in DIII [7] *for two resonance positions deduced from SXR signal. (P=0.65 MW, I=0.5 MA, $\bar{n}_e$=3.3-4.7x10^{19}m^{-3}, a=0.43m, HFS 60°).*

For finite $k_\parallel$, ΔX can be an order of magnitude larger than (2). The absorption line width observed in T10 [6] derived from the EC emission at $2\omega_c$ is shown in Figure 1. Of the launched power some 70 per cent was in the O-mode but the line width is substantially larger than (2) would suggest. Finite antenna pattern width and wall scattering of the X-mode component may be responsible for the broader line. Non-linear saturation of the absorption coefficient would have the same effect. The line width in DIII was measured for two resonance positions in terms of the change in the slope of the SXR diode signal [7]. These are shown in Figure 2 and are in reasonable agreement with expected doppler widths.

CONFINEMENT AS A FUNCTION OF POWER AND DENSITY

There are three density regimes for the effect of ECRH on energy confinement time (τ_E). At low densities ($n_e < 10^{19}m^{-3}$) in CLEO and TOSCA [8] a dramatic improvement in τ_E occurs coincident with an increasing distortion of the electron energy distribution. The reason for the improved confinement is not known although a significant population of hot electrons may have a beneficial effect on plasma stability [9]. At intermediate densities, where the energy distribution is close to Maxwellian, τ_E degrades in an L-mode like manner as illustrated by the results for central heating in T10 [2] (Figure 3). However with the resonance exactly on axis a very narrow peak in T_e has been observed in the centre of the plasma in PDX [10] and TFR [11] (Figures 4 and 5). In the PDX case the Thomson scattering measurement (TVTS) was taken at the top of the sawtooth and the rf power was just 75 kW. These observations suggest that at least close to the magnetic axis, the energy confinement is not significantly degraded by the heating and that very steep electron temperature gradients can be supported. At high densities ($\bar{n}_e > 5.5 \times 10^{19}m^{-3}$) in DIII [12] an improvement in τ_E over the value in the OH plasma is

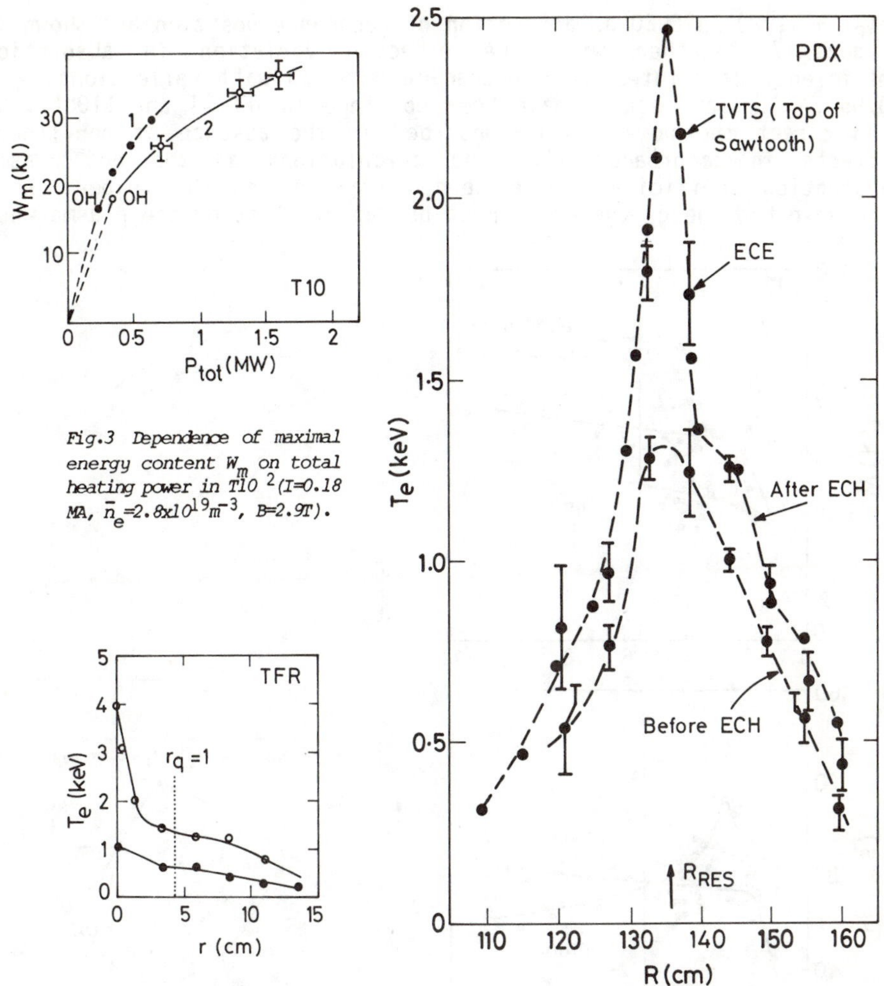

Fig.3 Dependence of maximal energy content W_m on total heating power in T10 [2] (I=0.18 MA, $\bar{n}_e$=2.8x10^{19}m^{-3}, B=2.9T).

Fig.4 T_e profiles in PDX [10] before and after the start of ECRH with resonance on axis. (P=75 kW, I=0.26 MA, $n_e(0)$=1.2x10^{19}m^{-3}, a=0.4 m, LFS 90°).

Fig.5 T_e profiles in TFR [11] for ohmic heating (●) and ECRH on axis (o). ($\bar{n}_e$=0.8x10^{19}m^{-3}, LFS 90°).

seen when the heating occurs outside r/a = 0.7. The expanded boundary divertor configuration was used and rf power was launched from the high field side. The improvement in τ_E (Figure 6) occurred at densities above the cut-off value and this prevented the wave reaching the centre of the plasma. Measurement of the time for the heat to diffuse through the plasma [12] confirmed that the heating occurred at the plasma edge and not in the centre.

VARIATION OF RESONANCE POSITION

The heating location can be moved across the plasma by varying the toroidal field. Results for the variation in ΔW(J) and

$\Delta(\beta_p + \ell_i/2)$ in CLEO as a function of resonance position are shown in Figure 7 together with the expected variation in absorption efficiency calculated by ray tracing with 15 wall reflections (η = 0.95). Similar results have been obtained in TFR [11] and T10 [13]. We can expect the power to be absorbed in the absence of non-linear effects in accordance with the calculations as the hot plasma absorption coefficients have been confirmed in other experiments. The absorbed energy therefore must be rapidly lost by the plasma when

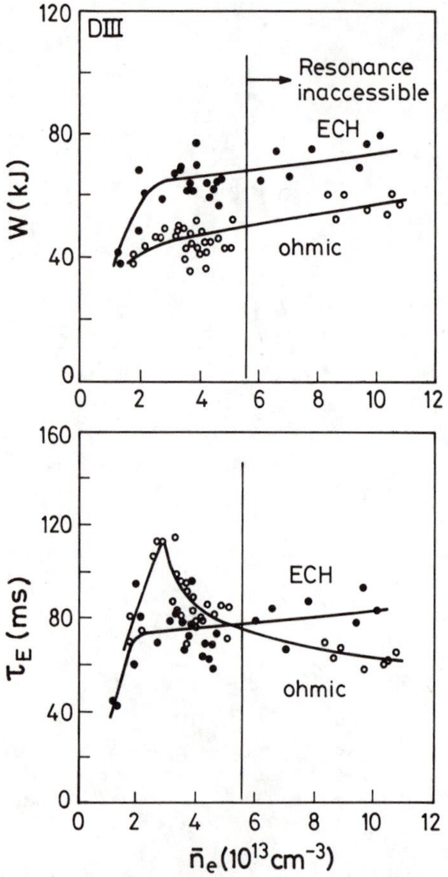

Fig.6 Total plasma energy W and gross energy confinement time τ_E for ohmic heating and ECRH in DIII [12] as a function of $\bar{n}_e$. (P=0.4 to 0.7 MW, I=0.5 MA, B=2.15T, HFS 60°).

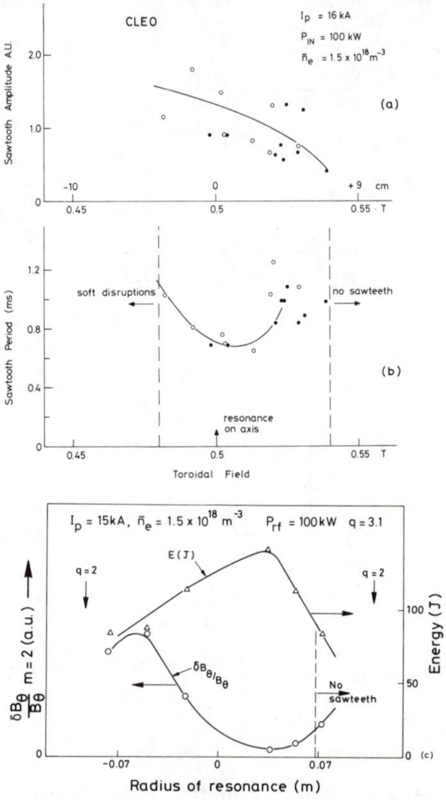

Fig.8 (a) Sawtooth amplitude (b) sawtooth period [8] and (c) stored energy E(J) and m=2 mode amplitude[18] in CLEO as a function of resonance position.

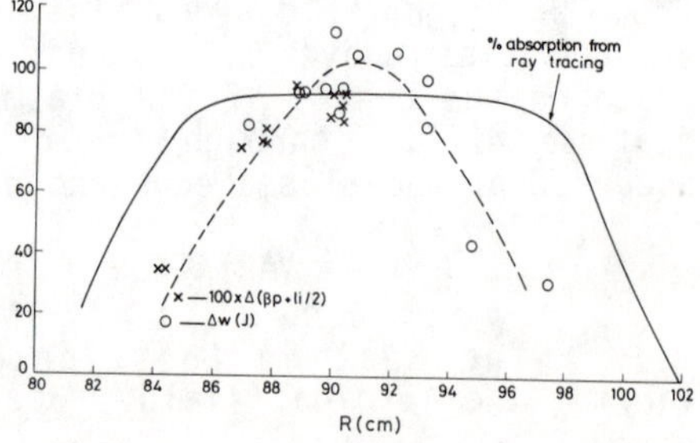

Fig.7 Dependence of ΔW(o) and $\Delta(\beta_p+\ell i/2)$(x) in CLEO on resonance position with ray tracing result for % absorption (solid curve) (P < 160 kW, I=20 kA, $\bar{n}_e$=0.4x10^{19}m^{-3}, LFS 90° 2ω_{ce}).

the resonance is in or beyond the region where $1 < q < 2$. This differs from the results at high density in DIII discussed above where an improvement in confinement was observed when the power was deposited outside $r/a = 0.7$. It may be that in the DIII case the 'turnover' in τ_E from the usual OH linear dependence on $\bar{n}_e$ is being recovered by the edge heating.

DENSITY DROP

A feature of ECRH is the decrease in $\bar{n}_e$ at the start of the rf heating pulse, which typically can be 30 to 50 per cent in a time $\tau \lesssim \tau_E$. The effect is pronounced under clean torus conditions but can be masked by enhanced recycling. On CLEO, where strong titanium gettering was used, ohmic discharges with $n_e > n_{cut-off}$ had their density reduced to below $n_{cut-off}$ [14]. The CLEO results are echoed by studies in THOR [15] where with central heating the density falls to about 0.5 $n_{cut-off}$ independent of the target density between about 0.3 and 1.5 times $n_{cut-off}$. Results from T10 [2] and elsewhere show that the density profile broadens and particles are lost during the density drop. For off-axis heating the profile becomes more peaked temporarily [2]. The increase in T_e and consequent fall in V_L reduces the neoclassical pinch but 1-D code modelling shows that this is not sufficient to account for the observed density changes. The rf wave may be affecting the transport directly [16] and of course changes in β_e or T_e during the heating may be affecting the underlying transport processes. Certainly the mechanism causing the enhanced diffusion is not yet understood.

MODE STABILISATION AND PROFILE CONTROL

It has been shown in CLEO [8], T10 [17], DIII [12] and TFR[11] that the sawtooth amplitude and frequency can be altered by ECRH. Heating on- or off-axis can enhance or reduce the sawtooth amplitude respectively. Thus in T10 an outward shift of r_{res} by $\sim$ 13 cm effectively suppressed the sawteeth for the duration of the heating. Results on control of sawteeth in CLEO are shown in Figure 8, [8, 18] for a moderate β_p, $q_a = 3$ discharge. The sawtooth amplitude and period, the stored energy and the amplitude of the m = 2 oscillation in B_θ are shown as a function of resonance position. Moving the resonance outwards to beyond +7cm stabilised the sawtooth. However, when the resonance was displaced far enough on the inside soft global disruptions occurred as the m=2 mode activity became too strong. Profile broadening presumably led to a too steep q gradient near the edge. By placing the ECR heating zone at slightly larger radii than the q=2 surface the m = 2 mode has been stabilised in T10 [19], CLEO [20] and TFR [21]. In the case of experiments on TFR, stabilisation is observed when the resonance is located at rather than outside the q = 2 surface [22]. The T10 results show that when the resonance is on the high field side ECRH stabilises the mode only over a narrow range of r_{res} in the neighbourhood of the q=2 surface. Insufficient data is available on the profile changes around the q=2 surface for changes in mode stability to be calculated although if the plasma is at a marginally stable point only small changes in profile may be required. Calculations have shown that quite small changes in plasma current ($\delta I/I \sim 3\%$) are all that are required for JET [23-24]. The off-axis heating in T10 referred to above produces some change in the T_e profile but recent results [2] show that the change is not permanent, the profile relaxes to a shape similar to that for on-axis heating. Strong heating off-axis in CLEO has also had only a small effect on the SXR profile [20]. This tokamak behaviour contrasts strongly with that in the WVIIA stellarator where quite dramatic T_e profile changes were observed [25] between on- and off-axis heating.

ELECTRON CYCLOTRON CURRENT DRIVE

Experiments which have studied current drive with ECRH are listed in Table 1 where the efficiency is defined as

$$\eta_{ECCD} = I_p(kA)\bar{n}_{20}(m^{-3})\,R(m)\,P_{RF}^{-1}(kW)$$

Table 1 ECCD Experiments

Machine	freq (GHz)	resonance	power (kW)	η_{ECCD} expt	η_{ECCD} theory	η_{LHCD} expt
TOSCA	28	$2\omega_c$	180	1.8×10^{-4}		
CLEO	28	$2\omega_c$	160	~ 0		
CLEO	60	$2\omega_c$	185	10^{-3}	3.10^{-3}	
TFR	60	ω_c	600	~ 0		
WT-2	35.6	ω_c	60	10^{-4}	10^{-2}	6×10^{-3}
WT-3	56	ω_c	120	10^{-3}		1.5×10^{-2}

Fig.9 Loop voltage V_L, plasma current Ip and electron density $\bar{n}_e$ in WT-2 [26] as a function of time during ECCD experiment; full curves with and dotted curves without ECCD. (P=50 kW, a=9cm, B=1.2T, LFS 48°).

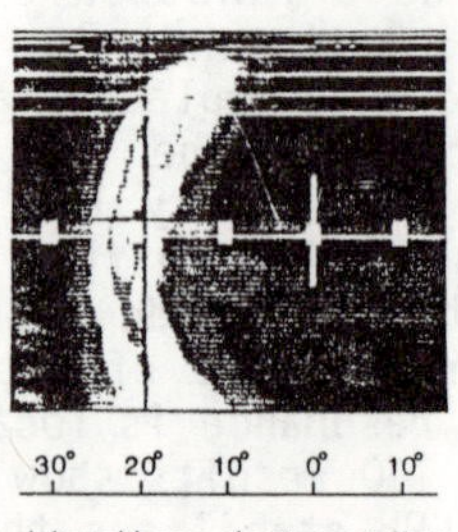

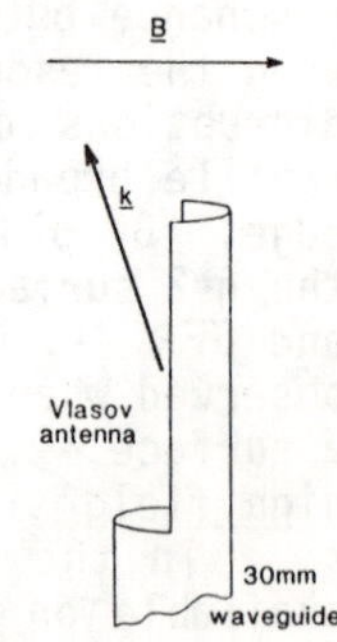

Fig.10 Far-field infra-red image of antenna pattern used for current drive experiments in CLEO [27].

In the WT-2 and WT-3 tokamaks [26] plasma currents of 3 and 10 kA respectively were totally sustained by ECCD (V_L = 0) as shown in Figure 9 for WT-2. The line average electron density was < $10^{18}m^{-3}$ and in WT-2 for example the plasma consisted of cold and hot components with T_e = 70 eV and 15 keV respectively, the bulk of the energy being in the hot component. Current drive experiments in CLEO [27] were carried out at $\omega_{rf} = 2\omega_c$, at 28 and 60 GHz. Attempts to observe ECCD at 28 GHz proved unsuccessful, a result which is not yet understood. In the 60 GHz experiments the antenna launched a single crescent-shaped pattern, at about 70° to the magnetic field (Figure 10). Experiments were carried out at $\bar{n}_e \sim 3.5 \times 10^{18}m^{-3}$ and with essentially Maxwellian central temperatures of 1.3 keV at I_p = 10 kA and 1.5 keV at I_p = 15 kA as deduced from SXR spectra. Reversing the servo-controlled plasma current or the antenna direction produced a consistent difference in loop voltage (ΔV_L). The theoretical assumption that the ECCD has the same efficiency for each current direction allows its magnitude to be unfolded from ΔV_L, with the result shown in Fig 11. The deduced RF current increases with plasma current from 2.5 kA at 5 kA to 5 kA at 15 kA. Theoretical calculations of the RF driven current were made, based on a current drive model allowing for trapping effects and including the weakly relativistic resonance condition [27]. With the resonance in the hot core the driven current profile has a half width of about 3 cm, with an efficiency of typically 60 A/kW. This leads to predicted currents approximately three times the observed values as shown in Figure 11. The bootstrap current is expected to flow in the same sense as the total plasma current, and for $\beta_p \gtrsim 2$, $\nu_e^* \sim 0.1$ and with the inferred profiles in CLEO it should be ~ 6 kA with a hollow distribution. It must be much smaller in the experiment as the average decrease in loop voltage corresponds closely to the rise in temperature if the resistivity is neoclassical.

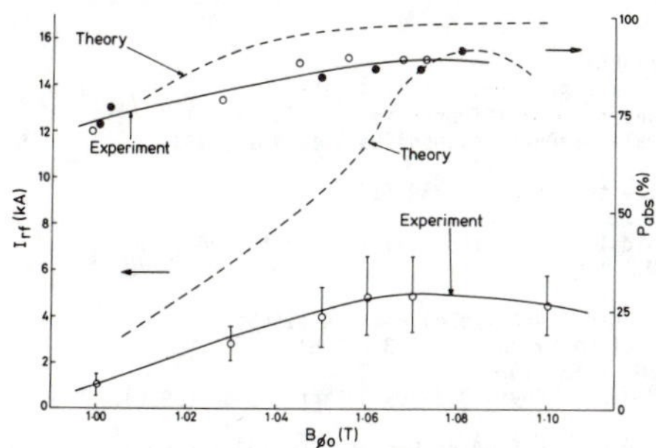

Fig.11 Deduced rf driven current in CLEO [27] as a function of toroidal field compared with theory and the observed % absorption compared with theory as a function of B, for I = 10 kA, P = 185 kW. Open and closed circles correspond to clockwise and counter clockwise plasma current respectively for the % absorption.

SYNERGISTIC EFFECTS WITH LHCD

An enhancement of LHCD efficiency when ECRH is applied has been observed in several experiments [1]. In WT-2 [28] with $\omega_{rf} \sim \omega_c$ the addition of ECRH to LHCD has resulted in a ramp-up in the plasma current. The strongest effect is observed with the ECR layer at the low field side of the plasma indicating down shifted resonance with hot electrons in the centre of the plasma. After switching off the

ECRH the current level remains higher than before ECRH was applied. When heating the bulk electrons, the current is reduced. A theoretical treatment of the combination of LHCD and ECCD (using top launch of the X-mode) suggests that the efficiency of both is increased by the action of the other [29].

TRAPPED ELECTRON EFFECTS

Cyclotron damping leads to an increase in $v_\perp/v_\parallel$ and therefore a localisation of hot electrons on the outer side of the plasma for appropriate resonance positions. This effect has clearly been seen as a local peak in the SXR emission profile during ECRH in TOSCA and T10 for example. An asymmetry in the variation of V_L and $\Delta W(J)$ with resonance position has been observed in CLEO which is consistent with the effect of trapped electrons on parallel resistivity. The decay time of the hot trapped electrons is classical and this has provided some insight into their contribution or otherwise to anomalous transport [30].

IN SUMMARY AND THE FUTURE

ECRH has been used successfully to perturb the electron component, bulk heat the electrons, stabilise MHD activity and drive current in tokamaks. Experiments at 60 GHz in DITE, COMPASS, TEXT, and DIII, 84 GHz in T10, 120 GHz in DIII, 150 GHz in FT-U and up to 300 GHz (using an FEL as source) in MTX are active or in preparation and will provide a strong programme for ECRH in the next few years.

ACKNOWLEDGEMENTS

The author gratefully acknowledges the work of M O'Brien for the ray tracing calculations and of B Lloyd for the CLEO confinement and current drive contributions.

1. A C Riviere, Plasma Phys and Contr Fus, 28, 1263 (1986)
2. V V Alikaev et al, Plasma Phys and Contr Nucl Fus Res (IAEA Kyoto) A-II-4 (1986) .
3. M W Alcock et al, Plasma Phys and Contr Nucl Fus Res (IAEA Baltimore) Vol II, 51 (1983)
4. M W Alcock et al, 11th Euro Conf on Contr Fus and Plasma Phys (Aachen) Part I, 401 (1983)
5. I Fidone et al, Phys Fluids, 29, 803 (1986)
6. V V Alikaev and V S Strelkov, Proc ICPP Lausanne,Invited Talks I, 259 (1984)
7. R Prater, APS Bulletin 29 1259 (1984)
8. A C Riviere et al, 4th Int Symp on Heating in Toroidal Plasmas (Rome) Vol II, 795 (1984)
9. M N Rosenbluth et al, Phys Rev Lett, 51, 1967 (1983)
10. A Cavallo et al, Nuc Fus 25, 325 (1985)
11. H P L de Esch et al, Plasma Phys and Contr Nucl Fus Res (IAEA Kyoto) F-III-4 (1986) .
12. R Prater et al, Plasma Phys and Contr Nucl Fus Res (IAEA Kyoto) F-III-3 (1986).
13. Yu V Esiptchuk et al, Plasma Phys and Contr Fus, 28, 1253 (1986)
14. T Edlington et al, 12th Euro Conf on Contr Fus and Plasma Phys (Budapest) Part II, 80 (1985).
15. A Airoldi et al,THOR report FP86/13 (1986)
16. Riyopoulos et al, University of Texas Institute for Fusion Studies Report IFSR#184-R (1985)
17. T-10 Group, 11th Euro Conf on Contr Fus and Plasma Phys (Aachen) Part I, 289 (1983)
18. D C Robinson et al, Proc of Conf on Phys Proc in Toroidal Confinement (Varenna) Vol II, 511 (1985)
19. V V Alikaev and K A Razumova . Wkshp App of RF Waves to Tokamak Plasmas (Varenna) Vol I, 377(1985)
20. T Edlington et al CLEO team private communication (1986)
21. TFR and FOM Groups 5th Int Wkshp on ECE and Elec Cyc Htg, GA Tech, San Diego) EC-5, 154 (1985)
22. TFR and FOM Groups, 13th Euro Conf on Contr Fus and Plasma Heat (Schliersee) Part II, 207 (1986).
23. M O'Brien, et al 13th Euro Conf on Contr Fus and Plasma Heat (Schliersee)Part II, 270 (1986).
24. T Hender et al, Plasma Phys and Contr Nucl Fus Res (IAEA Kyoto) A-V-3 (1986).
25. V Erckmann et al, Plasma Phys and Contr Nucl Fus Res (IAEA London) Vol III, 419 (1984).
26. S Tanaka et al, Plasma Phys and Contr Nucl Fus Res (IAEA Kyoto) F-II-6 (1986).
27. D C Robinson et al, Plasma Phys and Cont Nucl Fus Res (IAEA Kyoto) F-III-2 (1986).
28. T Maekawa et al, Nucl Fus, 23, 242 (1983)
29. I Fidone et al, Report No EUR-CEA-FC-1310 (1986)
30. D C Robinson, Proc of Workshop on Turbulence and Transport (Corsica) 21 (1986)

ELECTRON CYCLOTRON HEATING USING THE FUNDAMENTAL EXTRAORDINARY MODE LAUNCHED FROM THE LOW FIELD SIDE ON DIII–D.*

R. Prater, S. Ejima, R.W. Harvey, J.Y. Hsu, R.A. James† , K. Matsuda, J. Lohr, M.J. Mayberry, C.P. Moeller, T.C. Simonen†, and B.W. Stallard†
GA Technologies Inc., San Diego, California 92138, U.S.A.

ABSTRACT

Electron Cyclotron Heating experiments on the DIII–D tokamak using outside launch of the extraordinary mode have shown effective bulk heating at the fundamental resonance, in contradiction of the theory of wave propagation. This result may be explained by an efficient process of mode conversion from the extraordinary mode to the ordinary mode upon reflection at the vessel wall of waves reflected from the right hand cutoff in the plasma. The resulting heating has the characteristics expected for heating with the ordinary mode.

INTRODUCTION

The primary objective of the current set of ECH experiments on DIII–D, which use the outside (low field side) launch of the waves, is to study the effect of the electron temperature edge profile and/or pedestal on energy confinement. This work is motivated by previous results from Doublet III on ECH edge heating[1,2] and by ASDEX results which connect the H–mode to a threshold in edge temperature.[3] In order to accomplish edge heating, the antennas were installed in the extraordinary mode (X–mode) polarization in accordance with the theoretical prediction that the X–mode absorption is much stronger than ordinary mode (O–mode) absorption at the second harmonic.[4] Heating with the fundamental resonance placed near the plasma axis is not expected to be effective because the theory of wave propagation shows that the wave will be reflected from the right hand cutoff very near the outer edge of the plasma.[5] Nevertheless, when fundamental heating was tried the heating was found to be quite effective.

EXPERIMENTAL CONDITIONS

In this study, the ECH power incident on the plasma was between 0.45 MW and 0.70 MW. The applied frequency was 60 GHz, for which the gyromagnetic resonance corresponds to a magnetic field of 2.14 T. Eight ECH antennas are installed on the outer vessel wall (low magnetic field side) of the DIII–D tokamak, with each antenna connected to a single 0.2 MW gyrotron. The antennas are distributed vertically over a height 0.36 m centered on the midplane, and they are directed ±17° with respect to the radial direction. The antenna assembly consists of a vacuum barrier window of fused

---

*Supported by the United States Department of Energy under contracts DE–AC03–84ER51044 and W–7405–ENG–48.

†Permanent address: Lawrence Livermore National Laboratory.

quartz of diameter 3.8 cm which carries the TE_{01} mode, followed on the vacuum side by a taper to 1.9 cm, a mode converter to the TE_{11} mode, and a second converter to the HE_{11} mode. The HE_{11} waveguide radiates with an approximately gaussian radiation pattern of characteristic angular half width 11°. The output is linearly polarized with the electric field in the vertical direction; for the 17° orientation this provides over 90% of the incident power in the X–mode. About 68±5% of the generated power is launched as a pure X–mode and 8% is launched as an O–mode.

The DIII–D tokamak was operated in the expanded boundary divertor configuration,[6] with elongation of about 1.8. Plasma currents of 0.75 to 1.75 MA were run, with the toroidal field set at 2.14 T to place the fundamental resonance at the plasma axis. Most of the data were taken with deuterium as the working gas, but little difference in confinement time or other heating signatures was found when hydrogen was used. In the experiments described here, plasma density was measured with an interferometer which looks along a vertical chord.

EXPERIMENTAL OBSERVATIONS

The experiments were performed by applying ECH during the current flat-top, starting at least 100 msec after the start of the sawtoothing phase of the discharge. The ECH was applied for typical periods of 300 msec, by which time the plasma had attained its new steady state. Figure 1 shows data from such a discharge with plasma current of 0.75 MA, in which 605 kW of ECH was applied compared to 475 kW of ohmic power during the ECH. Due to the heating effect, the plasma conductivity increased and the loop voltage dropped from 0.85 V before ECH to 0.63 V during ECH, corresponding to an average temperature increase of 25%, assuming that Z_{eff} did not change. This behavior is shown in Fig. 1b. (The large excursions in loop voltage shown in Fig. 1b following the ECH are transient effects due to switching of the ohmic heating coil current as the current passes through zero.)

Unlike many previous ECH experiments, the line-averaged density was not affected by the ECH. Figure 1c shows no changes in $\bar{n}_e$ at the start or end of the heating pulse. In this discharge and all others discussed the gas injection rate was constant from well before to well after the ECH.

Clear evidence of electron heating was seen in the soft X–ray emission (SXR) and in the electron cyclotron emission (ECE). Figure 1d shows emission from an SXR chord through the plasma center. Figure 1e shows absolutely calibrated ECE measurements of T_e for a location near the plasma center, where the temperature increases from 1.15 keV to 1.60 keV, or about 40%. A 15% increase in Z_{eff} during ECH can resolve the increase in T_e and the loop voltage decrease. The total plasma kinetic energy W_{MHD} was determined from analysis of the MHD equilibrium, Fig. 1f, and it increased from 70 kJ to 96 kJ during the ECH, also about 40%. A very similar result was obtained from measurements of plasma diamagnetism. These results, taken together, suggest that only a small fraction of the plasma energy can be ascribed to a non-Maxwellian tail on the electron distribution function, or there would not be such good agreement between plasma conductivity, emissivity, and energy measurements.

The plasma heating did not appear to be highly localized. Figure 2 shows the evolution of the electron temperature determined from ECE for the same discharge as that of Fig. 1. When the ECH was applied, dT_e/dt was large both near the center of the plasma and at the edge, with a dip in between. The final profile is not much different from the initial profile, except for the increase near the edge. Analysis of the delay between the start of the ECH and the local increases in SXR emissivity for different chords through the plasma also does not show localized heating.

In order to appraise the effect of the plasma cutoffs on the heating, the density was scanned over the range $1.2 < \bar{n}_e < 4.6\times10^{19}$ m^{-3}. This range covers the O–mode cutoff, which is estimated to occur at the plasma center when the line–averaged density exceeds about 2.5 to 3.0×10^{19} m^{-3} (i.e., at a local density of 4.4×10^{19} m^{-3}). In order to increase the span of density available, the data were gathered for plasma currents of 0.75 and 1.0 MA; these data are shown in Fig. 3, in which $\Delta(\bar{n}_e T_{e0})$ and ΔW_{MHD} are plotted as a function of density. It is clear that plasma heating occurs for densities below the O–mode cutoff, but that above 3.0×10^{19} m^{-3} the heating effectiveness falls off dramatically. This behavior is what would be expected if a large fraction of the incident ECH power were polarized in the O–mode.

DISCUSSION

The experimental results can be explained by the presence of an efficient mechanism for conversion of the X–mode to the O–mode. Such a process is believed to occur even on a single specular reflection from a smooth wall.[7] As the linearly polarized wave propagates from the vacuum region near the antenna obliquely into the plasma, the phase velocities of the X–mode and the O–mode become nondegenerate, and the wave bifurcates into right–hand and left–hand elliptically polarized components. The small O–mode component (left–hand polarized) propagates into the plasma and is directly absorbed. The dominant X–mode component (right–hand polarized) is reflected by the right–hand cutoff back toward the vessel wall. When this wave reflects from the wall, its handedness is reversed, and even for slightly oblique propagation (17° from perpendicular) the conversion to O–mode is of order 50%. Only a small number of bounces between the cutoff and the wall can therefore result in effective conversion of most of the power to the O–mode. The optics of such a process cannot keep the power concentrated in the vertical direction, so broad heating across the plasma profile may be expected.

The gross energy confinement time τ_E decreases during ECH under the (unreasonable) assumption that all of the incident ECH power is absorbed in the plasma. Under this assumption, τ_E falls from 110 msec before ECH to 90 msec during ECH for the discharge of Fig. 1. However, if the hypothesis regarding the mode conversion process is correct, then several bounces of the rays are required for full conversion, and about 10% of power is lost on an average bounce on the wall. It also appears from Fig. 2 that a significant fraction of the power is being absorbed very close to the plasma edge where confinement may be poor. If the absorbed power is two-thirds of the gross incident power, then no degradation from ohmic confinement is taking place. Determinations of dW/dt at the start of the ECH indicate that $70\pm10\%$ of the incident power is absorbed, so little degradation in confinement may be taking place. At the ECH power levels used so far, which are about 4 times less than the minimum neutral beam power level required to obtain the H–mode,[6] no indication of the H–mode has been observed.

REFERENCES

1. R. Prater, S. Ejima, R.W. Harvey, A.J. Lieber, K. Matsuda, C.P. Moeller, Proc. Eleventh Int. Conf. on Plasma Physics and Controlled Nuclear Fusion Research (Kyoto, 1986) (IAEA, Vienna, 1987), paper FII–3.

2. S. Ejima and R. Prater, "Interpretation of Electron Cyclotron Heating Results in Overdense Plasma in Doublet III," GA Technologies Report GA–A18535 (1986); to be published in Nuclear Fusion.
3. M. Keilhacker, et al., Plasma Physics and Cont. Fusion **28**, 29 (1986).
4. M. Bornatici, R. Cano, O. DeBarbieri, and F. Engelmann, Nucl. Fusion **23**, 1153 (1983).
5. Thomas Howard Stix, *Theory of Plasma Waves* (McGraw-Hill, New York, 1962).
6. J. Luxon, et al., Proc. Eleventh Int. Conf. on Plasma Physics and Controlled Nuclear Fusion Research (Kyoto, 1986) (IAEA, Vienna, 1987), paper AII–3.
7. J.Y. Hsu and C.P. Moeller, "Polarization Change of the Electron Cyclotron Waves by Reflection," this conference.

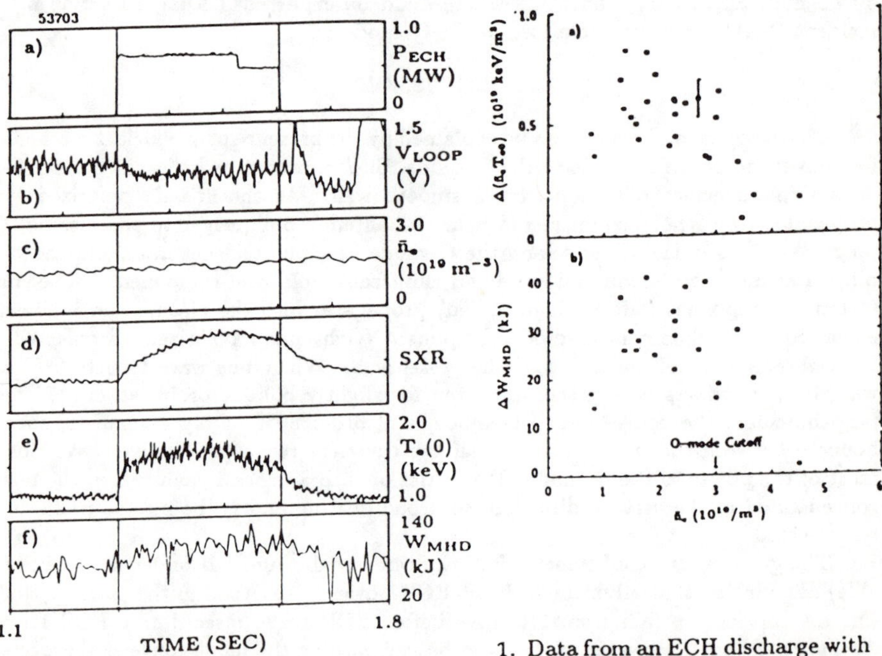

Electron Cyclotron Emission Shot 53703

1. Data from an ECH discharge with $I_P =$ 0.75 MA, B_T=2.14 T, and divertor configuration, in deuterium.

2. $T_e(r,t)$ evolution for the discharge of Fig. 1, from ECE.

3. Changes in the plasma energy due to ECH for discharges with $I_P =$ 0.75 MA (filled squares) and with 1.0 MA (hollow squares), as a function of line-averaged density. P_{ECH} is 0.45–0.63 MW and $B_T =$ 2.14 T.

POLARIZATION CHANGE OF ELECTRON CYCLOTRON WAVES BY REFLECTION

JANG-YU HSU and CHARLES P. MOELLER
GA Technologies Inc., San Diego, CA 92138

ABSTRACT

Given that the wave numbers of X and O modes merge as they approach the plasma boundary, a pure mode reflected from the plasma by a cut off may excite both modes due to polarization change when it is reflected off the vacuum vessel wall since the tangential components of the electric field change sign. This may explain the good central heating in DIII-D ECH experiments in which the fundamental X-mode is launched from the low field side.

In recent DIII-D ECH experiments, the X-mode was launched from the low field side of the torus in order to do second harmonic edge heating. When the toroidal field was raised to give a fundamental resonance at the plasma center, however, good central heating was observed [1] despite the expected right hand cut off and the evanescent layer beyond. Since the tunneling to the upper hybrid layer is unlikely, given the relatively short wavelength mode describable by Budden's equation, the good central heating observed may have to depend on a conversion from the X-mode to the O-mode.

If the incident X-mode, after reflection from the cut off, is assumed to leave the plasma, still as the X-mode, then this process might be thought to happen by random scattering off wall irregularities, limiters, etc. However, such processes would also randomize the propagation angle as well as the polarization, and might lead to uniform but inefficient heating over all flux surfaces, contrary to the observation. This paper suggests that a plane conducting surface can produce effective mode conversion in a single reflection with no randomization of propagation angle.

For the purpose of this study, we assume that if an X-mode is launched at the plasma edge by the proper choice of elliptical polarization for a specified launch angle, it will continue as an X-mode in the plasma interior, unless a mode conversion point is encountered, where the X-mode and O-mode have the same wave numbers. Since the perpendicular wave numbers $k_\perp^x$ and $k_\perp^o$ of the two modes diverge further inside the plasma, even as the right hand cut off is smoothly approached, where $k_\perp^x$ vanishes, the X-mode will simply turn around, as suggested in Fig. 1, with no mode conversion so that the exit polarization is the same as the incident polarization. At the wall, however, the reflection is abrupt. The tangential electric field components are reversed, so that the handedness of the ellipticity of the reflected wave is reversed. Such a depolarization could also occur at a sharp plasma boundary, along with the generation of a reflected wave, but that will not be treated here.

To launch exclusively an extraordinary mode at an oblique angle, a specific polarization is required at the plasma edge. This can be shown from the wave equation

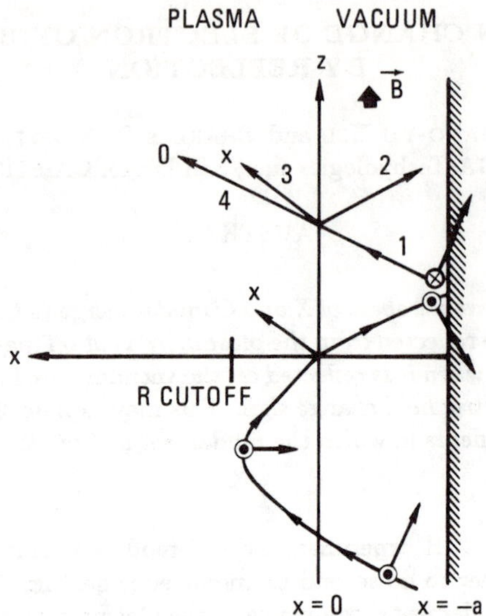

FIG. 1. Paths for an X-mode reflected from the right hand cut off.

$$\begin{pmatrix} S - n_\parallel^2 & -iD & n_\perp n_\parallel \\ iD & S - n^2 & 0 \\ n_\perp n_\parallel & 0 & P - n_\perp^2 \end{pmatrix} \begin{pmatrix} E_x \\ E_y \\ E_z \end{pmatrix} = 0 \quad , \tag{1}$$

where P, D, S are given by the cold plasma dielectric tensor as in Stix [2]. The plasma is located at $x \leq 0$ with the conducting wall at $x = -a$ (*cf.* Fig. 1). The electric field polarization may be defined [3] by $E_H/E_V = \vec{E}\cdot\hat{y}\times\vec{n}/\vec{E}\cdot\hat{y} = (n_\parallel E_x - n_\perp E_z)/E_y = E_x P n_\parallel/(P - n_\perp^2)E_y$, where $\hat{y}$ is the unit vector along the y-coordinate. In the low density region, the values of E_H/E_V are well separated for the two modes given by $n_\perp^2 = (B\pm\sqrt{B^2 - 4\,AC}\,)/2\,A$, where the dispersion equation is cast into $An_\perp^4 - Bn_\perp^2 + C = 0$ with $A = S, B = PS+RL-n_\parallel^2(P+S), C = P(n_\parallel^2-R)(n_\parallel^2-L)$. Expanding in terms of the small parameter $\delta \equiv \omega_p^2/\omega^2 \ll 1$, we find $B^2 - 4\,AC = \delta^2[(1-a)^2(1-n_\parallel^2) - 4\,a(1-a)n_\parallel^2] + O(\delta^3)$, where ω_p is the electron plasma frequency, $a \equiv \omega^2/(\omega^2 - \Omega^2)$, Ω is the electron cyclotron frequency. It follows that $E_x/E_y = (n^2 - S)/iD = -(\omega/2\,\Omega)\Big[(1- n_\parallel^2)(a-1) \pm \sqrt{(1-a)^2(1-n_\parallel^2) - 4\,a(1-a)n_\parallel^2}\,\Big]$. Therefore, the polarizations are given by [3]

$$\frac{E_H}{E_V} = \frac{i\Omega}{2\,\omega n_\parallel}\left[\left(1 - n_\parallel^2\right) \pm \sqrt{\left(1 - n_\parallel^2\right)^2 + 4\,n_\parallel^2\,\frac{\omega^2}{\Omega^2}}\,\right] \equiv iR_\pm \quad , \tag{2}$$

where $a/(a-1) = \omega^2/\Omega^2$ has been used, and only the lowest order in δ is retained. By defining the unit vectors $\hat{e}_o = (R_+,1)/\sqrt{1+R_+^2}$, and $\hat{e}_x = (R_-,1)/\sqrt{1+R_-^2}$ on the plane of (E_H, iE_V), it is straightforward to show that $\hat{e}_x \cdot \hat{e}_o = 0$, *viz.*, they form an orthonormal set. A pure mode must have its (E_H, iE_V) pointing along the principal axes, say an X-mode along $\hat{e}_x$, and an O-mode $\hat{e}_o$.

Taking a pure X-mode reflected from the right hand cut off, its polarization as shown in Fig. 1 is supposedly unchanged and can be given by $E_o\hat{e}_x$. After reflecting off the wall, the tangential components and $n_\perp$ change sign so that the polarization is given by $\vec{E} = E_o(R_-,-1)/\sqrt{1+R_-^2}$. The wave emerges with a fraction of the rf power in the O-mode given by $P_o = |\vec{E}_r \cdot \hat{e}_o|^2/E_o^2 = 1 - |\vec{E}_r \cdot \hat{e}_x|^2/E_o = 4R_-^2/(1+R_-^2)^2$. Figure 2 shows the values of R_+, R_-, and P_o in terms of $n_\parallel^2$ and the launch angle $\theta = \sin^{-1} n_\parallel$ for the inverse aspect ratio $\epsilon \simeq 0.4$. With a launch angle of 17 deg, as in the DIII-D ECH experiment, P_o is 44%, while 56% is relaunched as the extraordinary mode. The P_o quickly increases to 90% at 40 deg but vanishes at the perpendicular launch ($\theta = 0$), as expected. It is interesting to note that if a pure O-mode shines on the wall, a P_o fraction of the energy will be converted to the X-mode. This can easily be shown with the identity that R_+R_- =-1.

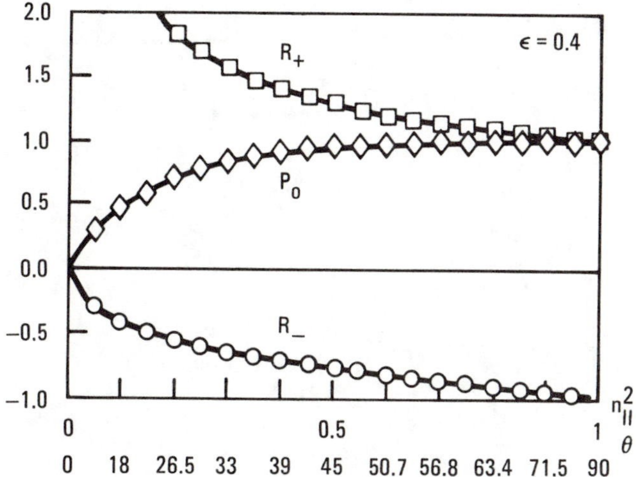

FIG. 2. The polarization parameters R_+, R_-, and the fraction of O-mode conversion after reflection as function of $n_\parallel^2$ and the launch angle.

In summary, we have shown that reflection off a smooth conducting wall may allow mode conversion of the cyclotron waves by polarization mixing. While this explains the central heating when an X-mode is launched from the low field side, it would be interesting to verify this mechanism in the experiment by raising the central density above the O-mode cut off to see whether the central heating goes away. Direct measurement of the rf intensity in the vacuum region may reveal the transmission coefficient, since after n bounces off the wall, the remaining rf energy should be reduced to $\sim (1/2)^n$.

This is a report of work sponsored by the U.S. Department of Energy under Contract No. DE-AC03-84ER53158.

References

[1] R. Prater *et al.*, this conference.
[2] T.H. Stix, *The Theory of Plasma Waves* (McGraw-Hill, New York, 1962).
[3] C. Moeller *et al.*, Phys. Fluids **25**, 1211 (1982).

EXPERIMENTAL INVESTIGATION OF A LOCALIZED ELECTRON TEMPERATURE SPIKE PRODUCED BY COLLISIONLESS ELECTRON CYCLOTRON DAMPING*

R.R. Mett, S.W. Lam and J.E. Scharer
University of Wisconsin, Madison, WI 53706

ABSTRACT

A spacially localized axial temperature spike due to electron cyclotron resonance absorption in a steady-state plasma column has been observed by B.W. Rice and J.E. Scharer[1]. Recent investigations have focused on its dependence on neutral pressure, plasma source input power, antenna excitation frequency and magnetic field gradient using Langmuir probes, emissive probes, diamagnetic loops and rf probes. The axial shape of the spike has been observed to strongly depend on the microwave coupling to the rf plasma source. The position, height and width of the spike vary with antenna excitation frequency in a localized mirror field to form a smooth spacial envelope. Measurements indicate a substantial temperature anisotropy near the resonance region.

INTRODUCTION

Detailed spatial measurement and understanding of large amplitude electron cyclotron wave propagation, damping and nonlinear effects is important for the analysis and design of electron beam devices such as the gyrotron and FEL, analysis of whistler wave propagation in the ionosphere and heating of fusion plasmas. We present experimental results for a controlled steady state experiment in which detailed wave, plasma temperature and density measurements are made in the vicinity of the electron temperature spike. Conditions which allow the presence of the spike are discussed.

EXPERIMENTAL APPARATUS

The physical apparatus used for these experiments is shown in Fig. 1 below. The glass column is 7.6 cm in radius and is covered with copper screen acting as a microwave shield and wave radial boundary condition. The plasma source is an open-ended microwave cavity residing at the cyclotron resonance as shown on the left side of Fig. 1. A magnetron provides 50-300 W of power to this source at a frequency of 2.45 GHz. The source produces a quiescent argon plasma 6 cm in radius at neutral pressures 1-9 × 10^{-4} Torr (λ_{en}~1m) plasma densities of 1-8 × 10^{10} cm^{-3}, electron temperatures of 2-7 eV (λ_{ee}~10m) and ion temperatures of ~0.1eV. The state of the plasma is greatly affected by the magnetron output power and by the state of a waveguide tuner that matches the magnetron to the plasma source. The electron cyclotron wave is launched at the other end of the plasma column by a helical beam antenna in a high magnetic field region of the system. A TWT provides 0-15W of microwave power that is tunable in the range of 1.8-2.7 GHz. A RHCP

*Research supported by NSF Grant ECS-8514978

wave is launched from the antenna and is absorbed near cyclotron resonance a few centimeters before the magnetic field minimum.

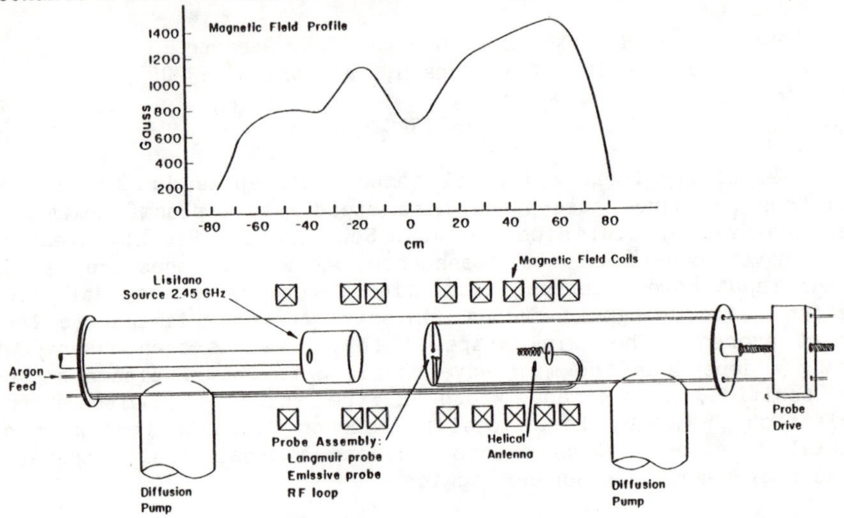

Fig. 1. Wave propagation facility

Data acquisition is controlled by a microcomputer that allows simultaneous monitoring of Langmuir, RF, and emissive probes. Part of a typical data set is shown in Fig. 2 below.

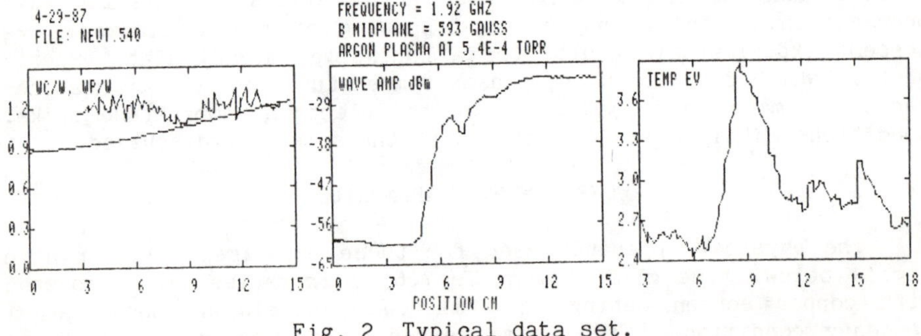

Fig. 2 Typical data set.

The RF probe picks up the time varying magnetic field at the antenna excitation frequency. Its signal is sent to an interferometer and spectrum analyser to obtain the plasma wave amplitude and phase. The interferometer indicates a steady shortening of the wavelength before the damping of the wave near resonance as the wave comes in from the right hand side on the plots shown in Fig. 2. The temperature spike and a corresponding dip in plasma density (ω_p) just where the wave begins to die away may also be seen. The axial dependence of the magnetic field (at r = 0) over the range of interest is shown by the monotonic curve in the left hand box as $\omega_{ce}/\omega \propto B_o$. Emissive probes have consistently obtained < 0.5 V variation of plasma potential across the resonance zone, consistent with that obtained with the Langmuir probes.

EXPERIMENTAL RESULTS

Dependence of the spike on neutral pressure is complex. This is because the neutral pressure affects the plasma source operation and thus the plasma density and temperature. The spike may only be observed under certain source coupling conditions for any given neutral pressure and magnetic field and where the antenna coupling is good (P_f = 15 W, $P_r \lesssim 3$ W). Under very similar conditions a spike may or may not exist. A spike has not yet been observed at neutral pressures below 2×10^{-4} Torr.

The dependence of the spike on antenna excitation frequency is shown in Fig. 3. Here all experiment parameters were held fixed except the frequency which did not affect the quiescent operation of the system. Antenna coupling remained good (P_f = 14-16 W, P_r = 1.5-3.3 W) and showed no correlation with temperature spike height. Correlation of the movement of the resonance as frequency is increased is $\lesssim 3\%$. The peak is always a few centimeters before the cold resonance point in agreement with hot plasma theory.

An estimate of the local temperature anisotropy was furnished by disc Langmuir probes facing perpendicular and parallel to the magnetic field. This preliminary measurement shows a steady state anisotropy of about three.

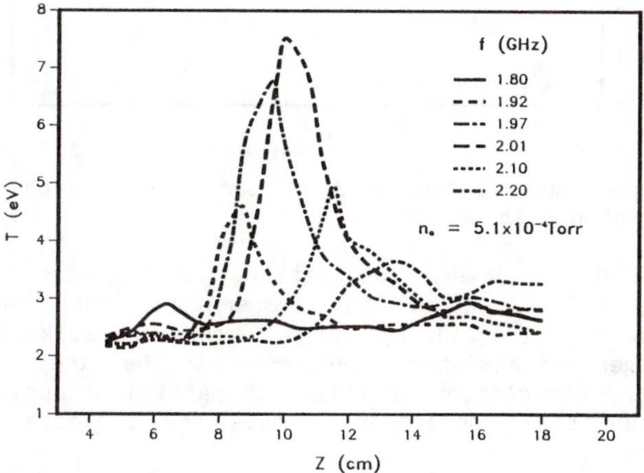

Fig. 3. Dependence of temperature spike on antenna excitation frequency.

DISCUSSION AND CONCLUSIONS

The evidence presented here substantiates our theory that the temperature spike is due to a particular velocity distribution coming from the source such that there is a large class of elections that have turning points near the resonance zone. These particles pick up substantial transverse energy from the wave and scatter off neutrals and into the loss cone before they reach the other side of the well due to the tight helical motion of the turning electrons. The class of turning particles roughly corresponds to the Maxwellian-like envelope of temperature peaks shown in Fig. 3.

Since the temperature spike is due primarily to perpendicular energy, the disc probe facing parallel to the field does not respond to this class of particles as shown in Fig. 4. The alignment of the disc must be quite precise otherwise the probe shows a temperature spike. The data in Fig. 4 was taken during one pass through the system. Pulsed RF experiments have shown that the Langmuir probes are not significantly affected by the resonant RF signal since the electron temperature was shown to decay in times long compared with the RF turnoff. This is corroborated theoretically in that most of the energy of the RHCP whistler is in the particles and RF magnetic field rather than in the electric field.

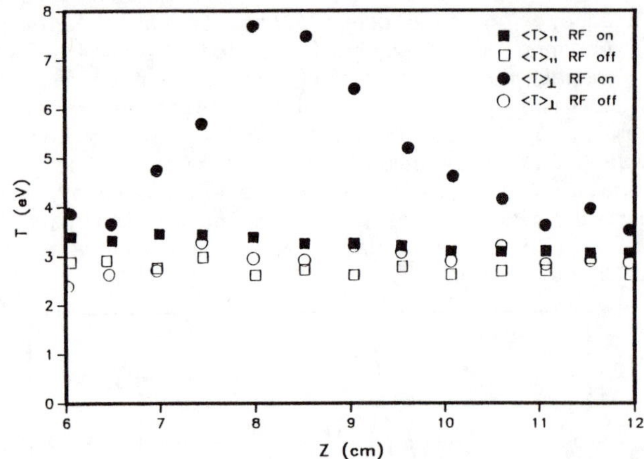

Fig. 4. Temperature estimates from parallel and perpendicularly oriented disc probes.

A correlation between oscillations in the wave amplitude as resonance is approached and the temperature spike phenomenon is apparent (see Fig. 2). When a large temperature spike exists these dips are larger whereas when no spike exists they are smaller or are not apparent. Correlation of this with particle trapping, sideband production and other nonlinear effects are subjects for further investigation.

Future investigations will also include quantification of diamagnetic loop data to measure the local anisotropy and other pulsed experiments to measure the transient response and verify the existence of trapped particles. Previous diamagnetic loop measurements on the system [2] led to the conjecture that a large temperature anisotropy exists in the resonance zone.

REFERENCES

1. B.W. Rice, J.E. Scharer, IEEE Trans. Plasma Science PS-14, 17 (1986).
2. T.L. Owens and J.E. Scharer, "Thermal Effects on Electron Cyclotron Resonance Heating," Plasma Physics 18, 663 (1976). T.L. Owens, Masters Thesis, "Hot Plasma Electron Cyclotron Heating," Univ. of Wisconsin (1974).

ECH IN THE MICROWAVE TOKAMAK EXPERIMENT*

B. W. Stallard, G. R. Smith, R. A. James, and K. I. Thomassen
Lawrence Livermore National Laboratory, University of California
Livermore, CA 94550

A. H. Kritz
Hunter College/CUNY, New York, NY 10021

M. Makowski, T. Samec, and R. Yamamoto
TRW
Redondo Beach, CA 90278

ABSTRACT

The Microwave Tokamak Experiment (MTX) at LLNL will investigate electron heating in the MTX tokamak (formerly Alcator-C) at high density (up to $6 \times 10^{20}\ m^{-3}$) and high power by using a free electron laser (FEL). Parameters of the FEL are a peak power up to 8 GW and 50 ns duration, with average power 1 to 2 MW, at a frequency of 250 GHz. The planned input driver for the FEL is a gyrotron oscillator. The FEL output will be transported quasi-optically, inside a 50 cm evacuated pipe, to the input port of the tokamak by means of a four-mirror system. Launch polarization is the ordinary mode. This experiment will test the FEL technology at short wavelength and high peak and average power levels. Important physics issues to be explored are the effects of intense pulse heating (electric field up to 500 kV/cm) on nonlinear wave absorption and bulk heating, plasma confinement, plasma impurities, and parametric instabilities. Because the FEL technology is scalable to higher frequency and power, success of these experiments has importance for next-generation tokamaks.

INTRODUCTION

Electron cyclotron heating (ECH) of tokamak plasmas has many attractive features for present and future machines: (i) controlled and localized heating, (ii) high power density requiring limited port access, and (iii) simple antennas located away from the plasma boundaries. At LLNL, experiments in ECH using the new FEL technology will be carried out in the MTX tokamak. The first major objectives of these experiments are to (1) demonstrate FEL power generation and transport at high frequency, high peak power, and high average power, and (2) investigate plasma heating efficiency at high density (cutoff density $7.7 \times 10^{20}\ m^{-3}$). During these experiments we will study confinement scaling (at $P_{aux} >> P_{ohmic}$), plasma transport,

*This work was performed under the auspices of the U.S. Department of Energy by the Lawrence Livermore National Laboratory under contract number W-7405-ENG-48.

MHD control, plasma profile shaping (combined with pellet fueling), and applications to current drive.

HEATING SYSTEM

The ECH system consists of the FEL amplifier and associated driving source and a quasi-optic transmission system. The FEL consists of a 5-m wiggler with 8-cm period and a tapered magnetic field in the final portion of the wiggler to maximize energy extraction. The e-beam injected into the 1-in.-diameter wiggler waveguide has a nominal energy of 7 MeV at 3 kA current. The FEL will produce rf pulses at 250 GHz with 50 ns duration and a repetition rate of 5 kHz. FEL code calculations (FRED) predict a peak power exceeding 8 GW with corresponding average power up to 2 MW. Previous experiments at lower frequency (35 GHz) have demonstrated e-beam power to rf power conversion efficiencies of 34%, in good agreement with FRED calculations.[1] The input driver source for the FEL will be a gyrotron. The expected output mode is the $TE_{11,2}$ with nominal output power of 10 kW for 1 μs. The output mode will be linearly polarized for coupling into the wiggler waveguide using a Vlasov antenna. FRED calculations predict that 50 W input power is sufficient to drive full output power for the FEL.

The transmission system to transport rf power from the FEL output to the MTX input port is a four-mirror quasi-optic design enclosed within a 50-cm-diameter evacuated pipe. To calculate microwave transport through the system from mirror to mirror, we use scalar diffraction theory (applying Huygen's principle) for each vector component. The final mirror design can be varied to change the power profile launched into MTX (port dimensions: 4 cm × 30 cm). Figure 1(a) shows the calculated field contours for a beam elongated to fill the input port and plasma cross section. For more localized heating the beam can be circularized as shown in Fig. 1(b). In both cases the power at the walls in the narrow dimension of the input port is very small. A preliminary mirror design (unoptimized) predicts an overall transport efficiency of 91% for a TE_{01} mode in square waveguide (E horizontal) launched from the FEL output.

A possible concern for intense pulse transmission is microwave breakdown. We have considered multipacting, gas breakdown and avalanche, and arcing on surfaces. Multipacting and avalanche are not expected to be important, since dimensions are large (fd > 100 GHz-cm) and ionization times greatly exceed the rf pulse length. On mirror surfaces, the peak electric field ε < 300 kV/cm. Within the MTX port,
$\varepsilon \sim 1$ MV/cm at the beam center, but the field at the wall is much less. The maximum electric field at a wall will occur in the wiggler, where $\varepsilon \sim 1$ MV/cm. Even here breakdown should not occur on a clean, conditioned surface. Tests at 35 GHz have shown breakdown at 3.6 MV/cm. At 250 GHz, the threshold is expected to be much higher.

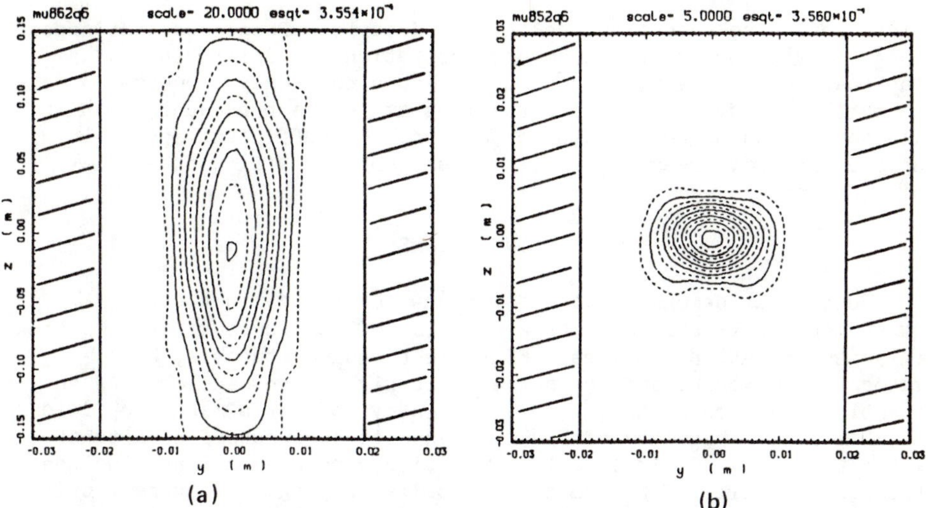

Figure 1. Beam electric field contours (linear) at center of TMX input port: (a) beam filling port, vertical scale ±15 cm, last contour -20 dB power; (b) circularized beam, vertical scale ±3 cm, last contour -22 dB power.

POWER ABSORPTION ISSUES

For the intense electric field of the FEL beam, nonlinear effects are expected to be important for absorption.[2] Linear theory predicts strong, single-pass absorption for the ordinary mode at the fundamental. For example, for $T_e = 1$ keV and $n = 2 \times 10^{20}$ m^{-3}, ray tracing calculations using TORAY predict 90% absorption. At very high electric fields, nonlinear effects reduce absorption since relativistic effects can cause the electrons to become "detuned" from resonance as they transmit the ECH beam. The nonlinear effects have been incorporated into TORAY, and calculations to address this issue are underway. A calculation at $T_e = 2$ keV and $n = 2 \times 10^{20}$ m^{-3} at 5 GW power predicts 82% absorption of the total beam. This is reduced from the 99% calculated by using linear theory. During startup, temperatures are expected to be in the 1-keV range and absorption will be less. However, as T_e increases because of heating, strong single-pass absorption is expected.

Since not all electrons experience resonance during the short pulse of the ECH, thermalization of these electrons with the bulk has been examined. We find that at the high densities in MTX there is sufficient time for thermalization between ECH pulses. Hence, the electron distribution is expected to remain Maxwellian.

Parametric instabilities are a possibility for intense pulse heating. These effects could actually increase absorption. Conversely, they could lead to back scattering of the microwave power. The required thresholds and frequency and wavenumber matching conditions have been examined in some detail.[3] Processes that appear marginally important for coherent excitation are Brillouin backscatter. However, effects that reduce coherency increase thresholds. Edge density fluctuations at levels previously seen in Alcator-C plasmas are predicted to produce a sufficient spread in $k_{\parallel}$ of the microwave beam that backscatter probably will not be significant.

CONCLUSIONS

Successful demonstration of the new technology in these experiments will show the attractiveness of ECH for auxiliary heating in next-generation machines. FELs are frequency tunable, have moderate bandwidth, and are scalable to higher frequencies. As an example, the second harmonic extraordinary mode at a magnetic field of 10 T and a frequency of 560 GHz can heat to a cutoff density of 1.9×10^{21} m^{-3}. Other important benefits already mentioned include the possibilities of MHD control, profile shaping, and current drive.

REFERENCES

1. T. J. Orzechowski et al., Phys. Rev. Lett. 57, 2172 (1986).
2. W. M. Nevins, T. D. Rognlien, and B. I. Cohen, submitted to Phys. Rev. Lett., Lawrence Livermore National Laboratory, Livermore, CA, UCRL-95006.
3. M. Porkolab and B. I. Cohen, to be submitted to Nucl. Fus.

ECRH ON RTP

P. Manintveld and A.G.A. Verhoeven
Association Euratom-FOM, FOM-Instituut voor Plasmafysica,
Rijnhuizen, 3430 BE Nieuwegein, The Netherlands

ABSTRACT

A design is presented of the Electron Cyclotron Resonance Heating system to be used on the Rijnhuizen Tokamak Petula (RTP), which will be dedicated to plasma transport studies.

The RTP-ECRH facility will use three 60 GHz Varian gyrotrons. Each of the gyrotrons can deliver more than 200 kW of microwave power during pulses up to 100 ms. The output power of each gyrotron is controlled by switching on the anode voltage in two steps. This enables very fast rise times. Modulation of the ECRH power by feedback control using diagnostic signals from the plasma is discussed.

The power will be launched into the RTP plasma both from the low magnetic field side through ports in the equatorial plane and from the high-field side through top ports. The envisaged launching structures are presented.

First ECRH experiments on RTP are scheduled for 1989.

TRANSPORT STUDIES IN RTP

The aim of the research to be carried out on RTP is to analyze in detail transport phenomena in tokamaks [1].

The experimental work will focus on determining the radial and poloidal plasma equilibrium and on turbulent transport of plasma particles and energy. In addition to analyzing plasmas with ohmic and ECRH heating in steady-state conditions, a large effort will be made to study the evolution of relevant plasma parameters for transient states, generated as a response to well-defined perturbations by local electron cyclotron heating.

The main parameters of the RTP tokamak, formerly in operation in Grenoble as PETULA are: a major radius of 0.72 m, a minor radius up to 0.18 m and a toroidal magnetic field up to 2.6 T.

APPLICATION OF ECRH ON RTP

With ECRH it is possible to inject power into the tokamak in a very efficient way through small antennae without introducing additional impurities. The absorption can be very high, close to 100%, and power deposition can be rather well localized [2,3].

Rijnhuizen has over 600 kW of ECRH power available at 60 GHz and 100 ms pulse time. This ECRH facility was used on the TFR tokamak and has recently been moved to Rijnhuizen. The magnetic field value at fundamental resonance is 2.14 T for 60 GHz. When higher magnetic fields will be accessible, it becomes possible to study absorption also at a down-shifted resonance frequency. In that case ECRH can also be used to affect the velocity distribution [4].

TRANSMISSION LINES

Two top launchers and two outside launchers are envisaged (Fig. 1). Since three gyrotrons are available, the outside launcher (I) and the inside launcher (II) will be permanently connected to a gyrotron. Depending on the experimental program, the third gyrotron will be connected to either an inside (III^a) or an outside launcher (III^b).

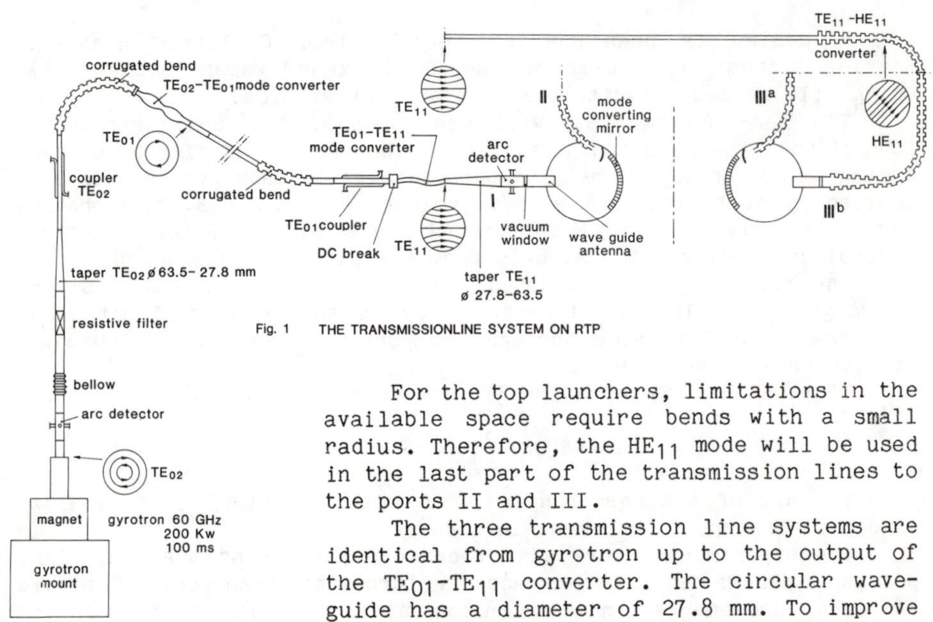

Fig. 1 THE TRANSMISSIONLINE SYSTEM ON RTP

For the top launchers, limitations in the available space require bends with a small radius. Therefore, the HE_{11} mode will be used in the last part of the transmission lines to the ports II and III.

The three transmission line systems are identical from gyrotron up to the output of the TE_{01}-TE_{11} converter. The circular waveguide has a diameter of 27.8 mm. To improve power measurements, most probably k-spectrometers will be used instead of TE_{01} directional couplers. Corrugated wall filters will be inserted to improve mode purity [6]. At the end of the transmission line, a tapered waveguide and a commercially available vacuum window are placed to transmit the power to launcher I. For ports II and III, the system is completed by a TE_{11}-HE_{11} converter.

LAUNCHERS

The launcher in port I is an open-ended smooth waveguide with a 63.5 mm diameter radiating the O-mode. At the opposite torus wall a grooved mirror will be mounted to convert the unabsorbed power to the X-mode. The foregoing also applies to the launcher III^b. In that case, the waveguide wall has to be corrugated to prevent decomposition of the HE_{11} mode. For the same reason, corrugated inserts will be mounted into the available smooth-walled vacuum window. The top launchers consist of a bent corrugated waveguide with an inner diameter of 27.8 mm (Fig. 2). A quartz window will be mounted at the end of this guide. An elliptical mirror will direct and focus the microwave beam. The angle of the beam is about 45° in the toroidal direction. The adjustment requirements of the mirror position are studied by means of ray tracing calculations and radiation pattern analysis.

For optimal coupling of the beam energy to the plasma, the radiation has to be elliptically polarized. We are investigating whether this elliptical polarization will be accomplished by an elliptical waveguide [7] or by corrugations in the focusing mirror [5].

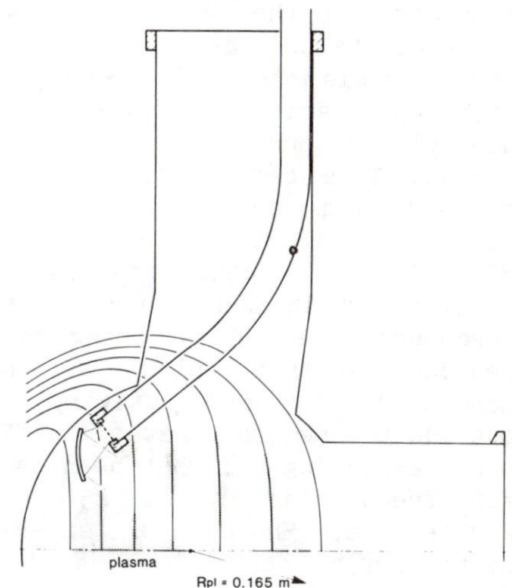

Fig. 2 Top-launcher with focusing mirror.

A stabilizing effect of ECRH on sawteeth was observed in TFR for heating at the q = 1 surface. This effect will be further explored in RTP, not only by using the O-mode from the outboard side, but also with the X-mode from the inside. This can be more efficient if suprathermal effects are important. As the position where X-mode absorption takes place can be varied using the launching mirror, it will be possible to heat around the q = 1 radius with one or two gyrotrons while at the same time the center of the plasma is heated with the other gyrotron(s). If it is indeed possible to stabilize the sawtooth, then the additional heating on axis may further increase the central electron temperature.

THE POWER SUPPLY SYSTEM

The output of the high-voltage transformer-rectifier which will be connected to the 10 kV public grid, serves as the input for the modulator regulator, see Fig. 3. Here a series tube stabilizes the high voltage to exactly 80 kV between the collector and the cathode of the gyrotrons. For each gyrotron 2 switches will be installed for 2 separate anode voltage levels to enable 2 different RF power output levels during ons 100 ms pulse.

MODULATION OF RF POWER

Square wave modulation of the RF power with a frequency up to 100 kHz can be achieved for modulation depths up to 100 percent. Even higher modulating frequencies are possible since the rise time from half-to-full power is only 3 μs [8]. The possibilities to modulate the power can be used for example to react upon periodic plasma phenomena. This can be achieved by pulsing at 100% RF output in case the plasma activity is in the center of the resonance area and switching to a reduced gyrotron output at the moment the plasma activity is outside the resonance area. At TFR a phase-locking unit has been used for frequencies up to 25 kHz. Furthermore, MHD-control experiments can be done by means of a feedback control unit. This unit switches on gyrotrons automatically in case of increasing plasma activity [9].

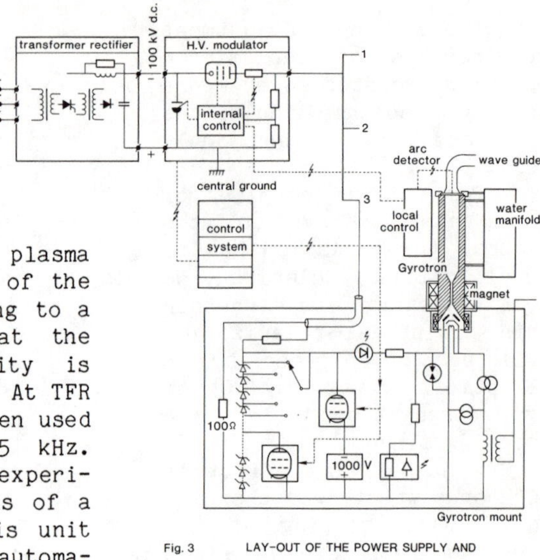

Fig. 3 LAY-OUT OF THE POWER SUPPLY AND CONTROL SYSTEM FOR ECRH ON RTP

ACKNOWLEDGEMENT

This work was performed under the Euratom-FOM association agreement with financial support from ZWO and Euratom.

REFERENCES

1. "Application for Euratom Preferential Support for the Rijnhuizen Tokamak PETULA (RTP), Phase I and II", FOM-Instituut voor Plasmafysica, Rijnhuizen, Nieuwegein, April 1987.
2. A.G.A. Verhoeven, Proc. 14th Symp. on Fusion Technology, Avignon (1986) p. 75-84. Invited Lecture.
3. G. Tonon, A.G.A. Verhoeven and F. Wesner, Proc. 14th Symp. on Fusion Technology, Avignon (1986) p. 107-112.
4. C.A.J. van der Geer et al., "Study of profile control and suprathermal electron production with electron cyclotron waves", Rijnhuizen Report 86-171 (1986).
5. C.A.J. van der Geer et al., Proc. 14th Symp. on Fusion Technology, Avignon (1986) p. 751-757.
6. W. Kasparek et al., Proc. 14th Symp. on Fusion Technology, Avignon (1986) p. 829-834.
7. J.L. Doane, Int. J. Electron. **61** (1986) 1109-1133.
8. A.G.A. Verhoeven et al., Proc. 11th Symp. on Fusion Engineering, Austin, Texas (1985) p. 724-727.
9. A.G.A. Verhoeven et al., Proc. EC-5, 5th Int. Workshop on EC and ECH, San Diego, Cal. (1985) p. 303-312.

INITIAL OPERATION OF ECRH HEATING EXPERIMENTS ON THE VERSATOR II TOKAMAK

S.C. Luckhardt, K.I. Chen, R. Kirkwood,
M. Porkolab, D. Singleton, J. Squire, J. Villasenor
MIT Plasma Fusion Center, Cambridge, Mass. 02139

Z. Lu
Southwestern Inst. of Plasma Physics, Leshan, P.R.C.

ABSTRACT

Operation of a 35GHz electron cyclotron heating experiment has begun on Versator II with gyrotron power of 100kW. The EC antenna is located on the high magnetic field side of the plasma and launches linearly polarized radiation in the HE11 hybrid mode with externally controllable polarization and parallel index of refraction. The transmission system provides mode conversion from the TE01 output mode of the gyrotron to the HE11 mode and polarization control. The mode transformation characteristics of the transmission system were measured by means of a computer controlled two dimensional scanning system, and contour plots of the far field radiation pattern of each transmission system element were made and compared with theory. Overall the transmission system is found to be approximately 95% efficient with mode patterns in generally excellent agreement with theory.

INTRODUCTION

The purpose of the Versator ECRH experiment is to provide perpendicular heating to the high energy electron component produced by lower-hybrid current drive. Such perpendicular heating is predicted to improve the current drive efficiency of lower hybrid waves.[1] The relativistic cyclotron resonance condition, $1 - n_{\parallel} v_{\parallel}/c = \omega_{ce}/\omega\gamma(v)$ permits interaction with the desired region of electron velocity space through control of the n-parallel spectrum of the launched waves and the magnetic field strength of the experiment.

ECRH SYSTEM

The ECRH transmission system, Fig. 1, launches power from the high magnetic field side of the torus using the HE11 hybrid waveguide mode. Variation of the launching angle with respect to the magnetic field, and therefore n-parallel, is achieved by means of a rotatable plane mirror, and the polarization angle of the launched power is continuously adjustable with a rotatable TE01/TE11 mode convertor, Fig. 1. Interaction with energetic electrons in the energy range of 0-40keV is expected for a launch angle of approximately 15 degrees with respect to the perpendicular in the direction of current flow. In this configuration the antenna launches a power spectrum extending from n-parallel= 0.1 to 0.4.

The transmission system provides for mode conversion from the TE01 circular output mode of the 35GHz NRL gyrotron to the TE11 mode, and finally to the HE11 hybrid mode, Fig. 1. The output power passes through a resistive wall mode filter section and then a down taper to WC109 circular waveguide. A mode selective directional coupler then allows forward and reflected power to be monitored in the TE01 mode. Mode conversion from the TE01 to the TE11 mode is accomplished with a sinuous ripple type mode convertor which also provides polarization control by rotation about its axis. Thus the polarization can be varied on a shot-to-shot basis, and the polarization change introduced later by reflection at the plane mirror can be compensated. The final mode conversion from the TE11 to the HE11 hybrid mode is made by a reverse tapered corrugated mode convertor constructed from compressed stacked disks. The transmission system next undergoes a 45 degree bend consisting of a curved section of electro-formed copper and a straight vacuum feed-through section; both sections are in corrugated guide. The HE11 guide corrugations have depth and width of 0.135 cm and period of 0.270 cm, and the guide surface impedance is 0.587 in units of the free space impedance. The quartz vacuum window is located inside the toroidal vacuum vessel, Fig.1, and the transmission line is terminated with a four period corrugated wall horn section. The radiated microwave power is then incident on a plane rotatable mirror and is reflected back from the high magnetic field side at a poloidal angle of approximately 45 degrees. The function of the various mode converting elements has been verified by means of full two dimensional far field radiation pattern measurements carried out with a cw power source. The measurement apparatus consists of a computer controlled x-y translation stage on which is mounted a microwave power detector and microwave absorbing cowel. In operation the detector is scanned over a two dimensional plane, and the incident power is AD converted. Contour plots of far field power obtained with this system are shown in Fig. 1 with power in the y and x polarization components shown above and below respectively. The power in the crosspolarized component, x-component, is 15dB less than the power in the dominant polarization for the TE11 and HE11 patterns. The spatial distribution of the dominant polarization patterns is found to be in substantial agreement with calculations of the TE11 and HE11 patterns. The cross polarization pattern of the HE11 mode was found to have substantial asymmetry, Fig. 1, however, this affects less than 3% of the total RF power.

The transmission system is now initial operation at high power with up to 45kW injected into plasma, further conditioning of the gyrotron source is underway to raise the power to the 100kW level.

REFERENCE

1. I. Fidone et al., Phys. Fluids 27, 2469 (1984).

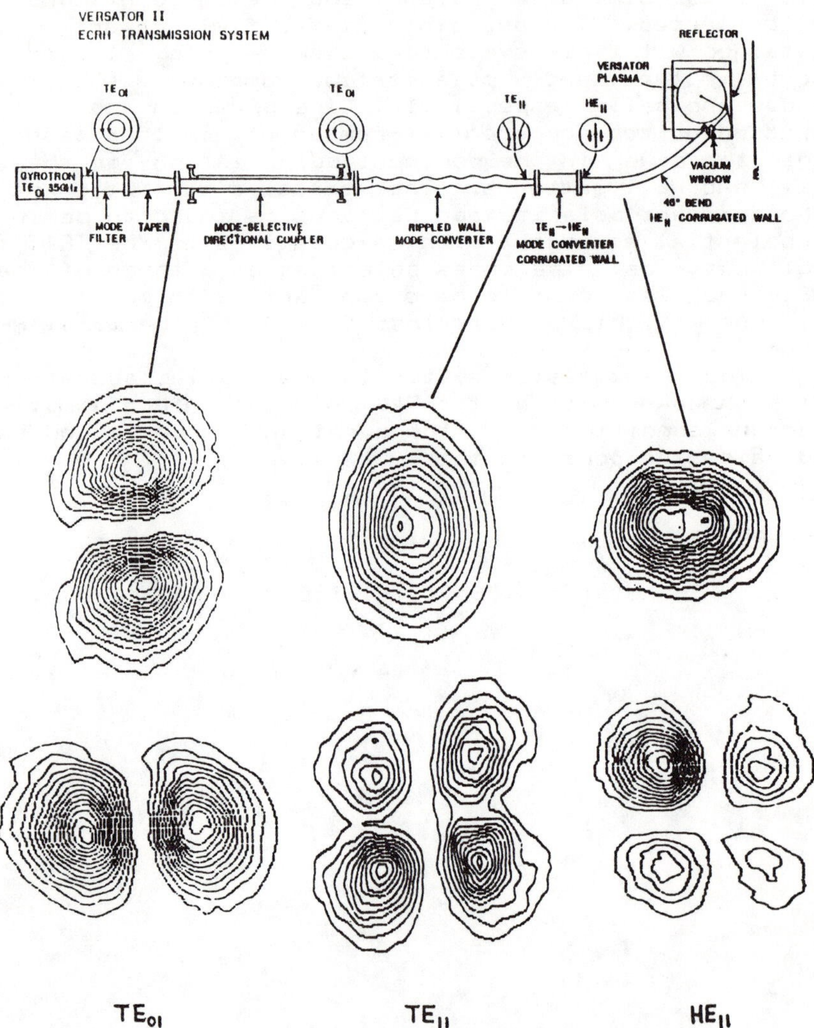

Fig.1. Versator II ECRH system and contour plots of the measured far field radiation patterns after each mode converting section, both polarizations shown. The crosspolarization patterns for the TE11 and HE11 modes (lower patterns) are approximately 15dB below the dominant polarization patterns (upper patterns).

ON APPLICATION OF ECH IN A TOKAMAK REACTOR

Hulbert Hsuan
Princeton University, Princeton, NJ 08544

ABSTRACT

Spatial localization of ECH energy depositon may be used tactically as a control of plasma parameters in addition to its usually assumed role of bulk heating and current drive in a tokamak reactor. As the fusion power amplification factor Q increases to and beyond "power breakeven," ECH may be used to tailor the α-particle slowing-down time and, thus, to control the initial time evolution of Q. Eventually, ECH when applied at rational magnetic surfaces can also be used as a control valve of power thruput, impurity concentration, and the extent of plasma edge regions.

INTRODUCTION

Electron Cyclotron Heating (ECH) has a number of advantages when applied to a tokamak reactor as the primary heating and current drive source. These advantages include strongly localized power deposition to bulk electrons; small launching structure with highest power flux; no direct wave absorption by alpha particles; and high-beta plasma heating with virtually no problem of density cut-off when power at higher harmonics (such as second and third) becomes available. These advantages and the convenient application of ECH have been generally recognized and discussed elsewhere.[1,2] In this article, we will propound an argument that ECH can be used naturally and effectively as a means of controlling the time evolution of the initial low values of fusion power amplification factor Q, as well as of moderating the plasma core region and the plasma edge region at high Q operation. This assertion arises from two observations: (1) ECH at the plasma center can provide a control of the slowing-down time of fast ions associated with neutral beam injection and minority ICRF heating before α-particle heating becomes dominant; and (2) ECH, when applied close to rational magnetic surfaces, may also be used as a control valve of power thruput and impurity transport.

ROLE OF ECH WHEN $Q \lesssim 1$

Recent experiments in TFTR with deuterium neutral beams have produced plasmas which project an equivalent deuterium-tritium (d-t) fusion power amplification factor, $Q_{dt} \simeq 0.25$.[3] These optimized results are obtained with low pre-injection densities and low plasma current and with a very low level of MHD activity at the plasma core. Detailed calculations indicate that the fusion neutron production source strenth is dominated by beam-plasma and beam-beam contributions, in spite of the fact that ions at the plasma core have already achieved the 20-keV range, in which the d-t thermal fusion cross section maximizes. This dramatizes the advantage of non-Maxwellian tokamak plasmas[4,5] as the TFTR-generation of tokamaks seek to achieve "scientific breakeven" with $Q_{dt} = 1$.[6]

Another successful plasma heating method is ICRF heating through direct minority ion damping of the wave. The formation of an energetic ion species has proven to be the source of plasma heating.[7] In fact, selective ICRF absorption by deuterons in a predominantly triton plasma can lead to a two-component velocity distribution suitable for the production of abundant fusion reactions -- similar to a neutral beam injection case.[8] A fundamental advantage of a tokamak reactor is that the 3.5 Mev α-particles of the d-t reaction can be confined effectively within the plasma volume. Thus, it is important to realize that some of the best currently proven plasma heating methods all involve the production of fast ion species, just as α-particles form a natural fast ion species in a fusion reactor.

The nature of the fast ion heating is such that T_e acts as an important parameter.[9] Indeed, neutral beam heating results in TFTR - to this point -- demonstrate clearly that the fusion-neutron production rate is a monotonically increasing function of the central electron temperature.[10] This is understandable, since the equilibrium fast ion density, n_f, is linearly proportional to the fast ion slowing-down time. As a result, the fusion neutron source strength is the sum of three terms:

the beam-plasma reaction rate = $n_f n_e \langle \sigma V \rangle_{bp} \propto \langle \sigma V \rangle_{bp} T_e^{3/2}$

the energetic-ion self reaction rate = $n_f^2 \langle \sigma V \rangle_{bb} \propto \langle \sigma V \rangle_{bb} T_e^3 / n_e^2$

the thermal ion reaction rate = $n_d n_t \langle \sigma V \rangle_{dt} = n_d n_t f(T_i)$

The importance of a higher value of T_e enters directly through the beam-plasma and the beam-beam contributions, and indirectly through the thermal plasma contribution due to the fact that the energy portion transferred to plasma ions is also a montonically increasing function of electron temperature.

Having established the importance of higher electron temperature, we shall briefly discuss the electron energy balance in the plasma core with emphasis on the functional dependence upon T_e. We shall use a simple zero-D electron power balance model for electron energy W_e in a core region:

$$\dot{W}_e + W_e/\tau_{Ee} = P_b(1 - G_{bi}) + P_\alpha(1 - C_\alpha T_e) + P_{\Omega o}\, T_e^{-3/2} + P_{ieo}\, T_e^{-3/2}(T_i - T_e) + P_{ECH} - P_{rad} \qquad (1)$$

where P_b is the beam heating power in the region; P_α is the α-particle heating power which is roughly $QP_b/5$ for d-t operation; $C_\alpha = (m_\alpha/m_e)^{1/2}(1 + m_\alpha/m_i)^{2/3}/E_\alpha$, $E_\alpha = 3.5$ MeV; $P_{\Omega o}$ a normalized ohmic heating power; P_{ieo} a normalized thermal ion-electron power transfer; P_{rad} the radiative power including the Bremsstrahlung radiation which is proportional to $T_e^{1/2}$; W_e the electron thermal energy; τ_{Ee} the transport related electron energy confinement time, assumed proportional to $T_e^{-\ell}$, where ℓ is a positive number that may possibly be 1/2; and G_{bi} is the portion of beam energy transferred to the thermal ions (a monotonically increasing function of T_e).

An important observation should be noted about Eq. (1). When the α-particle heating power is negligible, all the power inputs, except ECH, decline with increasing T_e (which also leads to a lower τ_{Ee}). Thus, we are confronted with the frustrating experimental fact of nearly no change in electron temperature during neutral beam heating.

Introduction of ECH power to the plasma core should make the electron energy balance in the core directly controllable. This is precisely what experimental evidence showed in the PDX-ECH experiment.[11] Localized heating by ECH can sometimes be achieved experimentally only by precise choice of the launching antenna pattern. This is shown most dramatically by ECH experiments in the Wendelstein VII-A stellarator.[12] ECH power can become dominant due to its direct localized heating of the bulk electrons and has been shown to give very peaked electron temperature in the plasma core which then excited MHD activity.[13] However, for the TFTR-generation of tokamaks, fast electron heating of the central core cannot be limited by sawtooth activity due to a much slower magnetic flux diffusion time.

ROLE OF ECH WHEN Q >> 1

When Q is negligibly small, ECH power can be used to raise the electron temperature, which in turn will lengthen the α-particle slowing-down time. However, as more α-particles accumulate in the plasma core, the α-particle-heated region grows until finally α-particle heating becomes dominant among the electron heating processes. At that point, ECH power should be turned off in the tokamak core. The electron temperature is determined by the balance between α-particle heating power and electron power loss terms. This may be ideal except that the thermonuclear self-heating phenomenon gives rise to problems as well as benefits. The possible onset of MHD modes, some possible thermal instability, and impurity transport in the central region may render it desirable to control the electron energy loss in the core region by creating a valve action. In addition, the radial excursion of α-particles will exceed the width of the birth profile and will affect the overall plasma profiles.[14] "Control valve" action may also be required in the outer region of plasma in order to avoid disruptions.

Heating by the α-particles is so large that the ECH power needed is insignificant except when being applied to some rational magnetic flux tubes. Interaction with MHD activity at rational magnetic surfaces becomes an attractive control method.[15] The TFTR-generation and future tokamaks are certainly to be operated with plasma parameters such that the optical depth for ECH waves (even up to third harmonics) is much greater than unity and permit the ulilization of ECH current drive. Stabilizing and deliberately destabilizing effects are easily obtained from non-linear magnetic island develoment.[16] Although control of modes with $m \geq 2$ requires rather low ECH power, the control of the m=1 mode is more difficult because of its proximity to ideal MHD instability. Further experimental work using ECH is necessary to gain insights.

REFERENCES

1. H. Hsuan, "ECH in a Tokamak Reactor," Princeton Plasma Physics Laboratory Report PPPL-2022 presented at the EC-4 Fourth International Workshop on ECE and ECH, Rome, March 1984; A.C. England and H. Hsuan in Wave Heating and Current Drive in Plasmas, ed. V.L. Granatstein and P.L. Colestock (Gordon and Breach Science Publishers, 1985) Chapter 10, pp. 459-509.
2. M. Firestone, T.K. Mau, and R.W. Conn, Comments on Plasma Phys. and Cont. Fusion 2, 149 (1985).
3. J.D. Strachan et al., Phys. Rev. Lett. 58, 1004 (1987).
4. J.M. Dawson, H.P. Furth, and F.H. Tenney, Phys. Rev. Lett. 26, 1156 (1971).
5. R.M. Kulsrud and D.L. Jassby, Nature 259, 541 (1976).
6. D.J. Grove and D.M. Meade, Nucl. Fusion 25, 1167 (1985).
7. J.C. Hosea et al., Phys. Rev. Lett. 43, 1802 (1979).
8. T.H. Stix, Nucl. Fusion 15, 737 (1975).
9. T.H. Stix, Plasma Phys. 14, 367 (1972).
10. H.W. Hendel, A.C. England, D.L. Jassby, A.A. Mirin, and E.B. Nieschmidt, J. Fusion Energy 5, 231 (1986).
11. H. Hsuan et al. Proc. Fourth Int. Symp. on Heating in Toroidal Plasmas 2, 809 (1984).
12. V. Erckmann et al., Proc. Sixth Topical Conf. on Radio Frequency Plasma Heating, ed. D.G. Swanson (AIP Conference Proceedings No. 129, 1985) p. 198.
13. H. Hsaun et al., Plasma Phys. and Contr. Fusion 26, 265 (1984); A. Cavallo et al., Nucl. Fusion 25, 335 (1985); A.H. Kritz et al., Proc. of Sixth Topical Conf. on Radio Frequency Plasma Heating, ed. D.G. Swanson (AIP Conference Proceedings No. 129, 1985) p. 235.
14. T.E. Stringer, Plasma Phys. 16, 651 (1974).
15. D.W. Ignat, P.H. Rutherford, and H. Hsuan, Course and Workshop on Applications of RF Waves to Tokamak Devices, ed. S. Bernabei, B. Casparino, and E. Sindoni, Vol. II, 525 (1985).
16. P.H. Rutherford, Phys. Fluids 16, 1903 (1973); R.B. White et al., Phys. Fluids 20, 800 (1977); D. Biskamp, H. Welter, Plasma Phys. and Cont. Nucl. Fusion Research Vol. I, 579, IAEA-CN-35/B2.2 (1977); R.B. White et al., Magnetic Reconnection and Turbulence (Editions de Physique, Orsay, 1985) ed. M.A. Dubois et al., p. 299.

HIGHLY CHARGED ION PRODUCTION IN ECRH PLASMA SOURCES FOR HEAVY-ION ACCELERATORS AND OTHER APPLICATIONS*

T. A. Antaya
National Superconducting Cyclotron Laboratory
Michigan State University, East Lansing, Mi. 48824

ABSTRACT

The design and status of three ECRH ion sources under development at NSCL are briefly discussed. The RT-ECR ion source, with two minimum B plasma stages and ECRH heating at 6.4 GHz, produces useable intensities of fully stripped light ions up to oxygen; for heavier species, charges such as Argon 14+, Krypton 20+, Iodine 25+ and Tantalum 29+ have been measured. The 6.4 GHz CP-ECR, just beginning operation, has a high temperature metal vapor oven replacing the first plasma stage, and will be used for metal ion production. Initial results for Lithium ions are presented. The SC-ECR, now in the design stage, has a superconducting magnet structure to allow first harmonic ECRH heating at 30-35 GHz. With a higher cutoff density, it is hoped that A≈200 ions with Q> 50+ will be realized.

ECRH PLASMA SOURCES FOR MULTIPLY CHARGED IONS

The general features of the NSCL RT-ECR ion source, shown in Fig. 1, are typical of ECR sources for high charge state heavy ions. The plasma axis is vertical and there are two minimum B plasma stages with ECRH heating at 6.4 GHz. A $SmCo_5$ hexapole composed of 8.7 kG B_{rem} pieces provides the radial field. The plasma is extracted axially at the bottom in a spherical pierce geometry. The power consumption, mostly due to the mirror coils, is 25-30 kW. Gas is fed into the upper plasma stage at the rate of a few standard cubic centimeters per hour; for gas feed operation there are no consumable source parts so that the time between hardware changes is long. The ECRH resonance is first harmonic at 6.4 GHz in both stages and typical power levels are 100 W in the first stage and 300-600 W in the main stage. (Obviously these parameters are far below the normal levels in mirror fusion experiments.) These sources operate at plasma densities of a few times 10^{11} cm^{-3}. The ECRH heating provides a source of keV energy electrons for electron impact ionization. Electron energies up to 200 keV have been measured in the RT-ECR. Table 1 shows the performance of the RT-ECR for gaseous feed.

The key to the production of highly charged ions is to decouple the main stage plasma from the neutral gas feed. In Fig. 2 full two stage operation is compared with main stage only operation for nitrogen ion production. In the two stage mode, the gas is partially ionized in a microwave cavity in the first stage at pressures in the range of 10^{-3} T to 10^{-4}T, while the main stage pressure is maintained by differential pumping at 5-8 10^{-7}T. The plasma diffuses across the magnetic mirror between the two stages. The pressure differential between the the two stages reduces substantially the charge exchange losses of high charge state ions in the main stage. As can be seen in Fig. 2, the intensity of successively higher charge states are raised a higher percentage in two stage mode over operation of the main stage only. The peak extracted nitrogen charge state is 4+. The large 1+ peak is due to first stage ions (dominantly 1+) that go through the main stage un-ionized. On the other hand, single stage operation shows a monotonic decline of intensity with charge from the nitrogen 1+ peak. This is because the plasma "source" for single stage operation can only be the neutral background, which must be increased to raise the density. The production of high charge states requires high density, but since the only knob is the neutral pressure, the high charge states are suppressed through increase charge exchange. In the two stage mode the first stage is the plasma "source" for the main stage-- high density can therefore be maintained with a low neutral background. The helium in the spectrum arises from a yet not explained effect- the intensities of high charge ions are enhanced by the additiion of a light gas to the plasma. Helium boosts high charge state nitrogen and oxygen ions, while nitrogen and oxygen work better for argon, and neon for krypton.

As can be seen in Fig. 2, for light ions and intermediate charge state heavier ions, the single stage mode can be utilized, and a less costly ECR ion source produced. Fig. 3 shows the new compact CP-ECR ion source at NSCL. This single stage source includes a high temperature oven that provides metal vapor for direct ionization in the plasma stage. This source is based on the single stage mode in the RT-ECR and is also smaller- having only 1/8 the confined plasma volume. Fig. 4 shows early lithium ion production. Fully stripped lithium was obtained.

The two stage mode is limited by the cutoff frequency for microwave reflection by the plasma. Recent experiments in Grenoble in which high charge state production at 16-18 GHz was compared to 10 GHz heating, demonstrated that the plasma density does scale with the square of the plasma frequency and that extracted currents of very high charge state ions are enhanced by the higher density.[1] The next step in frequency is to 28-35 GHz. Design work is in progress at NSCL for a superconducting magnet with a 5-35 GHz operating range (first harmonic) for the SC-ECR ion source, shown in Fig. 5. This project was recently approved for funding by the NSF and construction will start in April 1987. Initial operation will be at lower frequencies, but it is hoped that testing at 30-35 GHz will quickly be possible, though at present the high frequency transmitted is not funded. The design goal for this source is 50-60+ ions for M≈200.

REFERENCES

*Work supported by the US NSF under Grants PHY86-11210 and PHY82-15585.

1. R. Geller, et.al., 7th Int. Workshop on ECR Ion Sources, Julich, FRG, 187 (1986).

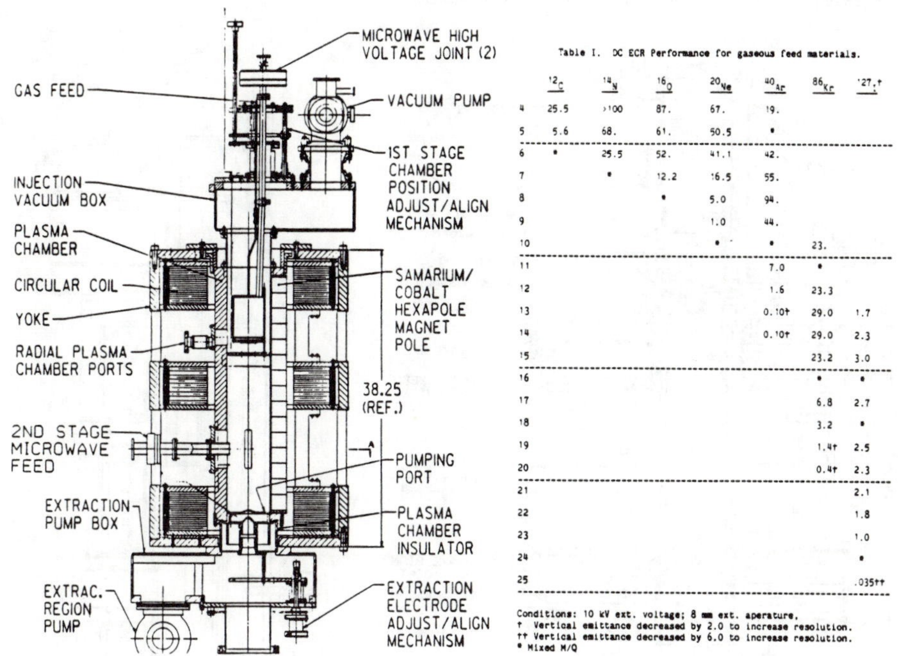

Figure 1.

Table I. DC ECR Performance for gaseous feed materials.

	^{12}C	^{14}N	^{16}O	^{20}Ne	^{40}Ar	^{86}Kr	^{127}I†
4	25.5	>100	87.	67.	19.		
5	5.6	68.	61.	50.5	*		
6	*	25.5	52.	41.1	42.		
7		*	12.2	16.5	55.		
8			*	5.0	94.		
9				1.0	44.		
10				*	*	23.	
11					7.0	*	
12					1.6	23.3	
13					0.10†	29.0	1.7
14					0.10†	29.0	2.3
15						23.2	3.0
16						*	*
17						6.8	2.7
18						3.2	*
19						1.4†	2.5
20						0.4†	2.3
21							2.1
22							1.8
23							1.0
24							*
25							.035††

Conditions: 10 kV ext. voltage; 8 mm ext. aperature.
† Vertical emittance decreased by 2.0 to increase resolution.
†† Vertical emittance decreased by 6.0 to increase resolution.
* Mixed M/Q

Table 1.

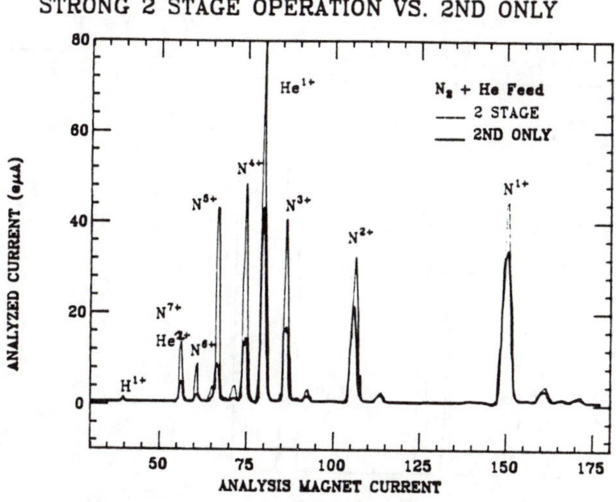

Figure 2.

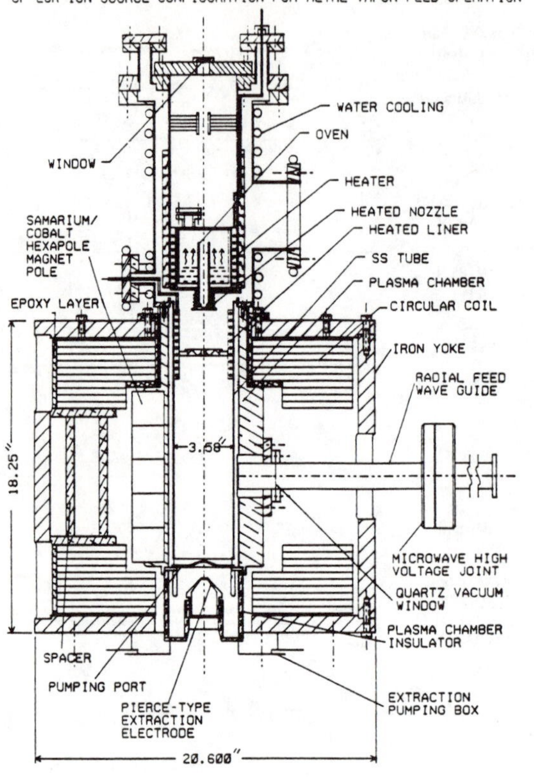

Figure 3.

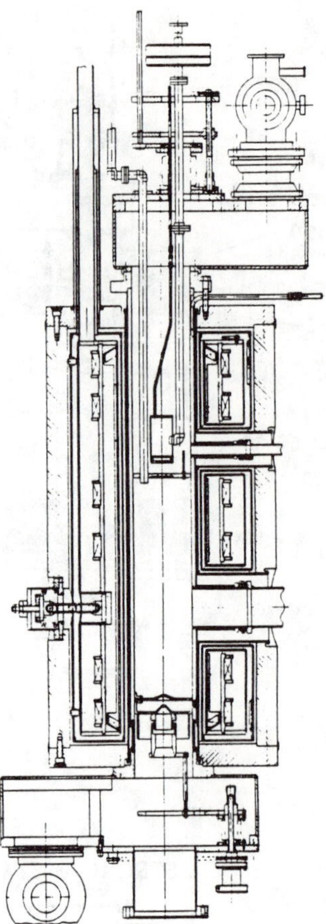

Figure 5.

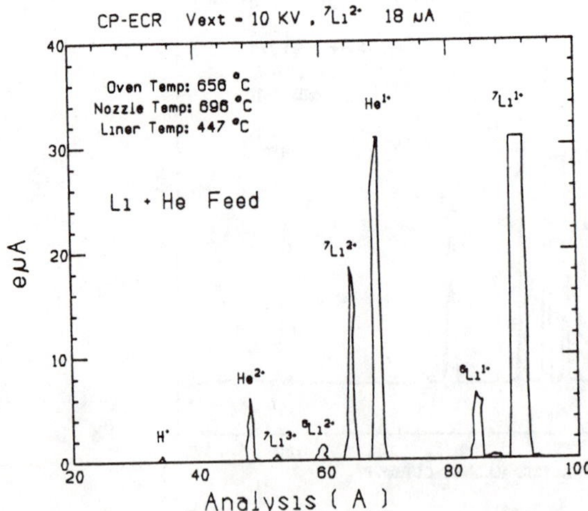

Figure 4.

STATUS OF THE 500 kW - 8 GHz GYROTRON OSCILLATOR DEVELOPMENT PROGRAM

G. Agosti, H.-G. Mathews
BBC Brown, Boveri and Company Limited, Baden, Switzerland

S. Alberti, P. Muggli, A. Perrenoud, M.Q. Tran, T.M. Tran
Centre de Recherches en Physique des Plasmas,
Association Euratom Confédération Suisse,
Ecole Polytechnique Féderale, Lausanne/Switzerland

ABSTRACT

In the frame of a development program aiming towards an 8 GHz, 500 kW, CW or long pulse (~10 s) gyrotron , we are testing a 200 kW prototype. Its principal characteristics are: cavity mode = TE_{01}, quality factor = 400, single anode magnetron injection gun insulated in a SF_6 high voltage tank, beam voltage = 80 kV, maximum beam current = 25A, ratio $p_\perp/p_\parallel \approx 1.9$. The computed electronic efficiency is around 50%. The prototype has been tested at low current ($I \leqslant 5A$) and in short pulse (5 ms). The power obtained with a beam current of 4.8 A was 200 kW, yielding an efficiency of 50%. Tests are under way to completely characterize the tube. The next prototype designed to achieve a higher output power ($\geqslant$ 300 kW) and longer pulse is under construction.

INTRODUCTION

Gyrotron oscillators are usually associated with high frequency ($f \geqslant$ 28 GHz) microwaves. However with the present need of high power sources (~1 MW) at X band (f = 8 - 12 GHz) for lower hybrid heating, gyrotron[1] oscillators (and, more preferably, amplifiers) have also been considered. In the general framework of high power microwave tube development at the Electron Tubes Department EKR of BBC Brown, Boveri and Company Ltd and at the Centre de Recherches en Physique des Plasmas (CRPP), an 8 GHz gyrotron has been designed and tested. The goal of this project is a long pulse or CW tube capable of delivering up to 500 kW of microwave power. As an intermediate step a 200 kW prototype was built and tested. The tube is a modified version of the one described at the 1986 IEDM.[2] Its design and the results of its tests are reported in this paper.

CHARACTERISTICS OF THE TUBE

The operating frequency of the tube has been fixed at 8 GHz. A schematic of the whole gyrotron with all the magnet coils is shown on figure 1. The Magnetron Injection Gun (M.I.G.) is a single anode gun. A relatively low maximum emission current of 3.75 A/cm^2 was chosen, at the maximum beam current of 25 A. The operating voltage is 80 kV and the ratio $\alpha = p_\perp/p_\parallel$ is around 1.9. A distinctive feature of the M.I.G. is its SF_6 high voltage insulation. Extensive tests both at factory and during operation on the test stand have shown that this approach does not present any insulation or thermal problem while it allows easier handling of the tube. The cavity operates at TE_{011} mode and its quality factor is 400 at the resonant frequency of 8.053 GHz. At the cavity output, an uptaper flares the diameter up to a standard C-18 waveguide. The output beam is spread by a series of collector coils: at full beam power (P_b = 2 MW) the thermal load is 420 W/cm^2. The window is a single Al_2O_3 edge cooled disk. The overall length of the tube is 2.3 m.

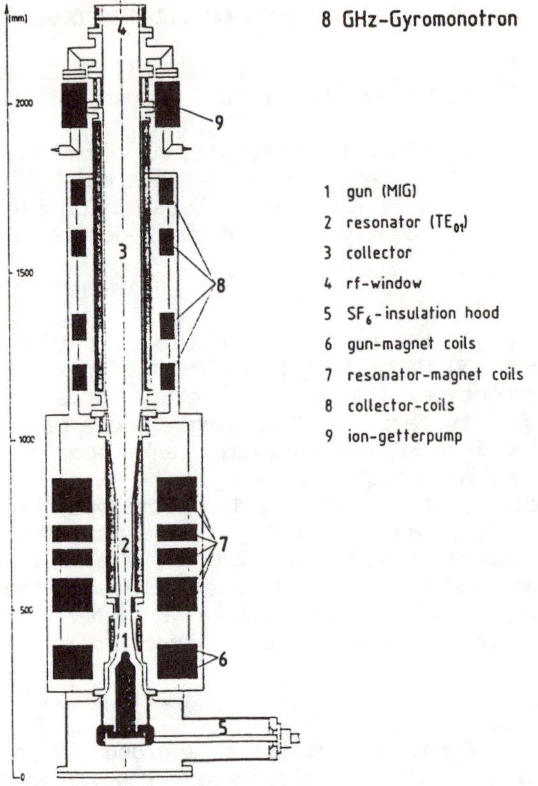

Fig. 1. Schematic of the gyrotron showing the main components.

EXPERIMENTAL RESULTS

The gyrotron was tested with the available HV modulator of the CRPP, which can deliver up to 10 A at 90 kV. The typical pulse length in this experiment is around 5 - 10 ms. Diagnostics include calibrated couplers and water load for power measurement, frequency discriminators and spectrum analyzer for frequency determination, liquid crystal paper and a k-spectrometer[3] for mode measurements. A CAMAC data acquisition system allows to store and handle all the important signals such as the beam voltage and current, the body and collector currents and the microwave signals. At each current the magnetic fields in the cavity and in the gun were optimized to yield the maximum output power. The relation between the optimized power and the beam current is shown on figure 2. One notes that 200 kW was obtained with a beam current of 4.8 A at 80 kV. The corresponding efficiency is around 50%. We have not yet investigated the domain of current above 5 A. Frequency measurements were also performed using both the spectrum analyzer and the frequency discriminators: the oscillation frequency is 8.056 GHz ±2 MHZ. The mode pattern as recorded by a liquid cristal sheet corresponds to the TE_{01} one (Fig. 3) indicating that there is no significant mode conversion in the uptapered section. Further measurements are in progress with the k-spectrometer to determine the mode purity. No oscillations at the second harmonic and at the TE_{02} mode were found. X-ray films were used to determine the impact zone of the spent beam on the collector. We found good agreement between the experiment and the prediction from the numerical code.[4]

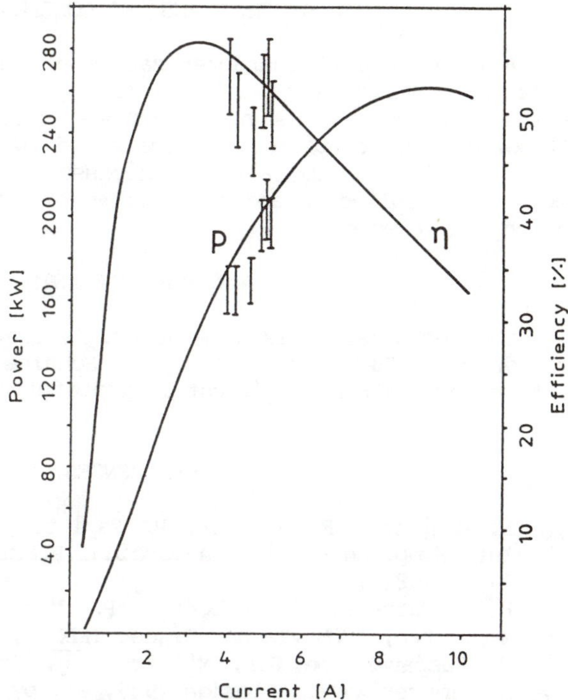

Fig. 2. Optimized power and efficiency versus beam current at 80 kV. The points are experimental data. The solid lines are the theoretical predictions with α = 1.89.

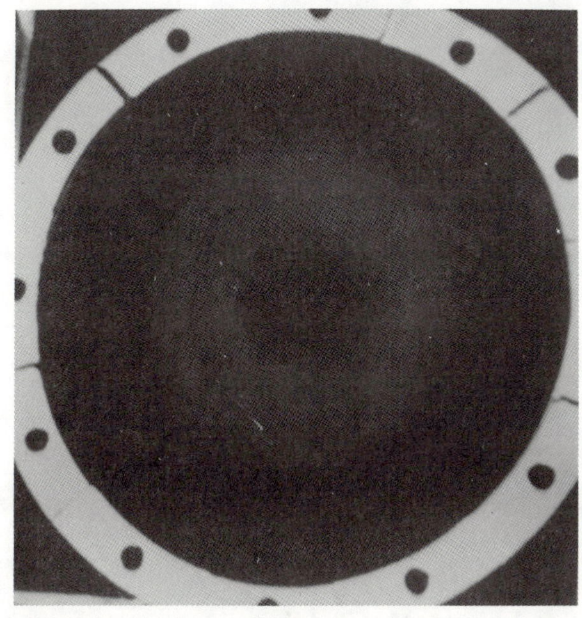

Fig. 3. Mode pattern of the output power. The bright zone shows the maximum of the E-field.

FUTURE PLANS AND CONCLUSIONS

A more complete characterization of the present prototype is under way. Measurements include mode purity, power and frequency versus beam current. A new prototype designed for operation at 300 kW will be completed and installed by summer this year. This tube will incorporate all the necessary cooling systems in the cavity, the collector and the window and shall be able to operate in long pulse mode.

ACKNOWLEDGEMENTS

The technical assistance of Dr. A. Lietti and Mr. P. Hostettler of the CRPP and Messrs. O. Schafheitle, B. Zigerlig and H. Kaser of the EKR Department is gratefully acknowledged.

REFERENCES

1. R. Andreani, F. Mirizzi, P. Papitto, M. Sassi, Proc. of 4th Int. Symp. on Heating in Toroidal Plasmas (Rome 1984) Vol. II, 1171 (1984).
2. H.G. Mathews, S. Alberti, P. Muggli, A. Perrenound and M.Q. Tran, IEDM, Techn. Digest IEDM86, 805 (1986).
3. W. Kasparek and G.A. Müller, Dig. Int. Conf. Infrared and Millimeter Waves, Florida 1985, ed. by R. Temkin, 238 (1985).
4. W.B. Herrmannsfeldt, SLAC Report SLAC-226 UC-28 (1979).

Third-Harmonic Electron Cyclotron Heating for Tokamak High-Beta Profile Control Studies

T. C. Simonen, K. Matsuda, R. Prater, and G. R. Smith**
GA Technologies Inc.
San Diego, CA 92138

ABSTRACT

Third-harmonic extraordinary-mode ECH could be employed for D III-D high-beta studies at reduced toroidal magnetic field. D III-D neutral beam injection increases the electron temperature above 4 keV. At 4×10^{19} m^{-3} density, the third-harmonic 120 GHz ECH power should be absorbed with greater than 80% efficiency. Absorption would increase at higher temperature unless the distribution function is too severely distorted by the high-power ECH. A 6 MW–120 GHz system could thereby allow well-controlled beta and profile control studies using third-harmonic ECH.

INTRODUCTION

The advent of high-frequency, high-power free-electron-laser microwave sources increases the suitability of electron cyclotron heating of high-field magnetic fusion devices. In addition, the FEL high-frequency capability allows heating at higher density and at cyclotron harmonics. The cutoff density is

$$n_{19} = \begin{cases} 0.5B_T^2 = 1.2\times10^{-3}f^2 & \text{O-mode} \\ B_T^2 = 2.4\times10^{-3}f^2 & \text{X-mode } \ell = 1 \\ B_T^2[\ell(\ell-1)] = 1.2\times10^{-3}f^2\,(1-1/\ell) & \text{X-mode } \ell \geq 2 \end{cases} \qquad (1)$$

where n_{19} is in units of 10^{19} m^{-3}, B_T in Tesla, f in GHz, and $\ell = \omega/\omega_{ce}$. Second- and third-harmonic heating, rather than the fundamental heating, increases the cutoff density by a factor of 2 and 6, respectively. Third-harmonic heating of a 10 T device such as the Compact Ignition Tokamak (CIT) is possible to densities up to 6×10^{21} m^{-3}.

In this paper we describe the application of third-harmonic ECH to the D III-D tokamak operating at reduced toroidal magnetic field for high-beta experiments. Third-harmonic ECH resonance occurs at 1.4 T with 120 GHz

* Permanent address: Lawrence Livermore National Laboratory, Livermore, CA 94550

allowing wave penetration up to densities of $1 \times 10^{20}\ \mathrm{m}^{-3}$. Third-harmonic absorption requires electron temperatures higher than obtainable with ohmic heating alone. The concept proposed here is to first increase the electron temperature above 4 keV with neutral beam injection, as has already been achieved.[1] Then third-harmonic ECH is well absorbed to provide localized heating for high-beta and profile control studies. When the electron temperature exceeds 10 keV the power deposition profile broadens and fourth-harmonic absorption sets in.

THIRD-HARMONIC ABSORPTION EFFICIENCY

The third-harmonic absorption efficiency can be evaluated in terms of the optical depth τ,

$$P_{\mathrm{abs}}/P_{\mathrm{incident}} = 1 - e^{\tau}, \tag{2}$$

where τ is given by Bornatici[2] for $3\omega_{ce}$ X-mode

$$\tau_3^{\mathrm{x}} = \frac{81\pi}{16}\left(\frac{\omega_{ce}}{c}\right)\left(\frac{kT_e}{mc^2}\right)^2 L_B \mu_3^{\mathrm{x}'} \ . \tag{3}$$

Here

$$\mu_3^{\mathrm{x}'} = \frac{4}{243}\alpha\left(\frac{12-\alpha}{8-\alpha}\right)^2\left(\frac{\alpha^2-18\alpha+72}{8-\alpha}\right)^{3/2}, \tag{4}$$

where $\alpha = \omega_{pe}^2/\omega_{ce}^2 = n_{19}/B_T^2$. For D III-D, $L_B = B_0/|dB_0/dr| = 1.67$ m. Evaluating Eq. (3) for D III-D parameters of $T_e = 4$ keV and $n_e = 4$ to $10 \times 10^{19}\ \mathrm{m}^{-3}$, we find $\tau = 2$, corresponding to 86% absorption efficiency. TORAY[3] ray tracing calculations confirm the above estimate and indicate that the absorbed ECH power is localized within an 8 cm zone when $T_e = 10$ keV for near normal launch.

EVALUATION OF HEATING AND DISTORTION OF THE ELECTRON DISTRIBUTION FUNCTION

We evaluate here the extent to which the electron distribution function is distorted from Maxwellian. If we first consider the heating on a global basis, we then estimate the electron temperature rise from 5 MW of ECH power for D III-D parameters to be

$$\Delta T_e = \frac{P_{ECH}\tau_E}{en_eV} = \frac{5 \times 10^6\ \mathrm{W} \times 0.2\ \mathrm{s}}{1.6 \times 10^{19} \times 6 \times 10^{19}\ \mathrm{m}^{-3} \times 20\ \mathrm{m}^3} = 5\ \mathrm{keV}\ .$$

The expected temperature increase is large.

Since the heating is expected to be localized to an 8 cm radial zone, the high power ECH could drastically distort the electron distribution function. This is particularly true since third-harmonic heating takes place on the Maxwellian tail of the distribution, for electrons with energy $4T_e$. If ECH increased the electron temperature to 10 keV, then such electrons would have an energy of 40 keV. The Coulomb electron-electron collision time τ_{ee} for 40 keV electrons is 0.5 ms. In comparison, the energy confinement time within the $\Delta r = 8$ cm heating zone can be roughly estimated as

$$\tau_E^{\text{zone}} = \tau_E^{\text{global}} \left(\frac{\Delta r}{a}\right)^2 = 0.2 \times \left(\frac{8\,\text{cm}}{60\,\text{cm}}\right)^2 = 3.5\,\text{ms} \quad .$$

Since $\tau_{ee} < \tau_E^{\text{zone}}$, we expect the electrons in the heating zone to remain Maxwellian. The temperature rise in the 8 cm heating zone depends on the major radius of the resonance. At $r = a/3 = 20$ cm, then, the heating volume is approximately $2\,\text{m}^3$. In this small volume, the electron temperature rise in the heating zone would be

$$\Delta T_e^{\text{zone}} = \frac{P_{ECH}\tau_E^{\text{zone}}}{en_e V^{\text{zone}}} = \frac{5 \times 10^6\,\text{W} \times 3.5 \times 10^{-3}\,\text{s}}{1.6 \times 10^{19} \times 6 \times 10^{19}\,\text{m}^{-3} \times 2\,\text{m}^3} = 0.9\,\text{keV} \quad .$$

These estimates thus indicate that the ECH power will be distributed radially as well as in velocity space. At lower densities, the distortions would be larger and the formation of a very energetic electron population might be established. Such a population could possibly be used to aid MHD stability, as has been proposed theoretically.

D III-D BETA STUDIES

In Table I we illustrate tokamak operating parameters using 60 and 120 GHz ECH for magnetic fields appropriate for D III-D. In all cases, we assume that a beta of 6% (3% average) is initially obtained by neutral beam heating. ECH is used as supplemental heating to increase beta locally. Such a scenario is analogous to the use of ECH to supplement ICRF on the CIT. Given in Table I are possible electron densities (n_e) and temperatures (T_e) for 6% beta, together with the calculated optical depth and percentage absorption. Also listed is the plasma current I_p for a Troyon parameter $I_p/aB = 2$ and the DITE parameter $n_e qR/B$ (evaluated for $q = 2.6$), which must be in the range of 10-30 for successful tokamak operation. For both 60 and 120 GHz, the third harmonic

Table I. Parameters for D III-D with 6% on-axis beta (3% average).

ECH frequency	60 GHz			120 GHz		
Harmonic	$2\omega_{ce}$	$3\omega_{ce}$	$4\omega_{ce}$	$2\omega_{ce}$	$3\omega_{ce}$	$4\omega_{ce}$
B (T)	1.05	0.7	0.53	2.1	1.4	1.05
n_{cutoff} ($10^{19}\,\text{m}^{-3}$)	2	3	3.4	8.8	11.8	13.2
T_e (keV)	4.5	1.8-3.6	4.2	5	1.8-3.6	3.4
n_e ($10^{19}\,\text{m}^{-3}$)	1.7	2-1	0.5	8	8-4	2
τ (optical depth)	85	0.43-1.33	0.08	189	1.5-2.9	0.1
% absorption	100%	35-74%	8%	100%	78-94%	10%
I_p (MA)	1.4	0.93	0.7	2.8	1.9	1.4
$n_e qR/B$	7	12-6	4.1	16.5	25-12	8.2

has significant absorption. For 120 GHz, second-harmonic heating is well suited for D III-D. In addition, third-harmonic heating with 120 GHz ECH power is an attractive possibility.

ACKNOWLEDGMENTS

The authors wish to acknowledge helpful discussions with V. S. Chan, W. M. Nevins, T. Taylor, R. A. James, and B. W. Stallard.

This work was performed under the auspices of the U.S. Department of Energy by the Lawrence Livermore National Laboratory under contract number W-7405-ENG-48 and DE-AC03-84ER51044 at GA Technologies Inc.

REFERENCES

1 J. Luxon et al., to be published in *Proc. 11th International Conference on Plasma Physics and Controlled Nuclear Fusion Research, Kyoto, 1986.*

2 M. Bornatici et al., *Nucl. Fusion* **23**, 1153 (1983).

3 A. H. Kritz et al., *Proc. 3rd Joint Varenna-Grenoble International Symposium on Heating in Toroidal Plasmas, Grenoble, 1982*, Vol. 2, 707 (1982).

3D BOUNCE AVERAGED FOKKER-PLANCK CALCULATION OF ELECTRON CYCLOTRON CURRENT DRIVE EFFICIENCY

R.W. HARVEY
GA Technologies Inc., San Diego, CA 92138

M.G. MCCOY and G.D. KERBEL
NMFECC, Lawrence Livermore National Laboratory, Livermore, CA 94550

Using the Kerbel-McCoy code to solve for the 2D (in velocity) relativistic bounce-averaged electron distribution at each point on a radial array, consistent with an ECH energy deposition model incorporating quasilinear damping, we obtain the toroidally averaged distribution everywhere in the tokamak. This is used to explore ECH current drive for several launching scenarios.

Construction of comprehensive models for tokamak electron cyclotron heating and current drive is motivated by the increasing level of experimental activity in this area [1]. We have assembled a code which predicts the profiles of heating and current drive, the global current drive efficiency, and diagnostic signals which depend upon profiles, such as soft x-ray (SXR) and electron cyclotron emission (ECE). By running an array of two dimensional, bounce averaged Fokker-Planck/quasilinear (FP/QL) codes on a set of flux surfaces parameterized by minor radius r we obtain the steady state, toroidally averaged electron distribution functions resulting from a balance between collisions and QL diffusion [2]. The incoming rf energy is damped consistent with the QL distortion of the distribution functions at each point in the tokamak. Heat thereby given to the electrons is removed by linearizing the Coulomb collision operator about a background Maxwellian distribution. Plasma density and temperature vary parabolically in the radial direction. Figure 1 depicts the physical situation which we are modeling. The vertically delimited channel of ECH energy may be injected from the inside as shown or from the outside; its frequency ω equals the electron cyclotron frequency $\omega_{ce}(R)$ at the cyclotron layer.

For a given location in the plasma cross-section, the relativistic resonance conditions determines the velocity space coordinates at which the wave-particle interaction occurs. From this condition

$$v_{\parallel} = \frac{\omega - \omega_{ce}/\gamma}{k_{\parallel}} \quad , \tag{1}$$

we can recognize the potential importance of relativistic effects even in moderate ($\sim$ keV) temperature plasmas. The resonance velocity $v_{\parallel}$ for significant interaction is $\sim 3\,v_{Te}$ $[v_{Te} \equiv (T_e/m_e)^{1/2}]$, and this velocity is the difference between two relatively large velocities of $\omega/k_{\parallel}, \omega_c/\gamma k_{\parallel} \sim 2\,c$, thus small changes in γ significantly affect $v_{\parallel}$.

Expressing Eq. (1) in $\bar{\boldsymbol{v}} \equiv \bar{\boldsymbol{p}}/m_e = \gamma\ \boldsymbol{v}$-coordinates, gives an ellipse in $(\bar{v}_{\parallel}, \bar{v}_{\perp})$-space [3] $(n_{\parallel} < 1)$. We generally work in coordinates giving the parallel and perpendicular velocity $(\bar{v}_{\parallel o}, \bar{v}_{\perp o})$ of collisionless particle orbits at the outer equatorial plane of a circular cross-section plasma. Figure 2 shows Eq. (1) in these coordinates for various minor radii of a wave with (fixed) positive $k_{\parallel}$ corresponding to EC injection on the equatorial plane at an angle 30 deg from $\perp$ to the toroidal direction. In this case the resonance is at $v_\parallel \gg v_{Te}$ for locations near the outside of the plasma ($r/a = \pm 0.8$) and $v_\parallel$ reduces to $\sim \pm 3\, v_{Te}$ as the resonance location approaches the cyclotron layer ($r/a = \pm 0.2$). (The cyclotron layer passes through the tokamak major radius.) As shall become evident, an important feature of these curves is that for $r/a < 0$ they intersect the trapped-transiting boundary but this boundary can be avoided from $r/a > 0$. These resonance effects are incorporated into the QL operator for the FP code.

The equations solved in the 3D code are the steady state of

$$\frac{\partial f_o}{\partial t}\ (\bar{v}_{\parallel o}, \bar{v}_{\perp o}, r, t) = \mathcal{C}(f_o) + \mathcal{Q}(f_o) \quad , \tag{2}$$

at each radial location r, and the rf energy transport equation

$$\nabla \cdot (\ v_g \mathcal{E}) = -P_{\rm ABS}\ (r, \theta_{\rm pol}) = -\int d\,\bar{v}\,\tfrac{1}{2} m v^2\, \mathcal{Q}(f) \quad . \tag{3}$$

Referring to Eq. (2), at each radius f_o is the electron distribution function at the equatorial plane, $\mathcal{C}$ is the collisional, and $\mathcal{Q}$ is the QL bounce-averaged operator. Numerical solution of Eq. (2) is described in Ref. 4. In Eq. (3) $\mathcal{E}$ is the rf energy density and v_g is the group velocity. We obtain the relation of $\mathcal{E}$ to rf electric field, the group velocity, and the warm plasma polarizations necessary for the $\mathcal{Q}$ using a modification of a code by Hsu *et al.* [5]. Presently the EC energy is approximated to enter the plasma in planes at constant vertical distance z. The vertical profile of radiation is chosen to correspond to that expected at the region of maximum power absorption. Figure 3 shows contours of the right-hand rf component obtained with the 3D code for a case of inside, X-mode launch of 1 MW power at 30 deg from $\perp$, into a DIII–D-like geometry (R_o = 167 cm, a = 67 cm) with parabolic density and temperature profiles — central values $n_{eo} = 7.5 \cdot 10^{12}\ \mathrm{cm}^{-3}$, T_{eo} = 2.5 keV. The abrupt decrease in the $|E^-|^2$ component at $r/a \sim -0.2$ is due to damping of most of the rf energy within a surrounding 10 cm region. Figure 4 gives contours of the perturbed distribution function at $r/a = 0.2$, which is Maxwellian for $\bar{v}_o$ to $\sim 3\, v_{Te}$, and has a marked QL plateau for $\bar{v}_o \sim 4$ to $6\, v_{Te}$. Examination of the velocity space flux $D \cdot (\partial f_o / \partial\, \bar{v}_o)$ reveals that the strongest components are due to the rf pushing particles at $\bar{v}_\parallel \sim -3\, v_{Te}$ from low $v_{\perp o}$ into the trapped region, a result which could be anticipated from the resonance curves of Fig. 2 with appreciation that EC QL diffusion is primarily in the $v_{\perp o}$-direction. This asymmetric trapping, first proposed as an EC current drive mechanism by Ohkawa [4], is in the opposite direction to the Fisch-Boozer asymmetric-resistivity current [6]. The Ohkawa trapping effect becomes significant in the QL regime.

Figure 5 (a) and (b) show results from the 3D code for the global current drive efficiency for inside launch and outside launch cases. At very low power, inside launch is marginally more efficient than outside launch. The marked reduction of efficiency for inside launch at higher input power is due to the previously mentioned trapping effect. With outside launch the efficiency increases with applied power due the QL enhancement of tail particle population yet avoiding the trapping effect, consistent with the resonance curves in Fig. 2 at $r/a > 0$.

Whether EC launch is from the outboard or inboard side, the rf generally penetrates until the minimum resonance velocity on the resonance curve at given r, $v_{\rm min}$ is

2 to 3 v_{Te}. In the linear regime, local current drive efficiency scales as $\sim T_e/n_e\,\bar{v}^2_{\parallel\,min}$ [6]. Rough analytic calculation of the local rf power density such that QL and collisional diffusion are equal, gives

$$p_{ABS}|_{D_{QL}=D_{coll}} \propto n_e^2 \left(\frac{v_{Te}}{\bar{v}_{min}}\right)^2 k_\parallel \left[1 - erf\left(\frac{\bar{v}_{min}}{2^{1/2} v_{Te}}\right)\right] \quad .$$

For $\bar{v}_{min}/v_{Te}$ fixed at $\sim$2 to 3, the onset of QL degradation of the current drive efficiency scales as n_e^2. Scaling from our results, the conclusion is that degradation will be an important factor in inside launch experiments at low density; at reactor relevant densities the inside launch degradation, and also the outside launch enhancement, are pertinent but not necessarily over-ridingly important.

This is a report of work sponsored by the U.S. Department of Energy under Contracts No. DE-AC03-84ER53158 and W-7405-ENG-48.

REFERENCES

[1] R. Prater *et al.*, GA Technologies Report GA-A18795 (1987).
[2] M.G. McCoy, G.D. Kerbel, and R.W. Harvey, this meeting.
[3] H.P. Freund *et al.*, Phys. Fluids **27**, 1396 (1984).
[4] G.D. Kerbel and M.G. McCoy, Phys. Fluids **28**, 3629 (1985).
[5] T. Ohkawa, GA Technologies Report GA-A13847 (1976); V.S. Chan *et al.*, Nucl. Fusion **6**, 787 (1982).
[6] N. Fisch and A. Boozer, Phys. Rev. Lett. **45**, 270 (1980).

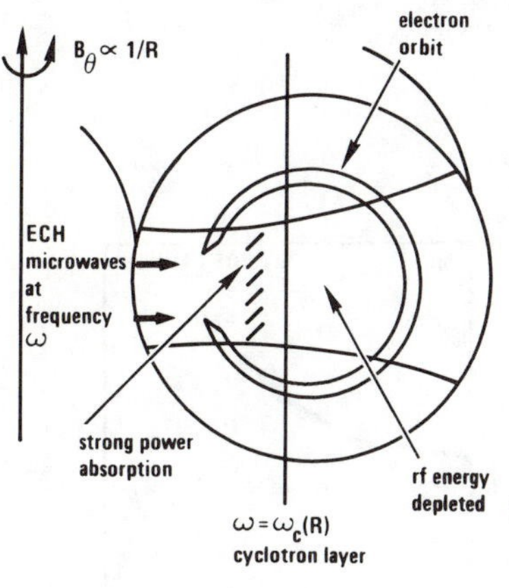

FIG. 1. ECH heating geometry.

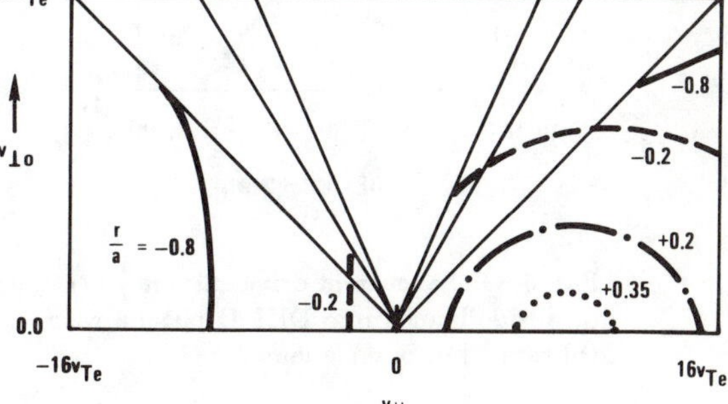

FIG. 2. Resonance condition projected onto outer equatorial plane.

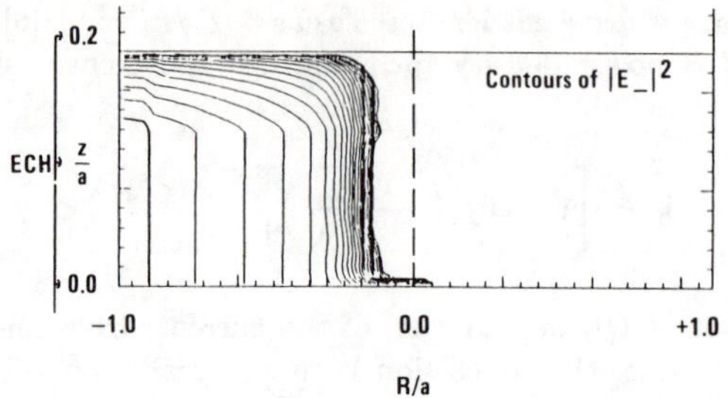

FIG. 3. Contours of $|E^-|^2$ as a function of R and z.

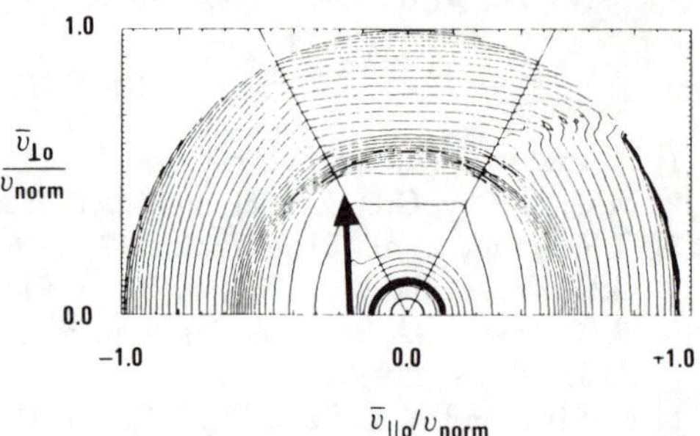

FIG. 4. Contours of f_{eo} as a function of $\bar{v}_{\| o}$ and $\bar{v}_{\perp o}$. v_{norm} is 1.3×10^{10} cm/sec, which is 16 v_{Te}.

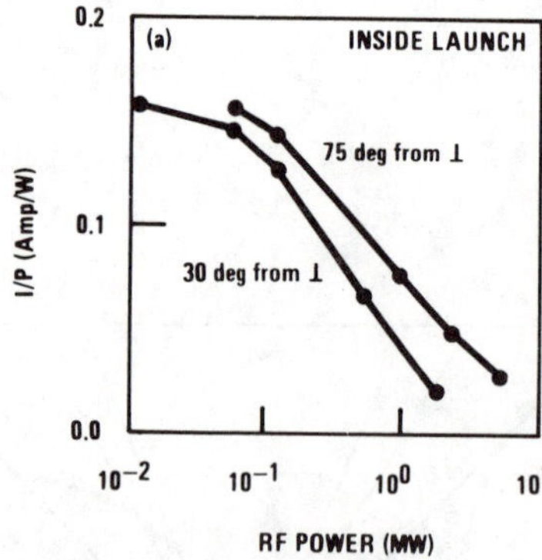

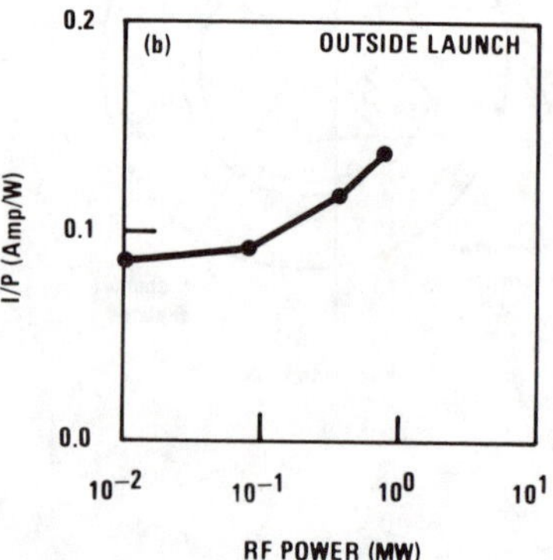

FIG. 5. Global current drive efficiency (I/P) as a function of input power P: (a) inside launch into DIII–D plasma with n_{eo}=$7.5\cdot10^{12}$ cm^{-3}, T_{eo}=2.5 keV; (b) same, but outside launch.

Finite Larmor radius effects on plasma absorption of electron cyclotron waves

E. Mazzucato, I. Fidone[a], and G. Granata[a]
Princeton University, Plasma Physics Laboratory
Princeton, N. J. 08544

Abstract

Finite Larmor radius effects on the plasma absorption of electromagnetic waves with oblique propagation and frequency below the electron cyclotron frequency are investigated. It is shown that the strong reduction of the damping of the extraordinary mode, caused by high order Larmor radius terms for perpendicular propagation and large values of $(\omega_p/\omega_c)^2$, diminishes significantly at large values of $N_{\|}$.

I. Introduction

Recently[1-3], the linear theory of electron cyclotron waves has been used, to the lowest significant order in the Larmor radius, for the study of plasma absorption of extraordinary waves with oblique propagation and frequency ω significantly smaller than the electron cyclotron frequency ω_c. This type of wave absorption offers several advantages: it allows control of the power deposition in both the physical and the velocity space, which can be exploited for profile control or current drive in tokamaks; it maximizes the plasma density for a given value of ω; it makes possible the use of existing high power sources in present large tokamaks.

The damping of electron cyclotron waves in hot plasmas, as those considered in Refs. 1-3, can be affected by finite Larmor radius (FLR) effects. It has been shown previously[4] that the plasma absorption of the extraordinary mode with perpendicular propagation ($N_{\|} = 0$) is significantly weakened by FLR effects when $(\omega_p/\omega_c)^2 \approx 1$. In this article we address the same issue for arbitrary values of $N_{\|}$. Our conclusion is that FLR effects become less important at large value of $N_{\|}$ and that they do not modify significantly our previous results[1-3].

II. The dispersion relation

The relativistic dispersion relation for a Maxwellian plasma has been discussed in several articles. Here we use a form derived recently[5], which is particularly useful for investigating the plasma absorption of electron cyclotron waves with oblique propagation.

The dielectric tensor ϵ is expanded in powers of the Larmor radius keeping up to third order terms. In the orthogonal frame with basis $(\mathbf{e}_1, \mathbf{e}_2, \mathbf{e}_3)$, where

---

[a]Permanent address: Association Euratom-C.E.A., C.E.N.-Cadarache, France.

the magnetic field is $\mathbf{B} = B\mathbf{e}_3$ and the complex refractive index is $\mathbf{N} = N_\perp \mathbf{e}_1 + N_\parallel \mathbf{e}_3$, the anti-hermitian part of ϵ has the form

$$\epsilon''_{ij} = a_{ij} + N_\perp^2(b_{ij} + c_{ij}) + N_\perp^4(d_{ij} + f_{ij} + g_{ij}) \ . \quad (1)$$

This expression contains the contribution of the first (a_{ij}, b_{ij}, d_{ij}), the second (c_{ij}, f_{ij}), and the third (g_{ij}) harmonic of the electron cyclotron frequency, and is therefore valid whenever the effect of the forth harmonic is negligible, i.e., for thermal plasmas with $T_e \leq 50\ keV$ when $\omega < \omega_c$.

Each coefficient in Eq. (1) contains a factor $F(p) = \exp[-\mu(\gamma - 1)]$, with $\mu = mc^2/T_e$, $\gamma = (1 + p^2)^{1/2}$, and p a solution of the resonant equation

$$(1 + p^2)^{1/2} - n\ \omega_c/\omega = N_\parallel p \ .$$

It is this factor that makes the use of the relativistic theory essential for a correct description of the wave absorption, even when $mc^2 >> T_e$.

III. Numerical results

Our previous results[1-3] were obtained with a reduced dispersion relation where only the lowest order terms (a_{ij}) were used in the dielectric tensor. In this section we shall investigate the validity of this approximation.

The imaginary parts of $N_\perp$ given by the reduced and the full dispersion relation, respectively, are shown in Fig. (1) as a function of $(\omega_p/\omega)^2$ for the extraordinary mode with $\omega_c/\omega = 1.3$, $T_e = 20\ keV$, and $N_\parallel = 0$. In agreement with Ref. 4 we find that the high order FLR terms cause a reduction of the wave damping which increases with plasma density.

The ratio of the two results (full/reduced) is plotted as a function of $(\omega_p/\omega)^2$ in Fig. (2) for different values of $N_\parallel$. This shows that while for perpendicular propagation there is a substantial FLR reduction of the wave damping, especially at large values of ω_p/ω, the two results become rapidly very similar as the value of $N_\parallel$ increases. The same phenomenon is shown in Fig. (3) where the ratio is displayed as a function of $N_\parallel$ for several values of T_e. It appears that the ratio is a fast growing function of $N_\parallel$. Also, for perpendicular propagation the FLR effect is not very sensitive to the value of T_e.

Finally, we found that the ordinary mode is only slightly affected by the high order terms of Eq. (1) when $\omega_c < \omega$.

IV. Conclusion

In conclusion we have investigated the contribution of high order Larmor radius terms in the dispersion relation of the extraordinary mode around the electron cyclotron frequency.

We have found that the previously known[4] reduction of the wave absorption, caused by FLR effects at high densities and perpendicular propagation, diminishes significantly at large values of $N_\parallel$.

Acknowledgment

This work was supported by U.S. Department of Energy Contract No. DE-AC02-76-CHO-3073.

REFERENCES

[1] Fidone, I., Giruzzi, G., Mazzucato, E., Phys. Fluids **28** (1985) 1224.

[2] Mazzucato, E., Fidone, I., Giruzzi, G., Krivenski, V., Nucl. Fusion **26** (1986) 3.

[3] Fidone, I., Giruzzi, G., Krivenski, V., Ziebell, L. F., Mazzucato, E., Phys. Fluids **29** (1986) 803.

[4] Bornatici, M., Engelmann, F., and Lister. G., Phys. Fluids **22** (1979) 1664.

[5] Fidone, I., Giruzzi, G., Krivenski, v., and Ziebell, L., Nuclear Fusion **26** (1986) 1537.

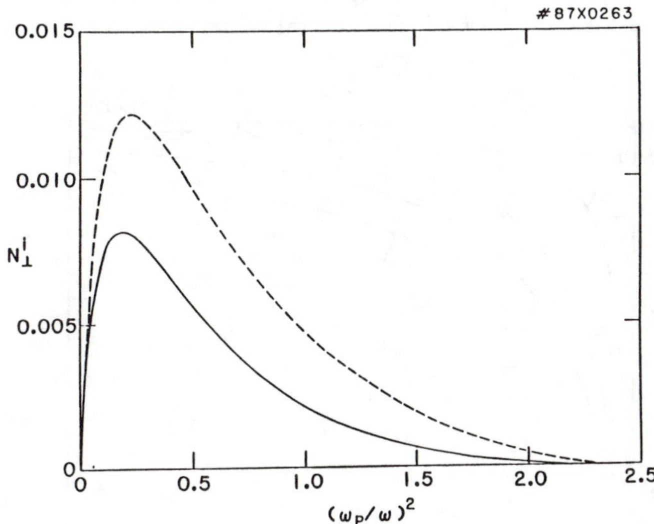

Fig. 1. Imaginary part of $N_\perp$ with (full) and without (dotted) high order FLR terms for the extraordinary mode with $\omega_c/\omega = 1.3$, $T_e = 20\ keV$, and $N_\parallel = 0$.

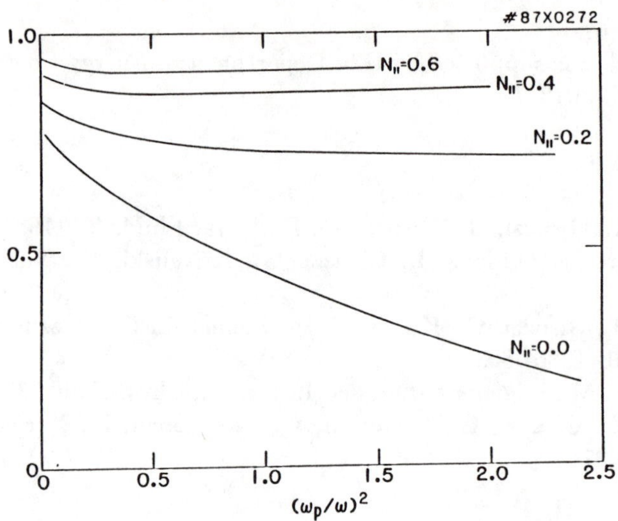

Fig. 2. Ratio of the imaginary parts of $N_\perp$ obtained with and without high order FLR terms for the extraordinary mode with $\omega_c/\omega = 1.3$, $T_e = 20$ keV, and several values of $N_{\|}$.

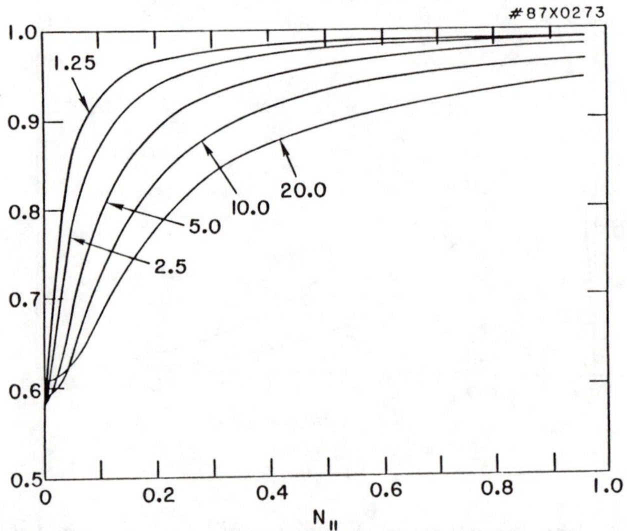

Fig. 3. Ratio of the imaginary parts of $N_\perp$ obtained with and without high order FLR terms for the extraordinary mode with $\omega_c/\omega = 1.2$, $(\omega_p/\omega)^2 = 0.5$, and several values of T_e.

Electron Cyclotron Heating in the Compact Ignition Tokamak

E. Mazzucato, I. Fidone[a], and R. L. Meyer[b]
Princeton University, Plasma Physics Laboratory
Princeton, N. J. 08544

Abstract

A new form of the dispersion relation of electron cyclotron waves in hot plasmas is used for the study of auxiliary heating in the Compact Ignition Tokamak (CIT). It is shown that the extraordinary mode with a frequency of 190 *GHz* can be used in CIT plasmas with a central electron density of $1 \times 10^{15}\ cm^{-3}$ and a toroidal magnetic field of 105 *kG*.

I. Introduction

The Compact Ignition Tokamak (CIT) is a small high field tokamak experiment designed to study the physics of α particle heating[1]. It is expected that its large toroidal magnetic field ($\approx 100\ kG$) will allow the values of plasma density ($10^{15}\ cm^{-3}$) and current ($10\ MA$) which are required for reaching thermonuclear ignition in a small device. Auxiliary heating is expected to be necessary, as well, but transport studies[1] indicate that it will be also necessary either to avoid the deterioration in confinement caused by auxiliary heating (H-mode), or/and to control very carefully the density and temperature profiles.

ICRF has been selected[1] for CIT. Unfortunately, the use of this type of auxiliary heating under CIT plasma conditions faces two major obstacles:

(1) the plasma coupling of the fast magnetosonic wave is extremely sensitive to the plasma edge conditions which, in turn, are completely dominated by the H-mode;
(2) it is difficult to control the rf power deposition profile.

In this article we suggest that ECH at downshifted frequencies[2-4] be also considered a viable candidate for the first phase of CIT. In fact, this type of plasma heating is not sensitive to edge plasma conditions and allows control of the power deposition in both the physical and velocity spaces, which is crucial for profile control of tokamaks.

II. ECH in CIT

The CIT magnetic configuration has been modelled by solving the Grad-Shafranov equation for the poloidal magnetic flux ($2\pi\psi$)

$$\nabla \cdot (\nabla\psi/R^2) = -dp/d\psi - (F/R^2)dF/d\psi,$$

with plasma pressure (p) and poloidal current flux ($2\pi F$) given by $p = p_1\psi^2 + p_0$, and $F^2 = f_1\psi^2 + f_0$, respectively. Figure 1 shows an example of a CIT magnetic

---

[a]Permanent address: Association Euratom-C.E.A., C.E.N.-Cadarache, France.
[b]Permanent address: Université de Nancy, Nancy, France.

configuration with a vacuum toroidal magnetic field $B_o = 105\ kG$ at $R = 125\ cm$, and a volume average $<\beta>= 0.05$. With a plasma current $I = 8.0\ MA$, the value of the safety factor q is 1.0 on axis and 3.5 on the outmost magnetic surface shown in Fig. 1. If the central electron density $n_e(0) = 1 \times 10^{15}\ cm^{-3}$, we get $T_e(0) = T_i(0) = 24\ keV$.

The absorption of electron cyclotron waves has been studied with a ray tracing code[3], where the power deposition along the ray is obtained from a new form of the relativistic dispersion relation[5] in which high order Larmor radius terms are taken into account for a correct description of wave absorption in high temperature plasmas.

Some example of wave trajectories with the extraordinary mode and a frequency of 190 GHz are shown in Fig. 1. The rays are labelled with their initial toroidal ($N_t = sin\alpha$) and poloidal ($N_p = sin\beta$) components of the refractive index. The rays of Fig. 1 have $\alpha = 12^o$.

The absorbed power density profile, obtained with a Gaussian beam with a full half intensity width of 5°, is shown in Fig. 2 for the top launching case of Fig. 1. Here the power density is averaged over the magnetic surfaces and is plotted versus the volume within the surface. The wave energy is absorbed in a central plasma region, even though the central value of the electron cyclotron frequency is more than 50% larger than the wave frequency. The position of the absorbing region can be controlled by adjusting the value of β.

To obtain strong absorption at lower values of T_e one must either increase the angle α (i.e., $N_{\parallel}$) or lower the magnetic field B_o. Here we consider this second scenario (Fig. 3) where we assume that the rf power is turned on when an ohmic plasma with $B_o = 80\ kG$, $T_e(0) = 4\ keV$, and $n_e(0) = 7 \times 10^{14}\ cm^{-3}$ has been produced. As the temperature rises, both the plasma density and the toroidal magnetic field are incresed to the values of Fig. 1 in such a manner that a total single pass absorption is obtained with a constant value of α.

To couple to the extraordinary mode, the wave must be launched with the proper elliptical polarization. In the above scenario the wave is injected almost perpendicularly to the magnetic field and therefore its polarization is very close to linear. In this case if a linearly polarized wave is launched with the electric field perpendicular to the local magnetic field, less than 5% of the energy is coupled to the inefficient ordinary mode. This simplifies enormously the launching structure. One can envision a simple antenna, where by bouncing the wave on a mirror whose position can be changed during the plasma pulse, one obtains a simple and efficient way of controlling the plasma temperature profile.

III. Conclusion

In conclusion, by using a new form of the relativistic dispersion relation we have shown that electron cyclotron waves with a frequency of 190 GHz can be used for heating CIT plasmas with a central density of $1 \times 10^{15}\ cm^{-3}$ and a toroidal magnetic field of 105 kG.

A new heating scenario has been described, where by injecting a linearly polarized wave from a top launcher, one can bring a CIT ohmic plasma to ignition. Control of the temperature profile can be achieved by varying the direction of the incident wave with a simple moveable mirror.

Acknowledgment

This work was supported by U.S. Department of Energy Contract No. DE-AC02-76-CHO-3073.

REFERENCES

[1] Schmidt, J. A., et al.,in Plasma Physics and Controlled Nuclear Fusion Research 1986 (Proc. 11th Int. Conf., Kyoto,1986), IAEA, Vienna(1987).

[2] Fidone, I., Giruzzi, G., Mazzucato, E., Phys. Fluids **28** (1985) 1224.

[3] Mazzucato, E., Fidone, I., Giruzzi, G., Krivenski, V., Nucl. Fusion **26** (1986) 3.

[4] Fidone, I., Giruzzi, G., Krivenski, V., Ziebell, L. F., Mazzucato, E., Phys. Fluids **29** (1986) 803.

[5] Mazzucato, E., Fidone, I., and Granata, G., these proceedings.

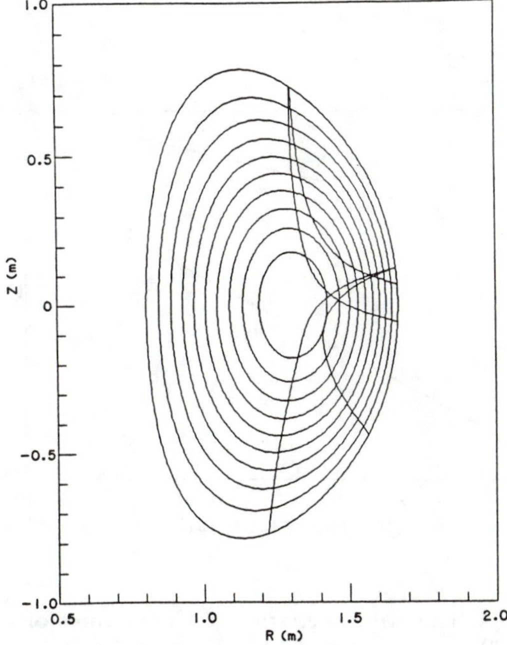

Fig. 1. CIT configuration with $B_o = 105\ kG$, $I = 8\ MA$, $n_e(o) = 1 \times 10^{15}\ cm^{-3}$, $<\beta>= 0.05$. Shown are also some ray trajectories with $f = 190\ GHz$, and $\alpha = 12^o$.

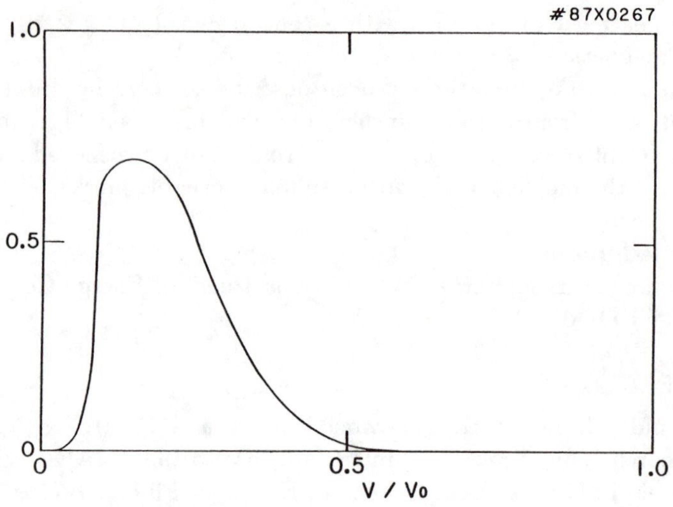

Fig. 2. Absorbed power density profile (surface average in arbitrary units) versus the normalized magnetic surface volume for the case of top launching of Fig. 1.

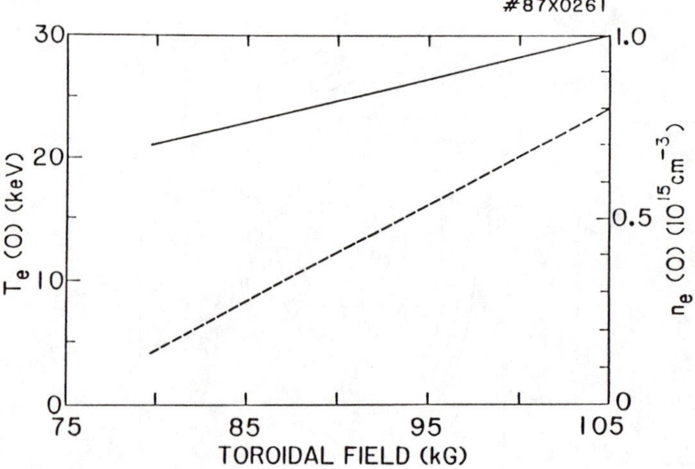

Fig. 3. An ECH heating scenario for CIT: values of $n_e(o)$ and $T_e(o)$ (dotted) versus the toroidal magnetic field for complete single path absorption in the case of top launching of Fig. 1.

Ray-Tracing Study of Electron-Cyclotron Heating and Current Drive in D III-D*

A. H. Kritz,†
Hunter College/CUNY, New York, NY 10021

Gary R. Smith and W. M. Nevins
Lawrence Livermore National Laboratory
University of California, Livermore, CA 94550

ABSTRACT

Ongoing and planned experiments in D III-D utilize X-mode power for heating and current drive. For X-mode power launched from the low-field side, the magnetic field is reduced so that the power is deposited at the second-harmonic resonance near the center of the plasma. It is found that when the central density is $\hat{n} = 1.8 \times 10^{13}\,\mathrm{cm}^{-3}$ and the eight available antennae are used to launch the power, the power deposition is not localized. For X-mode power launched from the high-field side, the increase in ECH driven current with decreasing density and increasing toroidal angle is illustrated. For the plasma parameters considered here it is found that current-drive efficiencies of the order of 0.8 Amps/Watt can be achieved in D III-D.

OUTSIDE LAUNCH OF SECOND-HARMONIC X-MODE POWER

A cluster of eight launchers are available for introducing ECH power from the low-field side into D III-D. We examine the propagation of second-harmonic X-mode power from the eight launchers into a D III-D plasma where $\hat{n} = 1.8 \times 10^{13}\,\mathrm{cm}^{-3}$ and $\hat{T}_e = 1.5\,\mathrm{keV}$. The density, temperature and magnetic field profiles are obtained from the flux surface information produced by an MHD equilibrium code. Poloidal and toroidal views of the ray paths of the central rays associated with the eight antennae are illustrated in Fig. 1.

It is seen that for the D III-D plasma parameters that were used (parameters that are typical of D III-D operation) the rays are strongly refracted. Refraction, together with the spread in launch directions of the eight launchers, causes power to be deposited over a range of 60 cm in both the vertical and toroidal directions. Because the ECH power is strongly refracted, determination of the range in normalized poloidal flux over which deposition occurs, requires a more complete analysis using the Gaussian beam pattern associated with each launcher. However, the results obtained here would indicate that it is reasonable to expect that deposition would occur over a range $\Delta\psi = 0.25$, where normalized ψ varies from 0 at the plasma center to 1 at the edge.

---

* Performed by LLNL for USDoE under Contr. W-7405-ENG-48.
† Supported by USDoE under Contr. DE-FG02-84-ER5-3187.

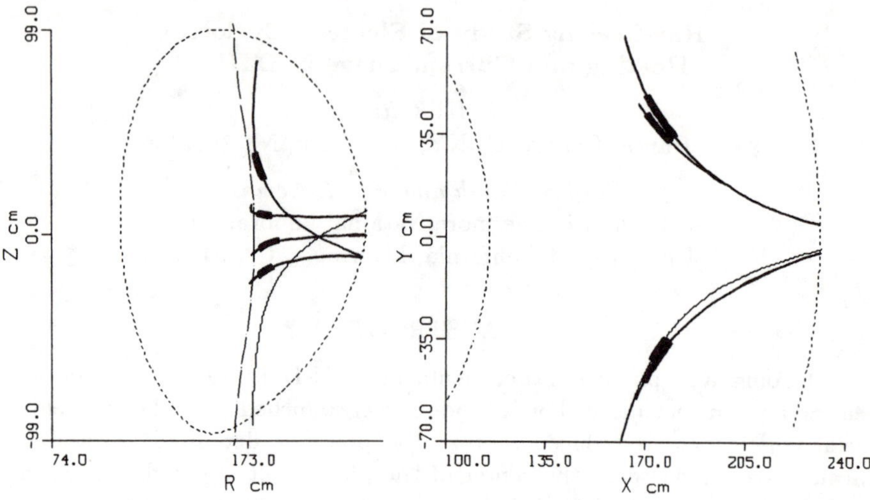

Fig. 1. Ray paths illustrating the propagation and deposition of second harmonic X-mode power incident from the eight launchers on the low-field side; $\hat{n} = 1.8 \times 10^{13}\,\mathrm{cm}^{-3}$ and $\hat{T}_e = 1.5\,\mathrm{keV}$.

INSIDE LAUNCH OF FUNDAMENTAL X-MODE POWER

Launch of X-mode power from the high-field side of the D III-D plasma is planned because this mode has the largest absorption and, possibly, the largest current drive. Also, penetration of electron-cyclotron power can occur at higher density ($\approx 6 \times 10^{13}\,\mathrm{cm}^{-3}$) than is possible with low-field side O-mode waves. Within the assumptions of the model used here, determination is made of the optimum launch angle of X-mode power to maximize current drive.

The calculations were performed with the TORAY code using a simple concentric-circle equilibrium model with the major and minor radii, axial magnetic field, density and temperature profiles ($\hat{T}_e = 4\,\mathrm{keV}$), and Z_{eff} as suggested by experimental results.

The calculations of absorption and current-drive efficiency assume that the electron distribution function is only slightly distorted from a Maxwellian by the electron-cyclotron power. This approximation may not be adequate for the power levels planned for D III-D. In particular, R. Harvey[1] has obtained Fokker-Planck code results that suggested strong distortion and reduced current-drive efficiency at high power due to significant heating of electrons with both signs of parallel velocity. Further Fokker-Planck code calculations are required to address this issue. The computation of the current-drive results presented below are carried out using a relativistically correct module developed by R. H. Cohen.[2]

The incident ECH power originates from a point ($R = 1$ m, $Z = 0$), and the beam direction is parallel to the $Z = 0$ plane and is represented by a single ray. The driven current is then obtained as a function of the toroidal launch angle θ, the angle with respect to perpendicular incidence, and as a function of central density, $\hat{n}$. The results of 12 TORAY runs at four values of θ and three values of $\hat{n}$ are summarized in Fig. 2, where the radial integral of the driven current density, normalized to the incident power,

$$\frac{I_{\text{driven}}(r)}{P_{\text{incident}}} \equiv \frac{2\pi \int_0^r J_{\text{driven}}(r)\, r\, dr}{P_{\text{incident}}} \quad ,$$

is plotted as a function of plasma radius. Note that in no case is current driven near $r = 0$, because resonant electrons with finite $k_{\parallel}v_{\parallel} < 0$ absorb the waves at higher field than at $\Omega = \omega \approx \Omega_{\text{axis}}$. If D III-D were run with a relatively flat density profile and with $\Omega_{\text{axis}} > \omega$, more current could be driven because absorption and current drive would occur at higher T_e and closer to the magnetic axis, where trapped electrons cause less degredation of the current-drive efficiency. From Fig. 2, it is seen that larger θ results in driven-current profiles that are peaked at larger r because of the larger Doppler $|k_{\parallel}v_{\parallel}|$ shift.

The qualitative trends for current drive, as seen in Fig. 2, can be understood as follows. Increasing the launch angle θ makes $\tan^{-1}(N_{\parallel}/N_{\perp})$ larger, which increases the absorption rate of the waves.[3] A larger absorption rate implies that a typical resonant electron is located farther from the surface $\Omega = \omega$ and has larger parallel velocity, which implies higher current-drive efficiency. The decrease of the driven current as density increases is understood in terms of decreasing absorption rate for X-mode waves near the fundamental cyclotron resonance with increased density.[3]

In summary, the driven current increases rapidly as density decreases. Also, the driven current is largest at the largest angle $\theta = 75°$ with respect to perpendicular incidence except when severe refraction does not allow significant power deposition. The results indicate that current- drive efficiencies of the order of 0.08 Amps/Watt can be achieved in D III-D with inside X-mode launch.

REFERENCES

1 R. W. Harvey, M. G. McCoy, and G. D. Kerbel, poster at APS-DPP meeting in Baltimore, Nov. 1986.

2 R. H. Cohen, "Effect of Trapped Electrons on Current Drive," Lawrence Livermore National Laboratory Report No. UCRL-95813 Rev. 1, submitted to *Phys. Fluids*, Mar. 1987.

3 M. Bornatici, R. Cano, O. DeBarbieri, and F. Engelmann, *Nucl. Fusion* **23**, 1153 (1983), Fig. 7 on p. 1201.

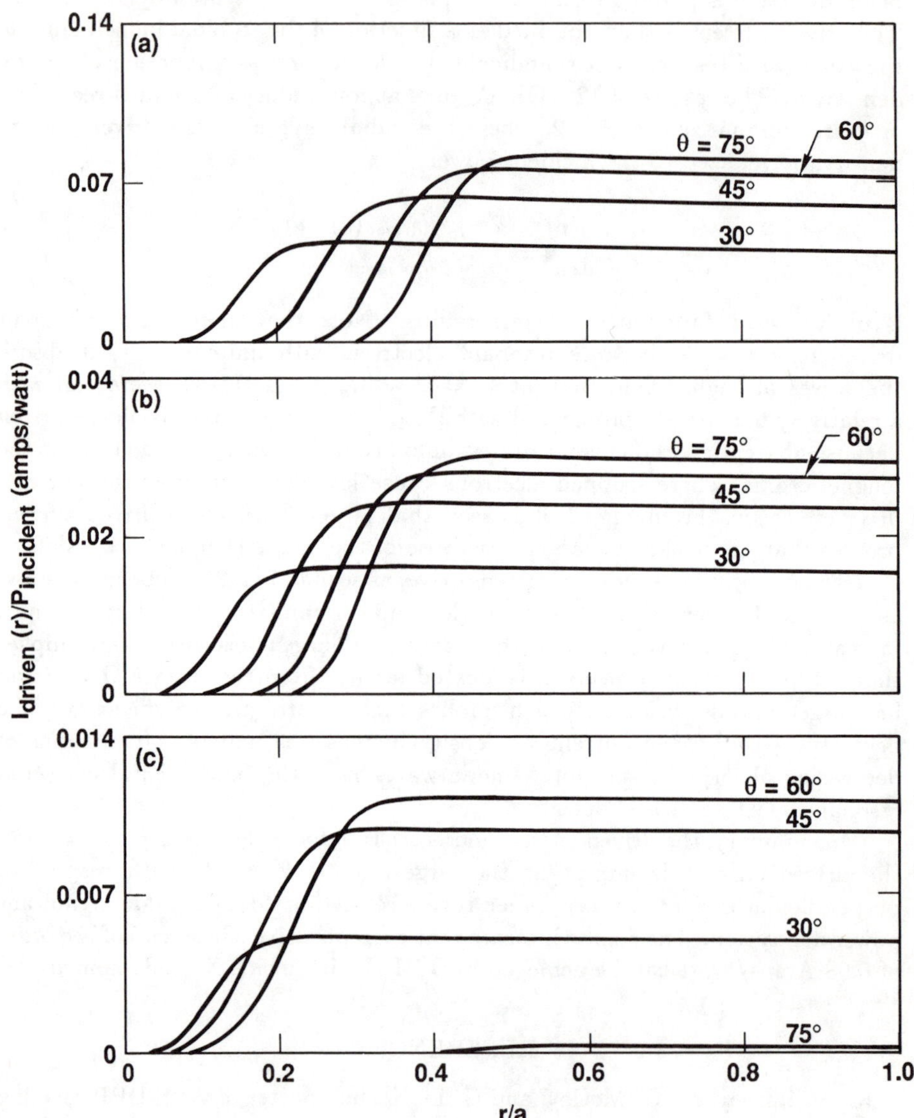

Fig. 2. Amps of current driven $I_{\text{driven}}(r)$ within radius r, divided by the incident power P_{incident}, for various launch angles θ and peak densities $\hat{n}$. (a) $\hat{n} = 2 \times 10^{13}\,\text{cm}^{-3}$. (b) $\hat{n} = 4 \times 10^{13}\,\text{cm}^{-3}$. (c) $\hat{n} = 6 \times 10^{13}\,\text{cm}^{-3}$.

SECOND HARMONIC ELECTRON CYCLOTRON HEATING CALCULATIONS FOR THE ATF TORSATRON*

R. C. Goldfinger and D. B. Batchelor
Oak Ridge National Laboratory, Oak Ridge, Tennessee 37831

ABSTRACT

The heating of the Advanced Toroidal Facility now under construction at Oak Ridge National Laboratory has been investigated using ray tracing techniques. The effect of electron cyclotron waves at the second harmonic resonance is studied using the RAYS geometrical optics code assuming anticipated steady state parameters. A comparison is made of the heating efficiency using two launching schemes: a linearly polarized antenna providing a narrow beam and a TE_{02} waveguide. The first pass ray tracing calculations are incorporated in a wall reflection/power balance model which includes the effects of wall reflected rays and wall losses. It is shown that the power absorbed in the plasma after multiple reflections is preferentially absorbed near the center of the plasma. A significant fraction of the incident power is predicted to ultimately be deposited in the plasma, rather than lost to walls and ports.

I. INTRODUCTION

The ray tracing code RAYS[1,2,3] has been used to investigate electron cyclotron heating (ECH) in the Advanced Toroidal Facility (ATF). Steady state parameters anticipated for ATF operation are used in all the calculations ($f_{\text{gyro}} = 53.2\text{GH}_z, n_e = 8 \times 10^{12}\text{cm}^{-3}, T_e = 750\text{eV}, B_0 = 1\text{T}$); the separate problem of plasma breakdown and startup is not addressed in this paper. During the design of the ECH launching system for ATF, several options were considered including a TE_{02} waveguide (circular polarization, broad null on-axis) and a carefully designed antenna giving rise to a narrow, linearly polarized beam. The absorption rate of a given ray is small unless the density and the magnetic scale length in the resonance region are large; the magnetic geometry of ATF therefore dictates that the rays should pass through resonance near the saddle point in the field (see Figs. 1 and 2). Thus the broad beam, superposition of X- and O-mode waves emanating from the TE_{02} waveguide is a cause for concern about whether significant ECH power will be centrally deposited in the plasma. To complete the picture, an elementary treatment of the power that is not absorbed in the first pass through the plasma is provided.

II. X-MODE NARROW BEAM ANTENNA VERSUS TE_{02} WAVEGUIDE

A RAYS calculation has been done assuming a 3° beam (full width) with the linearly polarizing antenna oriented so as to give pure X-mode at the center. The position of the resonance at the central saddle point region, with the beam

---

* Research sponsored by the Office of Fusion Energy, U.S. Department of Energy, under contract DE-AC05-84OR21400 with Martin Marietta Energy Systems, Inc.

passing through the region, results in 100% single pass absorption within 7 cm of the magnetic axis.

Figures 3A, 3B, and 3C show a run simulating the TE_{02} waveguide. This mode has circular polarization with the first peak at 11% off the waveguide axis. The circular polarization means that the portion of the beam which lies in the toroidal direction (i.e., up and down the torus) will be predominantly X-mode, whereas the orthogonal part of the beam is O-mode, since the magnetic field is essentially toroidal. To simulate this in the RAYS code, we launch a cone of rays with half angle of 11° and pure X-mode polarization. Figure 3C shows that 21% of the beam is absorbed on the first pass through the plasma at approximately 10 cm off axis. The remaining power impinges on the wall and is the subject of discussion in the next section.

III. WALL REFLECTIONS/POWER BALANCE MODEL

The power in the beam that is not absorbed in the plasma during the first pass strikes the wall and is partially lost due to the metallic wall and through holes for diagnostic ports. The rest of the beam is depolarized and scattered back into the plasma. The fate of this scattered power is treated statistically in Ref. 3 as emanating isotropically from a homogeneously radiating wall surface. Figure 4 shows the resulting histograph of one such run for the case of second harmonic, X-mode. Due to the complicated interaction of longer magnetic scale lengths near the center with preferential absorption at perpendicular incidence, it turns out that the X-mode reflected power suffers 8% absorption per bounce within 10 cm of the axis. Since approximately one half of the wall scattered power is O-mode (which is very weakly absorbed), the net power absorbed in the plasma is one half of the X-mode absorption, or 4% per bounce.

The power remaining in the waves after the ith transit through the plasma, γ_i, can be expressed as

$$\gamma_1 = 1 - F_1^P \tag{1}$$

$$= \gamma_{i-1}\left[\left(1 - F_i^P\right)\left(1 - F_i^W\right)\right] \;\; ; \; i \geq 2 \tag{2}$$

where F_i^P is the fraction of remaining power absorbed in the plasma during the ith pass of the waves through the plasma; F_i^W is the fraction of remaining power lost to walls and diagnostic ports at the ith wall reflection; and the initial power, γ_0, is taken to be unity. Using the statistical assumption described above, we take F_i^P constant for $i > 1$ and F_i^W constant for all i. We now recognize Eq. 2 to be a geometric series which can be summed to give the total power deposited in the plasma, P_∞^P, and total power lost to the wall, P_∞^W:

$$\gamma_i = \left(1 - F_1^P\right)\left[\left(1 - F^W\right)\left(1 - F^P\right)\right]^{i-1} \tag{3}$$

$$P_\infty^P = F_1^P + \frac{F^P\left(1 - F^W\right)\left(1 - F_1^P\right)}{1 - \left(1 - F^W\right)\left(1 - F^P\right)} \tag{4}$$

$$P_\infty^W = 1 - P_\infty^P \tag{5}$$

where the subscripts have been dropped from F^P and F^W. Taking into account the surface resistivity of the stainless steel walls and the area of the diagnostic ports, the fractional wall loss at each reflection is found to be 2.45%.[4] Using the wall reflection calculation described above, $F^P = 4\%$, we obtain

$$P_\infty^P = .385F_1^P + .614 \tag{6}$$

$$P_\infty^W = .385\left(1 - F_1^P\right) \tag{7}$$

This means the power deposited in the plasma can range from 61% to 100%, depending on the efficiency of the wave launcher in achieving first pass absorption; the remaining power (38% or less) is lost to the walls. For the case of the TE_{02} waveguide described in Sec. II, $F_1^P = 21\%$. Thus the power going into the plasma and walls is 69% and 31%, respectively.

CONCLUSIONS

Second harmonic ECH gives 100% central heat deposition when a narrow beam polarized to give X-mode is used with the steady state parameters envisioned for ATF. For the case of the TE_{02} waveguide (circularly polarized with on-axis null), first pass absorption drops to 21%. A power balance expression has been derived for the power fraction deposited in the walls and plasma after multiple reflections. For the TE_{02} waveguide, this expression says that 69% of the original power ends up in the plasma and 31% in the walls.

REFERENCES

1. Batchelor, D. B., Goldfinger, R. C., RAYS: A Geometrical Optics Code for EBT, Oak Ridge National Laboratory Rep. ORNL/TM-6844 (1982).
2. Batchelor, D. B., Goldfinger, R. C., Weitzner, H., IEEE Trans. Plasma Sci. **PS-8** (1980) 78.
3. Goldfinger, R. C., Batchelor, D. B., Nucl. Fusion **27** (1987) 31.
4. White, T. L., Analysis of Mixed Mode Microwave Distribution Manifolds, Oak Ridge National Laboratory Rep. ORNL/TM-8127 (1987).
5. Houlberg, W. A. Attenberger, S. E., Lao, L. L., Computational Methods in Tokamak Transport, Oak Ridge National Laboratory Rep. ORNL/TM-8193 (1983).

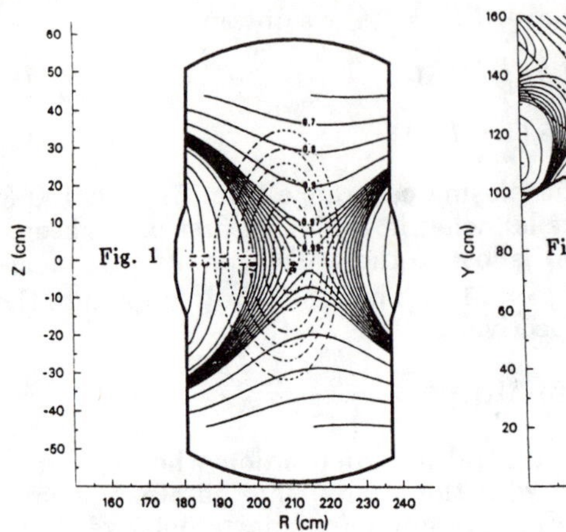

FIG. 1. Mod-B and toroidal flux contours in the $\phi = 0°$ plane with the outline of the ATF vacuum vessel superimposed. The value of $|\mathbf{B}|$ on each contour is indicated, normalized to unity at the magnetic axis. The flux contours (dashed lines) parameterize the density and temperature and were determined by following field lines.

FIG. 2. Mod-B and toroidal flux contours in the equatorial plane $(Z=0)$.

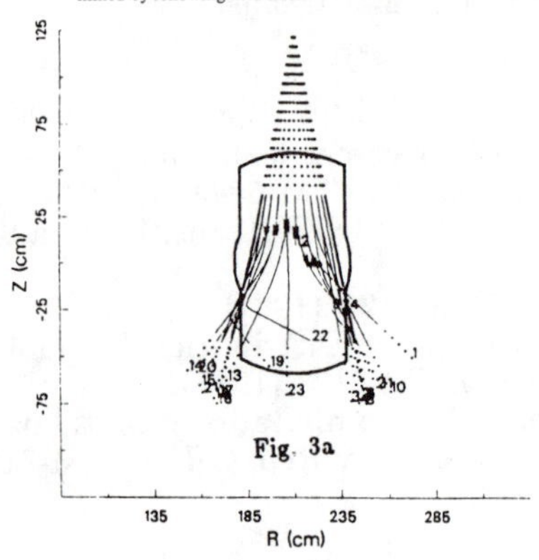

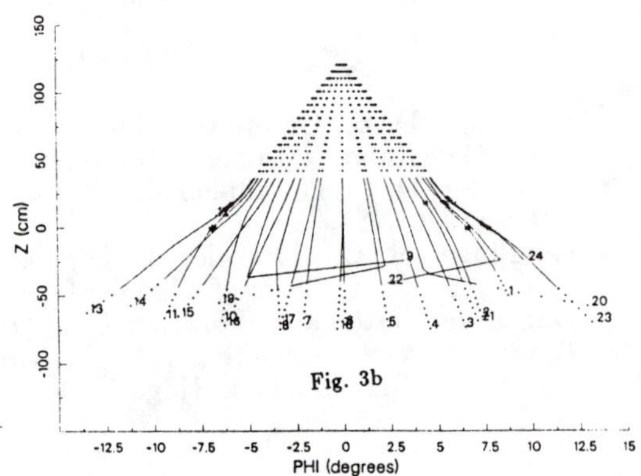

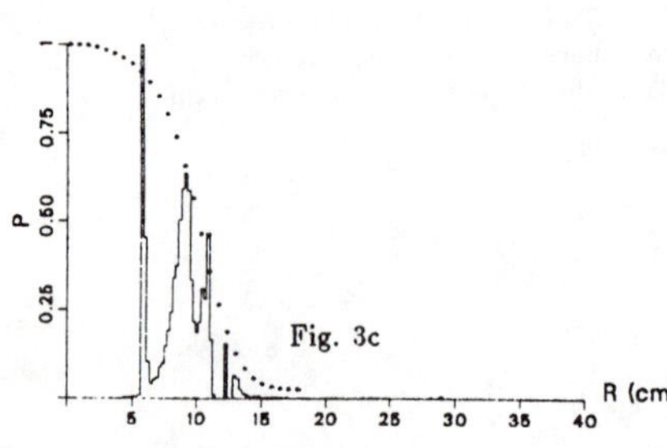

FIG. 3. Ray trajectories in the R–Z plane (plot 3a) and $\phi - Z$ plane (plot 3b) for the TE_{02} waveguide. The dotted curve represents ray propogation outside the plasma while the asterisks indicate location of second harmonic ECH absorption. In 3c the power absorption histograph for this run is displayed with the density profile superimposed (dashed line).

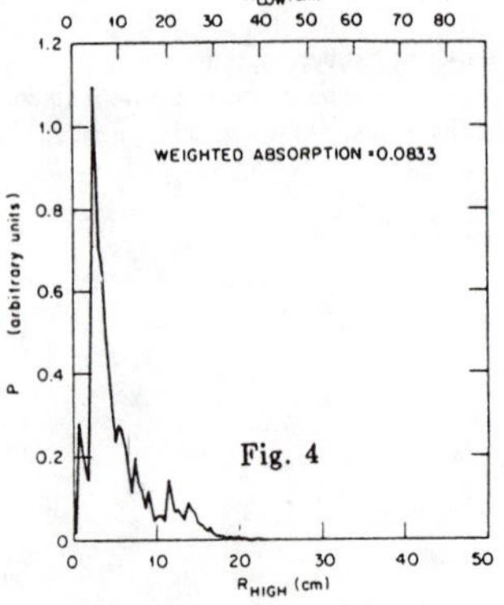

FIG. 4. Power absorption histograph for ensemble of wall reflected rays. A large number of second harmonic rays are launched from the wall to simulate the vacuum chamber as a homogeneous, isotropic radiator of power back into the plasma.

Localized Electron-Cyclotron Heating and Current Drive in the TIBER-II Reactor Study*

Gary R. Smith and B. G. Logan
Lawrence Livermore National Laboratory
University of California, Livermore, CA 94550

A. H. Kritz†
Hunter College/CUNY, New York, NY 10021

ABSTRACT

A scenario is shown for launching 10 MW of 450 GHz extraordinary-mode electron-cyclotron waves into a TIBER-II equilibrium. Localized, high-efficiency current drive near but outside the $q = 2$ surface causes substantial reduction in the current gradients that may play a role in major disruptions and anomalous transport. The same launch geometry shows promise for heating the plasma core during startup.

INTRODUCTION

In a high-temperature tokamak reactor, broad electron-cyclotron resonances lead to constraints on current-drive efficiency at the peak T_e achieved during burn in the plasma core.[1] At lower T_e, however, the injection of electron-cyclotron power can serve two useful purposes. First, during plasma startup, localized heating of the core can help initiate the burn. Second, to minimize transport and suppress disruptions, one can control the radial profiles of T_e and plasma current near and outside of the $q = 2$ surface. This paper reports initial work towards these goals.

LOCALIZED, HIGH-EFFICIENCY CURRENT DRIVE

High-efficiency current drive by electron-cyclotron waves is achieved by selectively heating high-energy electrons that have parallel velocity of one sign. The relativistic, wave-particle resonance condition is satisfied for such electrons if the parallel index of refraction $|N_\parallel| = \mathcal{O}(1)$ and if $\ell\Omega/\omega < 1$, where ℓ is the harmonic number, Ω is the cyclotron frequency, and ω is the wave frequency. Typical values for these parameters are $|N_\parallel| = 0.6$ and $\ell\Omega/\omega = 0.9$.

Localization of power deposition and current drive to a small range of flux surfaces is accomplished by choice of the launch geometry and of the polarization. Wave absorption must be strong and occur where propagation

* This work was performed under the auspices of the U.S. Department of Energy by Lawrence Livermore National Laboratory under contract No. W-7405-Eng-48.

† Supported by USDoE under Contr. DE-FG02-84-ER5-3187.

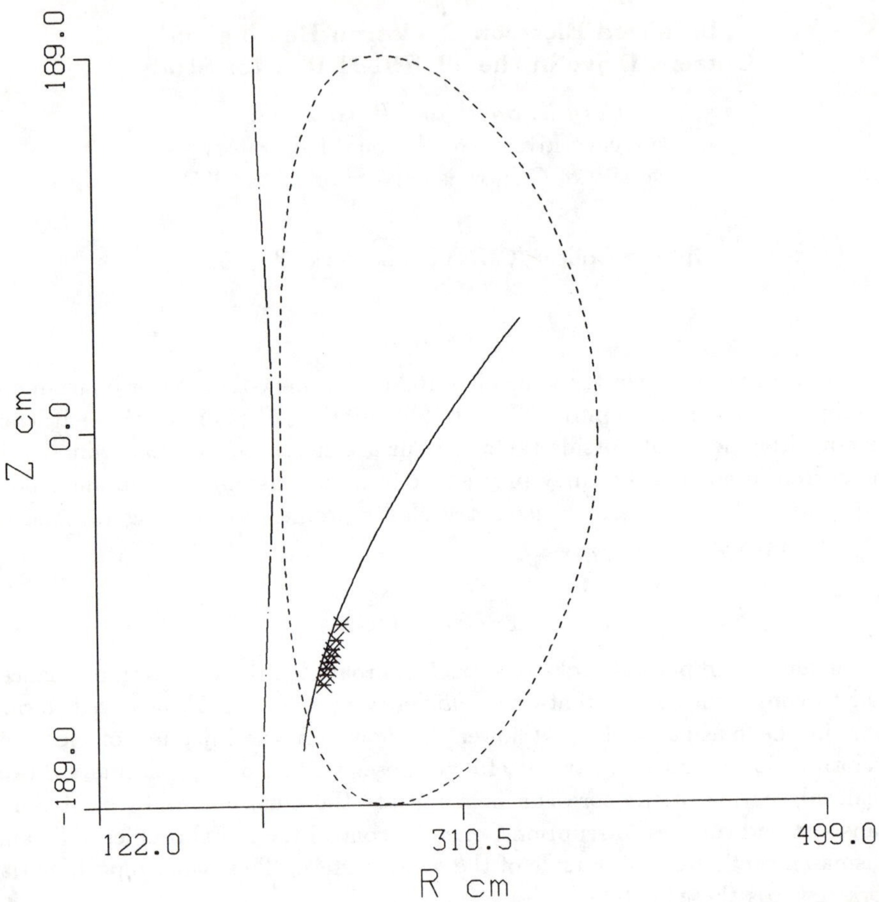

Fig. 1. Poloidal projection of a ray trajectory that deposits power and drives current outside the $q = 2$ surface in a TIBER-II equilibrium. The dashed curve is the $\tilde{\psi} = 0.96$ flux contour. The chain-dot curve is the second-harmonic cyclotron resonance. The asterisks indicate the location of wave absorption.

is at a relatively small angle with respect to the flux surfaces. For any harmonic ℓ, the extraordinary mode has the stronger absorption.

An attractive launch geometry for the TIBER-II study is illustrated in Fig. 1. The launch location is just outside the separatrix of the diverted equilibrium. Extraordinary-mode power at 450 GHz is launched at polar angle $\theta = 50°$ and at azimuthal angle $\phi = 285°$ in a cylindrical coordinate system coaxial with the tokamak axis of symmetry. Power absorption is limited to the region outside the $q = 2$ flux surface and amounts to 99% of the incident power. Second-harmonic cyclotron resonance absorbs the waves in a region where magnetic-field and plasma parameters have the values $B \approx 7.2\,\mathrm{T}$,

$n \lesssim 0.46 \times 10^{20}\,\mathrm{m}^{-3}$, and $T_e \lesssim 10\,\mathrm{keV}$. Third-harmonic absorption is not shown here but becomes significant farther along the ray, just after the second-harmonic absorption decreases; no power reaches the center of the plasma. These results are calculated with the TORAY ray-tracing code.[2,3] Absorption is calculated for a Maxwellian electron distribution function.

Second-harmonic extraordinary-mode absorption is used here, but an alternative scenario for localized power deposition and current drive utilizes the ordinary-mode polarization and the fundamental cyclotron resonance with $\omega \approx 225\,\mathrm{GHz}$. With either scenario, for accurate results the wave damping must include the warm-plasma effects on wave polarization. In this paper, cold-plasma polarization is used, and the results may be quantitatively inaccurate but are probably qualitatively correct.

The calculation of current-drive efficiency presented in Ref. 4 is used here to obtain the current driven in conjunction with second-harmonic power absorption. This calculation includes the effects of the relativistic resonance condition and of trapped electrons, both of which can alter current-drive efficiency significantly.

Figure 2 shows the effect of 10 MW of electron-cyclotron power on the current flowing in the TIBER-II plasma. The plots show the toroidal current enclosed within each flux surface, where the surfaces are parametrized by the normalized poloidal flux $\tilde{\psi} = (\psi - \psi_{\mathbf{axis}})/(\psi_{\mathbf{edge}} - \psi_{\mathbf{axis}})$. The integrated equilibrium current I_p (dashed curve) is characterized for $0 < \tilde{\psi} < 0.6$ by nearly constant slope, which implies that $dI_p/d\tilde{\psi}$ is almost flat. For $\tilde{\psi} > 0.6$ and, in particular, near the $q = 2$ surface at $\tilde{\psi} = 0.73$, $dI_p/d\tilde{\psi}$ decreases with increasing $\tilde{\psi}$, indicating a gradient in the current profile that may be associated with major disruptions. The solid curve, which is calculated from the sum of the equilibrium and the driven currents, maintains the nearly constant slope to larger $\tilde{\psi}$. With electron-cyclotron current drive, therefore, the current gradient is reduced near the $q = 2$ surface, which may help to control disruptions and minimize transport.

HEATING DURING STARTUP

Ideally, the chosen launch geometry should also serve to heat the plasma core during startup and help to initiate the burn. Typical values at the magnetic axis for density and temperature are $\hat{n} = 5 \times 10^{20}\,\mathrm{m}^{-3}$ and $\hat{T}_e = 8\,\mathrm{keV}$. The ray trajectory is similar to that shown in Fig. 1, because refraction is small both during startup and during steady state. Second-harmonic absorption is much reduced from that shown in Fig. 1, because the cooler plasma has very few electrons at the $\ell = 2$ resonance. The third-harmonic resonance occurs close to the magnetic axis, and localized absorption may occur there due to the strong T_e^2 dependence of third-harmonic damping. Future work will examine these questions quantitatively.

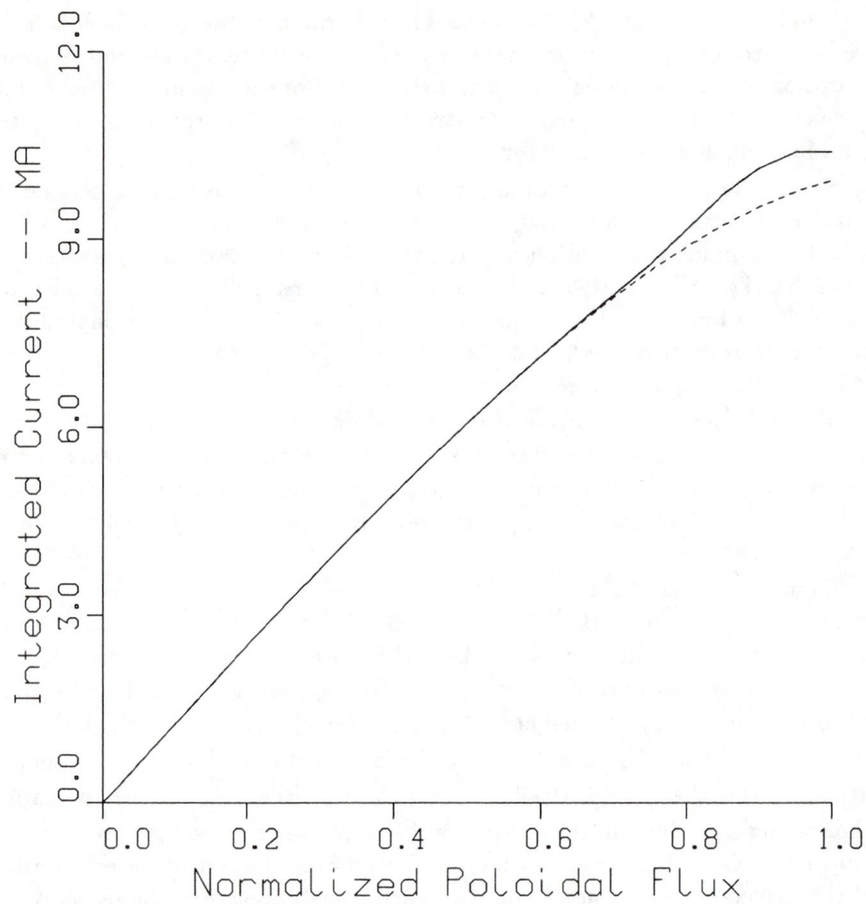

Fig. 2. Enclosed current as a function of normalized poloidal flux $\tilde{\psi}$. The solid curve differs from the dashed curve due to 10 MW of second-harmonic X-mode power.

REFERENCES

[1] G. R. Smith, R. H. Cohen, and T. K. Mau, LLNL Report No. UCRL-96364, submitted to *Phys. Fluids*, March 1987.

[2] A. H. Kritz, H. Hsuan, R. C. Goldfinger, and D. B. Batchelor, in *Proc. 3rd Joint Varenna-Grenoble International Symp. on Heating in Toroidal Plasmas*, (Commission of the European Communities, Brussels, 1982), Vol. II, p. 707.

[3] G. R. Smith, W. M. Nevins, R. H. Cohen, and A. H. Kritz, *Bull. Amer. Phys. Soc.* **31**, 1516 (1986).

[4] R. H. Cohen, "Effect of Trapped Electrons on Current Drive," LLNL Report No. UCRL-95813 Rev. 1, to be published in *Phys. Fluids*, 1987.

Third Harmonic Electron—Cyclotron Heating in Dense Tokamak Plasmas

R.L. Meyer , I. Fidone [a)] , and G. Granata [a)]

Laboratoire de Physique des Milieux Ionisés U.A. C.N.R.S. 835 , Université Nancy 1 BP 239 54506 Vandœuvre – les – Nancy France

and

E. Mazzucato

Plasma Physics Laboratory , Princeton University , Princeton N.J. 08544 U.S.A.

Abstract

Third harmonic electron – cyclotron heating in overdense plasmas is investigated using a ray tracing code for propagation from a top position . It is shown that the X – mode is strongly absorbed for $T_e > 1$ keV and $\omega_p^2 / \omega_c^2 \lesssim 6$.

I . Introduction

Electron cyclotron heating in overdense ($\omega_p^2 / \omega_c^2 \gg 1$) tokamak plasmas appears to be difficult for wave frequencies ω near the fundamental gyrofrequency ω_c and the second harmonic . For instance , for $B \approx 15$ kG and $n_e \approx 10^{14}$ cm^{-3} one obtains $\omega_p^2 / \omega_c^2 \approx 5$ and wave penetration at $\omega \approx \omega_c$, $2\,\omega_c$ becomes impossible . In present – day devices , an overdense plasma is generally used for high – β studies . In this case , heating at $\omega \approx 3\omega_c$ appears more appropriate since the cut – off density of the X – mode is given by $\omega_p^2 / \omega_c^2 = 6$. We show that fot $T_e > 1$ keV , strong X – mode damping generally occurs for wave launching from a top position . For this special launching direction , wave damping is sensitive to the values of $3\omega_c / \omega$ along the ray path and the effective absorption is computed with a ray – tracing code coupled with the equilibrium configuration of the tokamak magnetic field . We present here results for propagation at an angle nearly normal to the toroidal magnetic field for a tokamak with a circular cross – section and large aspect ratio . For the equilibrium configuration we use the well – known Shafranov[1] configuration for $a / R \to 0$.

II. Wave damping along the ray path

Damping along the ray path is described by the factor

$$\eta = 1 - \exp\left[(2\omega/c) \int (D'' / |\partial D'/\partial N_\perp|) \, dl \right],$$

where D' and D" are the real and imaginary parts of

$$D = N_\perp^2 - (\varepsilon_{11}\varepsilon_{12} + \varepsilon_{12}^2)/\varepsilon_{11},$$

thus ,

$$D'' = -|1 - i\,\varepsilon_{12}/\varepsilon_{11}|^2 \varepsilon_{11}'', \quad |\partial D'/\partial N_\perp| = 2N_\perp = 2N_x$$

For $\omega \approx 3\omega_c$,

$$\alpha = |1 + \omega_c \omega_p^2 / [\omega(\omega^2 - \omega_c^2 - \omega_p^2)]|^2$$

$$\times\ \pi^{1/2} (27/35) (\omega_p^2/\omega_c^2) N_x^3 (T_e/mc^2)\, \xi^{7/2} \exp(-\xi),$$

where ,

$$\xi = mc^2 (3\omega_c - \omega)/\omega T_e \quad \text{and} ,$$

$$N_x^2 = (6 - \omega_p^2/\omega_c^2)(12 - \omega_p^2/\omega_c^2)/9(8 - \omega_p^2/\omega_c^2),$$

For propagation in the equatorial plane

$$\tau = \int \alpha \, dl = 16 (\omega_c R/c)(\omega_p^2/\omega_c^2) N_x^3 (T_e/mc^2)^2 .$$

The function $(\omega_p^2/\omega_c^2) N_x^3$ is maximum at $\omega_p^2/\omega_c^2 \approx 2.86$, thus

$$\tau_{max} \approx 22.4 (\omega_c R/c)(T_e/mc^2)^2 .$$

For $R = 140$ cm , $T_e = 1.5$ keV , and $B = 15$ kG , we obtain $\tau_{max} \approx 0.16$

III Numerical results

Much stronger absorption is obtained for vertical propagation . We consider a circular cross – section. The wave launching position x_o is referred to the vertical diameter and the initial direction of propagation is parallel to the reference axis. Toroidal deviation is generally negligible ($N_{//} \approx 0$) and all rays lie in the poloidal cross – section . The local values of n_e, T_e, and B are obtained from Shafranov equilibrium equations[1] for $a / R \to 0$. In Fig. 1, we present three rays , i.e. , $x_o = 0 , \pm 2$ cm , for $n_e(0) = 10^{14}\ cm^{-3}$, $B_0 = 15$ kG , and $\omega_{c0} / \omega = 0.338$, where $\omega_{c0} = eB_0 / mc$, and B_0 is the vacuum magnetic field at $x_o = 0$. We consider a case in wich the radial profiles of $n_e(r)$ and $T_e(r)$ are approximately described by $(1 - r^2 / a^2)$ and $(1 - r^2 / a^2)^2$, respectively. It appears from Fig. 1 that strong bending of the ray trajectory occurs in the poloidal direction. This is due to the combined effects of refraction and the 1/R dependence of B . Wave damping along the ray path is shown in Fig. 2 for the conditions of Fig. 1(b) and $T_e(0) = 1.5$ keV . Nearly total absorption in the first transit is found but the power deposition is not localized in space. In Fig. 3, we present the values of B along the ray path for the conditions of Fig. 1. It appears the B(l) is a sensitive function of the ray trajectory and explains the variation of the wave absorption with x_o . In Fig. 4, we show the fraction of the wave energy deposited in one transit versus x_o for the conditions of Fig. 1 and $T_e(0) = 1 , 1.5 , 2$ keV .

IV Conclusions

In conclusion , we have shown that third harmonic heating can be envisaged in overdense ($\omega_p^2 / \omega_c^2 \approx 6$) tokamak plasmas using the X – mode launched from a top position in the tokamak poloidal section .

Acknowledgments

R.L. Meyer thanks the Conseil Général de Meurthe et Moselle for his financial support .

[a] Permanent address: DRFC , Association EURATOM – CEA CADARACHE 13108 Saint – Paul – lez – Durance , France

[1] V.D. Shafranov , in Reviews of Plasma Physics , edited by M.A. Leontovich (Consultants Bureau , New – York , 1966) Vol.2 p.103

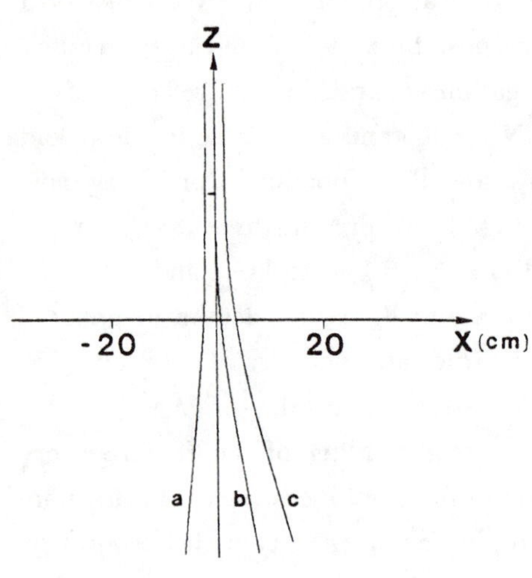

Fig.1
Ray path in the poloidal cross-section for $x_o = 0, \pm 2$ cm , $B_0 = 15$ kG , $n_e(0) = 10^{14}$ cm^{-3} and $f = \omega / 2\pi = 124.3$ GHz

Fig.2
$\eta(\ell) = 1 - \exp[-2(\omega/c)\int N''.d\ell]$ versus ℓ for the conditions of Fig.1(b) and $T_e(0) = 1.5$ keV .

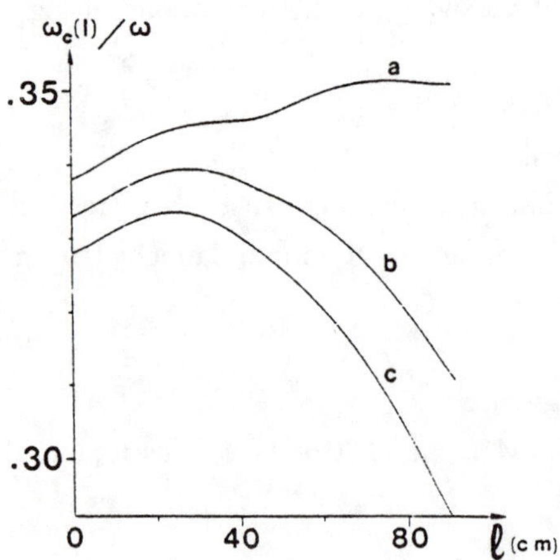

Fig.3
$\omega_c(\ell) / \omega$ versus ℓ for the conditions of Fig.1 .

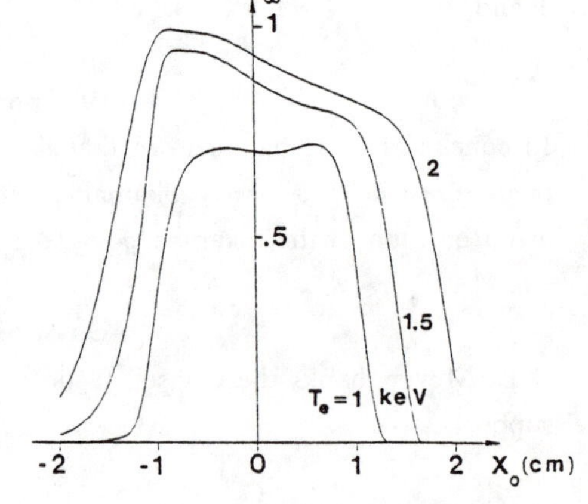

Fig.4
Total absorption η_∞ versus x_o for $T_e = 1$, 1.5 , and 2 keV and the conditions of Fig.1 .

THREE-DIMENSIONAL MODEL OF ELECTRON CYCLOTRON HEATING*

M.G. McCoy and G.D. Kerbel, National Magnetic Fusion Energy Computer Center, Lawrence Livermore National Laboratory, Livermore, CA 94550

R.W. Harvey, GA Technologies, San Diego, CA 92138%

ABSTRACT

To address heating problems in tokamaks, we have implemented a 3-D Fokker-Planck/rf code. This code uses a 2-D bounce-averaged Fokker-Planck package to solve for the distribution on a radial mesh. The rf energy density ξ_k is determined as a solution to a transport equation involving the local power absorption and the group velocity of the wave. Once ξ_k is determined it is used to update the quasi-linear diffusion coefficients and resume the Fokker-Planck calculation. A fast procedure for simulating a tokamak diagnostic, the soft-x-ray analyzer, is presented.

INTRODUCTION

The two dimensional (v=momentum, θ=pitch angle) bounce-averaged Fokker-Planck code CQL (for Collisional QuasiLinear) was developed to simulate a multispecies magnetized plasma whose evolution is governed by an equation with dominant diffusive mechanisms due to small angle Coulomb collisions and wave particle interactions. The model presupposes a symmetric magnetic well in which there occurs nearly recurrent motion of frequency ω_B, with which a gyro-center executes its orbital motion in the varying field structure:

$$\frac{2\pi}{\omega_B} = \tau_B = \oint \frac{ds}{v\cos\theta} = \oint \frac{ds}{v_\parallel}, \qquad (1)$$

Here ds is the element of arc length along the magnetic field line associated with the gyro-center motion. The magnetic moment $\mu = mv^2\sin\theta/2B$ and the energy $mv^2/2$ are invariants of the unperturbed motion. Defining $\psi = B(s)/B(0)$, the invariance of μ can be restated as $\sin^2\theta = \psi\sin^2\theta_0$ where θ_0 is the pitch angle coordinate at a fixed point s = 0, the bottom of the symmetric magnetic well. We adopt the notation that a subscript $_0$ shall adorn quantities evaluated at s = 0.

The bounce-averaged Fokker-Planck equation may be expressed in particle conservation form

*Work performed under the auspices of U.S.D.O.E. by LLNL under contract No. W-7405-ENG-48.

%Work performed under the auspices of U.S.D.O.E. by GA Technologies, Inc., under contract No. DE-AC-03-84ER53158.

$$\left(\frac{\delta\lambda f_0}{\delta t}\right)_{cql} = - \nabla_{v_0} \cdot \underline{\Gamma}_0$$

$$=\gamma\left[\frac{1}{v_0^2}\frac{\partial}{\partial v_0}(A_0+B_0\frac{\partial}{\partial v_0}+C_0\frac{\partial}{\partial\theta_0})+\frac{1}{v_0{}^2\sin\theta_0}\frac{\partial}{\partial\theta_0}(D_0+E_0\frac{\partial}{\partial v_0}+F_0\frac{\partial}{\partial\theta_0})\right]f_0 \quad (2)$$

Here $\gamma = 4\pi z^4 e^4/m^2$ and the coefficients A_0 - F_0 represent the sum of particle-particle, wave-particle and small amplitude steady electric field effects, appropriately bounce-averaged. The quantity $\lambda = v_0\cos\theta_0\tau_B$ is a geometric length factor.

Many heating problems in tokamaks are inherently four dimensional, involving not only the velocity space coordinates (v_0,θ_p), but the spatial coordinates (r,θ_p) as well. We have generalized CQL to CQL3D by adding effects involving these coordinates. The new code solves the Fokker-Planck equation at a number of flux surfaces in parallel and computes the local power absorption P_{abs} as a function of minor radius r and poloidal angle θ_p. Coupling between the various flux surfaces is achieved through the solution of a transport equation for ξ_k, the RF energy density

$$\nabla\cdot(\underline{v}_g\xi_k) = -P_{abs} \quad (3)$$

where $\underline{v}_g$ is the group velocity of the wave.

NUMERICAL FEATURES OF CQL

The Fokker-Planck equation (2) requires the generation of the six coefficients A_0 - F_0 which represent the accumulated effects of all relevant physical processes. Kerbel and McCoy[1,2] have discussed the generation of the quasilinear coefficients $B_{0_{ql}}$ - $F_{0_{ql}}$. In that case, the relevant bounce-averages can be evaluated analytically in the asymptotic limit $\Omega\tau_B \rightarrow \infty$ since the region of wave-particle resonant interaction is localized. On the other hand, collisions occur everywhere along a particle trajectory and for non-linear calculations the Fokker-Planck coefficients A_{0_c} - F_{0_c} must be computed through the numerical bounce-average of the local Fokker-Planck coefficients.

These coefficients are relativistic[3] and are accurate to order $\mu^2\mu'^2/c^4$ where μ and μ' are the velocities of interacting particles in the test and the field distributions respectively. The error except for tail-tail effects is small and the representation is accurate to all moments in the sense that as many terms in the Legendre expansion of the Rosenbluth potentials may be included as required.

The differencing of Eq. (2) is fully implicit. This allows the choice of a large (non-physical) time step - and for linear problems steady state is achieved in one iteration. The large sparse matrix, (frequently of order 50000) is inverted by a highly optimized Gaussian elimination routine written for the CRAY-2. A matrix of order 10000 can be inverted in less than three seconds.

THREE DIMENSIONAL GENERALIZATION - CQL3D

There are two major components to the three-dimensional version CQL3D. The first is the computation by CQL, run in parallel at a number of neighboring flux surfaces, of $f_0(v,\theta,r;t)$ and $P_{abs}(r,\theta_p)$. This latter power is that RF power absorbed in a given poloidal neighborhood. The second component provides the linkage between the flux surfaces themselves through a solution of the transport equation for RF energy density ξ_k. The quantity ξ_k may be used to determine the components of the electric field. This allows for the recalculation of the four quasilinear coefficients and thus feeds back into CQL.

For simplicity, we consider only the component of the group velocity in the $\hat{e}_R$ (major radial) direction: Thus, rays propagate parallel to the equatorial plane. RF power flowing into horizontal ray channels is constant from $z = 0$ to $z = z_1$ and then decreases linearly to zero at $z = z_2$. Up/down symmetry is assumed. This model avoids some of the complexities that would arise if the group velocity $\underline{v}_g$ were general.

The total RF antenna power is specified as is the mode type, i.e. O-mode or X-mode. Also required are the local background electron and ion temperature and density, the parallel wave refractive index and the wave frequency. A warm plasma dispersion relation subroutine[4] is used to determine the group velocity $\underline{v}_g$, the energy density factor

$$E_f = \frac{1}{16\pi}\left\{\frac{\underline{B}\cdot\underline{B}^*}{|\underline{E}|^2} + \frac{\underline{E}}{|\underline{E}|}\cdot\frac{\partial\omega\underline{\underline{K}}}{\partial\omega}\cdot\frac{\underline{E}^*}{|\underline{E}|}\right\} \qquad (4)$$

and the polorizations $|E_{\sim}/E|^2$. Here $_{\sim}$ may mean +, - or $\parallel$ and $\underline{K}$ is the warm plasma dispersion tensor. The transport equation for energy density (3) is now employed to obtain ξ_k. In the current approximation $\nabla\cdot \equiv \partial/\partial R + 1/R$: Then using the results of the wave characteristics package we obtain

$$E_{\sim}^2(R,z) = \frac{1}{E_f(R,z)}\xi_k(R,z)\frac{|E_{\sim}|^2}{|\underline{E}|^2}(R,z) \qquad (5)$$

Coupling with the kinetic equation solved by CQL is achieved through the updating of the coefficients Bo_{ql} - Fo_{ql}. The electric field $E_{\sim}(R,z)$ is interpolated onto the (r,θ_p) mesh. Then, as described by Kerbel and McCoy[2], CQL can update the quasilinear coefficients and recompute $P_{abs}(r,\theta_p)$ in order to bring these quantities closer to self consistency. The power is reinterpolated to (R,z) and the entire solution procedure for $E_{\sim}$ is repeated to convergence.

Among the diagnostics employed by the code is a module designed to compute the bremsstrahlung contribution to the soft-x-ray spectrum. The photon energy flux may be represented as

$$\Phi = k \int_0^L ds \int d\underline{v}\, d\underline{v}'\ f_s(\underline{v}')\ f_e(\underline{v}) \frac{\partial^2\sigma}{\partial k \partial\Omega}\ |\underline{v}-\underline{v}'| \tag{6}$$

where k is the photon energy, f_s the distribution of scatterers, f_e the electron distribution, and $\frac{\partial^2\sigma}{\partial k\partial\Omega}$ is the differential scattering cross section for photons into the detector per steradian per photon energy. The integral over s is along the sightline. The numerical complexity involved in evaluating Eq. (6) can be reduced by assuming that the scattering distribution is Maxwellian with density n(s) and that $|\underline{v}-\underline{v}'|=|\underline{v}|$. After some manipulations, Eq. (6) may be written as

$$\Phi(k,\chi) = 2\pi k \int_0^L n(s)ds \sum_{l=0}^{\infty} P_l(\cos\chi) \int_0^\infty v^3 dv\ r_l(v)\ G_l(v,k) \tag{7}$$

Here χ is the angle between $\underline{B}$ and the line of sight, r_l (v) is the l'th coefficient in the Legendre expansion of f_e. Finally,

$$G_l(v,k) = \int_0^\pi \sin\theta\ P_l\ (\cos\theta)\partial^2\sigma/\partial k\partial\Omega(v,\theta,k)d\theta \tag{8}$$

Here θ is the angle between $\underline{v}$ and the line of sight. Since G_l can be computed at initialization, evaluation of Eq. (7) is rapid.

CONCLUSION

We have described a program designed to address a global simulation of current ECRH experiments[5] in that the evolution of the electron distribution can be followed in the presence of RF excitation along an entire cross section. Further enhancements of the model will concentrate on an improved wave damping package and the coupling of CQL3D to an advanced transport code.

REFERENCES

[1]G.D. Kerbel and M.G. McCoy, Phys. Fluids 28 (12), 3629 (1985).

[2]G.D. Kerbel and M.G. McCoy, Comp. Phys. Comm. 40, 105 (1986).

[3]M. Franz, UCRL-96510, submitted to Journal Comp. Phys. (1987).

[4]J.Y. Hsu, V.S. Chan, and F.W. McClain, Phys. Fluids, 26, 11 (1983).

[5]R.W. Harvey, M.G. McCoy and G.D. Kerbel, 7th APS Topical Conference on Applications of RF Power, (1987).

CROSS-EFFECT ON ELECTRON-CYCLOTRON AND LOWER-HYBRID CURRENT DRIVE IN TOKAMAK PLASMAS

I. FIDONE - G. GIRUZZI[a] - V. KRIVENSKI - E. MAZZUCATO[b]
ASSOCIATION EURATOM-CEA sur la FUSION, DRFC-CADARACHE
13108 - SAIN PAUL LEZ DURANCE, FRANCE

Abstract

Electron cyclotron resonance current drive in a tokamak plasma in the presence of a lower hybrid tail is investigated using a 2D Fokker-Planck code. For an extraordinary mode at oblique propagation and down-shifted frequency, it is shown that the efficiency of electron cyclotron current drive becomes equal or greater than that of the lower hybrid waves and much greater than the corresponding efficiency of a Maxwellian target plasma at the same bulk temperature.

1. INTRODUCTION

Current drive by the combined system of LH and EC waves[1,2] is of interest because it optimizes the current efficiency and becomes very flexible for current profile control. Specifically, we show that the efficiency of EC current drive can be comparable or greater than that of the LH waves and much greater than the corresponding one of a Maxwellian distribution at the same bulk temperature. The mechanism of the current enhancement is as follows. The EC waves decrease the collision frequency of electrons in a given range of the tail distribution. Next, the combined effects of pitch angle scattering and reduced resistance to the LH parallel pusher result in the increase of the energetic part of the tail population. Since most of the current is carried by the fast part of the tail, the efficiency of the additional current is enhanced.

2. KINETIC EQUATIONS

In order to investigate current drive for the combined system we solve the equation for the distribution function f given by

$$\frac{\partial f}{\partial \tau} = \left(\frac{\partial f}{\partial \tau}\right)_{LH} + \left(\frac{\partial f}{\partial \tau}\right)_{cy} + \left(\frac{\partial f}{\partial \tau}\right)_{Coll} \qquad (1)$$

where $\tau = \nu_e t$, $\nu_e = 2\,\pi e^4\, n_e\, \Lambda / m^{1/2}\, T_e^{3/2}$,

$$\left(\frac{\partial f}{\partial \tau}\right)_{LH} = \frac{\partial}{\partial u_{\parallel}} D_{\parallel} \frac{\partial f}{\partial u_{\parallel}}$$

$$\left(\frac{\partial f}{\partial \tau}\right)_{coll} = \frac{\gamma(Z+1)}{u^3 \sin\theta} \frac{\partial}{\partial \theta} \sin\theta \frac{\partial f}{\partial \theta} + \frac{2}{u^2} \frac{\partial}{\partial u} \left(\frac{\gamma^3}{u} \frac{\partial f}{\partial u} + \gamma^2 f\right)$$

$$\left(\frac{\partial f}{\partial \tau}\right)_{cy} = \frac{1}{u_\perp} \left(\frac{\omega_c}{\omega} \frac{\partial}{\partial u_\perp} + \frac{u_\perp}{\mu^{1/2}} \frac{\partial}{\partial u_\parallel} n_\parallel \right) u_\perp D_{cy} \left(\frac{\omega_c}{\omega} \frac{\partial}{\partial u_\perp} + \frac{n_\parallel u_\perp}{\mu^{1/2}} \frac{\partial}{\partial u_\parallel}\right) f$$

where $\omega_c = eB/mc$, $\vec{B}$ is the tokamak magnetic field, $\mu = mc^2/T_e$, Z is the ion charge, $\vec{u} = \vec{p}/(mT_e)^{1/2}$, u and θ are polar coordinates, $\gamma = (1 + u^2/\mu)^{1/2}$, D is the lower hybrid diffusion coefficient,

$$n_\parallel = \mu^{1/2} (\gamma - \omega_c/\omega)/u_\parallel ,$$

D_{cy} is the quasilinear electron-cyclotron diffusion coefficient, and $(\partial f/\partial \tau)_{coll}$ is the relativistic Fokker-Planck collision operator for $u >> 1$. We consider a tokamak of major radius R = 225 cm, minor radius a = 70 cm, plasma density $n_e = 5 \times 10^{13} (1 - r^2/a^2) cm^{-3}$, electron temperature $T_e(r) = 3(1 - r^2/a^2)^{3/2}$ keV, and a magnetic field, at r = 0, B(o) = 45 kG. For the LH spectrum we assume D = 0.8 = const in space and non-zero for values of the phase velocities in the range $v_1 = 3.5(T_e(o)/m)^{1/2} \simeq 0.25$ c and $v_2 = 7(T_e(o)/m)^{1/2} \simeq 0.5$ c. The EC wave is an extraordinary mode launched from a top location in the poloidal section at frequency $f = \omega/2\pi = 100$ GHz. The launching direction is determined by the projections of the wave vector over the normal to $\vec{B}$ and the inner normal in the poloidal section, or by the angle ψ and Δθ, respectively. Note that $N_\parallel = \sin\psi$. We consider two cases, i.e., ψ = -45° (Δθ = 15°) and ψ = -15° (Δθ = -15°). Figure 1 shows the projection of the ray paths in the poloidal section for the two angles.

3. NUMERICAL RESULTS

We first solve Eq. (1) for $(\partial f/\partial \tau)_{cy} = 0$. At steady-state, a current channel is generated with a bell-shape profile. The total current is $I_{LH} = 500$ kA for a power absorbed of $W_{LH} = 2.8$ MW and one obtains $W_{LH}/I_{LH} = 5.6$ (W/A). Next we apply the EC wave-packet at an average angle of ψ = -45° or ψ = -15°, $\Delta N_\parallel = 0.12$ (Δψ = 7°) and initial power of $W_o = 2$ MW. A new steady-state situation is established in which the generated current is greater than I_{LH}. In Fig. 2, we present the spatial profile of I for the LH only (dashed) and the combined system (full). It appears that the maximum of the EC power deposition occurs at r = 30 cm and r = 15 cm for ψ = -45° and ψ = -15°, respectively. We define the additional current and power absorbed due to EC waves by $I_{cy} = I - I_{LH}$ and $W_{cy} = W - W_{LH}$ and we obtain $W_{cy}/I_{cy} = 6.7$ (W/A) and 4(W/A) for ψ = -45° and

ψ = -15°, respectively, to be compared with W_{LH}/I_{LH} = 5.6. It appears that W_{cy}/I_{cy} is close or smaller than W_{LH}/I_{LH}. The difference is due to the value of the velocity of the electrons resonating with the EC waves. The effect of this selective EC heating is shown in Fig. 3, where we present the parallel distribution $F(u_{\parallel}) = 2\pi \int du_{\perp}u_{\perp}f$ at r = 30 cm (a) and r = 15 cm (b). It appears that at r = 30 cm, the EC waves interact with the low velocity part of the tail. The combined effects of pitch angle scattering and reduced resistance to the LH parallel diffusion result in an increase of the high velocity tail population. Since the predominant part of the current is carried by the high velocity side of the tail, the efficiency of the additional current is close to that of the LH waves. In the case of r = 15 cm, the velocity of the resonant electrons lies in the energetic part of the tail and the result of EC waves is to flatten and extend further the far end of the preexisting tail. The additional current is now carried by electrons with velocity $v_{\parallel} > v_2$ and the efficiency is better than that of the LH only.

4. CONCLUSIONS

We have shown that the efficiency of electron-cyclotron current drive in the presence of the LH tail becomes comparable or better than that of LH waves. This contrasts with current drive by EC waves only. In particular, in the presence of LH waves the effect of electron trapping induced by EC waves is minimized since the predominant part of the current is always carried by electrons with large values of p .

REFERENCES

(1) I. Fidone, et al., Phys. Fluids 27, 2468 (1984) ; V. Krivenski et al., Nucl. Fusion 25, 127 (1985)

(2) K. Hoshino et al. 12th European Conf. on Controlled Fusion and Plasma Physics, Budapest, Sept. 2-6, 1985 (EPS, Geneva, 1985) 9F, II, 184 ; A. Ando et al., Nucl. Fusion 26, 107 (1986)

a) Present address : Association EURATOM-FOM, FOM-INSTITUUT voor Plasmafysica, "Rijnhuizen", NIEUWEGEIN, The NETHERLANDS

b) Permanent address : Plasma Physics Laboratory, P.O. BOX 451, PRINCETON, N.J., 08544, USA

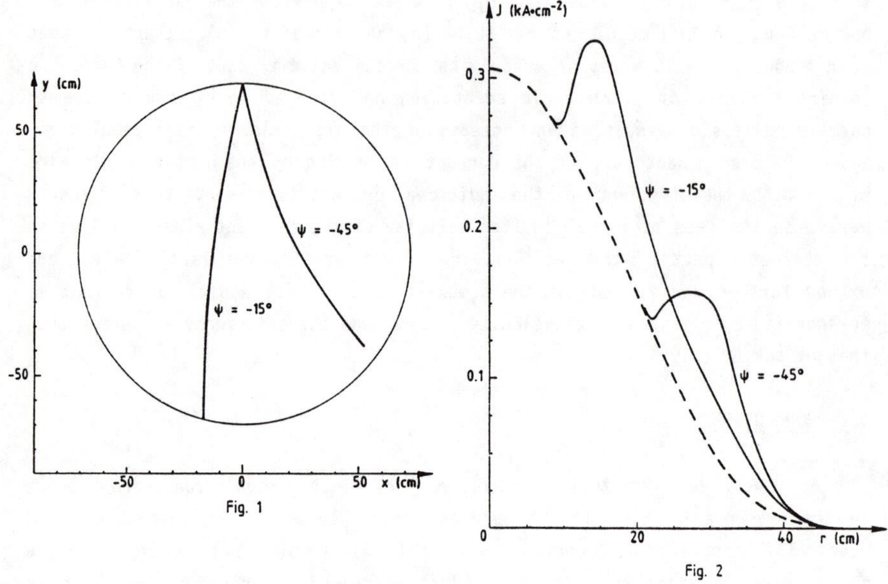

Fig. 1

Fig. 2

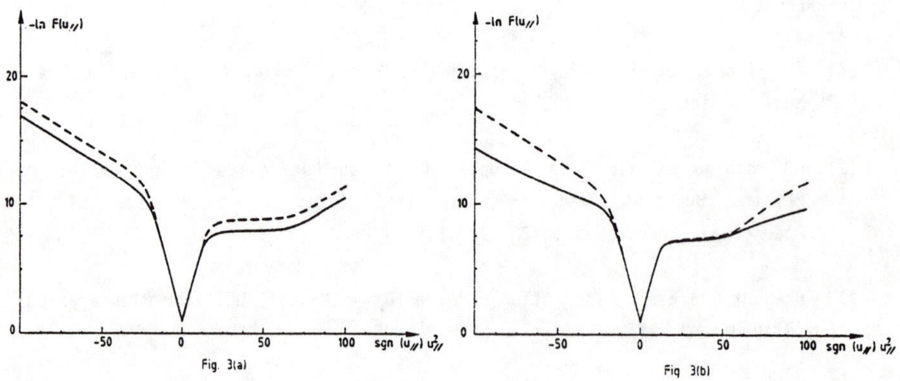

Fig. 3(a)

Fig 3(b)

"APPLICATIONS OF COMPUTER MODELS IN THE STUDY OF LOWER HYBRID RF PLASMA HEATING AND CURRENT DRIVE"

Paul T. Bonoli
Plasma Fusion Center, M.I.T., Cambridge, MA 02139

I. INTRODUCTION

In recent years lower hybrid (LH) waves have been successfully utilized for electron and ion plasma heating, to sustain toroidal plasma current, and to stabilize sawteeth in tokamaks.[1-8] In order to understand the underlying physics of these experiments, make detailed comparisons between theory and experiment, and to extrapolate these results to the reactor regime, a considerable effort has been devoted to the development of simulation models for lower hybrid current drive (LHCD) and lower hybrid heating (LHH). Such models incorporate a $1-D$ radial transport code, Fokker-Planck code, and a toroidal ray tracing calculation. Physics issues critical to the understanding of LHRF experiments can then be studied such as the "spectral-gap" problem [3,4] associated with LH wave propagation and absorption, the confinement and spatial diffusion of suprathermal tail electrons, and the apparent discrepancy between theoretical and experimental values of the current drive figure of merit. The objective of this paper is to review the basic structure of these computer models, pointing out the assumptions and major uncertainties in each model component. Comparisons between model predictions and experiment will be given for the PLT and Alcator C LHRF experiments.

II. TRANSPORT

The major uncertainty in the transport calculation is the confinement properties of the suprathermal tail electrons produced by the LHCD and the bulk electron plasma transport during RF injection. A common approach taken in order to infer these transport coefficients is to couple a $1-D$ radial transport code to a Fokker-Planck - toroidal ray tracing package. Certain theoretically derived forms for the bulk electron (ion) thermal diffusivities $\chi_e(\chi_i)$ are then chosen and the magnitudes are adjusted so as to obtain agreement with the experimentally observed temperature rises. Possible choices for χ_i and χ_e are [9] $\chi_i = M_i\chi_i^{HH}$ and [10] $\chi_e = M_e\chi_e^{TANG}$ where $M_i \simeq 3-6$ is determined by matching the ion temperature of the ohmic plasma, $M_e = 1$ for the ohmic plasma, and $M_e \geq 1$ is determined by matching the RF plasma.

Conservation equations are solved for the time evolution of the bulk plasma quantities n_e, n_i, T_e, T_i and ψ, where $B_\theta = \partial\psi/\partial r$. Although not necessary, n_e and n_i are sometimes taken to be specified functions of time and radius.[11,12] An evolution equation for ψ is formulated [11] from Faraday's Law, Ampere's Law, and a modified Ohm's Law $E_\parallel = \eta_\parallel(J_\parallel - \tilde{J}_{RF})$, where $\eta_\parallel$ is the Spitzer resistivity, $J_\parallel$ is the toroidal current density, and $\tilde{J}_{RF}$ is the "diffused" RF current density (discussed further in Sec. III). Source terms, $S_d^{(e)}(r)$ and $S_d^{(i)}(r)$ for the electron and ion transport equations are calculated in the wave propagation and absorption codes (Secs. III and IV).

Some information on how to model the confinement properties of suprathermal tail electrons is available from the experimental measurements of runaway electron confinement times in PLT [13], where the results could be parameterized by a form $\tau_L = \tau_o\gamma^n$, where $\gamma^2 = (1-v^2/c^2)^{-1}$ is the relativistic factor, $\tau_o \propto \tau_e$, τ_e is the bulk electron energy confinement time, and $n \geq 3$. Fast tail electrons are therefore better confined than thermal electrons.[13] In Sec. III it will be seen that τ_o influences the magnitude of the RF current as well as the fraction of RF power carried by the electron tail which eventually goes into stopped electrons (that have collisionally equilibrated with the bulk plasma).

III. FOKKER-PLANCK CALCULATION

The starting point for the Fokker-Planck analysis is an equation of the form

$$\frac{\partial}{\partial t}f_e = \frac{\partial}{\partial p_{\parallel}}D_{rf}(r,p_{\parallel})\frac{\partial}{\partial p_{\parallel}}f_e + C(f_e) - eE_{\parallel}\frac{\partial}{\partial p_{\parallel}}f_e + \frac{1}{r}\frac{\partial}{\partial r}r\chi_{ef}\frac{\partial f_e}{\partial r} + \Gamma_s\delta(p_{\parallel}) \qquad (1)$$

where $f_e = f_e(p_\perp, p_\parallel)$, $p_\parallel = \gamma m_e v_\parallel$, $E_\parallel$ is the DC electric field, D_{rf} is the quasi-linear diffusion coefficient [14], due to the RF waves, χ_{ef} is a diffusion coefficient for the fast electrons, Γ_s is a source of particles at low $p_\parallel$, and $C(f_e)$ is the Landau limit of the Balescu-Lenard collision opreator.[15,16] The $(t, r, p_\parallel, p_\perp)$ solution of Eq.(1) is computationally prohibitive within the context of a transport-Fokker-Planck-ray tracing calculation. However, exact $(t, p_\parallel, p_\perp)$ solutions of Eq.(1) have been obtained by neglecting Γ_s and the spatial diffusion term, assuming a constant D_{rf} for some range $p_1 \leq p_\parallel \leq p_2$ and $D_{rf} = 0$ for $p_1 < p_\parallel$ and $p_\parallel > p_2$.[16-22] Some of these studies [16-21,23] indicate the presence of a large effective temperature "$T_\perp$" in the perpendicular electron distribution due to enhanced pitch-angle scattering of electrons at high $p_\parallel$ into perpendicular momenta. This finding is in good agreement with experimental observations of $T_\perp \simeq (100-200)$ keV in PLT [23] and Alcator C.[24] Some of this work included the effect of a DC electric field [19,22,25,26,27] which was shown to be important in LH current ramp-up [19,25,27] and could lead to the anomalous Doppler instability [26] if $E_\parallel \gtrsim E_r$, where E_r is the runaway electric field.

However $E_\parallel \approx 0$ in steady-state LHCD. Also $\partial f_e/\partial t \simeq 0$ since the transport time scales (τ_e) are usually much longer than the quasi-linear diffusion time scales. Finally by assuming $f_e(p_\perp, p_\parallel) = F_e(p_\parallel)/(2\pi m_e T_\perp)\exp(-p_\perp^2/2m_e T_\perp)$, Eq.(1) can be integrated over perpendicular momentum to obtain

$$\frac{\partial}{\partial p_{\parallel}}D_{rf}(r,p_{\parallel})\frac{\partial}{\partial p_{\parallel}}F_e + \tilde{C}(F_e, T_\perp, p_\parallel) - eE_\parallel\frac{\partial F_e}{\partial p_\parallel} - F_e/\tau_L(p_\parallel) + \Gamma_s\delta(p_\parallel) = 0, \qquad (2)$$

where the spatial diffusion term has been replaced by a phenomenological loss. $\tilde{C}$ is a complicated collision operator which is evaluated for abitrary $T_\perp$ in the region of RF waves.[28] Equation (2) can be solved at a number of radial grid points and at each time step of the transport calculation. The dissipated RF power density $S_{rf}(r)$ is calculated by taking the energy moment $[n_e m_e c^2(\gamma-1)]$ of the RF quasi-linear term in Eq.(2). The power density that is collisionally equilibrated from the electron tail to the bulk plasma $S_d^{(e)}(r)$, is given by the energy moment over the collision plus electric field terms in Eq.(2). Similarly the power density $S_L(r)$ due to the tail loss is found by taking the energy moment over the loss term.

It should be noted that D_{rf} is a function of the quasi-linear wave damping, $\partial F_e/\partial p_\parallel$. A possible method [11,12,28] for solving Eq.(2) is to calculate $D_{rf} \equiv D_{rf}^{(0)}$ by first assuming that $F_e \equiv F_e^{(0)}$ is Maxwellian. Next, $D_{rf}^{(0)}$ is used in Eq.(2) to generate the first approximation to the quasi-linear $F_e \equiv F_e^{(1)}$. Then $D_{rf} \equiv D_{rf}^{(1)}$ is calculated based on $F_e^{(1)}$. This iteration process is continued until successive approximations to F_e and D_{rf} no longer vary.

In lower density discharges, the slowing down time of fast electrons (τ_s) may be long enough so that $\tau_s \gtrsim \tau_e$ and the earlier assumption of $\partial f_e/\partial t = 0$ may not be quite accurate. A technique for treating the RF current density in this case was described in Ref.(11) where the moment $\pi\int_0^\infty dp_\perp^2 \int_{p1}^{p2} n_e e v_\parallel dp_\parallel$ is applied to Eq. (1) to obtain

$$\frac{\partial}{\partial t}\tilde{J}_{rf} = S_J(r) + \frac{1}{r}\frac{\partial}{\partial r}r\chi_{ef}\frac{\partial \tilde{J}_{rf}}{\partial r} - \nu_J\tilde{J}_{rf} \qquad (3)$$

Here p_1 and p_2 define the resonant region of RF waves where $D_{rf} \neq 0$ and S_J, χ_{ef}, and ν_J are calculated from the steady-state solution [Eq.(2)]. The source term $S_J(r)$ corresponds to the moment over the RF operator in Eq.(2), the effective rate of current destruction ν_J corresponds to the moment over the collision plus electric field terms, and the diffusivity χ_{ef} corresponds to the moment over the loss term. If $\chi_e(r) \propto a^2/\tau_o$, then $\chi_{ef} = \chi_e(r)(\int_{p1}^{p2} n_e e v_\parallel F_e/\gamma^n dp_\parallel)/J_{rf}$, where $J_{rf} = \int_{p1}^{p2} n_e e v_\parallel F_e dp_\parallel$ is the RF current moment corresponding to the steady-state solution [Eq.(2)]. This is to be distinguished from the diffused RF current density $\tilde{J}_{rf}$ given by Eq.(3).

IV. WAVE PROPAGATION AND ABSORPTION

Lower hybrid wave propagation in tokamak plasmas can be accurately treated in the limit of geometrical optics, [29-34] where $\lambda_\perp \ll L_n$. Here $k_\perp = 2\pi/\lambda_\perp$ is the perpendicular wavenumber and $L_n = \mid \frac{1}{n_e}\frac{dn_e}{dr} \mid^{-1}$. This condition is well-satisfied away from the plasma cut-off and the confluence point for the fast and slow wave.[34] The trajectory of a wave packet satisfying the local dispersion relation $D(\mathbf{x}, \mathbf{k}, \omega) = 0$ is given by the ray equations [35]

$$\frac{d\mathbf{x}}{dt} = -\frac{\partial D}{\partial \mathbf{k}} / \frac{\partial D}{\partial \omega}, \qquad (4a) \qquad\qquad \frac{d\mathbf{k}}{dt} = \frac{\partial D}{\partial \mathbf{x}} / \frac{\partial D}{\partial \omega}, \qquad (4b)$$

where $\omega, \mathbf{k}$, and D are taken to be real. If one utilizes the Hamiltonian nature of these equations [31] they take on a particularly simple form in toroidal geometry where $\mathbf{x} \equiv (r, \theta, \phi)$ and $\mathbf{k} \equiv (k_r, m, n)$ are the canonically conjugate momenta. Here r is the minor radial position, θ is the poloidal angle, ϕ is the toroidal angle, k_r is the radial wavenumber, $m = rk_\theta$ is the poloidal mode number and $n = (R_o + r\cos\theta)k_\phi$ is the toroidal mode number. $D(\mathbf{x}, \mathbf{k}, \omega)$ is the local WKB dispersion relation for LH waves including electromagnetic and warm plasma effects.[32,33,36] The great advantage gained by using Eqs.(4) is that the effects of the two-dimensional (r, θ) inhomogeneities of the tokamak equilibrium can be accounted for in the wave propagagtion, i. e. , $D = D(r, \theta; k_r, m, n, \omega)$. Clearly from Eq.(4b) m will now vary due to the θ dependence of the equilibrium whereas for toroidal (ϕ) symmetry n will be conserved (a constant of the wave motion). The parallel wavenumber, $k_\parallel = \mathbf{k} \cdot \mathbf{B} / \mid \mathbf{B} \mid$ will also vary due to the combined effects of magnetic shear and toroidicity (m changes), leading to possible modifications in wave accessibility [36,37] ($n_\parallel \geq n_a$) and electron Landau damping [37] ($n_\parallel \geq n_{eld}$). Here $n_\parallel = k_\parallel c/\omega$.

Ultimately one should solve the full wave equation in two dimensions (r, θ) with the appropriate hot conductivity operator and compare the results with the ray approach. Also one should include the effects of wave scattering from electron density fluctuations which have been shown to be an important effect for the LH wave.[33,38,39] However, the numerical implementation of these effects within a larger LHCD simulation model is computationally prohibitive.

An evolution equation for the power flowing along the ray path can be integrated simultaneously with the ray equations and has the form

$$\frac{dP}{dt} = -2\gamma_T P, \qquad (5)$$

where $\gamma_T = \gamma_e + \gamma_i + \gamma_\alpha + \gamma_c$ and $\gamma_T \ll \omega$. Here γ_e is the damping due to quasi-linear electron Landau resonance, γ_i is the linear ion Landau damping assuming unmagnetized ions, γ_α is the Landau damping assuming unmagnetized fusion-generated alpha-particles,[40] and γ_c is the nonresonant damping due to electron-ion Coulomb collisions. Note that focussing terms in Eq.(5) have been neglected so that P is not a power density.

An expression for the RF diffusion coefficient can be obtained [28] by assuming that RF power propagates according to the ray equations and Eq.(5) in a 'tube' of constant cross-section, with varying group velocity. The resulting D_{rf}, will in general contain contributions from many wavenumbers $k_\parallel^0$, representing the full range of the Brambilla power spectrum [41] assumed to be launched at the plasma edge.

V. SIMULATION RESULTS

A. ALCATOR C: STEADY STATE LHCD

Detailed code simulations have been carried out for discharges maintained purely by LHRF driven currents in Alcator C.[42,43] The parameters used were $a = 16.5$cm, $R_o = 64$cm, $B_\phi = 8$T, $I_p = 140$kA, $\bar{n}_e = (3-9)\times10^{13}$cm^{-3}, $n_{eo} = 1.5\times\bar{n}_e$, $Z_{eff} = 1.5$, hydrogen gas, $T_{eo} = (1.3-1.8)$ keV, $T_{io} = (0.6-0.8)$ keV, $M_i = 6$, relative waveguide phase $\Delta\phi = 90°$, $\tau_L = (3\text{msec})\gamma^3$, $T_\perp = 50\times T_e(r)$, and $P_{in} = (300-1000)$ kW. The results of these studies are summarized in Table I:

Table I

$\bar{n}_e(10^{13}\text{cm}^{-3})$	P_{in}(kW)	W_B (kJ)	W_T (kJ)	M_e	$\tau_E^{TOT}(ms)$	τ_E^{OH} (ms)
3	320	1.2	1.1	1	7.1	5.7
5.5	650	2.3	1.3	1.75	5.6	9.3
7.0	950	3.1	1.2	2.1	4.5	10.6

The confinement times are defined by $\tau_E^{TOT} = (W_B + W_T)/P_{in}$ and $\tau_E^{OH} = W_B/P_{OH}$. The current drive figure of merit is $\bar{\eta} = (0.084 - 0.066)A/W/m^2$, where $\bar{\eta} \equiv \bar{n}_e(10^{20}\text{m}^{-3}I_p$ (kA) R_o(m)/P_{in} (kW). The energy carried by the electron tail (W_T) is comparable to the bulk energy (W_B) and thus contributes significantly to the overall power in good agreement with experiment.[4,42-44] The power lost due to electron tail loss (P_L) decreases as $\bar{n}_e$ is raised from $P_L/P_{in} = 0.32$ to 0.15. However the power lost due to collisional damping (P_C) increases as $\bar{n}_e$ is raised from $P_C/P_{in} = 0.04$ to 0.2. It is interesting to note that $\chi_e^{rf} = M_e\chi_e^{TANG} \approx$ constant, as $\bar{n}_e$ increases. This is to be contrasted with ohmic discharges where $\chi_e^{OH} \propto n_e^{-1}$.

The profiles for RF deposition and diffused RF current density are shown in Figs. 1(a)-1(b), for $\bar{n}_e = 3\times10^{13}$cm^{-3}. Note that although S_{rf} is peaked off-axis at $r = 2.8$ cm, the profile for $\tilde{J}_{rf}$ calculated from Eqs.(2) and (3) is centrally peaked. At $\bar{n}_e = 7\times10^{13}$cm^{-3}, this peaking in $\tilde{J}_{rf}$ no longer occurs because fast electrons slow down and thermalize on a time scale which is faster than their spatial diffusion time. Figure 1(c) is a plot of F_e for positive and negative velocities as a function of $E = m_ec^2[n_\parallel/(n_\parallel^2 - 1)^{1/2} - 1]$, at a radial location corresponding to $r \simeq 2.8$ cm. Although most of the RF power in the initial Brambilla spectrum ($\Delta\phi = 90°$) is at $n_\parallel \leq 2$ and $E \leq 100$ keV, the toroidal variations in $k_\parallel$ have "spread" the RF power to $n_\parallel \geq 6$ and $E \leq 10$ keV, causing a significant tail population. Plots of $\rho/a, m, n_\parallel$, and P_N versus ϕ are shown in Figs. 2(a)-2(d) for the 3×10^{13}cm^{-3} case. Here $\Delta(r)$ is the Shafranov shift, P_N is a normalized wave power, and the quasi-linear wave damping is used. About 70% of the wave power is absorbed due to significant tail damping, on the first pass of the ray in and out of the plasma, with $n_\parallel \lesssim 1.7$. On subsequent passes of the ray at $\phi \simeq (9,17)$ rad., $n_\parallel$ increases to $2.8-4$ and causes the remaining wave power to damp at lower phase velocities, thus maintaining the raised plateau on F_e.

B. ALCATOR C: RAMP-UP

If the plasma is "overdriven" with LHRF it is possible to ramp-up the total current, I_p. An example of this is shown in Fig.3 for Alcator C.[44] The parameters are the same as those used in Sec. VA with $P_{in} = 460$ kW, $I_p = 120$ kA initially, $\tau_0 = 3$ msec, and $M_e = 1.1$. The ramp rate from Fig.3 is $\dot{I}_p \equiv dI_p/dt \simeq 125$ kA/sec, although $\dot{I}_p \sim 1300$ kA/sec in the first $10-20$ msec of the RF pulse. The ramp-up efficiency defined here [11] to be $-P_{OH}/P_{in}$ is 5%.

C. PLT WAVEGUIDE PHASING SCAN

The PLT Waveguide phasing experiments [3,8] have been studied using a computer model consisting of a Fokker-Planck-toroidal ray tracing package. [28] The parameters used were $a = 40$ cm, $R_o = 132$ cm, $B_\phi = 3-3.1$ T, deuterium gas, and $T_\perp = 50\times T_e(r)$. Four cases were considered. This first three correspond to the 6 element, 0.8 GHz waveguide array with

$\Delta\phi = 60°, 90°$ and $120°$, $T_{eo} = 1.5$ keV, $Z_{eff} = 4.0$, $I_p = 200$ kA, $\bar{n}_e = 3.75 \times 10^{12}\text{cm}^{-3}$ and $\tau_L = (7.5 \text{ ms})\gamma^4$. The fourth case corresponds to the 16 element, 2.45 GHz array with $T_{eo} = 5$ keV, $Z_{eff} = 2.5, I_p = 500$ kA, $\bar{n}_e = 1 \times 10^{13}\text{cm}^{-3}$, and $\tau_L = (10\text{ms})\gamma^4$. The 16 element array launched a Brambilla spectrum centered at $n_\parallel = 1.5$ with $\Delta n_\parallel = 0.5$, while the 6 element array produced spectra whose maxima shifted from $n_\parallel \approx 1.2, 2, 2.7$ as $\Delta\phi$ was changed to $60°, 90°, 120°$. These results are summarized below in Table II:

Table II

$\bar{n}_e(10^{13}cm^{-3})$	f_o(GHz)	$\Delta\phi$	P_{in}(kW)	P_L(kW)	$\tilde{\eta}(A/W/m^2)$
0.375	0.8	60°	100	53	0.10
0.375	0.8	90°	123	54	0.08
0.375	0.8	120°	450	123	0.02
1.0	2.45	-	595	325	0.11

It is clear from Table II that as the injected RF power is shifted from faster to lower phase speeds the corresponding figure of merit $\tilde{\eta}$ decreases. The profiles of $S_{rf}, \tilde{J}_{rf}$, and the distribution functions $F_e(r, p_\parallel)$ produced in these cases are similar to those shown in Figs. 1(a)-1(c). The tail loss is significant for these cases with $P_L/P_{in} \lesssim 0.5$. This is not surprising since n_e is low ($\lesssim 10^{13}\text{cm}^{-3}$) and most of the current is carried by fast electrons.

D. ALCATOR C: RF SAWTOOTH STABILIZATION

The transport-current drive model has also been applied to RF sawtooth stabilization experiments in Alcator C.[7] The parameters were $B_\phi = 6.2$ T, $\bar{n}_e = 1.1 \times 10^{14}\text{cm}^{-3}, n_{eo} = 1.27 \times \bar{n}_e, I_p = 265$ kA, $P_{in} = 600$kW, $\Delta\phi = \pm 90°, Z_{eff} = 1.5$, deuterium gas, $T_\perp = 20 \times T_e(r), \tau_L = (7.5\text{ms})\gamma^3, T_{eo} = (1.3 - 1.8)$ keV, and $M_e = 1.0$. The resulting profiles for $J_\parallel$ and J_{rf} are shown in Figs.4(a) and 4(b) for $\Delta\phi = 90°$. The ohmic $J_\parallel$ profile shown in Fig. 4(a) leads to the $q(r)$ profile in Fig. 4(c) where $q(0) < 1, q(r_1) = 1$ for $r_1 \simeq 2$cm, and the plasma is presumably unstable to the $m = 1$ tearing mode. It can be seen from Fig. 4(b) that RF current is generated at $r \gtrsim r_1(2 - 4\text{cm})$. This causes an increase in $J_\parallel$ at $r < r_1$ (because the discharge is being maintained at constant current). The resulting $q(r)$ profiles [Fig. 4(c)] have $q(0) > 1$ for $0 \le r \le a$ at 10 msec into the RF pulse and eventually $q(r)$ is "flattened" at $r \gtrsim r_1$, about (50-100) msec into the RF pulse. It has recently been found using a fully toroidal, resistive MHD code that such $q(r)$ profiles with $dq(r)/dr \simeq 0$ near $r \ge r_1$ can be stable to the $m = 1$ mode.[45]

For $\Delta\phi = -90°$, a negative RF current is generated at $r \gtrsim r_1$, resulting in an increase in $J_\parallel$ at $r < r_1$. The corresponding $q(r)$ profiles retain $q(0) < 1$ and would be unstable to the $m = 1$ mode. We are thus lead to a possible explanation (or at least partial understanding) of the experimental observation [7] that sawteeth can be stabilized using LHRF injection with a "current drive" phasing and that sawteeth are not stabilized by RF injection with an "anti-current drive" phasing.

VI. CONCLUSIONS

It can be seen from the examples of the previous section that simulation models which combine transport, Fokker-Planck, and toroidal ray tracing packages can provide us with details of the wave propagation and absorption physics of LHCD, LH ramp-up, LHH, and RF sawtooth stabilization. Yet some questions still remain in the modelling. A full-wave calculation in (r, θ) must still be done and compared with the multi-bounce toroidal ray tracing theory. The possible role of nolinear effects at ray caustics must be resolved.[46] Although the density limit for LHCD has been shown experimentally to scale with f_0^2[4,5,8,47], only recently has any work [48] been done to study the possible role of nonlinear effects (i.e., parametric decay of the LH pump wave [49]) in explaining this density limit. In addition, the transport of fast electrons and thermal electrons in present day experiments needs further theoretical investigation, although in the reactor regime one would expect fast electrons to thermalize before they are able to diffuse spatially. A calculation which combines a $2-D$ (velocity) Fokker-Planck analysis and a toroidal ray tracing calculation could provide important information on the self-consistent effect of the RF on $T_{\perp}$. (Note that in the modelling results presented here, $T_{\perp}/T_e = 50$ has been used based on $2-D$ calculations [21], which in turn assumed a form for D_{rf}.) Despite these uncertainties, the type of RF models described in this review can provide valuable interpretations of the present experiments and give direction to future LHRF work.

ACKNOWLEDGMENTS

It is a pleasure to acknowledge the contributions of Dr. R. C. Englade, Professor M. Porkolab, Dr. S. Knowlton, and Dr. Y. Takase. This work was supported by the U. S. Department of Energy under Contract No. DE-AC02-78ET-51013.

REFERENCES

1. S.C. Luckhardt, et al., Phys. Rev. Lett. **48**, 152 (1980).
2. T. Yamamoto, et al., Phys. Rev. Lett. **45**, 716 (1980).
3. S. Bernabei, et al., Phys. Rev. Lett. **49**, 1255 (1982).
4. M. Porkolab, et al., Phys. Rev. Lett. **53**, 450 (1984).
5. C.C. Gormezano, et al., Proc. 11th Eur. Conf. on Contr. Fusion and Plasma Phys., Aachen, 1983 (EPS, 1983) Vol. I, 325.
6. F.X. Soldner, et al., Phys. Rev. Lett. **57**, 1137 (1986).
7. M. Porkolab, et al., Proc. 13th Eur. Conf. on Contr. Fusion and Plasma Heating, Schliersee 1986 (EPS, 1986) Vol. II, 445.
8. S. Bernabei, et al., Proc. 11th Int. Conf. on Plasma Phys. and Contr. Nucl. Fus. Res., (Kyoto, 1986) IAEA-CN-4/F-II-1.
9. F.L. Hinton and R.D. Hazeltine, Rev. Mod. Phys. **48**, 239 (1976).
10. W.M. Tang, Nucl. Fusion **26** 1605 (1986).
11. R. Englade and P.T. Bonoli, in *Radiofrequency Plasma Heating, (AIP, NY, 1985) 151.*
12. F.W. Perkins, et al., in Plasma Phys. and Contr. Nucl. Fus. Res. 1984 (IAEA, Vienna, 1985) Vol. I, 513.
13. H.E. Mynick and J.D. Strachan, Phys. Fluids **24**, 695 (1981).
14. C.F. Kennel and F. Engelmann, Phys. Fluids **9**, 2377 (1966).
15. B.A. Trubnikov in *Reviews of Plasma Physics, (Consultants Bureau, NY, 1970), Vol. I, 105.*
16. K. Hizanidis and A. Bers, Phys. Fluids **27**, 2673 (1984).
17. C.F.F. Karney and N.J. Fisch, Phys. Fluids **22**, 1817 (1979).
18. C.F.F. Karney and N.J. Fisch, Phys. Fluids **28**, 116 (1985).
19. C.F.F. Karney and N.J. Fisch, Phys. Fluids **29**, 180 (1986).
20. D. Hewett, K. Hizanidis, V. Krapchev, and A. Bers, in Proc. of IAEA Tech. Comm. Mtg. on Non-Ind. Current Drive in Tokamaks, (EUR/UKAEA, Abingdon, 1983) Vol. II, 124.
21. V. Fuchs, et al., Phys. Fluids **28**, 3619 (1985).
22. V.S. Chan and F.W. McClain, Phys. Fluids **26**, 1542 (1983).
23. J. Stevens, et al., Nucl. Fusion **25**, 1529 (1985).

24. S. Texter, S. Knowlton, M. Porkolab, and Y. Takase, Nucl. Fusion **26**, 1279 (1986).
25. N.J. Fisch and C.F.F. Karney, Phys. Rev. Lett. **54**, 897 (1985).
26. C.S. Liu, V.S. Chan,D.K. Bhadra, and R.W. Harvey, Phys. Rev. Lett. **48**, 1479 (1982).
27. C. S. Liu, V.S. Chan, and Y.C. Lee, Phys. Rev. Lett. **55** 583 (1985).
28. P.T. Bonoli and R.C. Englade, Phys. Fluids **29**, 2937 (1986).
29. Yu. F. Baranov and V.I. Fedorov, Soviet Phys. Tech. Phys. Lett. **4**, 322 (1978).
30. J.L. Kulp, Bull. Am. Phys. Soc. **23**, 789 (1978); J.L. Kulp, Ph.D. dissertation, MIT, 1978.
31. J.M. Wersinger, E. Ott, and J.M. Finn, Phys. Fluids **21**, 2263 (1978).
32. D.W. Ignat, Phys. Fluids **24**, 1110 (1981).
33. P.T. Bonoli and E. Ott, Phys. Fluids **25**, 359 (1982).
34. M. Brambilla, Plasma Phys. **24**, 1187 (1982).
35. S. Weinberg, Phys. Rev. **126**, 1899 (1962).
36. T.H. Stix, *Theory of Plasma Waves (McGraw-Hill, NY 1962)*.
37. M. Brambilla, in Phys. of Plasmas Close to Thermonuclear Conditions (CEC, 1980), Vol. I. 291.
38. P.T. Bonoli and E. Ott, Phys. Rev. Lett. **46**, 424 (1981).
39. P.L. Andrews and F.W. Perkins, Phys. Fluids **26**, 2537 (1983).
40. P.T. Bonoli and M. Porkolab, to be published in Nucl. Fusion; MIT Report PFC/JA-86-63.
41. M. Brambilla Nucl. Fusion **16**, 47 (1976).
42. S. Knowlton, et al., Phys. Rev. Lett. **57**, 587 (1986).
43. Y. Takase, et al., Nucl. Fusion **27**, 53 (1986).
44. Y. Takase, S. Knowlton, and M. Porkolab, Phys. Fluids **30**, 1169 (1987).
45. T.C. Hender, D.C. Robinson, and J.A. Snipes, Proc. 11th Int. Conf. on Plasma Phys. and Contr. Nucl. Fus. Res., (Kyoto, 1986) IAEA-CN-47/A-V-3.
46. E. Barbato, A. Cardinali, and F. Santini, in Proc. of 4th Int. Symp. on Heating in Toroidal Plasmas (Int. School of Plasma Physics, Varenna, 1984) Vol. II, 1353.
47. M. Mayberry, et al., Phys. Rev. Lett. **55**, 829 (1985).
48. C.S. Liu, V.K. Tripathi, and V.S. Chan, Phys. Fluids **27** 1709 (1984).
49. M. Porkolab, Phys. Fluids **20**, 2058 (1977).

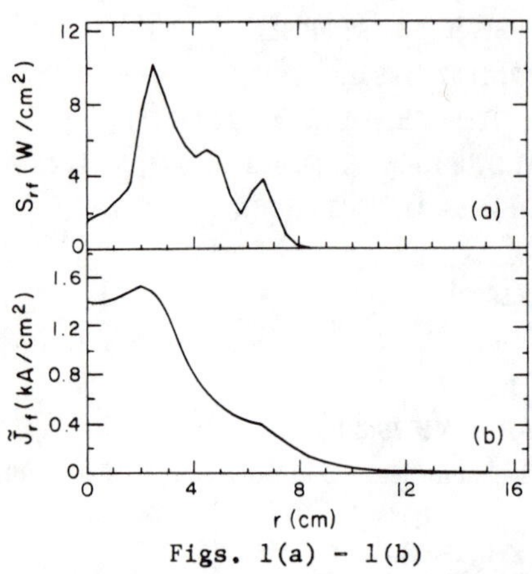

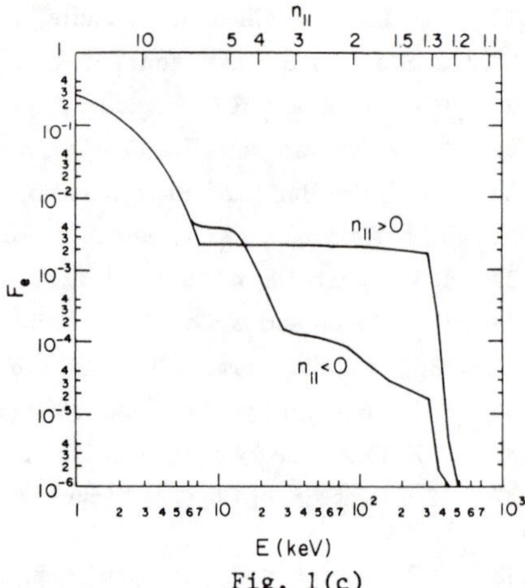

Fig. 1. Model results for Alcator C, LHCD parameters of Sec. V A; $\bar{n}_e = 3\times10^{13}$ cm^{-3}, $B_\phi = 8$T, and $P_{in} = 320$ kW. (a) Radial profile of RF power density. (b) Radial profile of diffused RF current density. (c) Electron distribution function at a radial location $r = 2.8$ cm versus parallel kinetic energy.

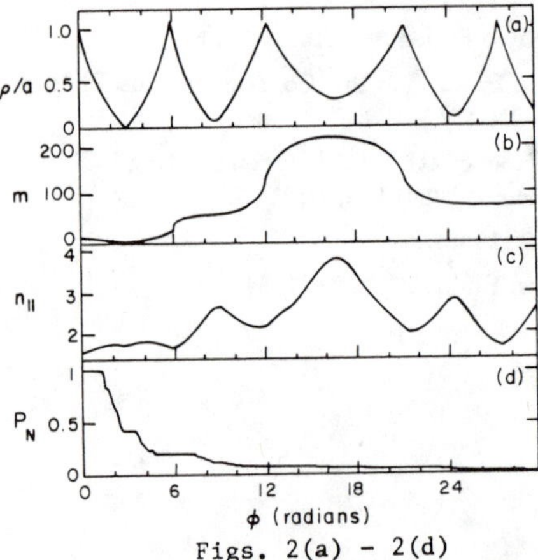

Fig. 2. Ray trajectory for Alcator C example shown in Fig. 1. (a) Variation in ρ/a versus ϕ. (b) Variation im m versus ϕ. Variation in $n_\parallel$ versus ϕ. Initially $n_\parallel = 1.55$. (d) Normalized wave power (P_N) that results from damping on quasi-linear distribution for $n_\parallel > 0$, versus ϕ.

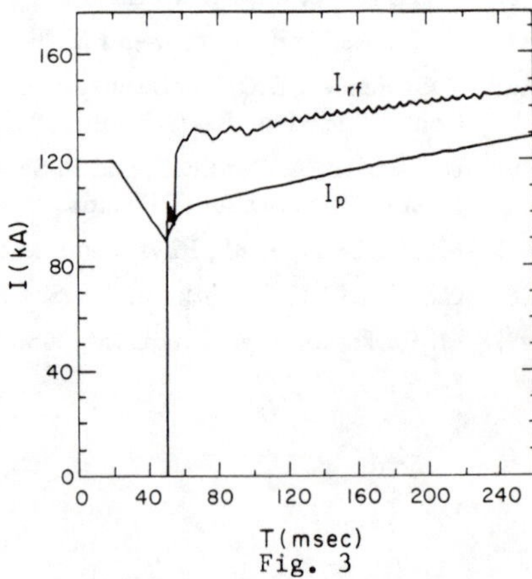

Fig. 3. Model results for Alcator C, LH ramp-up case of Sec. V B; $\bar{n}_e = 3\times10^{13}$ cm^{-3}, $B_\phi = 8$T, and $P_{in} = 460$ kW.

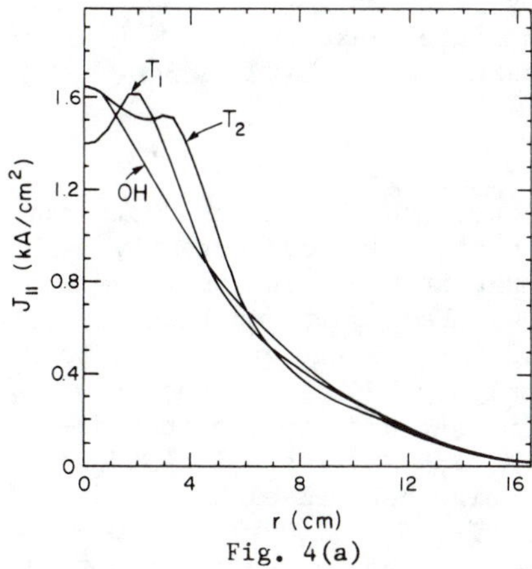

Fig. 4(a)

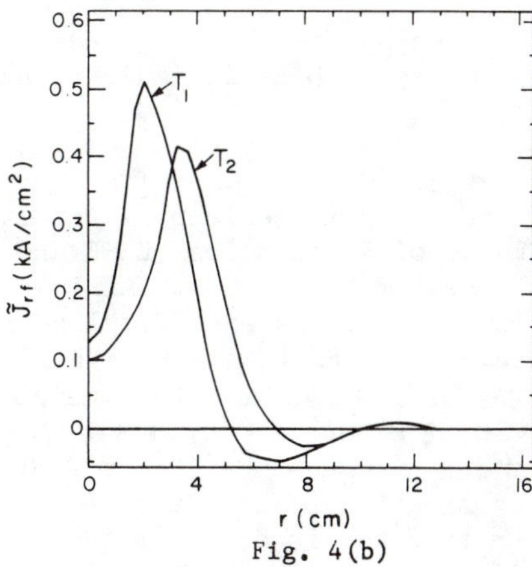

Fig. 4(b)

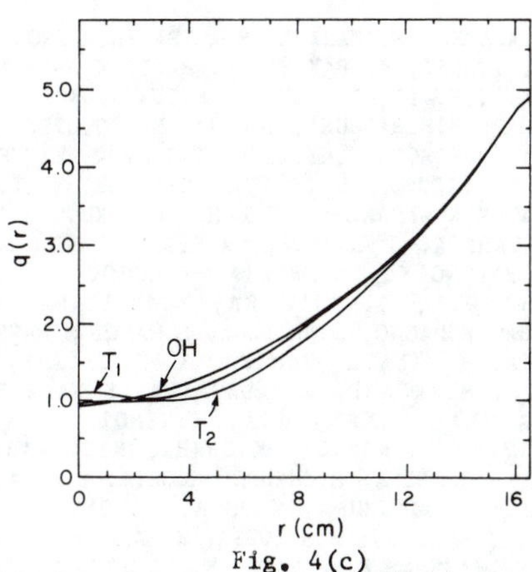

Fig. 4(c)

Fig. 4. Model results for Alcator C, LHRF sawtooth stabilization of Sec. V D; $\bar{n}_e = 1.1 \times 10^{14}$ cm^{-3}, $B_\phi = 6.2$T, and $P_{in} = 600$ kW. (a) Spatial profile of toroidal current density. (b) Spatial profile of diffused RF current density. (c) Spatial profile of the safety factor $q(r)$. The times $T_1(T_2)$ refer to 50 ms (90 ms) after the RF has been turned on.

RF HEATING AND CURRENT DRIVE EXPERIMENT ON JT-60

JT-60 TEAM* presented by T. Nagashima
Japan Atomic Energy Research Institute, Naka Fusion Research Establishment, Naka, Ibaraki-ken 311-02, Japan

ABSTRACT

Recent experimental results of Lower Hybrid and Ion Cyclotron Range of Frequencies (LHRF and ICRF) heating and current drive are presented on JT-60 at JAERI. Three LHRF at 2 GHz and one ICRF at 120 MHz systems are installed in JT-60. Each unit has launched 2.1 -2.4 MW of RF power into the JT-60, so far. Steady current up to 2 MA for 2.5 sec have been maintained only by LHCD at a density of $\bar{n}_e$ = 0.32 x 10^{19} m^{-3} with 3.1 MW. The current drive efficiency defined by $\eta_{CD} = \bar{n}_e$ $(10^{19}\ m^{-3})R(m)I_{RF}(MA)/P_{LH}(MW)$ reach 1.5-3.0 by combination of LHCD and NBI heating. High central electron heating up to 6 keV is demonstrated at the density $\bar{n}_e$ = 1.7 x 10^{-19} m^{-3}. Current profile control and improvement of energy confinement time via LHCD is observed with and without NBI heating. Optimization of the second harmonic ICRF heating is studied with 2 x 2 phased loop antenna. In combination heating of ICRF and NBI, remarkable beam acceleration is observed in the plasma core.

* T.ABE, H.AIKAWA, N.AKAOKA, H.AKASAKA, M.AKIBA, N.AKINO, T.AKIYAMA, T.ANDO, K.ANNOH, N.AOYAGI, T.ARAI, K.ARAKAWA, M.ARAKI, K.ARIMOTO, M.AZUMI, S.CHIBA, M.DAIRAKU, N.EBISWA, T.FUJII, T. FUKUDA, H.FURUKAWA, K.HAMAMATSU, K.HAYASHI, M.HARA, K.HARAGUCHI, H.HIRATSUKA, T.HIRAYAMA, S.HIROKI, K.HIRUTA, M.HONDA, H.HORIIKE, R.HOSODA, N.HOSOGANE, Y.IIDA, T.IIJIMA, K.IKEDA, Y.IKEDA, T.IMAI, T.INOUE,, N.ISAJI, M.ISAKA, S.ISHIDA, N.ITIGE, T.ITO, Y.ITO, A.KAMINAGA, M.KAWAI, Y.KAWAMATA, K.KAWASAKI, K.KIKUCHI, M.KIKUCHI, H.KIMURA, T.KIMURA, H.KISHIMOTO, K.KITAHARA, S.KITAMURA, A.KITSUNEZAKI, K.KIYONO, N.KOBAYASHI, K.KODAMA, Y.KOIDE, T.KOIKE, M.KOMATA, I.KONDO, S.KONOSHIMA, H.KUBO, S.KUNIEDA, S.KURAKATA, K.KURIHARA, M.KURIYAMA, T.KURODA, M.KUSAKA, Y.KUSAMA,, S.MAEBARA, K.MAENO, S.MATSUDA, S.MASE, M.MATSUKAWA, T.MATSUKAWA, M.MATSUOKA, N.MIYA, K.MIYATI, Y.MIYO, K.MIZUHASHI, M.MIZUNO, R.MURAI, Y.MURAKAMI, M.MUTO, M.NAGAMI, A.NAGASHIMA, K.NAGASHIMA, T.NAGASHIMA, S.NAGAYA, H.NAKAMURA, Y.NAKAMURA, M.NEMOTO, Y.NEYATANI, S.NIIKURA, H.NINOMIYA, T.NISHITANI, H.NOMATA, K.OBARA, N.OGIWARA, T.OHGA, Y.OHARA, K.OHASA, H.OOHARA, T.OHSHIMA, M.OHKUBO, K.OHTA, M.OHTA, M.OHTAKA, Y.OHUCHI, A.OIKAWA, H.OKUMURA, Y.OKUMURA, K.OMORI, S.OMORI, Y.OMORI, T.OZEKI, A.SAKASAI, S.SAKATA, M.SATOU, M.SAIGUSA, K.SAKAMOTO, M.SAWAHATA, M.SEIMIYA, M.SEKI, S.SEKI, K.SHIBANUMA, R.SHIMADA, K.SHIMIZU, M.SHIMIZU, Y.SHIMOMURA, S.SHINOZAKI, H.SHIRAI, H.SHIRAKATA, M.SHITOMI, K.SUGANUMA, T.SUGIE, T.SUGIYAMA, H.SUNAOSHI, K.SUZUKI, M.SUZUKI, M.SUZUKI, N.SUZUKI, S.SUZUKI, Y.SUZUKI, M.TAKAHASHI, S.TAKAHASHI, T.TAKAHASHI, J.E.STEVENS **, M.TAKASAKI, M.TAKATSU, H.TAKEUCHI, A.TAKESHITA, S.TAMURA, S.TANAKA, T.TANAKA, K.TANI, M.TERAKADO, T.TERAKADO, K.TOBITA, T.TOKUTAKE, T.TOTSUKA, N.TOYOSHIMA, H.TSUDA, T.TSUGITA, S.TSUJI, Y.TSUKAHARA, M.TSUNEOKA, K.UEHARA, M.UMEHARA, Y.URAMOTO, H.USAMI, K.USHIGUSA, K.USUI, J.YAGYU, K.YAMADA, M.YAMAMOTO, O.YAMASHITA, Y.YAMASHITA, K.YANO, T.YASUKAWA, K.YOKOKURA, H.YOKOMIZO, K.YOSHIKAWA, M.YOSHIKAWA, H.YOSHIDA, Y.YOSHINARI, R.YOSHINO, I.YONEKAWA, K.WATANABE

** Plasma Physics Laboratory, Princeton University, USA

1. INTRODUCTION

As the successful results of heating and current drive experiments of LHRF[1,2] and ICRF[3,4] on JAERI tokamaks, scaling up program of high power LHRF heating and current drive of 24 MW generator power with the combination of 6 MW ICRF generator power started in JT-60. The major objectives of JT-60 RF heating systems are production of a reactor grade plasma with the combination of 20 MW Neutral Beam Injection (NBI) heating and current drive in low and meadium density plasma in order to ascertain the scientific feasibility of stationary or quasi-steady operation of the tokamak reactor.

In this paper we report coupling results of array antennas, RF heating and current drive experiments during the last half year by the end of March, 1987 using LHRF and ICRF heating, both individually and in combined with NBI heating.

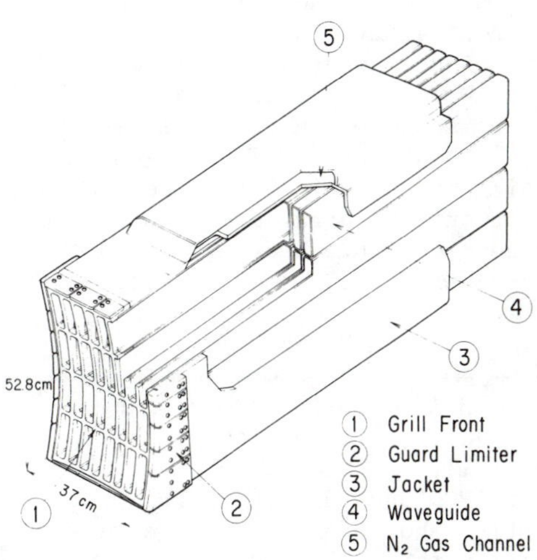

Fig. 1(a) 4 x 8 LHRF coupler.

2. SYSTEM DESCRIPTION

JT-60 is an outside poloidal divertor tokamak with major and minor radii of 3 m and 0.93 m. JT-60 presently operates at plasma current up to $I_p = 2.0$ MA with and without divertor configuration, toroidal fields up to $B_t = 4.5$ T, and plasma duration up to 10 sec.[5] Experiments described in this paper were performed in the hydrogen or helium plasma in the range of $\bar{n}_e = (0.2\text{-}4.2) \times 10^{19}\ m^{-3}$, $I_p = 0.5\text{-}2.0$ MA, $B_t = 4.0\text{-}4.5$ T, $q_{eff} = 3.7\text{-}11$.

NBI heating system is composed of 14 beam lines and has total injection power of 20 MW with hydrogen beam energy of 40-70 keV. Detailed descriptions of JT-60 are provided in the reference.[6]

LHRF heating system has three units. Each unit has 8 MW RF generators with 8 klys-

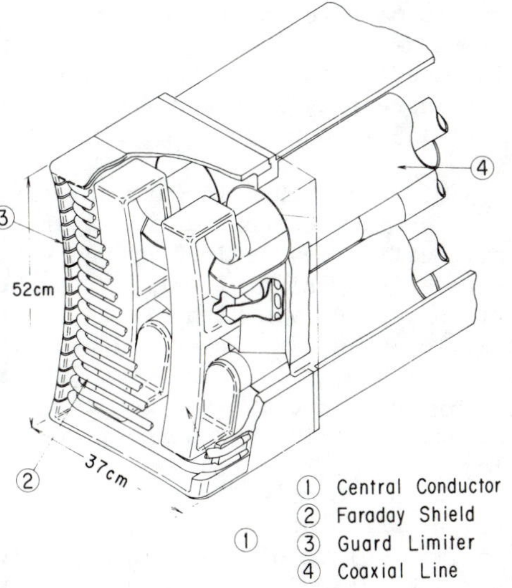

Fig. 1(b) 2 x 2 ICRF antenna.

trons of 1 MW at a frequency of 2 GHz. The LHRF waves were launched by three 4 x 8 waveguide couplers; one of these grills in Fig. 1(a) is designed for LHRF current drive (LHCD) with a $N_{\parallel}^{p}$ index of 1.7 for a phasing $\Delta\phi = 90°$ and other two grills for LHRF heating (LHH) with a $N_{\parallel}^{p}$ of 1.8 for $\Delta\phi = 180°$. One of the remarkable characteristics of JT-60 LHRF heating system is that 1 MW klystrons are stably operating without any output circulators to protect klystrons against reflective power from the plasma.[6,7] The vacuum windows are located far away from the grill mouth, the coupler conditioning being obtained mainly via RF conditioning in the vacuum and short pulse plasma discharges (TDC). RF power of 2.1-2.4 MW and power density up to 4.1 kW/cm^2 have been obtained allowing RF energies of up to 4.5 MJ to be launched into the plasma, so far.[8]

The fourth ICRF unit generates 6 MW at 120 MHz of RF power using eight chains of three-stage-tetrode amplifiers. The 2 x 2 loop antenna array has been developed for the optimization of the second harmonic heating of hydrogen plasmas, shown in Fig. (b).[6,9] RF power of 2.1 MW for 3 sec, so far, has been injected.

3. LHRF RESULTS

3.1 Coupling properties

Size of each element is 1.6 cm x 11.5 cm and 2.9 cm x 11.5 cm for LHCD and LHH. Coupling properties show qualitatively agreement with grill theory, if we assume a density gradient.[8]

Coupling efficiency depends on distance between the coupler and plasma edge and typically in the range of 0.85-0.75. Reflection coefficient stays constant with increasing RF power from 0.5 MW to 1.5 MW during 2 sec (Fig. 2).

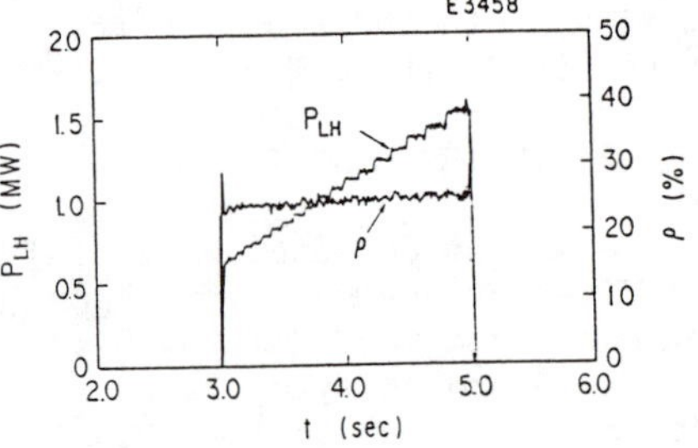

Fig. 2 Time dependence of reflection coefficient ρ with increasing RF power up to 1.5 MW.

3.2 Lower Hybrid Current Drive (LHCD)

The first current drive experiment on JT-60 shows the current drive efficiency defined by $\eta_{CD} = \bar{n}_e\ (10^{19}\ m^{-3})R(m)I_{RF}(MA)/P_{LH}(MW)$ is reletively high; 1.0-1.7 with the LHCD launcher.[10] Here, the grills were driven with a relative phase of $\Delta\phi = 90°$ between successive waveguides in order to excite an asymmetric spectrum of waves with $N_{\parallel}^{p} = 1.7$ for the LHCD launcher and $\Delta\phi = 120°$ with $N_{\parallel}^{p} = 1.3$ for the LHH one. η_{CD} decreased in case of LHH couplers because the accessibility and directivity is not so good as the LHCD coupler.

With a combined injection of one LHCD and Two LHH couplers, a driven current of 2 MA was achieved at $\bar{n}_e = 0.32 \times 10^{19}\ m^{-3}$ and $B_t = 4.5$ T in helium plasmas, as shown in Fig. 3. Total injected power is 3.1 MW at 2 GHz, where 0.8 MW was injected from the LHCD launcher. In this shot the loop voltage keeps zero or slightly negative during RF injection of 2.5 sec, which indicate the current was fully driven by LHCD without Ohmic power.

The efficiency η_{CD} of LHCD only (denoted as OH in the figure) and LHCD + NBI (denoted as NB) are summarized in $P_{LH}/\bar{n}_e$ dependency of the driven current by $I_{RF} = -I_p\Delta V_1/V_1$, shown in Fig. 4. The solid lines mean $\eta_{CD} = 1$, 2, and 3, respectively. In the case of LHCD only, η_{CD} lies from 0.9 to 2.0, however, η_{CD} is improved drastically to 1.5 to 3.0 when NBI heating is combined with LHCD. The range of peak electron temperature T_{e0} during LHCD on JT-60 is 2 - 3 keV for OH + LHCD case and 3 - 4 keV for OH + NBI + LHCD. The key to improve the current drive efficiency seems to be electron temperature which may dominate to enhance resonant electrons.

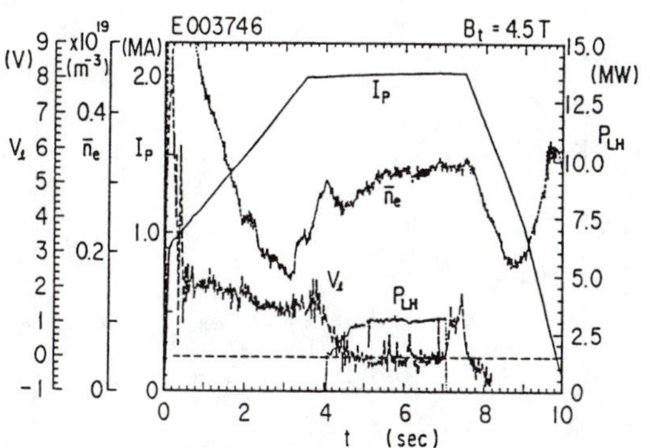

Fig.3 LHRF driven current of 2 MA.

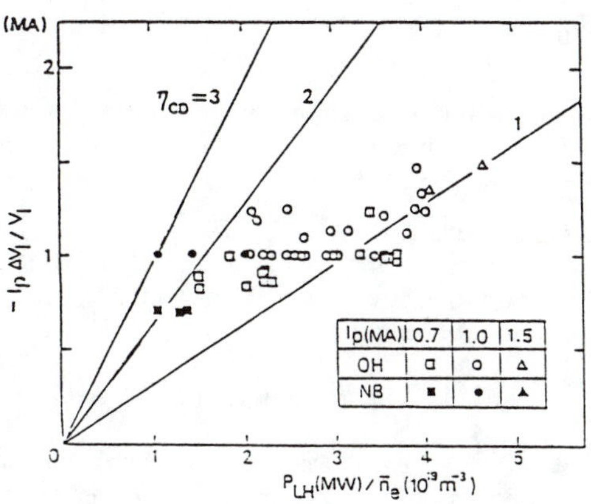

Fig.4 Current drive efficiency with and without NBI heating, where $|\Delta V_1|/V_1 > 1$ for OH.

3.3 LHRF Heating (LHH)

Density limit is expected to be 4×10^{19} m^{-3} for hydrogen and 8×10^{19} m^{-3} for helium at 2 GHz from other experiments. The electron heating experiment was carried out to get a high electron temperature with helium plasmas in the range of $\bar{n}_e = (1.0\text{-}4.0) \times 10^{19}$ m^{-3}. The two LHH couplers are used at $\Delta\phi = 180°$, usually accompanied by the LHCD coupler at $\Delta\phi = 90°$. The injected power from the LHCD unit was about half of the LHH units. In Fig. 5 the central electron temperature T_{e0} is plotted against $P_{abs}/\bar{n}_e$ for OH, NBI and LHEH (LHRF electron heating) in helium plasmas of 1.5 MA (P_{abs} means OH power plus total additional power, P_{LH} is the injected RF power). The heating efficiency η_e of LHEH is relatively high and about 2×10^{19} m^{-3}keV/MW. Up to now, the maximum T_{e0} of 6 keV was obtained at $\bar{n}_e = 1.7 \times 10^{19}$ m^{-3} and $P_{LH} = 2.4$ MW. The radial profile of the electron temperature measured by Thomson scattering is shown in Fig. 6, where $I_p = 1.5$ MA, $B_t = 4.5$ T, and $P_{LH} = 1.8$ MW from the LHH launcher and 0.6 MW from the LHCD launcher. Though T_{e0} decreases as $\bar{n}_e$ increases, the sig-

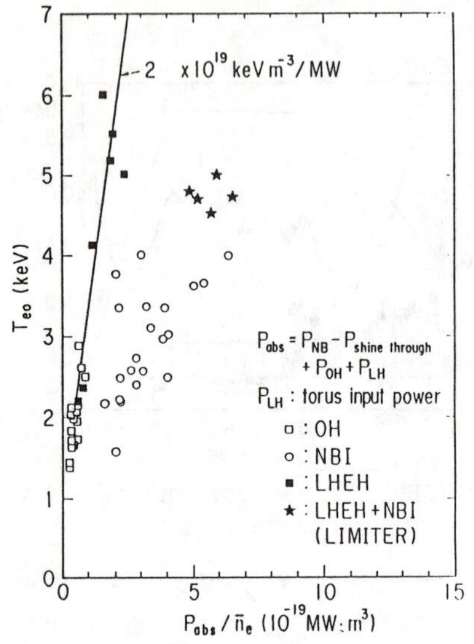

Fig. 5 Peak electron temperature vs $P_{abs}/\bar{n}_e$ in case of OH, NBI, LHEH, and LHEH + NBI.

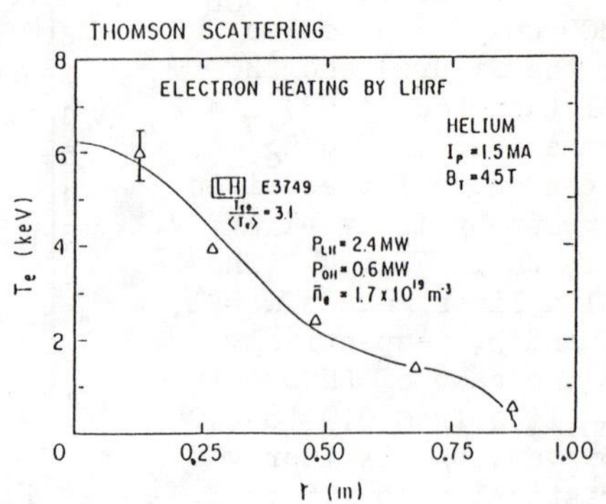

Fig. 6 Radial profile of the electron temperature by Thomson scattering, where I_p = 1.5 MA, B_t = 4.5 T, $\bar{n}_e$ = 1.7 x 10^{19} m^{-3}, and P_{LH} = 2.4 MW.

nals of hard X-ray and ECE (electron cyclotron emission) at 3 ω_{Ce} extend to $\bar{n}_e$ = 4 x 10^{19} m^{-3}. The increase of ion temperature up to 1 keV was observed in this density region, while no high energy ion tail was observed.

4. COMBINED LHCD PLUS NBI

4.1 Current Profile Control by LHCD

Current profile in LHCD seems to be broader than OH one.[10] Improvement of confinement in LHCD plasmas at low density is also reported in small and medium size tokamaks. To optimize the current profile: $\Delta(l_i/2) \sim \Delta\Lambda_{LH}$ for better confined plasma NBI heating, $\Delta\Lambda_{LH}/(P_{LH}(MW)/\bar{n}_e(10^{19}m^{-3})$ versus the phasing $\Delta\phi$ is studied for two types of couplers, LHCD and LHH, as shown in Fig. 7, where the Λ and $\Delta\Lambda_{LH}$ are defined as $\Lambda = \beta_p + l_i/2$ and $\Delta\Lambda_{LH} = \Lambda_{after\ LHCD} - \Lambda_{before\ LHCD}$. The coupler for LHH is not so effective to obtain a broad current profile.

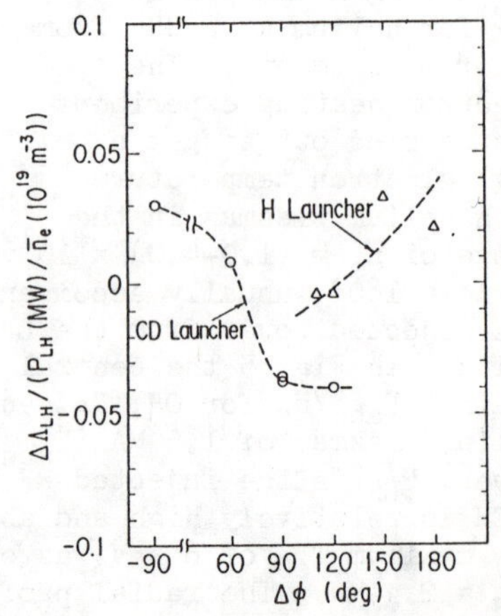

Fig. 7 LHCD properties of LHCD launcher and LHH launcher.

The change of the plasma internal inductance Δl_i measured by Faraday rotation also support current flattening during LHCD.

4.2 Improvement of τ_E

Further study is made using the LHCD launcher. Figure 8 shows the incremental energy confinement time $\tau_E^{INC} = \Delta W^*/P_{abs}$: the incremental stored energy divided by absorbed power, versus $\Delta\Lambda_{LH}$. The change of the current profile due to LHCD is the key parameter to improve the energy confinement time. Also shown in Fig. 9, $\Delta\Lambda_{LH}$ as a function of $P_{LH}/\bar{n}_e I_p$, for I_p = 0.7-1.5 MA. These experiment were done in the low density region of less than 1 x $10^{19} m^{-3}$. Optimization of $N_\parallel$ spectrum and high RF power after enough launcher aging is needed to obtain the well confined LHCD plasma in order to conduct NBI heating in the higher density region.

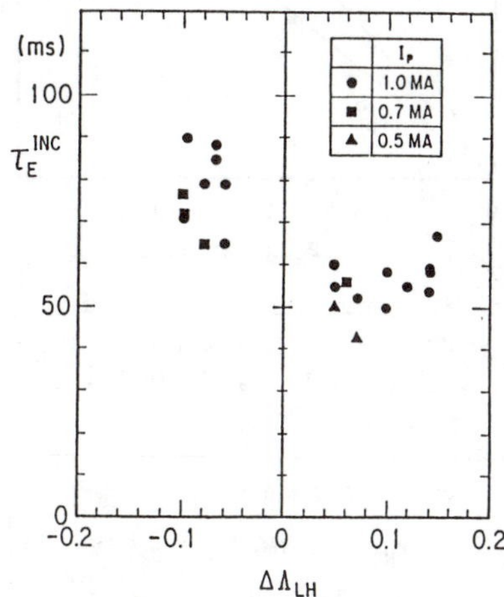

Fig. 8 τ_E^{INC} versus $\Delta\Lambda_{LH}$ in case of the LHCD launcher.

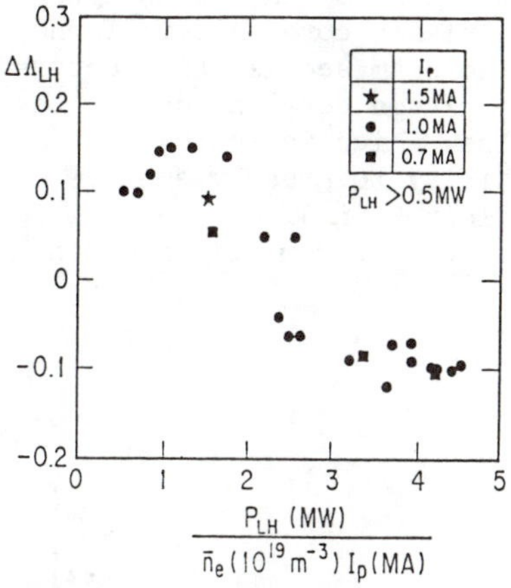

Fig. 9 $\Delta\Lambda_{LH}$ as a function of normalized RF power P_{LH} by $\bar{n}_e$ ($10^{19} m^{-3}$) I_p(MA) in case of the LHCD launcher.

5. ICRF RESULTS

5.1 Coupling Properties

Good coupling was obtained in spite of small size of antennas compared with the plasma size, where the length of each central conductor is 23 cm and minor radius of the plasma column is about 90 cm. The coupling resistance reached 6-10 Ω for (0,0) phasing and 2-4 Ω for (π, 0) phasing, where (Δϕ, Δθ) indicates the toroidal and poloidal phases, respectively.[9]

5.2 Heating Results with $k_\parallel$-phasing

The incremental stored energy ΔW_s^* evaluated from the magnetics versus the additional power P_{add} ($= P_{Ic} + \Delta P_{OH}$) is shown in Fig. 10. τ_E^{INC} for (π, 0) case is about 70 msec and comparable with that of NBI heating, but is almost 40 msec for (0, 0) case. Such large difference in heating effects between (π, 0) and (0, 0) may be explained by the power deposition profile, which comes from $k_\parallel$ spectrum.[11]

5.3 Beam Acceleration

In combination heating with NBI, significant beam acceleration above the beam injection energy was observed. Figure 11 shows the charge exchange (CX) spectra in the perpendicular direction. Fast ions, where the injected energy is 60 keV, are accelerated up to 150 keV. The region where the beam acceleration takes place was identified by chopping the NBI whose beam lines coincide with the line of the CX analyzer in the plasma core. It is found that the beam acceleration occurs in the plasma core and during the RF pulse, which indicates a possibility to greatly enhance equivalent Q in JT-60, where Q is the fusion power multiplication factor.

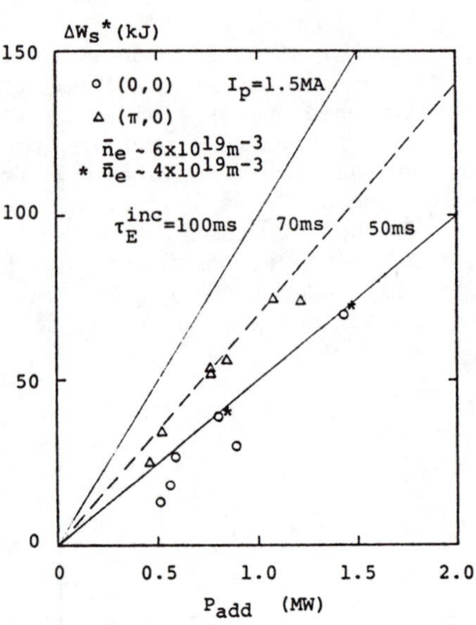

Fig. 10 Incremental stored energy versus additional power P_{add} for (π, 0) and (0, 0) phasing.

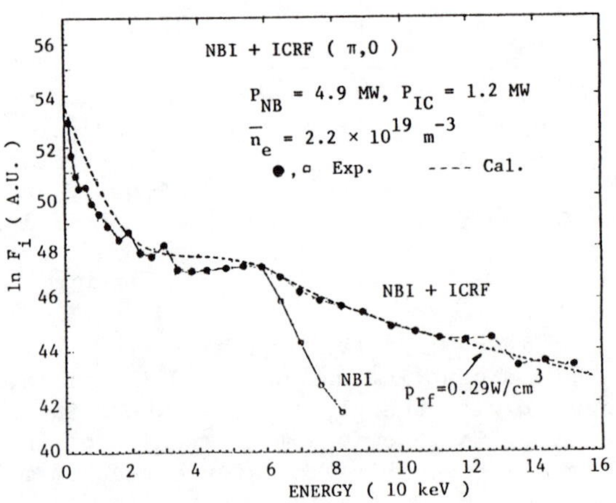

Fig. 11 Charge exchange neutral energy spectra during NBI + ICRF and NBI alone. The dotted curve is simulated by the Fokker Planck calculation. (I_p = 1.5 MA, B_t = 4 T, with divertor).

6. SUMMARY

(1) 2 MA steady current driven by LHCD is firstly demonstrated for 2.5 sec at the density of $\bar{n}_e = 0.32 \times 10^{19}$ m^{-3} with 3.1 MW.

(2) High efficient LHCD is achieved by combination of LHCD and NBI : η_{CD} = 1.5-3.0.

(3) Central electron temperature of 6 keV is obtained by LHRF electron heating at the dencity of $\bar{n}_e = 1.7 \times 10^{19}$ m^{-3}.

(4) Current profile modification and improvement of energy confinement time by LHCD is observed in a large tokamak with and without NBI heating. Optimization of $N_{\parallel}$ seems to be important.
(5) $k_{\parallel}$-shaping is important in optimization of the second harmonic ICRF heating.
(6) In combination heating of ICRF and NBI, remarkable beam acceleration above the injected energy is observed in the plasma core.

ACKNOWLEDGEMENTS

The authors wish to express thanks to Drs. K. Tomabechi, Y. Iso, and S. Mori for their continuous encouragements.

REFERENCES

1. T. Nagashima and N. Fujisawa, Proc. of Varenna-Grenoble Int. Symp. on Heating in Toroidal Plasmas, Vol.II (Grenoble, France, 1978), p. 281.
2. T. Yamamoto et al., Phys. Rev. Lett. 45, 716 (1980).
3. H. Kimura et al., Proc. of 9th Int. Conf. on Plasma Phys. and Controlled Nuclear Fusion Research, Vol. 1 (Baltimore, USA, 1982), p. 73.
4 K. Odajima et al., Proc. of 4th Joint Varenna-Grenoble Int. Symp. on Heating in Toroidal Plasmas, Vol. 1 (Rome, Italy, 1984), p. 243.
5. M. Yoshikawa et al., Proc. of 11th Int. Conf. on Plasma Phys. and Controlled Nuclear Fusion Research, (Kyoto, Japan, 1986), Paper-CN-47/A-I-1.
6 Nuclear Engineering and Design/Fusion (Special Issue), to be published in 1987.
7. T. Imai et al., this conference.
8. Y. Ikeda et al., this conference.
9. M. Saigusa et al., this conference.
10. JT-60 TEAM presented by T. Imai, Proc. of 11th Int. Conf. on Plasma Phys. and Controlled Nuclear Fusion Research, (Kyoto, Japan, 1986), Paper IAEA-CN-47/K-I-2.
11. H. Kimura et al., to be published in Proc. of 14th Europ. Conf. on Controlled Fusion and Plasma Phys., (Madrid, Spain, 1987).

Control of Plasma Profiles with Lower Hybrid Waves

F.X. Söldner
Max-Planck-Institut für Plasmaphysik, 8046 Garching, FRG

ABSTRACT

MHD modes can be strongly influenced by LH waves due to modifications of the current profile j(r). Sawtooth oscillations are suppressed with LH-current drive in a wide parameter range, also in combination with strong additional heating by NBI and ICRH. Broadening of j(r) with resulting q(0)>1 has been identified as one responsible mechanism in this case. Also stabilization of the m=2 mode could be achieved. By tailoring of the LH wave spectrum selective local profile shaping is attained.

INTRODUCTION

Active control of the plasma profiles is considered now a key to improvement of confinement and gross MHD stability of the tokamak. The importance of the form of the profiles, long known from the stability analysis of MHD modes, has become evident in recent experiments with strong additional heating where central electron temperatures were clamped by giant sawteeth. The observed operational limits in β and q(a) might be extended by skillful shaping of the pressure and current profiles. Experimental results on external control of these profiles by local heating and current drive with LH waves are examined in this paper.

THEORETICAL MODELS

Stability of the tokamak against kink and tearing modes depends critically on the form of the current profile j(r). Theoretical models have been developed for the search of stable profiles. An analytical theorem for stability comparison of different profiles was first proposed by Furth /1/. With numerical calculations on the basis of Δ' analysis, optimized current profiles were constructed which are stable against all resistive kink modes /2/. Such an optimized profile is shown in Fig. 1. The coupling of j(r) and $T_e(r)$ in an inductively driven tokamak, however, rules out direct external profile control. RF-current drive with LH waves was then investigated for stabilization of nonlinear tearing modes /3/. It was shown that local LH-current drive at the q=2 surface can suppress the nonlinear growth of the m=2/n=1 island. Various schemes of island suppression by rf waves were studied numerically /4/. The most promising method seems local LH-current drive in the island centre with pulsed wave launching synchronized with the island rotation. The rf current profile and the total plasma current density in this case are shown in Fig. 2. The LH-driven current has to be well localized and peaked inside the island.

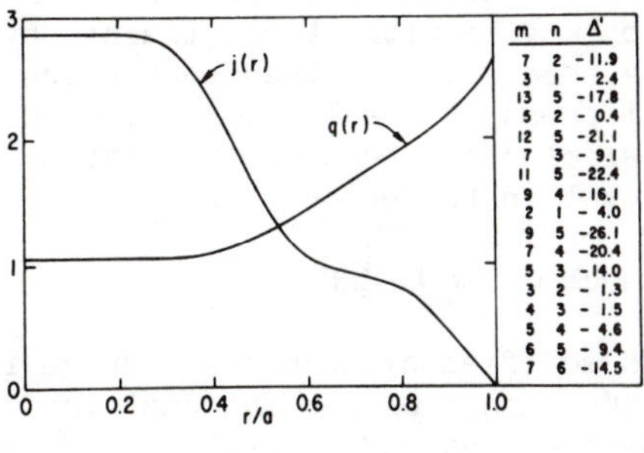

Fig.1. Optimized current profile, providing stability against all resistive kink modes. From Ref. /2/.

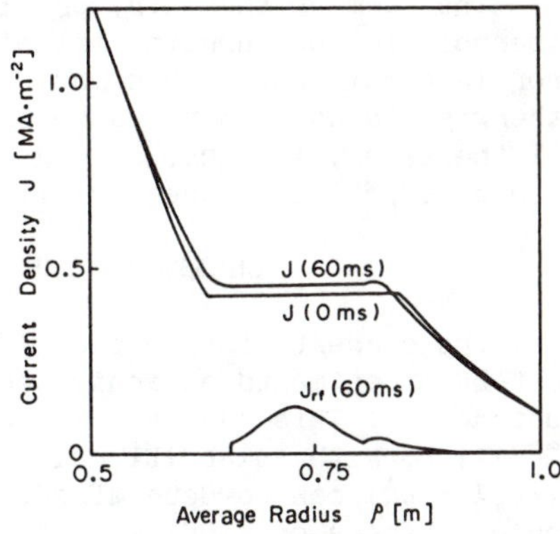

Fig.2. Current densities j(r) and $j_{rf}(r)$ in the case of island stabilization. From Ref. /4/.

CHANGE OF THE $T_e(r)$ PROFILE WITH LH

Strong electron heating is observed in many experiments upon injection of LH-waves at high power. In the suprathermal regime, the heating is most efficient for LH-spectra with the lowest $N_\parallel$ still accessible/5/. The variation of the $T_e(r)$ profile is shown in Fig. 3 for LH-current drive on ASDEX with P_{LH}=750 kW at different densities. Marked central heating with resulting peaking of $T_e(r)$ is obtained for $\bar{n}_e \leq 2\times10^{13}$ cm^{-3}. At higher densities $T_e(r)$ increases over the whole plasma region, as shown for $\bar{n}_e$=2.9x10^{13} cm^{-3} in Fig. 3d. A similar behaviour is found with symmetric LH-spectra. In both cases an increase of the central temperatures and peaking of the $T_e(r)$ profile are obtained in the whole density range where suprathermal electrons are generated by the LH. Above the density limit for formation of fast electron tails, thermal electron heating occurs over the entire profile.

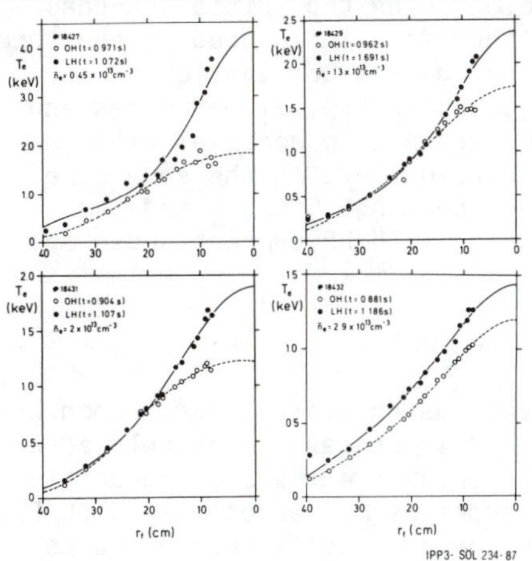

Fig.3. $T_e(r)$ profiles for LH-current drive with P_{LH}=750 kW at 4 densities.

The form of the $T_e(r)$ profile can therefore be varied considerably in the suprathermal electron regime without constraints by profile consistency. This indicates a change in the transport which otherwise is assumed to cause the stiffness of the $T_e(r)$ profile. On PLT there was, in fact, strong central electron heating correlated with a reduction in the turbulence level in the center /6/.

CHANGE OF THE J(R) PROFILE WITH LH

The current distribution j(r) is modified by LH-current drive in most LH experiments as indicated by the change of the internal inductance l_i. This can be deduced from the measured quantity $\beta_p^{equ}+l_i/2$. With an additional measurement of $\beta_p^{\perp}$, the combination $l_i+(\beta_p^{\parallel} - \beta_p^{\perp})$ can be determined. These signals are plotted in Fig. 4 for discharges on ASDEX with three LH spectra with same $\bar{N}_{\parallel} = 2$ but different directionality. The first increase in the magnetic signals at the begin of the RF-pulse is caused by the build-up of a high energy parallel component in the electron pressure. The anisotropy $\beta_p^{\parallel}-\beta_p^{\perp}$ stays then constant in time for heating (LHH) and increases slowly for opposite current. With normal current drive, a slow decay is observed which has to be attributed to a decrease of l_i. Owing to the different time constants, variations of l_i and of the pressure anisotropy can be separated in many cases. In Fig. 5 the change of the internal inductance Δl_i, as deduced from the magnetic signals and from Li-beam measurements /7, 8/, is plotted versus P_{LH} for ASDEX in the density range $\bar{n}_e$=0.5 - 1.6x10^{13} cm^{-3}. At low LH-power an increase of l_i is observed as also reported from other experiments. The Li-beam measurements show in this case a redistribution of j(r) resulting in an increase of the fractional current inside $r/a \approx 0.5$. This global peaking of the current profile may be related to the generation of a runaway distribution out of the RF produced fast electrons by the high dc electric field during the initial LH-phase /9/. The form of the j(r) profile is therefore influenced by the form of the electron distribution established by the combination of inductive and RF current drive. With increasing P_{LH}, l_i decreases and the current profile flattens. For two series of discharges with different parameters this is seen in Fig. 6 /10/. At the same time, T_e increases in the center and $T_e(r)$ is peaking. Current and temperature profiles are therefore decoupled with LH-current drive and can evolve independently.

SAWTEETH BEHAVIOUR WITH LH

The behaviour of sawtooth oscillations changes strongly upon LH injection at low $\bar{N}_{\parallel}$. The sawtooth period increases immediately after start of the RF, probably due to the increased electrical conductivity in presence of the LH generated suprathermal electrons. With P_{LH} above a threshold power P_{LH}^*, sawteeth are completely suppressed as found on Alcator C /11/, ASDEX /12/, Petula /13/ and PLT /5/. Only with directional current drive spectra sawteeth can be suppressed, not with symmetric spectra and not with opposite current drive.

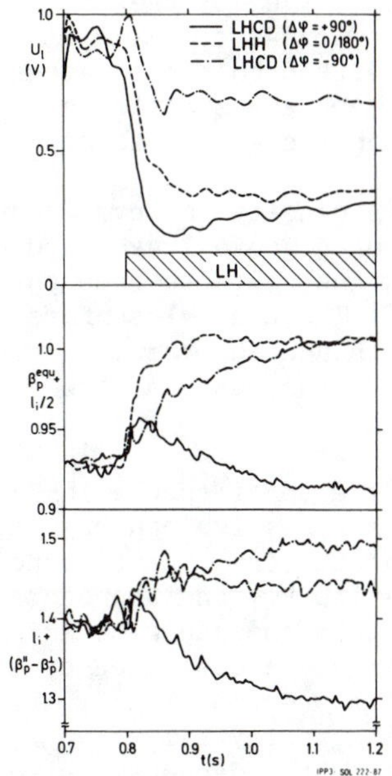

Fig.4. Temporal variation of U_l and the magnetic signals for $\beta_p^{equ}+l_i/2$ and the deduced $l_i+(\beta_p^{\parallel}-\beta_p^{\perp})$ for different LH spetra.

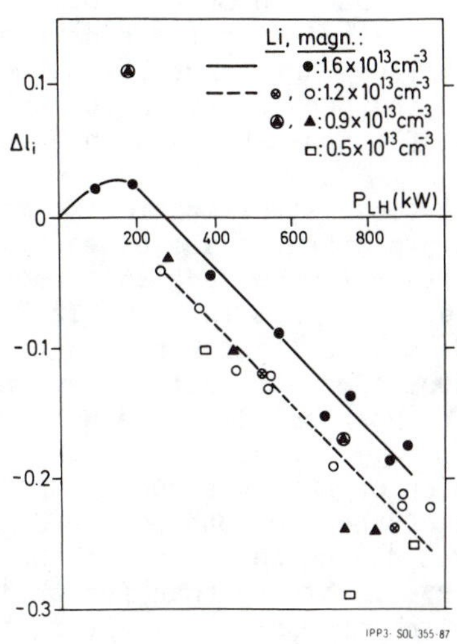

Fig.5. Variation of the change in l_i with LH-power.

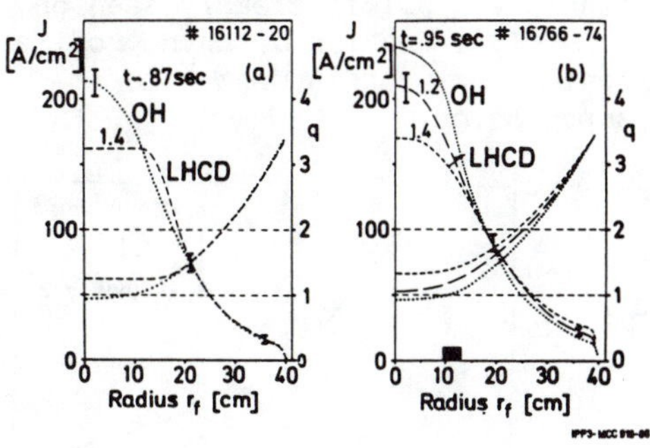

Fig.6. Current profiles for LHCD at
a) $\underline{n}_e$=0.7x10^{13} cm^{-3}, P_{LH}=400 kW, $N_{\parallel}$=1.65,
b) $\underline{n}_e$=1.2x10^{13} cm^{-3}, P_{LH}=870 kW, $N_{\parallel}$=1.9.

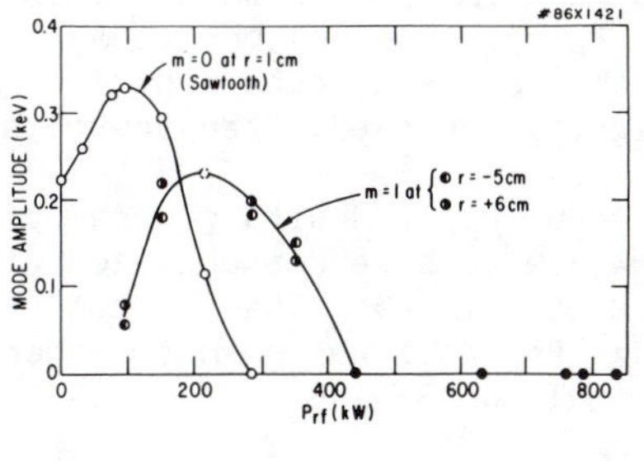

Fig.7. Amplitude of sawteeth and m=1 mode, Ref. /16/.

The threshold power P^*_{LH} for sawtooth stabilization is of order 1/3 - 2/3 of the power required for complete current drive. It increases with $\bar{n}_e$ and with I_p /14, 15/. The critical parameter therefore seems to be the fraction of the total plasma current driven by the rf. From parameter studies at low densities ($\bar{n}_e \leq 1.6 \times 10^{13}$ cm^{-3}) on ASDEX it was concluded that sawtooth suppression was related to the broadening of the j(r) profile by LH-current drive. Sawteeth are stabilized when j(r) is flattened to such an extent that q>1 in the whole plasma /12/.

Sawtooth stabilization at medium and high density, however, cannot be explained in all cases by an increase of q above 1 everywhere. Sometimes m=1 modes are still observed during the sawtooth-free phase /15, 16, 17/. This is shown in Fig. 7 for PLT. There a q=1 surface should still exist in the plasma while sawteeth are suppressed. The mechanism of stabilization remains unclear in these cases as the local current profiles are not known.

Injection of LH waves during NBI also leads to a considerable increase of the sawtooth period /14/. Sawtooth suppression can be achieved during NBI only with LH-current drive spectra. The threshold power P^*_{LH} is higher than for Ohmic plasmas at low P_{NI} and decreases then with increasing P_{NI}. At high P_{NI}, sawteeth disappear with NBI alone in divertor discharges in ASDEX. This might be explained by a flattening of the current profile in the central region already with NBI due to a broader $T_e(r)$ profile. In fact, the delay time τ_d between start of the LH and damping of the sawteeth, decreases with increasing P_{NI} (P_{LH}=const) as seen in Fig. 8. The change of l_i induced by the LH during this time drops with P_{NI}. Therefore j(r) must be already closer to a sawtooth-stable profile at higher P_{NI}. In contrast to the Ohmic case where the sawtooth amplitudes are small, the gain from sawtooth stabilization during NBI is clearly seen on the $T_e(r)$ and $n_e(r)$ profiles (Figs. 9, 10). Peaking of both profiles in the center is obtained. The global energy confinement time is slightly improved after sawtooth suppression.

During ICRH with powers up to 1 MW sawteeth were stabilized with LHCD on PLT /16/. The treshold power P^*_{LH} does not seem to depend strongly on P_{IC}.

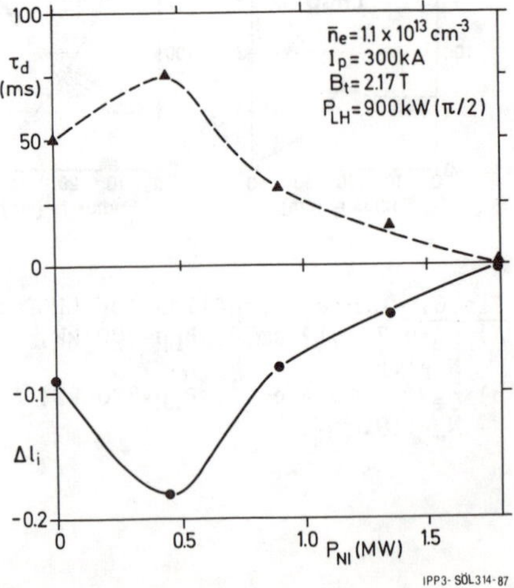

Fig.8. Delay time τ_d between start of the LH and sawtooth damping and change of the internal inductance Δl_i during τ_d for LHCD+NBI, versus injection power P_{NI}.

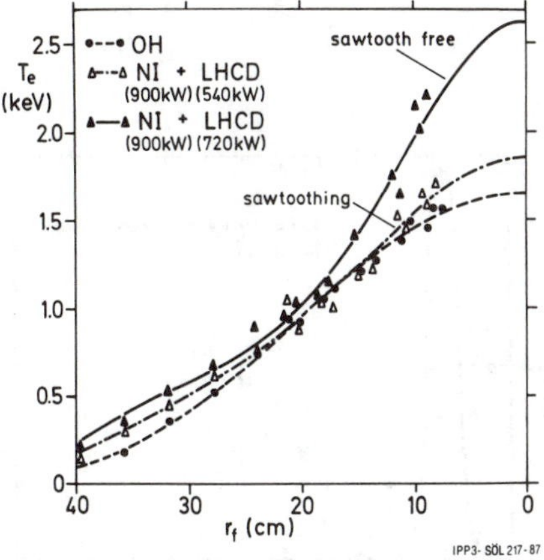

Fig.9. $T_e(r)$ in the Ohmic target plasma, during sawtoothing and sawtooth-free NBI-phases, both with additional LHCD.

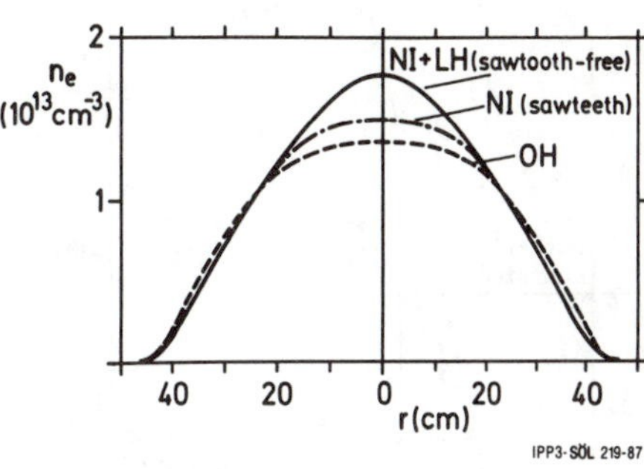

Fig.10. $n_e(r)$ during OH, after a sawtooth crash during NBI and during a sawtooth-free phase with NBI+LHCD.

BEHAVIOUR OF m=2 MODES WITH LH

Triggering of m=2 modes by LH waves occurs often briefly after begin of LH-current drive. Normally the modes saturate at a low level /18/. A correlation between changes in l_i and the behaviour of MHD modes can be seen in Fig. 11 from ASDEX. At low density, strong broadening of j(r) leads first to stabilization of the sawteeth but then later to the onset of m=2 modes. With increasing $\bar{n}_e$ the drop $-\Delta l_i$ is reduced and m=2 modes are no longer triggered while sawteeth can still be suppressed up to $\bar{n}_e$=1.6x10^{13} cm^{-3}. Therefore, the LH power has to be adjusted such that the j(r) broadening just establishes q>1 everywhere marginally. Stronger modification of j(r) might steepen $\nabla_r j(r)$ at the q=2 surface and trigger the m=2 mode. The margin for changes of j(r) diminishes with decreasing q(a). At q(a)≈2 the LH may initiate immediately after the start strong m=2 modes which cause major disruptions. Sawtooth stabilization cannot be obtained at such low q(a) with normal current drive spectra. Similar observations are made on other experiments /16, 18/.

Stabilization of m=2 modes could be achieved by combining OH and RF current drive /13/. In this case major disruptions caused by the m=2 tearing mode in OH operation, could be prevented by partial LH-current drive. But m=2 modes present before the RF can be damped by the LH only in a limited range around q(a)=3 (Fig. 12). At larger and smaller q(a) the m=2 mode absent in the OH phase is triggered by the LH. Feedback stabilization of the m=2 mode reduced the amplitude of the mode by a factor of 2, the same as in the unmodulated case /19/.

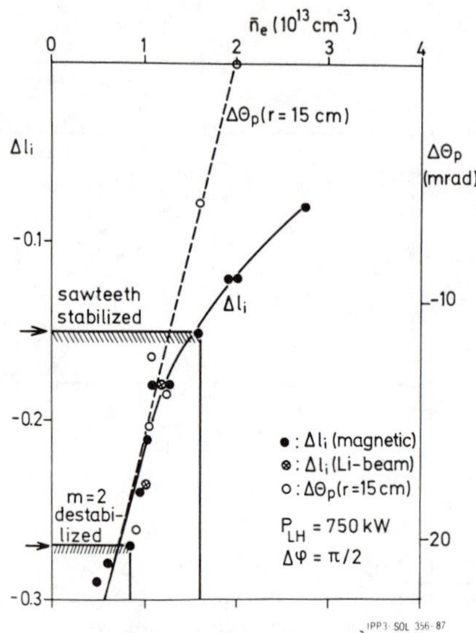

Fig.11. Δl_i and $\Delta\theta_p$(r=15 cm) from Li-beam diagnostics versus $\bar{n}_e$ for LH-current drive.

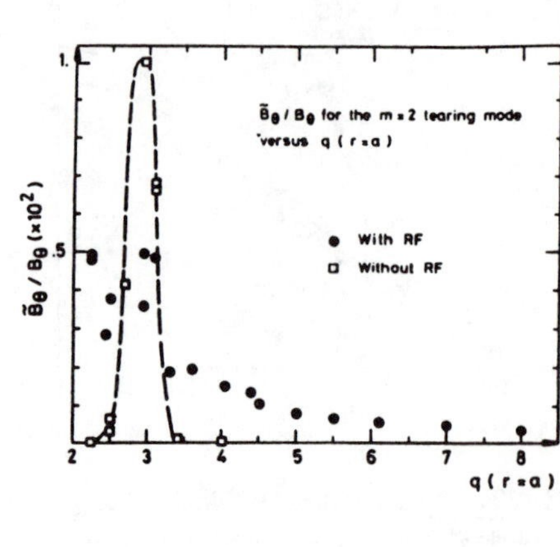

Fig.12. Amplitude of the m=2 mode versus q(a), from Ref. /18/.

The poor success of the attempts of stabilizing the m=2 mode may be related to a lack of well localized LH wave absorption. Suppression of m=2 modes could also be obtained by generating suprathermal electrons with fast LH-waves /20/.

PROFILE SHAPING WITH TAILORED LH-SPECTRA

Simultaneous stabilization of all MHD modes requires continuous control and adjustment of the wave deposition zone. The absorption zone varies with the plasma parameters as seen in Fig. 11 by comparing Δl_i and the fractional current change inside 15 cm as monitored by $\Delta\theta_p$(r=15cm) from the Li-beam diagnostics. With increasing $\bar{n}_e$, $\Delta\theta_p$ (15 cm) diminishes faster than the change of the total current distribution, Δl_i. Therefore it has to be concluded that at higher $\bar{n}_e$ j(r) is flattened closer to the plasma periphery and no modification is seen in the central region during the RF pulse times.

External control of the local deposition zone was achieved by adjustment of the LH wave spectra in ASDEX /21/. With gradual variation of the power in single waveguides of the grill antenna, $\bar{N}_\parallel$ could be varied in the range 1.9-3.2 for highly directional current drive spectra. With increasing $\bar{N}_\parallel$ the total RF driven current is reduced. The modification of j(r), however, as measured by Δl_i and

$\Delta\theta_p$(r=15cm) increases (Fig. 13b, c). The RF driven current therefore is shifted towards the plasma periphery with increasing $\bar{N}_{\parallel}$. This is confirmed by local j(r) measurements which show an increase of $-\Delta l_i$ but a decrease of $\Delta q(0)$ with increasing $\bar{N}_{\parallel}$. The shift of the LH wave deposition zone is also seen in a variation of the $T_e(r)$ profiles. With increasing $\bar{N}_{\parallel}$, the strong central heating is replaced by off-axis heating /22/. This demonstrates the feasibility of local modification of j(r) and $T_e(r)$ profiles by tailoring of the LH wave spectrum.

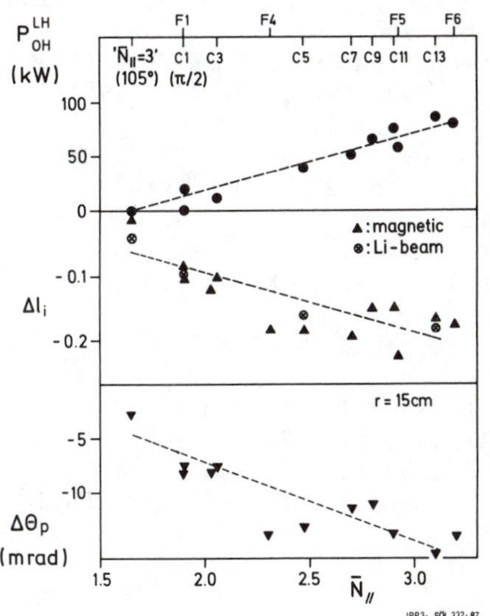

Fig.13. Ohmic power input, drop in l_i, and $\Delta\theta_p$(r=15 cm) versus $\bar{N}_{\parallel}$ for tailored LH-current drive spectra. All points of the C and F series are obtained with a grill antenna giving $\bar{N}_{\parallel}=4$ at 180° phasing. The spectrum at the lowest $\bar{N}_{\parallel}$ is produced with a different antenna which gives $\bar{N}_{\parallel}=3$ at 180° phasing.

SUMMARY

LH-current drive offers a large variety of applications of modifications of the plasma profile. Sawteeth could thereby be suppressed in a wide parameter regime, also with strong additional heating from NBI and ICRH. This results in improved particle and energy confinement in the central region. The plasma current profile can be decoupled from $T_e(r)$ by LH-current drive. This allows for independent control of j(r) by the LH and $T_e(r)$ by additional heating methods. Tailoring of the LH-wave spectrum is a way to local control of j(r) and therefore to optimized profiles stable against all MHD modes.

REFERENCES

/1/ H.P. Furth, et al., Phys. Fluids 16 1054(1973).
/2/ A.H. Glasser, et al., Phys. Rev. Lett. 38, 234 (1977).
/3/ A.H. Reimann, Phys. Fluids 26, 1338 (1983).
/4/ Y. Yoshioka, et al., Nucl. Fusion 24, 565 (1984).
/5/ J.E. Stevens, et al., 12th Europ. Conf. on Contr. Fusion and Plasma Physics, Budapest, 1985, Vol. II, p. 192.
/6/ T.K. Chu, et al., Nucl. Fusion 26, 666 (1986).
/7/ K. McCormick, et al., Ref. 5, Vol. I, p. 199.
/8/ K. McCormick, et al., 13th Europ. Conf. on Contr. Fusion and Plasma Physics, Schliersee, 1986, Vol. II, p. 323.
/9/ F.X. Söldner, et al., Internal report IPP III/111 (1986).
/10/ K. McCormick, et al., Phys. Rev. Lett. 58, 491 (1987).
/11/ M. Porkolab, et al., Ref. 8, Vol. II, p. 445.
/12/ F.X. Söldner, et al. Phys. Rev. Lett. 57, 1137 (1986).
/13/ D. vanHoutte, et al. Nucl. Fusion 24, 1485 (1984).
/14/ F.X. Söldner, et al., Ref. 8, Vol. II, p. 319.
/15/ D. vanHoutte, et al., Ref. 8, Vol. II, p. 331
/16/ S. Bernabei, et al., 11th IAEA Conf., Kyoto 1986, paper CN-47/F-II-1.
/17/ M. Porkolab, et al. Ref. 15, paper CN-47/F-II-2.
/18/ F. Parlange, et al., Ref. 5, Vol. II, p. 172.
/19/ F. Parlange, et al. Ref. 8, Vol. II, p. 393.
/20/ K. Toi, et al., Ref. 8, Vol.II, p. 457.
/21/ F. Leuterer, et al., Ref. 8, Vol II, p. 409.
/22/ F.X. Söldner, et al., 14th Europ. Conf. on Contr. Fusion and Plasma Physics, Madrid 1987.

ENERGY PROPAGATION AND FIELD STRUCTURE OF LH FAST WAVES IN THE UCLA CONTINUOUS CURRENT TOKAMAK (CCT)

K.F. Lai, T.K. Mau, B.D. Fried, R.J. Taylor
University of California, Los Angeles, CA 90024

ABSTRACT

3D field maps of the lower hybrid fast waves have been measured[1] in the UCLA CCT tokamak in both the current drive and test wave experiments. The waves was found to propagate at a relatively constant small angle (<20°) to the magnetic field lines. When launched from the edge, the plasma filters the antenna spectrum allowing only the component $N_{//} \approx N_{\perp}$ to penetrate. The wave damping cannot be accounted for either by Landau or collisional damping. The measured $N_{\perp}/N_{//} \approx 1$ suggests mode scattering between the fast and slow modes. The low $N_{//}$ component however can be excited when launch directly at the center of the plasma using a small dipole antenna. Substantial asymmetry of the propagated wave pattern was observed in both the up-down and in-out directions.

INSTRUMENTATION

Steady state lower hybrid fast wave current drive experiments are conducted in the UCLA CCT tokamak using an array of 8 large area antennas[2]. The antennas can be phased separately to control the $N_{//}$ spectrum imposed on the plasma. In the current drive mode, the penetration of the waves is sensitive to phasing[1]. As the plasma is sustained by the RF waves, the plasma also changes with the waves. In order to separate the wave physics form the plasma generation process, a continuous audio plasma has been used as the target plasma for our test wave studies. Typical audio plasma parameters are: n_e=1-4$\times 10^{12}$cm^{-3}, B=0.1-1kG and T_e=20eV which are similar to those of the RF sustained plasmas.

A specially constructed 3D magnetic probe is used to map out both the amplitude and phase profiles of the waves inside the plasma. A small Langmuir probe is also incorportated into this 3D probe. The density profile is routinely measured by the Langmuir probe and can be calibrated against the microwave interferometer. A tangential magnetic probe located toroidally 90° apart is also used to determine the $N_{//}$. (Fig.1)

Two different kinds of antennas, have been used for our studies. The large area antennas which surrounds the plasma has been used for our global field map measurements. Field maps of singly excited antenna module at various toroidal locations are first generated separately. By making use of the toroidal symmetry of the antennas, these separate maps can be combined to construct a global picture of the wave energy propagation in the tokamak. In the equatorial plane, areas up to 50% can be mapped out using this technique.

On the other hand, in order to pinpoint the wave trajectory and damping mechanisms, a tiny movable magnetic dipole launcher has also been used for our test wave experiments. Because of the small size of this launcher, it can be used as a point excitor of the fast waves and inserted into various locations of the plasma. The relatively simple geometry of the antenna enable us to compare measured near field structure with theories.

RESULTS

In CCT the dispersion of the LH fast waves has been measured and agreed reasonably well with the cold plasma dispersion. At f=21MHz, the observed wavenumbers are in the range of $N_{//} \approx 10-100$ and $N_{\perp} \approx 20-100$.

For all of our measurements, the energy of the LH fast waves was found to be propagating at a small angle along the magnetic field line, which is not as shallow as the cone angle of the LH slow waves ($\alpha \approx \omega/\omega_{ce}$). This angle agrees with prediction from cold plasma theory[3] and does not vary much over a wide range of parameters. This is especially clear in Fig.4 where the direction of energy (Mod(B)) propagation is very different from the phase angle of the waves.

From the global toroidal field maps (Fig.2), we can clearly see the group angle effect on the wave energy propagation. Although our antenna covers all poloidal angles, most of the wave energy was propagated from the inside portion of the antenna. This may be related to both the toroidal geometry and the tighter winding of the antenna (higher $N_{//}$) at the inside.

The fast waves was found to be moderately damped at 21MHz and lost most of its energy in less than half a transit along the toroidal direction. This cannot be accounted for either by Landau or collisional damping. Furthermore the plasma filters the antenna spectrum allowing only the high $N_{//}$ components ($N_{\perp}/N_{//} \approx 1$) to penetrate (Fig.3). The rapid mixing of the polarization of the fields from the antenna (as observed by the dipole antenna) suggests mode scattering between the fast and slow modes. Independent probe measurements[4] also confirms the existence of shorter wavelength electrostatic modes near the edge of the plasma.

By inserting the dipole antenna into the center of the plasma, the low $N_{//}$ waves can be excited (Fig.4). This indicates that the low $N_{//}$ waves are inaccessible to the center by refraction and mode scattering rather than cutoff due to the dispersion relation. This result is also confirmed by ray tracing analysis[5].

From the poloidal field maps (Fig.5), we have observed substantial asymmetry of the propagated wave patterns both in the up-down and the in-out directions. This asymmetry decreases with the increase of magnetic field as the wave restricted further along the field lines. No explanations for the asymmetry have been established, and further detailed measurements are required to resolve this puzzle.

CONCLUSIONS

The plasma filtering of the $N_{//}$ spectrum of the fast waves should be a major concern for designing future fast wave experiments.

The excited wave spectrum inside the plasma can be very different from what is imposed at the edge by the antennas. Although our frequency range of study is quite far away from the ICRF range, this may still suggest some clues to the modelling of ICRF heating. The verification of the anomalous damping of the LH fast waves by mode scattering will require close collaboration between theory and experiments in the future.

This work is supported by USDOE Contract DE-AM03-76SF-0010, Mod. A001.

REFERENCES

1. K.F. Lai, et al, Bull Am. Phys. Soc. 31 (1986) 1517.
2. R.J. Taylor, et al, Bull. Am. Phys. Soc. 30 (1985) 1623.
3. T.H. Stix, The Theory of Plasma Waves, (1962) 55.
4. R.J. Taylor, private communication.
5. T.K. Mau, et al, this conference.

FIGURE CAPTIONS

1. Top view of CCT showing the location of the RF antennas and magnetic probes.
2. Global toroidal field maps of Mod(B). (He, B=461Gauss, F=21MHz)
3. Comparison between observed wave spectrum and cold plasma theory predictions. The dominant excited mode in the plasma is observed at $N_{//}$ where the group angle is a maximum. (He, B=1kG, F=21MHz)
4. (a) Mod(B) and (b) Bz phase maps of the fast wave in the equatorial plane excited by a dipole antenna inserted at the center of plasma at ϕ=45°. (He, B=230 Gauss, F=21MHz)
5. Poloidal profile of the amplitude of the fast wave at two different magnetic fields (a) 230Gauss (b) 922Gauss (He, F=21MHz)

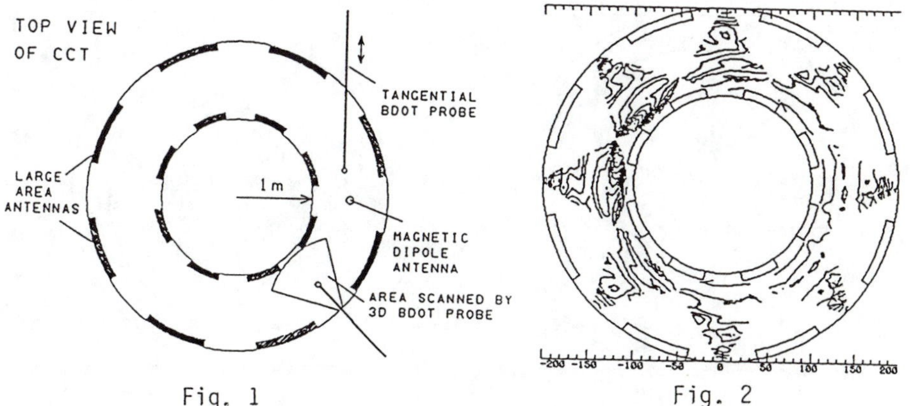

Fig. 1

Fig. 2

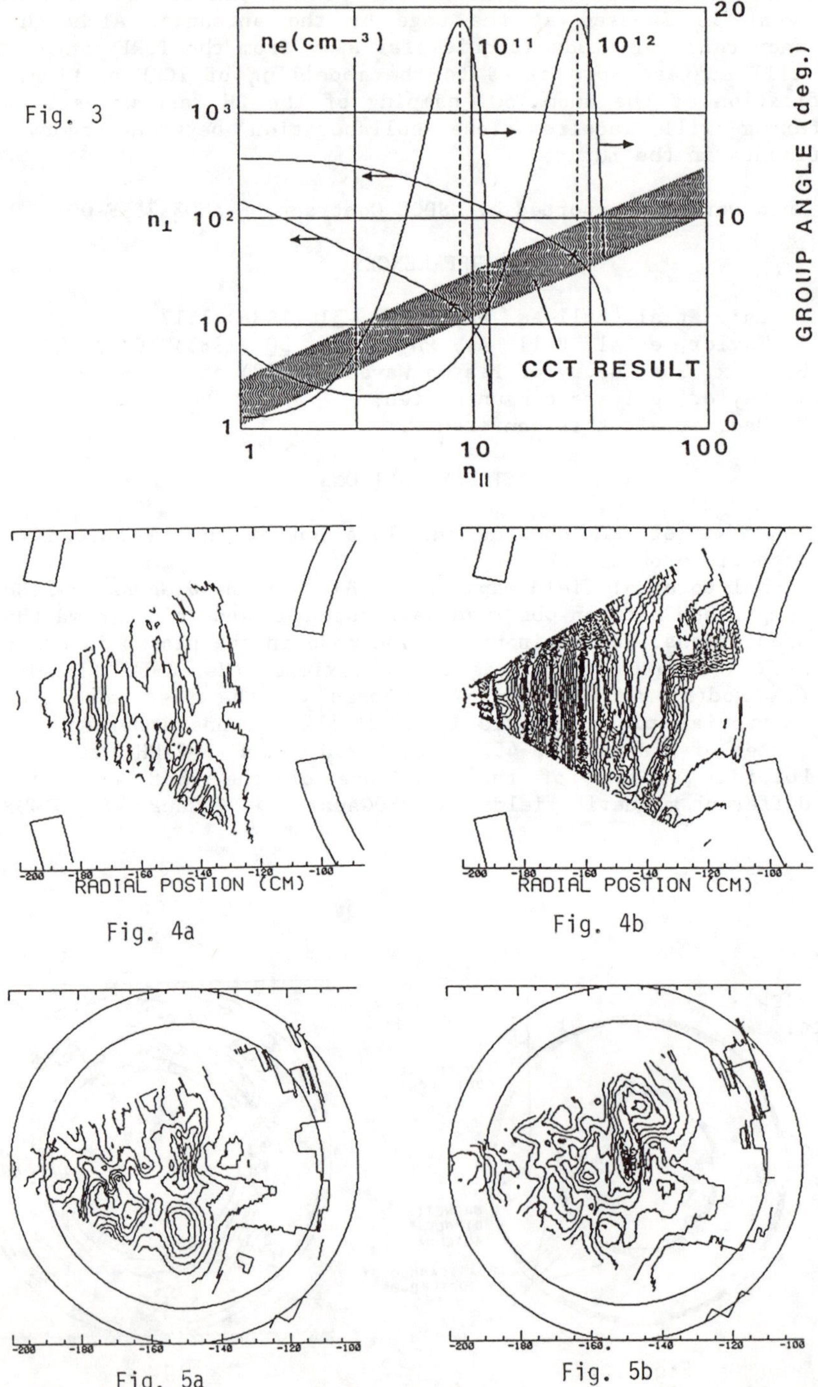

Fig. 3
$n_e(cm^{-3})$
10^{11}
10^{12}
$n_\perp$
$n_\parallel$
GROUP ANGLE (deg.)
CCT RESULT
RADIAL POSTION (CM)
Fig. 4a
Fig. 4b
Fig. 5a
Fig. 5b

RF Performance Test and Current Operations of LHRF and ICRF Heating Systems on JT-60

T. IMAI, T. NAGASHIMA, H. KIMURA, K. UEHARA, K. ANNOH, T. FUJII, M. HONDA, Y. IKEDA, T. KATO, K. KIYONO, S. KURAKATA, S. MAEBARA, M. SAIGUSA, K. SAKAMOTO, M. SAWAHATA, M. SEKI, H. SHIRAKATA, K. SUGANUMA, N. SUZUKI, M. TERAKADO, M. TSUNEOKA, K. YOKOKURA and JT-60 TEAM

Naka Fusion Research Establishment,
Japan Atomic Energy Research Institute,
Naka, Ibaraki, Japan.

ABSTRACT

Performance testing of JT-60 RF systems was completed in summer of 1986. In the integrated test of LHRF systems output power of eight klystrons was absorbed into water dummy loads at the level of 8 MW for 1 sec. at 2 GHz. In the ICRF case power test of each eight three-stage amplifier was conducted at the level of 750 kW for 10 sec. at 120 MHz. In the current operation with JT-60 plasma, each LHRF system has achieved 2.0-2.4 MW for seconds and the ICRF system has achieved 2.0 MW for 1 second.

1. INTRODUCTION

One of notable features of JT-60 is the huge RF power for heating and steady state current drive purposes. RF for JT-60 consist of three Lower Hybrid Range of Frequencies (LHRF, ~2 GHz) and one Ion Cyclotron Range of Frequencies (ICRF,~120 MHz) systems. Each has 6-8 MW RF Generator. The construction of the RF system started in 1983 and launching systems were installed at the end of 1985. RF performance test was completed in summer of 1986. Coupling test with JT-60 plasma was started in October 1986. One LHRF and ICRF systems were operated at first and successful initial results with multi-MA current drive and heating had been obtained[1,2]. All four systems have been operated from January of 1987. Recent results of RF experiments are reported in the companion papers[3-5]. The results of RF performance testing and present operation of the RF systems are described here.

2. OVER VIEW

The most notable feature of the RF system for JT-60 is the integration of complex components into an enormous system including 8 klystrons or tetrodes of MW class, which handles a wide variety of power levels from mW to MW, enabling very high power and long pulse operation with quite precise controllability. The system flexibility allows for launcher conditioning and wide varieties of experimental needs. The summary of RF system, design values and the results of RF performance test and current operation with plasma load is listed in Table I. Both LHRF and ICRF systems have already achieved full power output into a dummy load termination.

3. LHRF SYSTEM

A new 1 MW klystron has been developed for JT-60. By use of this, huge power generator without any output circulators are realized even with high reflection power from the plasma load.

Automatic Gain Control (AGC) and Automatic Phase Control (APC) circuits can stabilize the RF wave transmission to the grill mouth against the unstable plasma load. Amplifier chains and output microwave circuits to the launcher are shown in Fig. 1.

The key component of RF generator is 1 MW klystron. A typical waveform of 1 MW x 10 sec. operation with water dummy load at the test bed is shown in Fig. 2(a). The AGC circuit can make the complex amplifier chain of huge power as a linear amplifier with wide dynamc range of 30 dB, as shown in Fig. 2(b). The response time of the AGC circuit is a few tens μs and the gain instability of the klystron due to unstable reflection is stabilized by the AGC, which will be shown later.

Before launching RF power into JT-60 plasma, RF testing was performed to check the phase and power control, output microwave circuit, protection system and monitor system, in the real configulation. The phase control is the most important in LHRF system since it determines $N_{/\!/}$ spectrum radiated from launcher, where the $N_{/\!/}$ is refractive index parallel to the magnetic field. Careful adjustment of the geometrical and electrical length of the transmission lines of one klystron output, which is provided to 4 poloidal elements of the same line of 4x8 grill launcher, has been accomplished. The APC circuit can provide a stable and intentional phase difference setting among waveguide mouths even against the unstable load. The stable waveforms of output power and phase difference are shown in Fig. 3, where reflection power and phase are intentionally changed using a high speed power phase shifter placed between klystron and launcher. As a tool for protecting the klystrons from the steady high reflection power, the high speed power phase shifter, which optimize the reflection phase to the klystron, and the combination of magic tees and movable phase shifters as shown in Fig. 1, are employed.

Stable power and phase control is achieved with plasma load and reflection coefficient is constant in the power range up to ~2 MW. A knotching circuit in combination with various reflection alarms is also employed in order to condition the launcher efficiently. The arc detector for protecting launcher windows works well and the RF system is operated safely without worries of window cracking even in heavy duty breakdown during launcher conditioning. After conditioning of the launcher for several days, torus input power of ~2 MW/launcher for seconds has been achieved and successful experiments of 2 MA current drive, electron heating, and combination heating with NBI were carried out [1-3]. Max torus input power is 6.3 MW at present and seems to be limited by plasma wall interaction due to fast electrons from the measurement of impurity flux and edge [4].

4. ICRF SYSTEM

The ICRF system has similar amplifier chains of 8 lines but, instead of klystron, Eimac 8973 tetrode is used as a final tube. The grounded grid type amplifier with a 3/4 wavelength coaxial cavity is employed. A ferite ring efficiently suppreses parastic oscillations. Power of 750 kW/tube for 10 sec. in the frequency range of 110-130 MHz, which is the new operation regime of this tube, have been achieved in real system, that is shown in Fig. 4. Two amplifier chains are combined with 3 dB hybrid coupler and provided to one of the elements of the 2x2 loop array. A feedback control system of amplitude and phase is employed for the optimum power combining.

One of the notable features of the JT-60 ICRF system is programmable stub tunning control, which enables us to find the tunning point of complicated stub control for 2x2 loop antenna in one shot. The impedance matching is further refined with calculation of the tunning point based on the measurement of the antenna impedance. Phase control of the 2x2 loop antenna works well. Clear response is seen in coupling and heating with phase control [3,5].

A knotching circuit similar to LHRF system is also employed. Moreover, power regulation circuit is equiped in order to protect the launching system against high standing voltage. A typical example of knotching waveform and power regulation is shown in Fig. 5. The ICRF launching system employs an open Farady Shield, which works quite well and has achieved a power density of 1.1 kW/cm^2. The maximum power, so far, is 2.1 MW into torus.

5. SUMMARY

Full power RF performance has been tested with dummy load termination in both LHRF and ICRF systems. Phase and power control are stably attained by the AGC and APC circuits in both systems, even against the unstable tokamak plasma load. Very high power RF generators, with eight 1 MW klystrons, have been operated successfully without output circulators. A maximum RF power into torus of ~2 MW/launcher have been achieved in both systems.

ACKNOWLEGEMENTS

The authors wish to thank the menbers of JAERI, NEC and Toshiba corps., who have been contributed to the JT-60 RF project. We also wish to express our gratitude to Drs. S. Mori, Y. Iso, K. Tomabechi and M. Tanaka for their continued encouragement and support.

REFERENCES

[1] M. Yoshikawa and JT-60 Team, 11th Int. Conf. of Plasma Phys. and Cont. Nucl. Fusion Res., Kyoto, Japan (1986), IAEA-CN-47/A-I-1.
[2] JT-60 Team presented by T. Imai, ibid of Ref. 1 /K-I-2.
[3] JT-60 Team presented by T. Nagashima, this Conference.
[4] T. Ikeda, T. Imai, et al., this Conference.
[5] M. Saigusa, N. Kobayashi, H. Kimura, et al., this Conference.

Table I Summary of the JT-60 RF system performances.

	DESIGN VALUES		ACHIEVED VALUES			
				LHRF		ICRF
	LHRF	ICRF	I	II	III	IV
			LHH	LHH	LHCD	IC
FREQUENCY(MHz)	2000	120		2000		120
FRQ. RANGE(MHz)	-	-		1740 - 2230		110-130
RF POWER (MW)	8	6		8		6*1
DURATION (SEC)	10	10		10*2		10*1
HPA TUBES	KLYSTRON	TETRODE		LD4444/E3778		EIMAC8973
LAUNCHER	-	-		4X8 PHASED ARRAY		2X2 LOOPS
WINDOW	-	-		PILL-BOX		BELL SHAPE
TORUS INPUT(MW)	2.5-5.0	2.5-5.0	2.1	2.2	2.4	2.1
POWER DENSITY	3-4.5 kW/cm^2		2.0	2.1	4.1	1.1

*1: The tesing was done one tube by one at full power.
*2: The 10 sec operation of the klystrons at full power was done in the test bed.

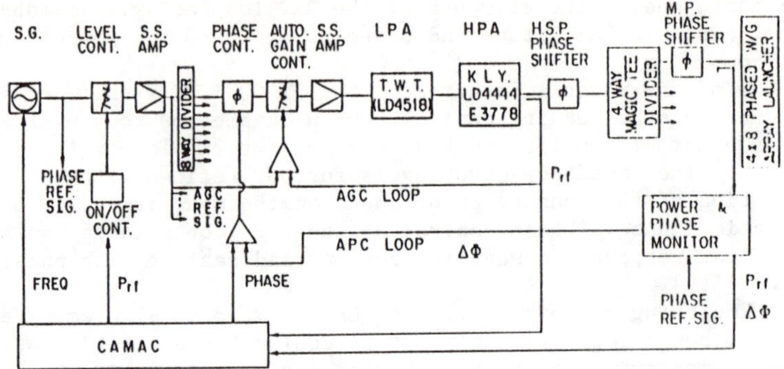

Fig. 1 Amplifier chains and output microwave circuits for JT-60 LHRF system.

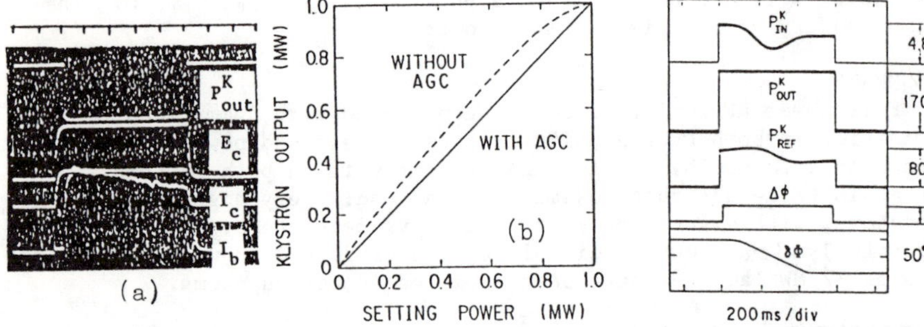

Fig. 2 (a) Time evolution of output power(P^K_{out}), collector voltage(E_c), collector current(I_c) and body current(I_b) of 1 MWx10 sec operation of the klystron for JT-60. (2.0sec/div) (b) Output characteristics with and without the AGC circuit.

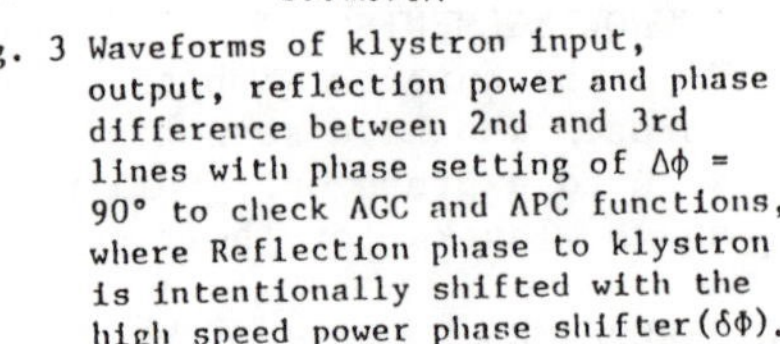

Fig. 3 Waveforms of klystron input, output, reflection power and phase difference between 2nd and 3rd lines with phase setting of $\Delta\phi$ = 90° to check AGC and APC functions, where Reflection phase to klystron is intentionally shifted with the high speed power phase shifter($\delta\Phi$).

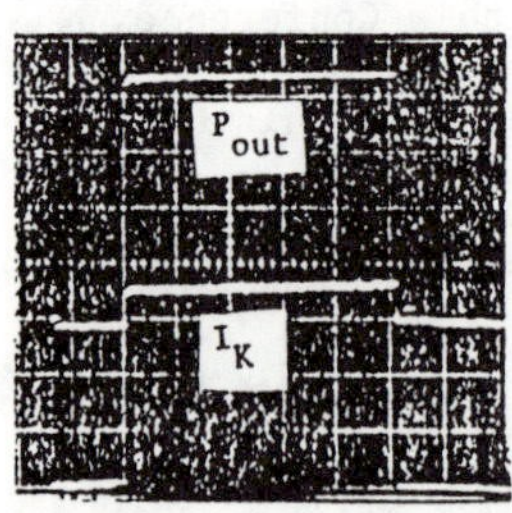

Fig. 4 Waveforms of output power(P_{out}) and cathode current(I_K) of 750 kW x 10 sec operation on JT-60 ICRF system. (2.0sec/div)

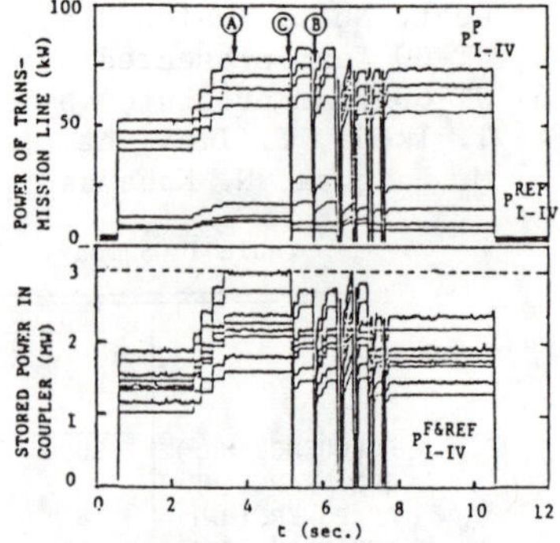

Fig. 5 Typical waveform of ICRF power during injection to JT-60 plasma with knotching and regulation. Arrows indicate the timings of standing wave voltage limit on(Ⓐ), off(Ⓑ) and reflection alarm knotching(Ⓒ).

Design Concepts and Inital Experimental Results of LHRF launcher in JT-60

Y.IKEDA, T.IMAI, K.SAKAMOTO, M.HONDA, T.KATO, S.MAEBARA, M.SAWAHATA, K.SUGANUMA, N.SUZUKI, M.TSUNEOKA, K.UEHARA, K.YOKOKURA and JT-60 TEAM

Japan Atomic Energy Research Institute, Naka, Ibaraki, Japan

ABSTRACT

The phased array launchers of 4x8 waveguide for current drive and heating in JT-60 have been constructed and have operated at a power level of ~ 2 MW for a few seconds. The transmitted power densities have reached 4.1 KW/cm^2 and 2.1 KW/cm^2 for current drive and heating launchers, respectively, after only a few weeks conditioning. The coupling efficiency of 80 ~ 85 % has been obtained at the power level of a few MW into 1 ~ 2 MA plasma and the coupling efficiency is explained well with Brambilla's theory.

1. INTRODUCTION

The JT-60 LHRF experiment is the first attempt to sustain plasma current by only RF power in a large tokamak device [1]. The electron and ion heating by LHRF is expected, too. The main parameters of the LHRF launcher are listed in Table I.

From the point of view of RF heating technology, the objectives for the launcher are as follows,

(1) to transmit high power long pulse without breakdown,

(2) to demonstrate the RF injection into a reactor grade plasma,

(3) to gain confidence with the use of very large number of waveguides.

This paper describes the design concepts of the launcher in JT-60. The launcher conditioning at the power level of 2 MW and the initial results of coupling properties are also mentioned.

2. DESIGN CONCEPTS

To attain these objectives, we have developed and designed the 4x8 phased array launcher, whose vacuum seal is a pillbox window located at about 7 m far from the plasma. The differential pumping system has been employed in this launcher to keep high vacuum in the evacuated waveguide and Carbon has been coated to suppress the unipole multipactoring discharge at $2f_{ce}$ zone due to the leakage of the poloidal field [2]. A sketch of the launcher is given in Fig. 1. This type launcher has many advantages as follows.

(a) The differental pumping system is expected to suppress the ECR and gaseous breakdown and also to make RF conditioning quicker by evacuating the released gas from the wall of the waveguide and the window during RF injection.

(b) Radiation damage of the window due to the plasma can be avoided and easy access and cooling of the window is expected.

(c) The launcher mouth is a simple and rigid structure because of no ceramic breazing.

We have adopted other options to inject high power into a large tokamak and installed many protection systems to suppress

RF breakdown. The specifications of the launcher have been shown in Ref[3].

3. LAUNCHER CONDITIONING

Three launchers were installed in JT-60 in December 1985, and the launcher conditioning started effectively in November 1986 with the current drive launcher and then in January 1987 with all, three launchers. At first, we carried out RF conditioning into the main plasma with the current drive launcher. The conditioning was done smoothly untill P_{in} = 1.25 MW after only 70 plasma shots. But RF breakdown frequently occured at the launcher mouth at the powers over 1.25 MW. A short pulse conditioning with a TDC plasma, whose pulse width was 30 ~ 50 msec. at every 5 sec, was very effective to suppress this breakdown. The transmitted power into main plasma was improved from 1.25 MW to 1.35 MW after only 3 hours of TDC conditioning.

Judging from the good result of the RF conditioning with TDC plasma, the two heating launchers were conditioned with TDC plasma at the start. The gas puffing and the plasma position were optimized to inject maximum power. The amplitude of the injected wave was modulated with ramp up shape and/or thin short pulse of a few ten μsec. An amplitude modulation of ramp up is expected to avoid RF breakdown at the leading edge of the RF pulse, and a thin pulse modulation can cut off the duration of the RF breakdown, too. Figure 2 shows the time history of the total klystron output power of the heating launcher for TDC conditioning. The maximum power reached about 3.2 MW and 5.5 MW in the ramp up and ramp up + thin pulse modulation cases, respectively after 30 hours. These values correspond to approximate 2 MW (1.9 KW/cm^2) and 3.6 MW (3.5 KW/cm^2) of the transmitted power (power density). After this TDC conditioning, the transmitted power of 2 MW for a few sec. could be injected successfully into 1.5 MA plasma. The current drive launcher was also conditioned at the power level of ~ 2 MW by the same means.

We usually carried out TDC conditioning at P_{in} ~ 2 MW for a few hours before LHRF experiments. The RF power during the experiment was increased gradually shot by shot. Up to now, a total power of 6.3 MW, 0.5 sec. has been injected successfully into plasma by three launchers. The maximum power has reached 2.4 MW (4.1 KW/cm^2), 0.65 sec. and 2.1 MW (2.1 KW/cm^2), 1.0 sec. for current drive and heating launchers, respectively. A long pulse operation of 8 sec. at P_{in} = 0.5 ~ 0.8 MW has been also carried out successfully for each launcher.

It has been also found that power seems to be limited by following reasons,

- breakdown at the vacuum window,
- breakdown at the launcher mouth,
- corruption of the periphery plasma due to the plasma wall interaction.

These phenomena are observed by a light emission monitor at the vacuum seal and by a periscope to watch the launcher mouth. Further RF conditioning and the optimization of the plasma conditions may be necessary to make a success of higher power injection.

4. COUPLING PROPERTY

One noteable characteristic of the launcher for JT-60 is its large number of the waveguides, which is 32 waveguides with 4 poloidal rows of 8 waveguide-grills. The coupling property must be discussed for each of the four poloidal rows. Figure 3 shows the reflection coefficients of each row against Δl for the current drive and heating launchers operated at $\Delta\Phi = 1/2\ \pi$, and $\Delta\Phi \fallingdotseq 2/3\ \pi$, respectively, in divator operation of I_p =1.5 MA, n_e = 1.0x1013 cm-3, where Δl is the distance between the separatrix and the limiter. The position of the launcher was 0.5 cm of the limiter towards the plasma in this shot. The solid and dotted lines indicates the reflection coefficients for each launchers calculated by the Brambilla's theory [4] on the assumption of the density gradient at the launcher mouth $\nabla n = \Delta L^{-1} \cdot n_s \cdot \exp(-\Delta l/\lambda)$, where n_s is the density at the separatrix, ΔL and Δl are the distance of launcher-limiter and limiter-separatrix, respectively, λ is the density decay length, and ΔL = 0.5 cm, λ = 0.5 cm, $n_s = 10^{12}$ cm^{-3} in this calculation. The experimental data agrees well with the Brambilla's theory on this density model. This data also indicates that the optimization of Δl is very important to gain good coupling efficiency for all rows due to the sharp density decay length in the divertor operation of JT-60.

On the contrary, when the plasma is operated on touching at high field side limiter (limiter operation), the distance between the plasma edge and the limiter is usually the large value of 5 ~ 10 cm. In spite of this large distance, the good coupling efficiency of ~ 85 % can be obtained at each row during limiter operation with NBI. Figure 4 shows the reflection ratio ρ of heating launcher, line density and NBI power. The solid and the dotted lines indicate the datas with/without NBI, respectively. The ρ is improved by NBI from 0.85 to 0.15.

5. SUMMANY

We have designed and constructed the LHRF launcher for JT-60. The launcher, whose vacuum seal is located far from the plasma, has been conditioned at the power level of ~ 2MW quickly after only a few weeks by injecting RF into a TDC plasma. Using three 4x8 waveguide launchers, up to 6.3 MW for 0.5 sec. of net power has been injected into a 1.5 MA plasma. The transmitted power densities have reached 4.1 KW/cm^2 and 2.1 KW/cm^2 for current drive and heating launchers, respectively. At present, the breakdown at the vacuum seal and/or at the launcher mouth limit the maximum power. The corruption of the periphery due to the plasma wall interaction may also affect the limit of the transmitted power. The coupling property is explained well by the grill theory. The coupling efficiency of 80 ~ 85 % could be obtained at the power level of a few MW into 1 ~ 2 MA plasma by optimizing the plasma-limiter distance and the density and by adding NBI.

ACKNOWLEDGEMENT

We would like to thank the members of JAERI who have contributed to the JT-60 project. We wish to express our gratitude to Drs. S. Mori, Y. Iso, K. Tomabechi and M. Tanaka for their continued encouragement and support. We also wish to thank J. Stevens for his excellent advices.

REFERENCES

1. JT-60 Team (presented by S. Tamura), " Status of JT-60 Experiments ", Eur. Phys. Soc. Schliersee (1986).
2. K.Sakamoto et al. IEEE Trans. Plasma Sci., vol. PS-14, P.548, 1986.
3. T.Nagashima, et al., to be published Nuclear Eng. and Design/Fusion
4. M. Brambilla; Nucl. Fusion 16 (1976) 47.

Table I Main Parameters of LHRF Launcher

Frequency	1.73 ~ 2.26 GHz
Number of Units	3 (one is for current drive) (two are for heating)
Launcher	4x8 phased array
Waveguide Size	11.5 x 1.6 cm for current drive 11.5 x 2.9 cm for heating

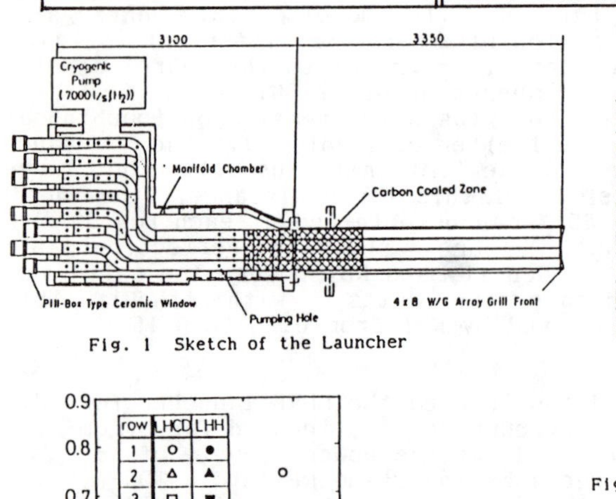

Fig. 1 Sketch of the Launcher

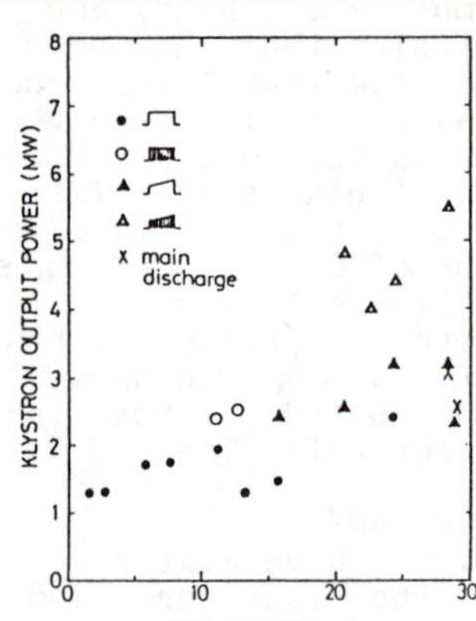

Fig. 2 History of Launcher Conditioning with TDC Plasma

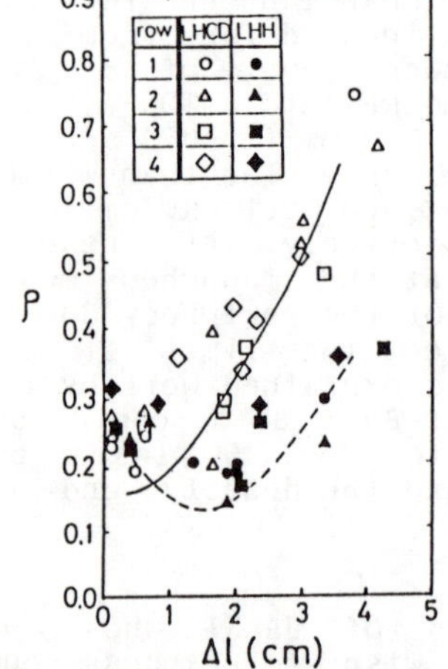

Fig. 3 Reflection Coefficiencies at each rows depending on the seperatrix-limiter distance

Fig. 4 Improvement of the Coupling by NBI

FAST-WAVE SLOTTED-WAVEGUIDE COUPLER ON VERSATOR-II

J. A. Colborn, R. R. Parker, K.-I. Chen, S. C. Luckhardt, M. Porkolab
Plasma Fusion Center, MIT, Cambridge, MA 02139

ABSTRACT

A slotted-waveguide fast-wave coupler has been constructed, without the use of dielectric, and used to drive current on the Versator-II tokamak. Up to 35 kW of net rf power at 2.45 GHz has been radiated into plasmas with 2×10^{12} cm^{-3} $\leq \bar{n}_e \leq 1.2 \times 10^{13}$ cm^{-3} and $B_{tor} \approx 1.0$ T. The launched spectrum has a peak near $N_{\|} = 2$ and a large peak near $N_{\|} = -0.8$ Radiating efficiency of the antenna is roughly independent of antenna position except when the antenna is at least 0.2 cm outside the limiter, in which case the radiating efficiency slightly improves as the antenna is moved farther outside. When the coupler is inside the limiter, radiating efficiency improves moderately with increased $\bar{n}_e$. Current-drive efficiency is comparable to that of the slow-wave, and is not affected when the antenna spectrum is reversed.

INTRODUCTION

The principal reason for experimenting with the slotted-waveguide is to see if it couples to the fast-wave and, if so, how well it drives current. Others have performed Fast-Wave Current-Drive experiments using the dielectric-filled waveguide grill and the loop array.[1,2] Some of these experiments have been plagued by lack of fast-wave excitation, poor efficiency, or waveguide breakdown.

A slotted-waveguide fast-wave coupler is shown in Figures 1 and 2. The slotted-waveguide has several advantages over the grill and loop array. First, a slotted-waveguide with an acceptable $N_{\|}$-spectrum can be produced without the use of dielectric. This lowers the cost of the antenna enormously and avoids the plasma-impurity and waveguide-breakdown problems associated with large dielectric surfaces exposed to the plasma.[3] Second, the toroidal extent of the slotted-waveguide antenna can be much greater than that of a grill. This creates a much narrower $N_{\|}$-spectrum, enabling more wave power to fit into the "window" in $N_{\|}$-space between the cutoff and accessibility limits. Also, the longer source should interact with the plasma electrons in a more spatially distributed way, decreasing the local rf power density in the plasma and lessening the possibility of deleterious nonlinear effects. Third, because it is very difficult to excite evanescent modes in the slots with polarization orthogonal to that desired (for example the TE_{01}-mode), the polarization purity should be better than for the dielectric-filled grill, in which it is relatively easy to excite these modes.

ANTENNA DESIGN

The available theory for slots in waveguides radiating into free space gives the following for the conductance of a long, narrow, longitudinal, resonant slot in the broad face of a rectangular waveguide.[4]

$$g = \frac{480}{73\pi}\frac{a}{b}\cos^2\left(\frac{\pi\lambda}{2\lambda_g}\right)\sin^2\left(\frac{\pi x_1}{a}\right)$$

where a and b are the long and short guide-dimensions, λ is the free-space wavelength, λ_g is the guide wavelength and x_1 is the distance of the slot from the centerline of the broad face of the guide. The perimeter of a resonant slot (that for which the reactance of the slot is zero) is generally about one free-space wavelength. Dumbbell-shaped slots are used to maintain this perimeter with a shorter slot length, enabling closer slot spacing.

For a single row of slots that are not probe-fed, the interference maxima, or peaks in the $N_\|$-spectrum, are given by the following:

$$N_\| = \frac{2\pi m}{d\omega/c} + N_{\|g}$$

where $m = 0, \pm1, \pm2, \ldots$, d is the slot spacing, and $N_{\|g}$ is the ratio of guide wavelength to free-space wavelength. Note that the $m = 0$ solution always yields a radiation peak at $N_\| = N_{\|g}$ independent of d, and that there is a countably infinite number of radiation peaks in $N_\|$-space. The height of these peaks is modulated by the transform of the individual slot field, and the medium determines which will propagate.

The spectrum of the present antenna is shown in Figure 3. The much-improved spectrum of a planned coupler incorporating probe-fed slots is shown in Figure 4. A schematic diagram of the probe-fed coupler is shown in Figure 5.

EXPERIMENTAL RESULTS

Coupling data were taken at 2×10^{12} cm^{-3} $\leq \bar{n}_e \leq$ 1.2×10^{13} cm^{-3}. When the Versator circular limiter was used (radius of 13.00 cm), data were taken at antenna radial positions (ARPs) from 13.35 cm to 12.45 cm. When a straight "bar" limiter was used at a radial position of 12.5 cm, data were taken at ARPs from 12.2 cm to 13.2 cm.

At any given density, the radiating efficiency—defined as the radiated power divided by the total forward power in the waveguide—was roughly independent of density for ARPs inside the limiter (in which case the antenna was the limiter). As the antenna was pulled back from the limiter, radiating efficiency improved slightly. These data are shown in Figures 6 and 7. With the antenna 0.45 cm inside the 13.00 cm limiter, radiating efficiency improved moderately with increased $\bar{n}_e$, as shown in Figure 8.

Current-drive was accomplished with the spectrum both forward and reversed, that is, with the plasma current traveling in the plus and minus $N_\|$ directions (see Figures 9 and 10). Efficiency was approximately equal for both cases. This was unexpected because the spectrum that coupled to the plasma was thought to be asymmetrical. It is possible that the spectral peak near $N_\| = -0.8$ is broader than predicted and much of the power there has $|N_\|| > 1$ and is hence able to couple to the plasma. Loop voltage drops and bursting activity, indicative of the anomalous Doppler instability, were observed during the rf pulse with a forward spectrum (see Figure 9). The driven current was calculated assuming $I_{rf} = (\Delta I/\Delta t)I_{L/R}$. The current-drive efficiencies $\eta = nIR/P_{rf}$ for the forward and reverse cases were 0.007 and 0.0065, respectively, comparable to the flat-topping current-drive efficiency obtained by the Versator 2.45 GHz slow-wave grill of 0.0072.[5] This is encouraging considering the relatively poor spectrum of the slotted-waveguide presently in use.

DISCUSSION

No evidence has been obtained as yet that the present antenna actually launches the fast-wave. In order to determine this, means of detection of the fast-wave, such as microwave scattering, will be explored.

The relatively poor spectrum of the present antenna can be substantially improved with use of probe-fed slots, as shown in Figure 5. The improved antenna should couple much more power to the plasma and eliminate the ambiguity introduced by the power at $N_\| \approx -0.8$.

REFERENCES

1. R. I. Pinsker, et. al., 28th annual mtg. APS/DPP, Baltimore, November, 1986.
2. T. Watari, et. al., 11th Int. Conf. on Plasma Phys. and Controlled Nucl. Fusion Research, Kyoto, November, 1986.
3. R. I. Pinsker, private communication, February, 1987.
4. S. Silver, Microwave Antenna Theory and Design, Peter Peregrinus, London, 1984.
5. M. J. Mayberry, et. al., Phys. Rev. Lett. 55, 892 (1985).

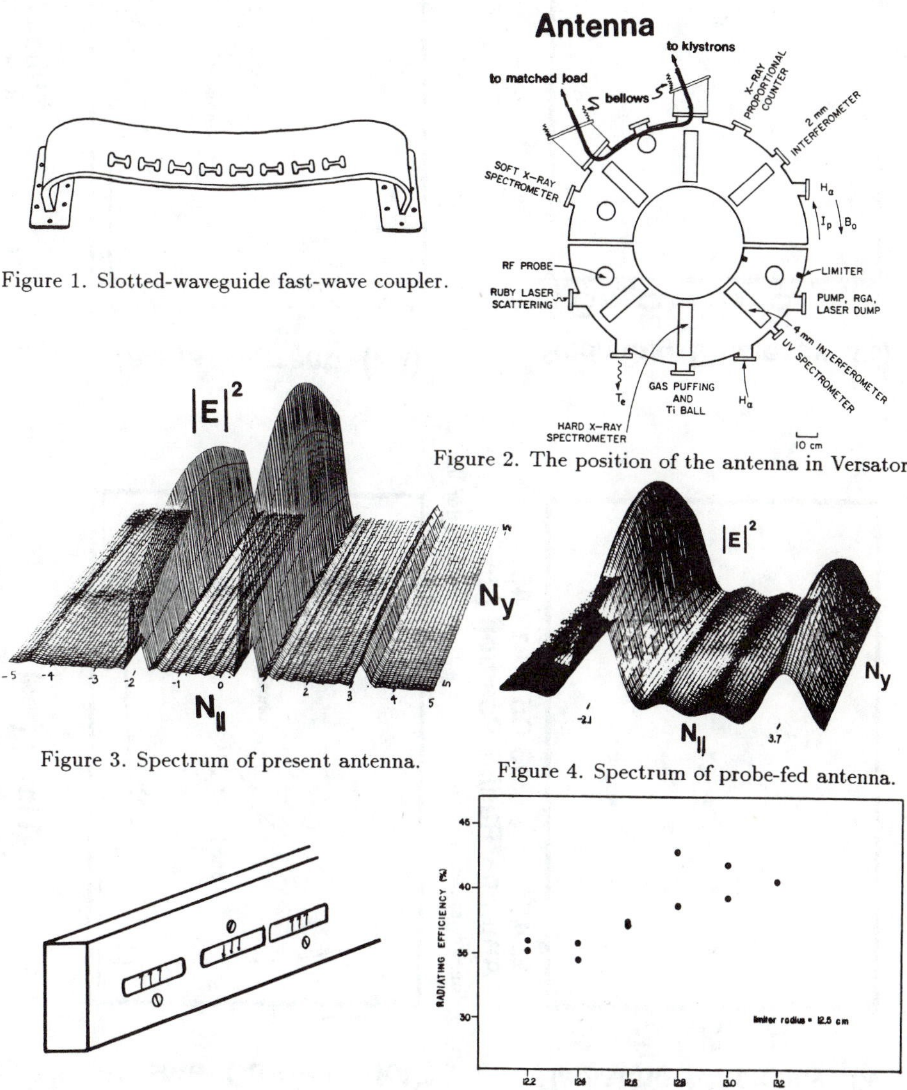

Figure 1. Slotted-waveguide fast-wave coupler.

Figure 2. The position of the antenna in Versator

Figure 3. Spectrum of present antenna.

Figure 4. Spectrum of probe-fed antenna.

Figure 5. Probe-fed antenna.

Figure 6. Coupling vs. antenna radial position with 12.5 cm bar limiter.

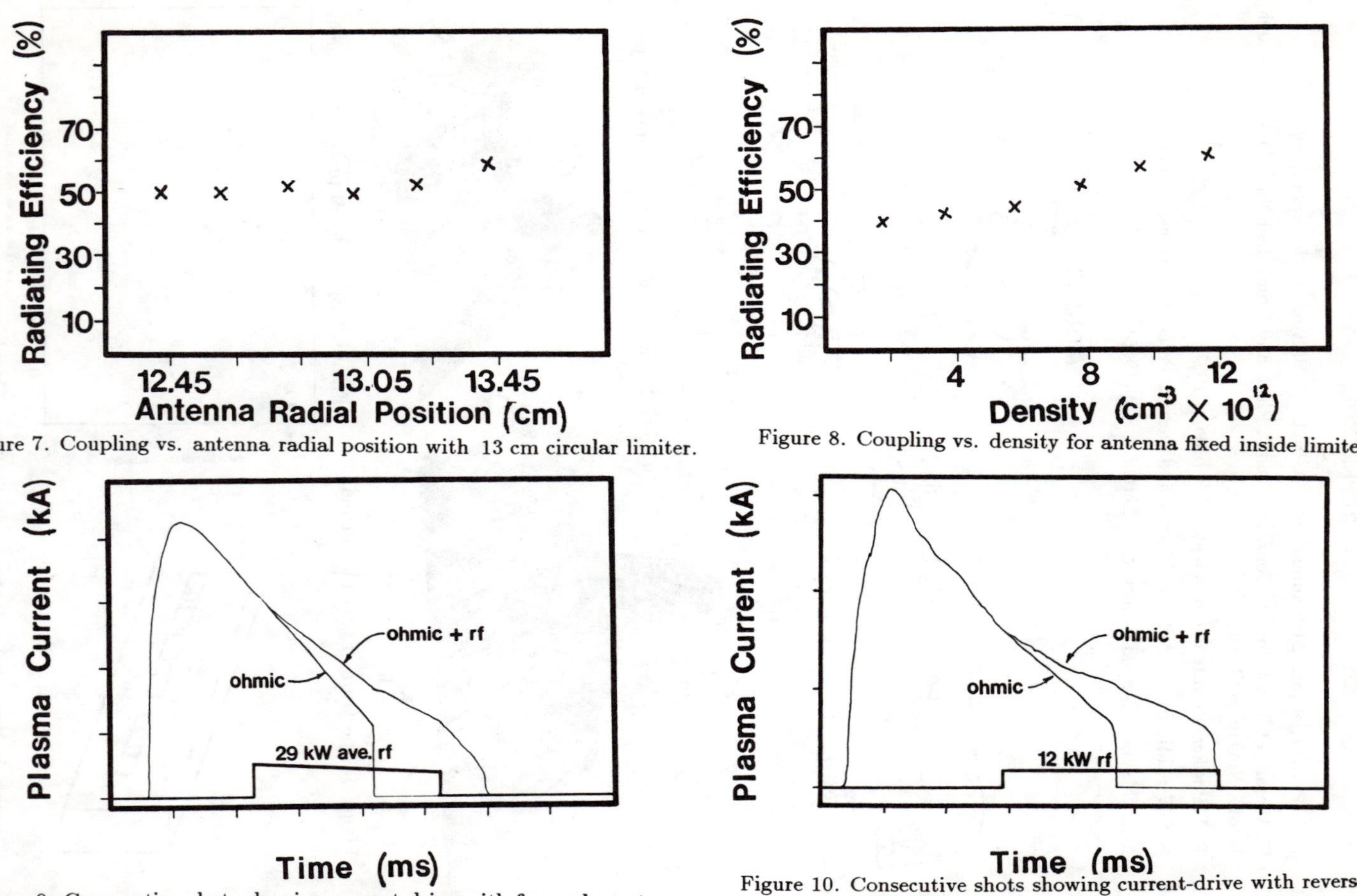

Figure 7. Coupling vs. antenna radial position with 13 cm circular limiter.

Figure 8. Coupling vs. density for antenna fixed inside limiter.

Figure 9. Consecutive shots showing current-drive with forward spectrum.

Figure 10. Consecutive shots showing current-drive with reversed spectrum.

TRANSMISSION AND COUPLING AT HIGH POWER DENSITY OF 8 GHz LOWER HYBRID ON FT

F. Alladio, E. Barbato, G. Bardotti, R. Bartiromo, M. Bassan, G. Bracco, F. Bombarda, G. Buceti, P. Buratti, E. Caiaffa, A. Cardinali, R. Cesario, F. Crisanti, R. De Angelis, F. De Marco, M. De Pretis, D. Frigione, R. Giannella, M. Grolli, S. Mancuso, M. Marinucci, G. Mazzitelli, F. Orsitto, V. Pericoli-Ridolfini, L. Pieroni, S. Podda, G.B. Righetti, F. Romanelli, D. Santi, F. Santini, S.E. Segre, A.A. Tuccillo, O. Tudisco, G. Vlad, V. Zanza

Associazione EURATOM-ENEA sulla Fusione, C.R.E. Frascati, C.P. 65 - 00044 Frascati, Rome (Italy)

INTRODUCTION

The possibility of injecting radio frequency power into the plasma with a grill structure makes the lower hybrid (LH) a very interesting additional heating system for high field tokamaks.

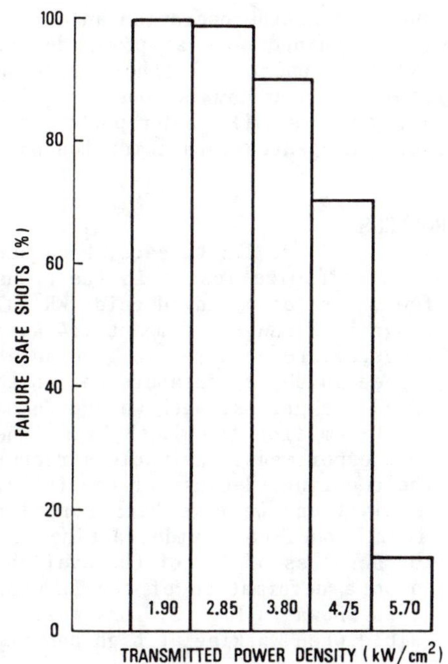

Fig. 1 Fraction of failure safe RF pulses vs transmitted power density from ~ 1000 FT shots at f=2.45 GHz

Starting from this consideration and taking into account the dependence of the density limit on the square of the launched frequency [1] we have chosen a frequency of 8 GHz for LH heating of the FTU machine [2]. To define the coupling structures a foundamental parameter to be known is the RF power density capability. In the preliminary design of the FTU grills a limit of 9 kW/cm^2 was assumed in order to be able to launch 8 MW through 8 ports. This limit seems to be quite conservative if compared with the best results [3] from previous LH experiments.

The experience on FT at f = 2.45 GHz indicates that one must distinguish between the best results and the routine operations. As shown in Fig. 1, at f= 2.45 GHz, a maximum transmitted power of

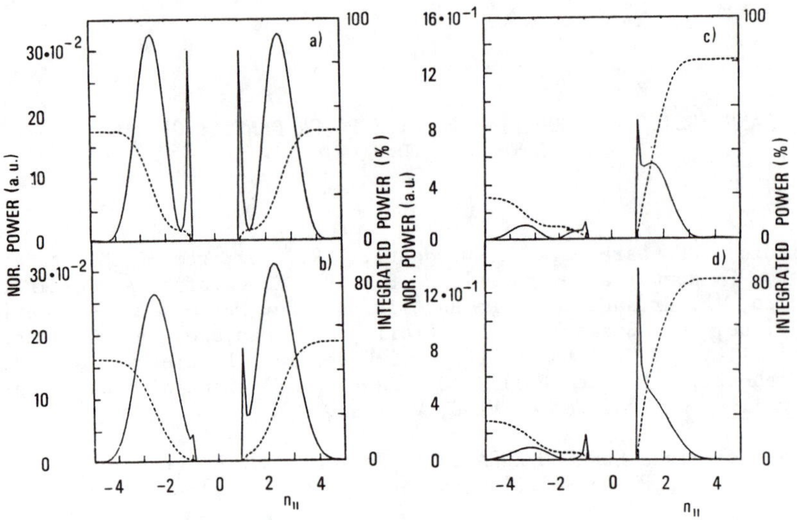

Fig. 2 Spectra from the 16 waveguides, 8 GHz FT grill at phasing π: a) 4 w.g., b) 3 w.g. and phasing 2/3 π: c) 4 w.g., d) 3 w.g.

about 6 kW/cm^2 has been achieved, but for routine operation an acceptable number of failure safe shots is obtained only at power density around 4.5 kW/cm^2. In order to test this point and other technical solutions a 0.5 MW, 1 s, 8 GHz system has been commissioned on FT at the end of February 87. Preliminary answers will be reported here together with the first physical results presented in a companion paper in this conference [4].

SYSTEM DESCRIPTION AND TECHNICAL PROBLEMS

Four Varian VKX-7879A tubes, with f = 8 GHz 125 kW each, have been installed directly in the torus hall to minimize losses in the transmission lines. These consist of a few meters of standard guide WR 137. Each klystron feeds a column of a 4x4 grill through a compact 1:4 power divider. The overall losses of the system are measured to give an attenuation of ~ 1.4 db mainly concentrated in the circulators and in the power dividers. As for the previous 2.45 GHz grills, each waveguide has its own ceramic window brazed at ~ 15 cm from the mouth, the inner dimensions being 5x35 mm in the present experiment. The whole structure can move radially 3 cm to optimize the coupling. Because of the failure of two out of five of the available klystrons we have been forced to work with only 3 columns and the fourth one, on a side, acting as a passive waveguide. In addition to the net loss of 1/4 of the available power, in this condition there is also a deformation of the Launched spectra, namely a shift to lower $n_{\parallel}$, as shown in Fig. 2, thus making a larger fraction of the power inaccessible when working at high density.

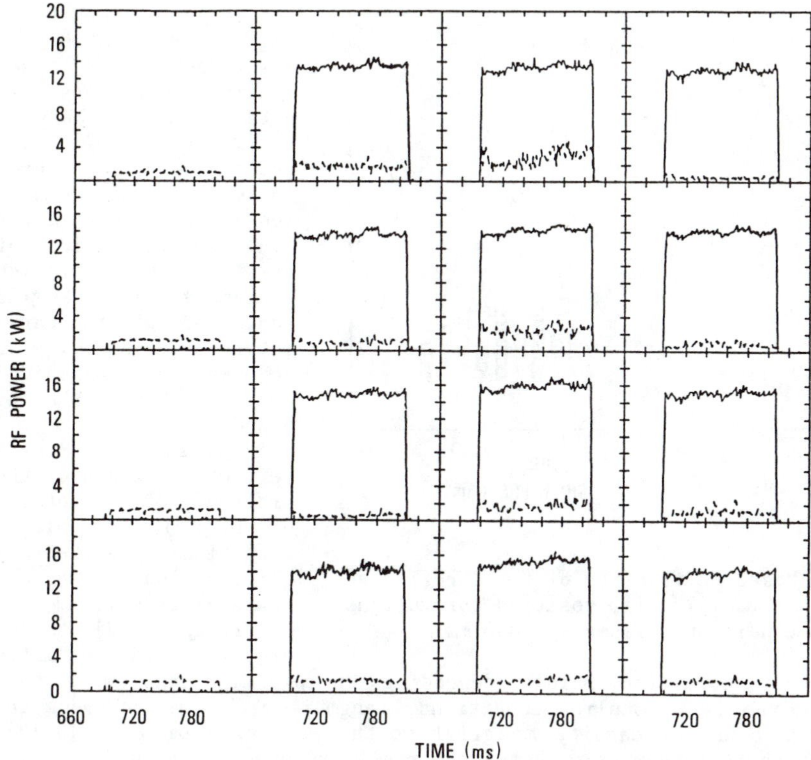

Fig. 3 Direct and reflected power signals from the FT grill for a RF pulse at ~ 9 kW/cm^2.

We hope to get again the system fully in operation during the fall of this year.

TRANSMISSION AND COUPLING

Due to the narrow size of the waveguides in the reduced section of the grill it is impossible to insert any system of guide conditioning. Thus the only remaining way is to fire the RF on the plasma. After the exclusion of the first column, the system was able to rise the transmitted power up to a level corresponding to a density power of about 9 kW/cm^2 without any particular care. A routinely obtained 100 ms RF pulse at this power level is reported in Fig. 3 where an image, from the outside, of the 16 waveguides, direct and reflected power is shown. Further rising the power, some spikes on the reflected power signals appear, triggering the switch off of the system which automatically reset itself after ~ 5 ms. Generally a few RF long pulses (~ 500 ms) at a sligthly inferior power are sufficient to avoid the tripping of the system at the next power step and so on. This behavior indicates the

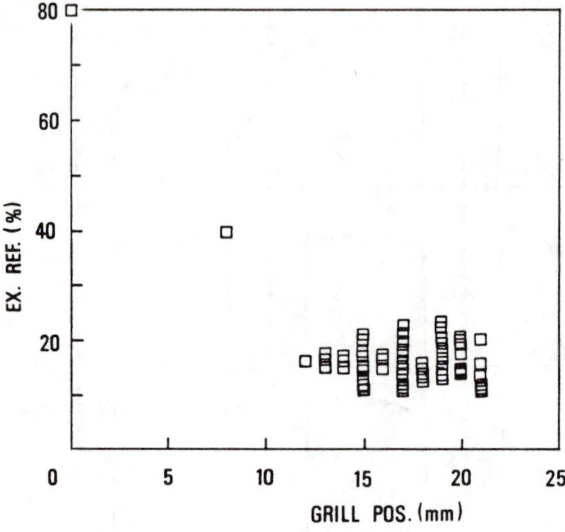

Fig. 4 Measured fraction of total reflected power vs radial grill position for various plasma condition (plasma at r=30 mm)

need of grill conditioning at least at higher power density. Up to now routine operations are obtained at 150 kW with ~ 100% of failure safe shots and only three columns. In this condition we have checked the coupling properties of the 8 GHz LH radiofrequency on a wide range of plasma parameters: B_T=4÷8 T, I_p ~230-450 kA, $\bar{n}_e$ ~ 4-20·10^{13} cm^{-3} for the two phasing π and 2/3 π.

As shown in Fig. 4 reflection values of the order of 10% are obtained for all the plasma conditions when the radial position of the antenna exceeds 15 mm.

Further adjustments are required to match the best condition mainly depending on plasma density.

In the best coupling conditions, Langmuir probes measurement indicate a boundary density at grill mouth, ranging from 2 to 10·10^{12} cm^{-3} quite in agreement with the theorical expectation [5] $n_b/N_{cutoff} \gtrsim n_{\parallel}^2$.

CONCLUSION

Despite the technical problems, good performances of the system have been obtained up to a power density at grill mouth of 9 kW/cm^2. There are indications that this value is not a limit to current operations but that higher power density requires a better grill conditioning. Good coupling has been obtained in all the investigated plasma conditions for boundary density in agreement with theorical expectations.

REFERENCES

[1] F. Alladio et al., Proc. 4th International Symposium on Heating in Toroidal Plasmas, Roma 21-28 March, (1984) Vol. 1, p. 546.
[2] R. Andreani et al., ENEA Report RT/FUS/85/12, Centro Ricerche Energia Frascati (1985).
[3] G. Tonon, Plasma Phys. Control. Fusion, 26, 1A (1984).
[4] F. De Marco, this conference.
[5] F. Santini, Course and Workshop on Applications of RF Waves to Tokamak Plasmas, Varenna 5-14 September, (1985).

PRELIMINARY RESULTS IN LOWER HYBRID HEATING IN FT AT 8 GHz

F. Alladio, E. Barbato, G. Bardotti, R. Bartiromo, M. Bassan, G. Bracco, F. Bombarda, G. Buceti, P. Buratti, E. Caiaffa, A. Cardinali, R. Cesario, F. Crisanti, R. De Angelis, F. De Marco, M. De Pretis, D. Frigione, R. Giannella, M. Grolli, S. Mancuso, M. Marinucci, G. Mazzitelli, F. Orsitto, V. Pericoli-Ridolfini, L. Pieroni, S. Podda, G.B. Righetti, F. Romanelli, D. Santi, F. Santini, S.E. Segre, A.A. Tuccillo, O. Tudisco, G. Vlad, V. Zanza

Associazione EURATOM-ENEA sulla Fusione, C.R.E. Frascati,
C.P. 65 - 00044 Frascati, Rome (Italy)

INTRODUCTION

The first Lower Hybrid Heating system on FT operated with f = 2.45 GHz, P $\leq$ 450 kW [1,2]. Good heating results were obtained in the electron regime, electron and ion temperatures increasing of about 1 keV and 0.5 keV respectively without significant enhancement of Z_{eff}. The evolution of the sawtooth was studied both with symmetric and progressive waves [3]. In the range of power used, ($P_{RF} \leq 3/4\ P_{OH}$) an increase of sawtooth period, reaching about a factor 4 at maximum coupled power, is observed. The available power was not sufficient for a full stabilisation.

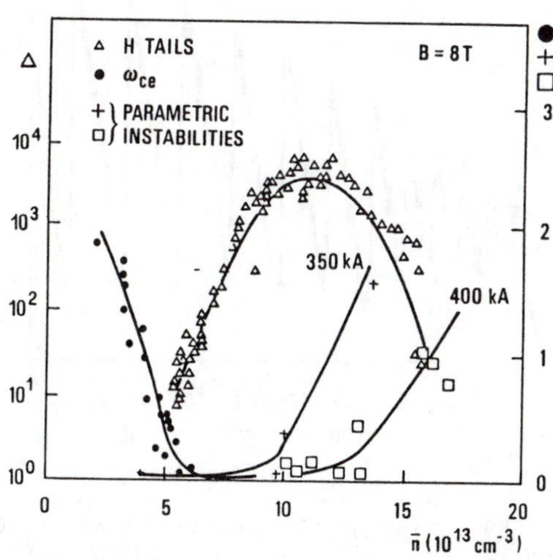

Fig. 1 - Overall scenario of the different heating regimes in FT at 2.45 GHz (D plasma B = 8 T).

About the density limit, the overall scenario is well represented in Fig. 1. The electron wave interaction, indicated here by the ECE signal, disappears above $\bar{n}$ = 5×10^{13} cm^{-3} because of ion absorption. As the density is further increased the onset of Parametric Decay Instabilities [4] strongly affects the wave propagation.

FT and the next Tokamak in Frascati FTU can work up to $\bar{n}$ ~ ~ 5×10^{14} cm^{-3}, therefore an experiment at f=8 GHz was prepared in order to heat the high density plasma of FT since it was demontrated

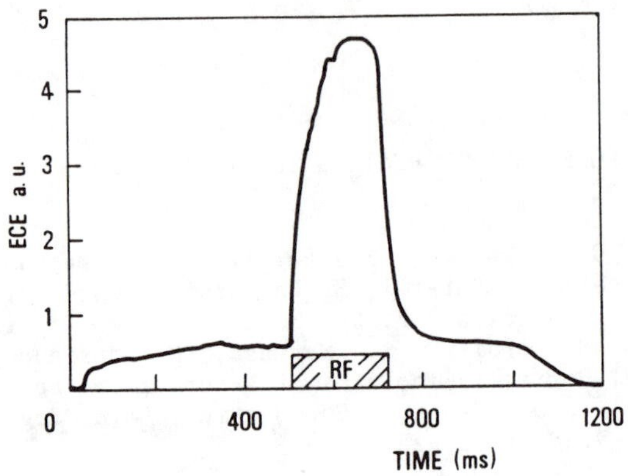

Fig. 2 Electron cyclotron emission (fourth harmonic) B_T=6T, I_p=280 kA, $\bar{n}$=5.5x10^{13} cm^{-3}, P_{RF}=110 kW, $\Delta\phi$=2π/3.

that the density limit increases with frequency [1]. The aim of the experiment is both physical (control of the density limit, heating at high density where the collisionality is stronger) and technological (control of the scaling of the specific power with frequency) in view of the FTU machine where 8 MW of RF power will be injected.

EXPERIMENTAL RESULTS

The RF system, its difficulties, the results on coupling are described at length in a companion paper [5]. The coupler is a 4x4 grill, each waveguide being 0.5x3.5 cm. For a series of failures [5] we were forced to work up to now with three adjacent columns, the fourth lateral column functioning as a dummy waveguide. The phasing was set at π or 2π/3. The plasma parameters were B_T= 6-8 T, I_p = 300--450 kA, $\bar{n}$ = 4-20x10^{13} cm^{-3}.

Operating at low density, the effects on the plasma were similar to the ones at 2.45 GHz as expected. Figure 2 shows the EC emission (4th harmonic) for the indicated plasma parameters. In Figure 3 the soft X-ray signal is reported showing the evolution of the sawteeth.

As the density increases, the enhancement of the 4th harmonic of the ECE signal becomes weaker (Fig. 4) presumably due to:

a) worse accessibility (notice that Fig. 4 is at B_T = 6 T)

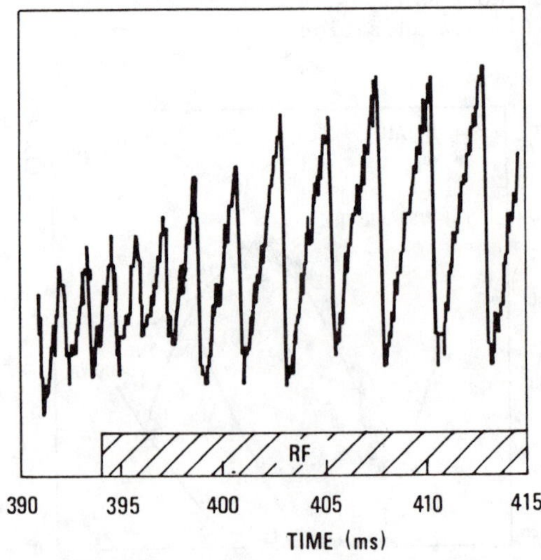

Fig. 3 Evolution of the sawteeth (SXR) B_T=6T, I_p=300 kA, $\bar{n}$=4x10^{13} cm^{-3}, P_{RF}=70 kW, $\Delta\phi$=π.

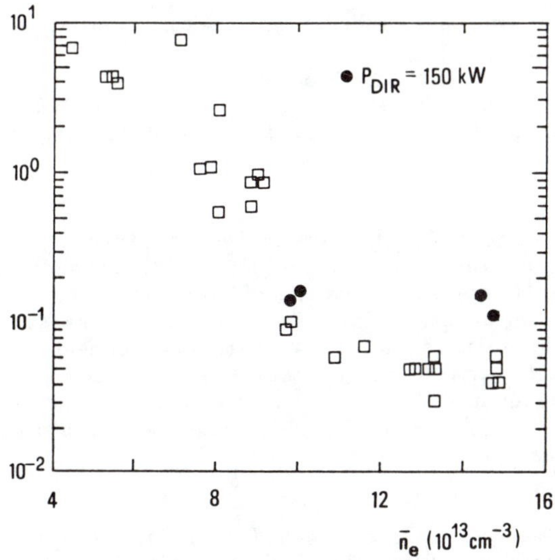

Fig. 4 Normalized enhancement of ECE signal (4th harmonic) vs density B_T = 6 T, P_{RF} ~ 100 kW (except as noticed).

b) larger drag progressively quenching the development of electron tails.

Signals on the EC emission due to RF power were clearly seen up to $\bar{n}$ = 1.5x10^{14} cm^{-3} in the range of plasma current and toroidal field values explored. At present it seems certain that this limitation is only an effect of insufficient power (P coupled $\leq$ 130 kW). Anyhow no ion tails have been observed even looking at ripple trapped particles. The spectrum of the signal collected by a RF probe in the limiter shadow shows only a weak broadening of the pump up to $\bar{n}$ = 1.7x10^{14} cm^{-3}. At this density in a low current (I_p = 300 kA, B_T = 6 T), quite cold discharge a first satellite was observed in the low frequency side of the pump (Fig. 5). This threshold

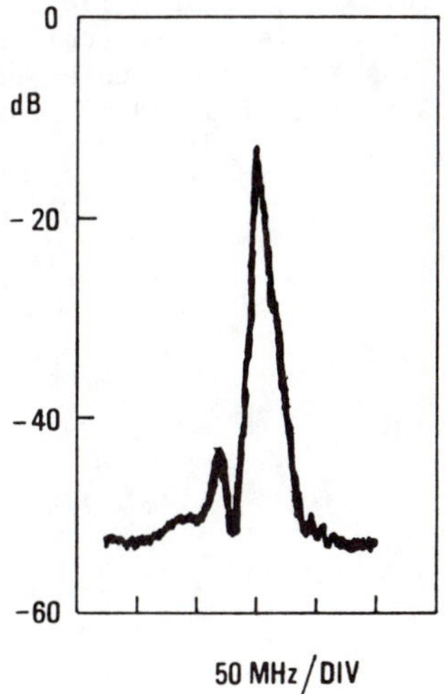

Fig. 5 RF spectrum detected by RF probes. B_T=6 T, I_p=300 kA, $\bar{n}$=1.7x10^{14} cm^{-3}, P_{RF}=140 kW, $\Delta\phi$=2π/3.

is expected to increase increasing plasma current and in a higher temperature discharge [4].

We had no time for an extended investigation of bulk effects. An increase of Te and Ti at the center of about 100 eV was detected in plasmas with B_T=8 T, I_p = 380 kA, $\bar{n}$ = $1x10^{14}$ cm^{-3}, P_{RF} = 90 kW, $\Delta\phi = \pi$.

CONCLUSIONS

The density limit for the wave-electron interaction has been investigated in the range $\bar{n} \sim 4\text{-}20x10^{13}$ cm^{-3}. Enhancement of the 4th harmonic of the EC emission have been observed up to $\bar{n} = 1.5x10^{14}$ cm^{-3} despite the worse accessibility due to the necessity of functioning with 3 waveguide columns out of 4. This enhancement is strong at low density ($\bar{n} \sim 5x10^{13}$ cm^{-3}) and becomes weaker as the density is increased. Probably the quoted limit is due to insufficient power (P coupled $\lesssim$ 130 kW) because no ion tails were observed even looking at ripple trapped particles.

REFERENCES

[1] F. Alladio et al., Proc. "4th International Symp. on Heating in Toroidal Plasmas", Roma, 21-28 March 1984, Vol. I, p. 546 (1984)
[2] F. Alladio et al., Proc. "12th European Conference on Controlled Fusion and Plasma Physics", Budapest, 2-6 September 1985, Vol. Ia, p. 179 (1986)
[3] F. Alladio et al., Proc. 13th European Conference on Controlled Fusion and Plasma Physics", Schliersee, 14-18 April 1986, Vol. 10c, Part. II, p. 327 (1987)
[4] R. Cesario, V. Pericoli-Ridolfini, ENEA Report RT/FUS/86/5, C.R.E. Frascati (1986), to be published on Nuclear Fusion
[5] A.A. Tuccillo and FT Groups, this Conference

OPTIMIZATION OF SLOW WAVE MULTIJUNCTION ANTENNAE FOR CURRENT DRIVE

D. Moreau*, C. David*
Association EURATOM-CEA Cadarache,
13108 St Paul lez Durance, France

C. Gormezano, S Knowlton+, R.J. Anderson, G. Bosia,
H. Brinkschulte, J.A. Dobbing, J. Jacquinot, A.S. Kaye, M. Pain
JET Joint Undertaking, Abingdon, Oxfordshire, OX14 3EA, UK
* Presently attached to JET Joint Undertaking
\+ Plasma Fusion Centre, MIT Cambridge, Mass 02139, USA

ABSTRACT

Having defined a directivity parameter based on current drive efficiency we quantitatively compare various $n_{//}$ spectra which can be launched by a multijunction coupler. We show how such couplers can be optimized on the basis of linear coupling theory by choosing the locations of the E-plane junctions.

INTRODUCTION

During the last few years, travelling slow waves above the lower hybrid frequency have been providing a very successful way of stabilizing sawteeth and in some cases improving energy confinement in tokamaks. Recent experiments on JT-60 have demonstrated that the ability of slow waves for generating fast current-carrying electrons was enhanced at high electron temperature[1]. It was also shown in PLT that current drive efficiency may be improved by launching a well defined narrow wave spectrum into the plasma[2].

Multi-megawatt launchers at 3.7 GHz are being designed for current drive and profile control in TORE SUPRA and JET and will produce the narrowest spectra ever achieved with 32 waveguides in each row. In both cases the antennae will be of the multijunction type[3,4]. A computer code (SWAN) based on linear plasma coupling theory has been developed for simulating such antennae and allows the launched parallel wave number ($n_{//}$) spectra to be properly shaped.

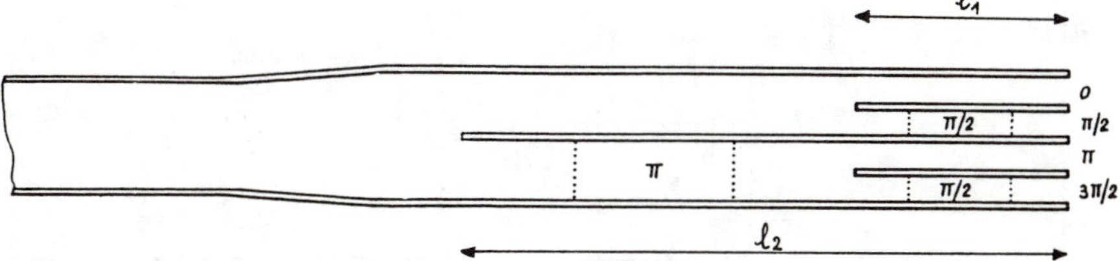

Fig. 1 Schematic drawing showing one of 8 juxtaposed multijunction grills (N = 4)

FLEXIBILITY OF THE N// SPECTRUM OF A PHASED SET OF "MULTIJUNCTION GRILLS"

The peak value of the $n_{//}$ spectrum launched by a multijunction antenna in which horizontal groups of N waveguides are fed and phased separately (Fig. 1) is given by

$$n_{//peak} = n_{//o}\left[1 + \frac{\delta\phi}{N\phi_o}\right] \quad (1)$$

where $n_{//o}$ is the nominal wave-number of each N-waveguide multijunction grill alone, ϕ_o is the fixed phase shift between secondary waveguides and $\delta\phi$ is the offset of the phasing between adjacent groups with respect to the nominal phasing ($n_{//} = n_{//o}$ for $\delta\phi = 0$). In the case of JET where $n_{//o} = 1.84$, N = 4 and $\phi_o = 90°$, $n_{//}$ can therefore vary between 1.38 and 2.3 if we allow the phasing to vary between -90° and +90°. However, due to the fact that reflected waves are sent back towards the plasma, the power spectrum can be distorted, especially for large phasings.

The multijunction power divider has therefore to be carefully designed by properly choosing the location of the various E-plane junctions. Such an optimization will reduce the parasitic peaks which can appear in the spectra e.g. an n// peak on the negative side of the spectrum which would tend to drive an opposing RF current, and also minimize the reflection coefficients and the field enhancement in the reduced waveguides.

Typical spectra obtained for $\delta\phi = -90°$ ($n_{edge} = 4.5\times10^{17} m^{-3}$), $\delta\phi = 0°$ ($n_{edge} = 7\times10^{17} m^{-3}$) and $\delta\phi = +90°$ ($n_{edge} = 1.4\times10^{18} m^{-3}$) are shown in figure 2.

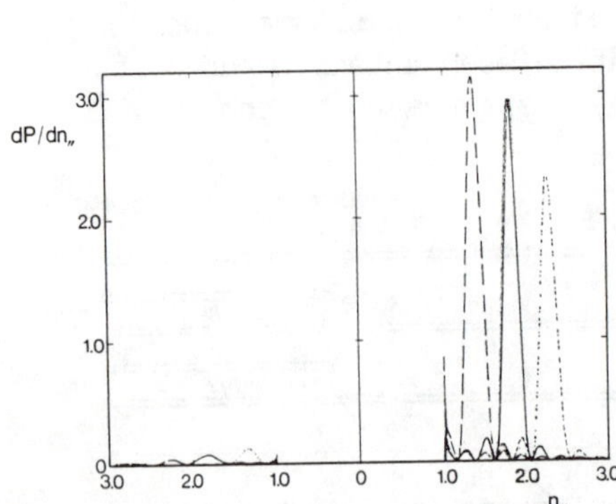

Fig. 2 Optimized n// spectra for $\delta\phi$ = - 90°, 0°, + 90°

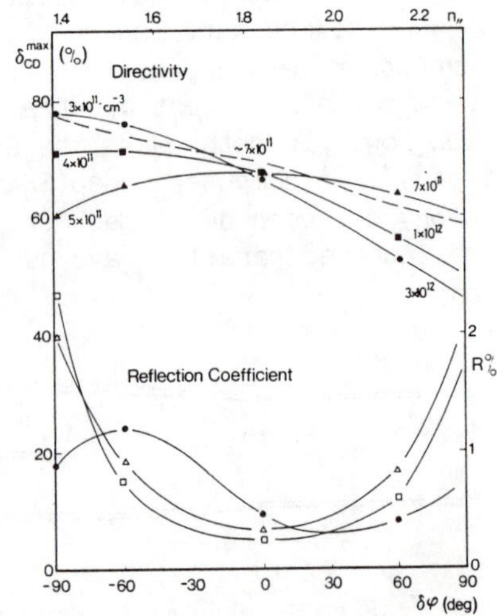

Fig. 3 Directivity and Reflection vs. phasing at optimum density

QUANTITATIVE CRITERION FOR OPTIMIZING THE SPECTRUM DIRECTIVITY

In order to compare various spectra we have defined a directivity for current drive based on the $n_{//}^{-2}$ theoretical scaling of current drive efficiency:

$$\delta_{CD} = (1-R)\, n_{//peak}^2 \left[\int_1^{+\infty} \frac{1}{n_{//}^2} \frac{dP}{dn_{//}}\, dn_{//} - \int_{-\infty}^{-1} \frac{1}{n_{//}^2} \frac{dP}{dn_{//}}\, dn_{//}\right] \quad (2)$$

δ_{CD} would then be unity if all the incident power was radiated into the plasma with a δ-function spectrum at $n_{//} = n_{//peak}$. R is the power reflection coefficient and $dP/dn_{//}$ the power density in $n_{//}$ space normalized to the total radiated power. Curves showing the directivity and reflection coefficient for the optimum density in front of the coupler are plotted versus phasing in figure 3 for three different possible locations of the E-plane junctions. The influence of the plasma density on both the directivity and the reflection coefficient is shown on figures 4 and 5 for $\delta\varphi = -60°$ ($n_{//} = 1.54$) on figures 6 and 7 for $\delta\phi = 0°$ ($n_{//} = 1.84$) and on figures 8 and 9 for $\delta\phi = +60°$ ($n_{//} = 2.13$) where we have assumed $n_e/\nabla n_e = 1$cm. Symbols refer to the same antennae as for figure 3 and we have indicated in dotted lines the performances which would be obtained with a 32-waveguide conventional grill at the expense of phasing all waveguides independently.

CONCLUSION

Slow wave multijunction antennae must be properly designed in order to obtain the best directivity in the range of $n_{//}$ values of interest and to minimize the HF fields in reduced waveguides. In our case the antenna corresponding to the square symbols on the figures is preferred because it has a good directivity at the lowest $n_{//}$ ($\delta\phi<0$) and for densities which are neither too high nor too close to the cut-off. It also has a small reflection coefficient ($\leq 2\%$) for $-90°<\delta\phi<+90°$.

REFERENCES

1. K. Ushikusa et al., Report JAERI-M 87-012, Feb. 1987
2. J.E. Stevens et al., 28th APS Ann. Meet. Div. of Plasma Phys., Nov 1986
3. D. Moreau and T.K. N'Guyen, Proc. 1984 Int. Conf. on Plasma Physics, Lausanne, Vol. 1., 216 (1984)
4. D. Moreau and T.K. N'Guyen, Report EUR-CEA-FC 1246, (1984)

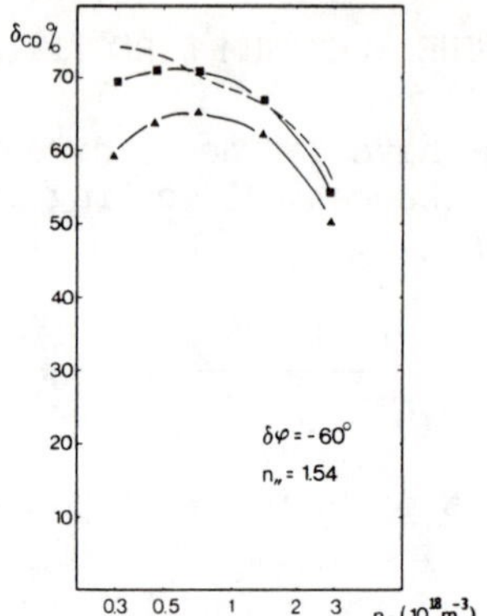

Fig. 4 Directivity vs. density for $\delta\phi$ = - 60°

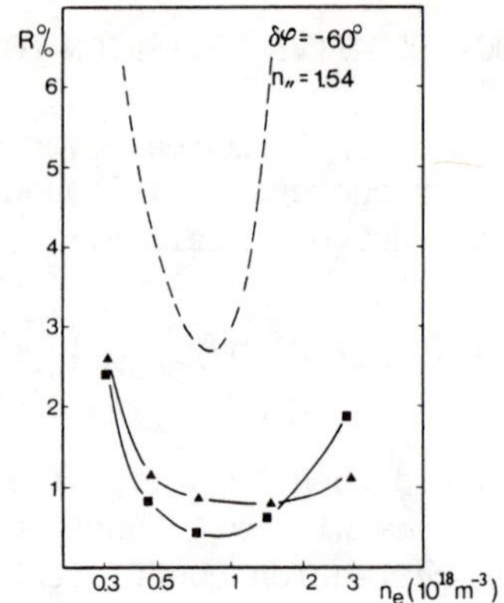

Fig. 5 Reflection vs. density for $\delta\phi$ = - 60°

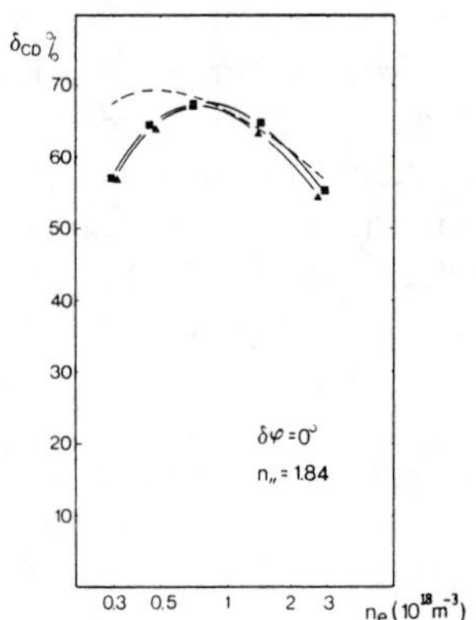

Fig. 6 Directivity vs. density for $\delta\phi$ = 0°

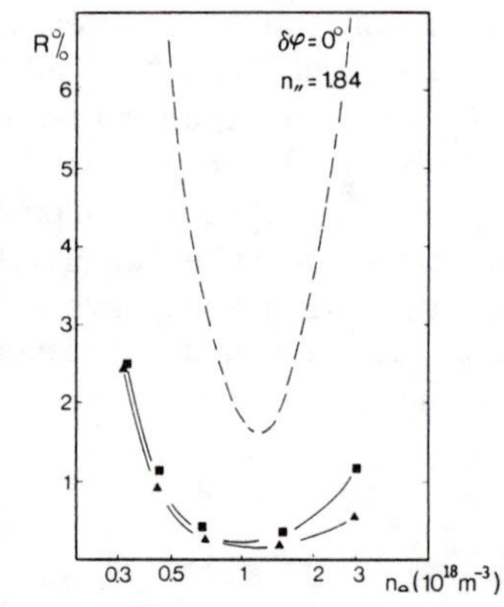

Fig. 7 Reflection vs. density for $\delta\phi$ = 0°

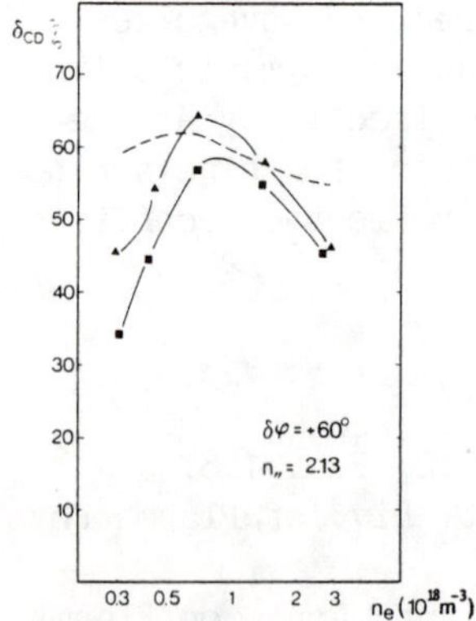

Fig. 8 Directivity vs. density for $\delta\phi$ = + 60°

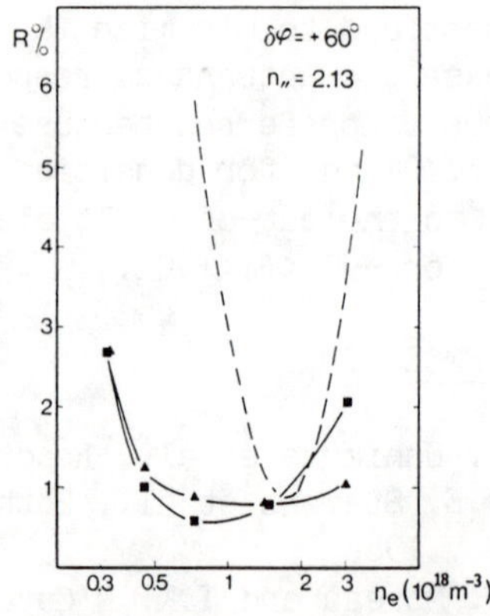

Fig. 9 Reflection vs. density for $\delta\phi$ = + 60°

POWER AND PHASE CONTROL IN THE JET LOWER HYBRID CURRENT DRIVE SYSTEM

G. Bosia, C. David**, C. Gormezano, J. Jacquinot, S. Knowlton*, D. Moreau**, M. Pain, T.J. Wade,

JET Joint Undertaking, Abingdon, Oxon, OX14 3EA, UK.
* From Plasma Fusion Centre, MIT, Cambridge, Mass. USA
** From EUR-CEA Association, CEN Cadarache, France

ABSTRACT

The JET Lower Hybrid Current Drive System, presently under construction, consists of a set of 24 klystrons (first stage: 5 klystrons), feeding in parallel a single launching structure (Grill) located at some 50 m from the generators. The "multijunction" technique has been adopted for the power splitting between a total of 384 reduced size waveguides. The total output power, at a frequency of 3.70 GHz is in excess of 10 MW, with an $n_{//}$ adjustable between 1.4 to 2.3, and a directivity better than 70%. Control of the JET current profile requires a careful adjustment of the central value of $n_{//}$ which implies an accurate control of the phase between modules. In particular the tritium phase of JET operation makes the use of conventional techniques [HF probes, coaxial cables, diodes] impractical near the JET machine. A discussion of the proposed techniques for the grill monitoring, and for a phase stable transmission of the detected signals is presented. The proposed scheme utilizes the power coupled between different modules at the Grill through the edge plasma, to measure the instantaneous electrical length differences between waveguides.

An overall phase resolution of ±5° is aimed for.

1. INTRODUCTION

The JET Lower Hybrid Current Drive (LHCD) experiment aims to produce in JET plasma currents up to 1.5 MA at maximum electron densities of $\langle n_e \rangle \approx 5 \times 10^{19}\ m^{-3}$, by launching the travelling wave by means of one single Grill launcher. A total of 12 MW of RF power, at 3.70 GHz will be produced by 24, 0.5 MW klystrons, connected to the Grill by separate waveguides. At the Grill, the power in each waveguide is further divided in 16 reduced size waveguides by three hybrid (H) junctions and 12 3dB (E) junctions, to form a Multijunction (MJ) pair[1], as sketched in Fig 1. The phases between elements of a MJ pair, are mechanically set, to follow the sequence 0 ; $\pi/_2$; π ;$3/_2\pi$.

The phase between adjacent multijunctions pairs will be instead electronically controlled at any value between $-\pi/2$ and $\pi/2$. This will allow a $n_{//}$ variation between 1.4 to 2.3. The computed "best" value of the launcher directivity is 70% at $\Delta\phi = 0$ ($n_{//}$=1.8) and $7 \times 10^{17}\ m^{-3}$ edge plasma density.

The computed module of the overall reflection coefficient for one MJ array of 4 waveguides is shown in Fig 2 ($|\sum S_{ij}|$) as a function of electron density. Also shown are the module of the diagonal terms of the scattering matrix ($|S_{ii}|$), the scattering terms of two adjacent MJ ($|S_{i\ i\pm1}|$) and the scattering coefficient between two non adjacent MJ ($|S_{i\pm2}|$). It is seen that for the selected array phase ($\Delta\phi$=0), a considerable phase cancellation takes place on the overall reflected wave.

Phase distortions in excess of 180° are expected in the waveguides/launcher system, during the application of the RF power, due to thermal expansions (ΔT up to 500°C). These effects require by themselves a full 4 quadrants phase control range.

The phase pattern should be ideally imposed at the mouth of the Grill. This is an inconvenient location where to install monitoring directional couplers, also in view of the fact that the LHCD System is planned to operate on JET during the Tritium Phase, when access to the Torus Hall for repairs and maintenance will be severely restricted or forbidden.

A method to monitor the phase at the mouth of the launcher, with measurements remotely performed, at the generators end is discussed. Experimental tests of the performances of the system will be carried out on JET, on a prototype launcher.

2. CONTROL METHOD

The simplified diagram of one of the 24 phase and power control loop is shown in Fig 3.

The driving frequency f_c is obtained by coherent frequency translation of an offset reference source (f_u) by means of an amplitude and phase modulated IF carrier produced by a local Voltage Controller Oscillator. The VCO is phase locked to a common IF phase reference and is used as a linear phase shifter. The frequency converter is used as (suppressed sideband) modulator of the RF frequency (f_u). The offset reference is also used to heterodine the monitored intelligence(s) at the output of the klystrons to the IF frequency. Both amplitude and phase control are therefore performed at the IF frequency for improved accuracy, linearity, and extended dynamic range. A proper choice of the location of the equipment, also allow the transmission of the IF frequency over the long distances, with an increase in phase stability of the overrall system against temperature variations typically of the order of $\frac{f_c}{f_i}$.
The selected IF frequency is 10.7 MHz.

The expected performances of the control loops are:

	Phase	Power
Linear Dynamic Range	700°	30 dB
Accuracy	± 2°	0.2 dB
Loop Step Response	100 µs	1 msec

3. PHASE CORRECTION FOR WAVEGUIDE/LAUNCHER ELECTRICAL LENGTH VARIATIONS

From Fig 2 it is seen that a considerable amount of power (in certain conditions in excess of the total reflection coefficient) is scattered back from one MJ to the adjacent ones. The wave scattered from MJ A to MJ B has the form:

$$b_{BA} = S_{BA}\ e^{-J\gamma(\ell_A+\ell_B)}$$

where S_{BA} is the (vectorial) scattering coefficient between A and B, (at the mouth); ℓ_A and ℓ_B are the electrical lengths of line + MJ A and line + MJ B.

If a third MJ C, symmetrical to B with respect to A is selected, it is now

$$b_{CA} = S_{CA}\ e^{-J\gamma(\ell_A+\ell_C)}$$

with $S_{BA} = S_{CA}$ because of symmetry.

The difference in electrical length between systems B and C can be computed as

$$\ell_B - \ell_C = \frac{1}{\gamma}\left(\angle b_{BA} - \angle b_{CA} \right)$$

Various other symmetries, can be predicted in the terms of the Grill scattering matrix, and used to establish relations between horizontally and vertically adjacent MJs. Some are sketched in Fig 4.

The problem is therefore reduced to make the individual scattered contribution recognizable. This is done by periodically shifting of some small amount (Δf = 3 to 5 MHz) the frequency in each waveguide, and by comparing the phases of the frequency shifted components, in that waveguide and two adjacent ones.

It should be noted that once all electrical length differences are known, the vectorial values of each scattering coefficients, at the mouth of the Grill can be computed.

A source of phase error can be due to the frequency broadening of the scattered waves, due to local non linear effects and/or to power scattering from parametric effects from the plasma. Experimental evidence at 2.45 GHz[2)] shows, however, power ratios between reflected pump and sidebands powers, in excess of 25 dB at HF power densities comparable with the ones in JET.

REFERENCES

1. D. Moreau, T.K. N'Guyen, Proc. 1984 Int. Conf. on Plasma Physics, Lausanne, Vol. 1, 216 (1984)
2. J.J. Schuss et al, Nucl. Fus. 21 No 4, page 427 (1981)

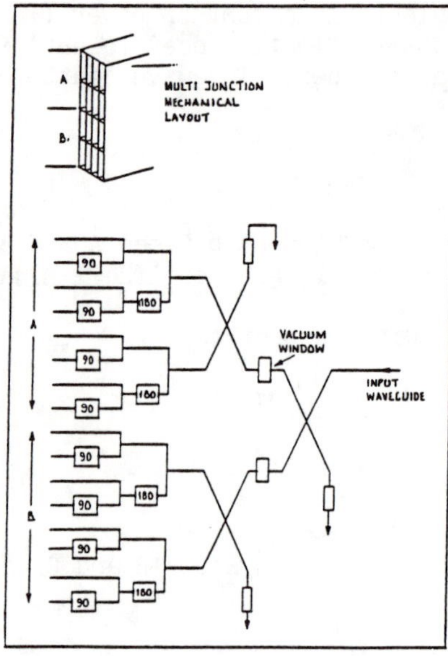

Fig. 1 Power division in a MJ pair

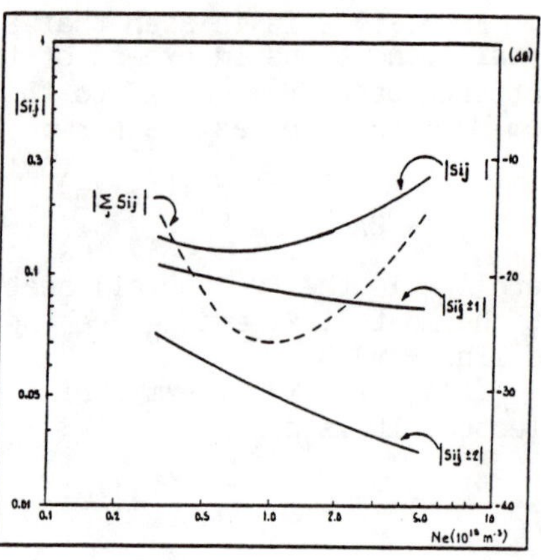

Fig. 2 MJ scattering matrix modules

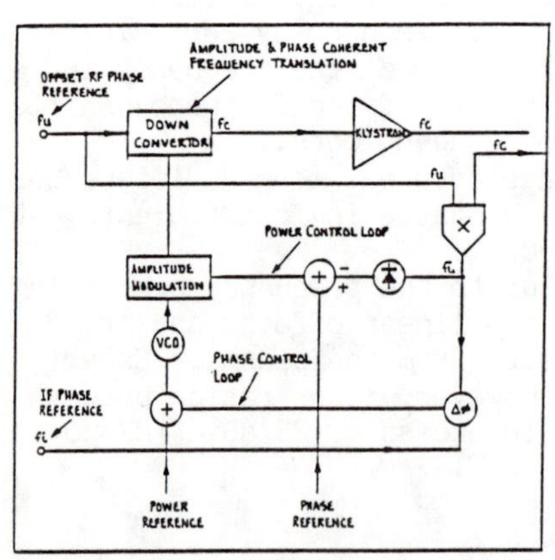

Fig. 3 Diagram of the control system

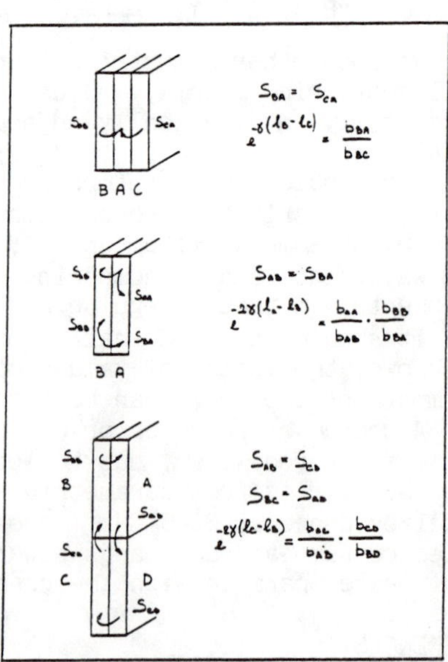

Fig. 4 Grill scattering symmetries

CURRENT DRIVE VERSUS THE LAUNCHED $N_{\parallel}$ SPECTRUM

J.Stevens, R.Bell, S.Bernabei, A.Cavallo, T.K.Chu,
R.W.Motley, and S.VonGoeler,
Princeton Plasma Physics Laboratory

ABSTRACT

Lower hybrid experiments on PLT with a narrow wavenumber spectrum have shown a 40% improvement in the current drive figure of merit η_{CD} over previous PLT results. The simple assumption that the electron tail energy is determined by the launched $n_{\parallel}$ spectrum gives a good qualitative description of all of the PLT current drive results.

EXPERIMENTS AND CALCULATION

Lower hybrid experiments at 2.45 GHz were carried out on the PLT tokamak with a 16-waveguide grill, whcih produces a narrow wavenumber ($n_{\parallel}$) spectrum ($\Delta n_{\parallel} = \lambda/L \approx 0.45$ FWHM). The current drive figure of merit, $\eta_{CD} = I\bar{n}R/P$ ($10^{13}Acm^{-3}m/W$), was measured in discharges where inductive current sources (dI_p/dt, $d\ell_i/dt$, dI_{OH}/dt) were nearly zero(figure 1). For the 16-waveguide grill $\eta_{CD} \approx 1.4$ versus ≈ 1.0 and ≈ 0.6 for previous grills having $\Delta n_{\parallel} = 1.55$ and 2.7 respectively. A simple estimate of η_{CD} can be made by assuming that the launched spectrum, upshifted slightly by first pass ray tracing effects, determines the tail energy. We approximate the launched Brambilla spectrum shape with the Gaussian function $P(n_{\parallel}) = 0.94/(\Delta n_{\parallel})\exp[-2.77(n_{\parallel} - n_{\parallel 0})^2/(\Delta n_{\parallel})^2]$, where $n_{\parallel 0}$ is the spectrum peak determined by waveguide phasing, and $\int P(n_{\parallel})dn_{\parallel} = 1$. Then, since η_{CD} for a narrow spectrum varies as $n_{\parallel}^{-2}$, the efficiency is calculated as $\eta_{CD} \approx 0.41 \int n_{\parallel}^{-2} P(n_{\parallel}) dn_{\parallel}$ where the lower limit of integration is determined by wave accessibility to $r/a \approx 0.5$ and where the factor of 0.41 is chosen to fit the data. The results of this calculation are plotted in figure 2 versus $n_{\parallel 0}$ for the three spectral widths used on PLT. A 10% upshift in launched $n_{\parallel}$, calculated from ray tracing for an outside launcher position, is included in the calculated curves. The symbols in figure 2 show the best values of η_{CD} obtained for the PLT grills for various values of peak $n_{\parallel 0}$ and $\Delta n_{\parallel}$. A good qualitative agreement exists between this simple model and the data. Specifically, this calculation duplicates the relative values of η_{CD} versus $\Delta n_{\parallel}$ for the various grills as well as the relative η_{CD} versus $n_{\parallel 0}$ for each grill.

The current drive efficiency from Fisch's theory for a narrow spectrum is $\eta_{CD} \approx 31[6/(5+Z_{eff})]\Lambda^{-1} n_{\parallel}^{-2}(n_0/\bar{n})^{-1}(1+T_{keV}/50)$. Estimating $Z_{eff} \approx 4$, $\Lambda \approx 20$, $n_0/\bar{n} \approx 1.5$, $T \approx 5keV$, $n_{\parallel} \approx 1.1 n_{\parallel 0}$ from ray tracing, and that the power in the main spectral lobe is $\approx 85\%$ of the launched power gives a theoretical value of $\eta_{CD} \approx 0.53 n_{\parallel 0}^{-2}$ for the PLT experiment in the limit of a narrow spectrum. This curve is plotted as the dashed line in figure 2. The numerical factor of 0.41 for the experimental data is ≈80% of the theoretical value of 0.53. The remaining power in the experiment presumably goes to filling the spectral gap or is lost by other mechanisms.

The role of accessibility in determining the lower value of $n_{\parallel}$ is illustrated by the sharp peak in η_{CD} versus $n_{\parallel 0}$ seen with the 16 waveguide grill (open squares in figure 2). The drop in η_{CD} as $n_{\parallel 0}$ goes from 1.5 to 1.2 demonstrates that there is a limit to η_{CD} at the point where a large fraction of the spectrum is less than $n_{\parallel a}=\sqrt{K_{\perp}+K_{x}}/\sqrt{K_{\parallel}} \approx 1.3$. The overall efficiency was slightly lower for this scan since it was performed just after a machine opening when the plasma conditions were not optimum. The density scan shown in figure 3 also shows the role of accessibility. In this case, 2/3 of P_{rf} was supplied by the 16-waveguide grill and 1/3 of P_{rf} was supplied by an 8 waveguide top grill. Applying the same calculation proceedure to these two spectra results in the curves shown in figure 3. The simple model again agrees qualitatively with the trend of the data which is towards reduced η_{CD} with higher density and with lower magnetic field.

DISCUSSION AND CONCLUSION

The experimental current drive efficiency is ≈80% of the theoretical value calculated from the assumption that the launched spectrum, modified by ray tracing, is damped on the first pass of the wave. This demonstrates that the initial launched spectrum is the major determining factor for the PLT lower hybrid experiments. This simple estimate works well because current drive efficiency η_{CD} is mainly determined by the high energy component of the tail, which is well understood, while several necessary but poorly understood physics details appear not to affect the overall lower hybrid current drive results. For instance, some of the remaining power presumably upshifts in $n_{\parallel}$ and automatically fills the spectral gap no matter how wide the gap. Also, radial variations in wave damping and current profile evolution have little effect on η_{CD} as long as the wave is accessible to $r/a \lesssim 0.5$. Finally, even though poor tail energy confinement might be expected because of the long slowing down time ($\tau_{s} \gtrsim \tau_{E}$), the tail losses appear to be a small fraction of P_{rf}. Improvement in η_{CD} should result from $Z_{eff} \rightarrow 1$, $T_{e} \rightarrow 10$keV, and better accessibility.

ACKNOWLEDGEMENTS

We thank the PLT technical staff for their excellent work on this experiment. The support of W.Hooke, J.Hosea, K.Bol, D.Meade, H.Furth, and P.Rutherford is greatly appreciated. J.S. thanks the members of the JT-60 rf heating team for their comments and discussions on the subject matter of this paper. This work is supported by the U.S.D.O.E. contract No. DE-AC02-76-CH03073.

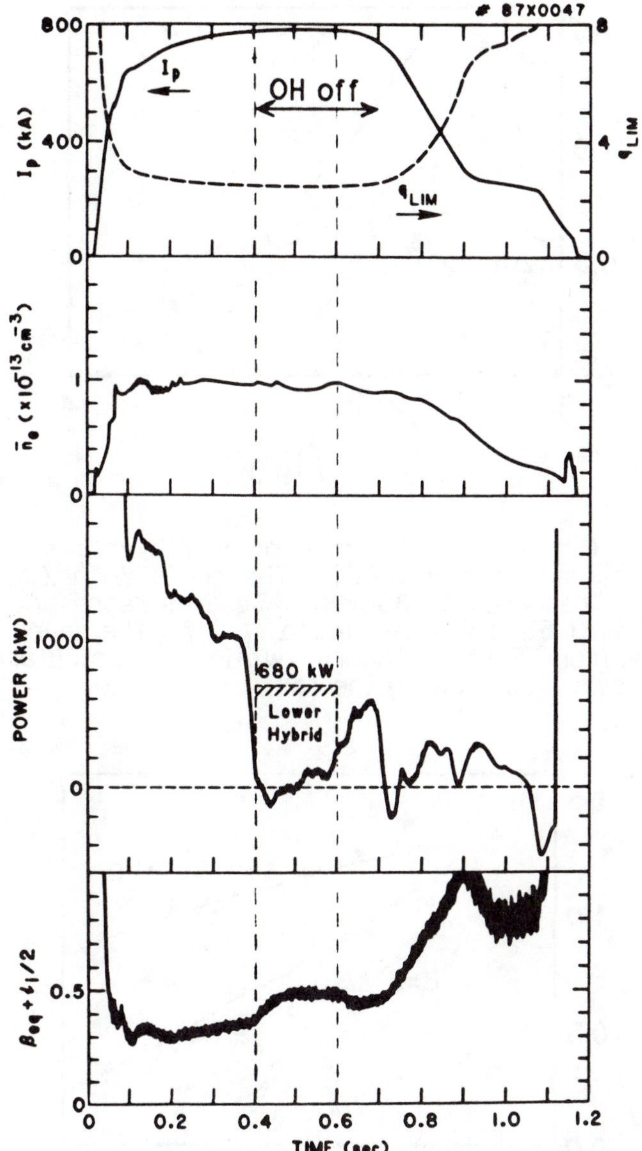

Fig.1 - Plasma current of 770kA maintained by 680kW of lower hybrid power from 0.4-0.6sec. with $\bar{n}_e$=0.95x10^{13}cm^{-3}. Intentional impurity injection at 0.45 sec. caused Z_{eff} to increase, which reduced η_{CD} and resulted in some Ohmic input power for t>0.5 sec.

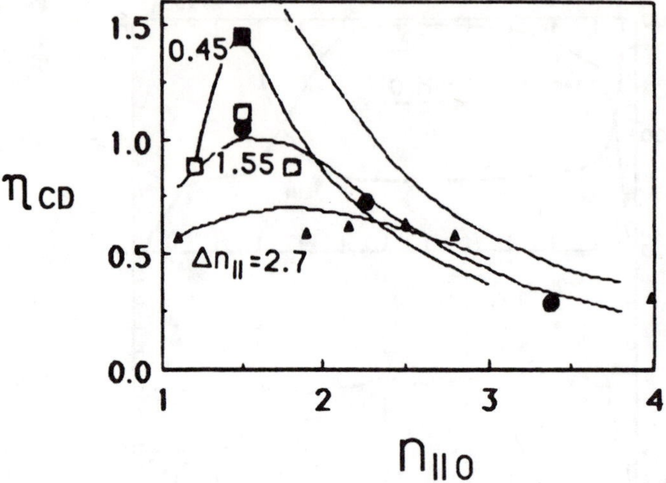

Fig.2 - Current drive figure of merit versus peak of the launched $n_{\parallel}$ spectrum for PLT. The grills with $\Delta n_{\parallel}=1.55$ and 2.7 used a frequency of 800MHz. Squares represent $\Delta n_{\parallel}=0.45$, circles $\Delta n_{\parallel}=1.55$, and triangles $\Delta n_{\parallel}=2.7$. The upper line is the maximum from Fisch's theory, while the calculation of the other lines is discussed in the text.

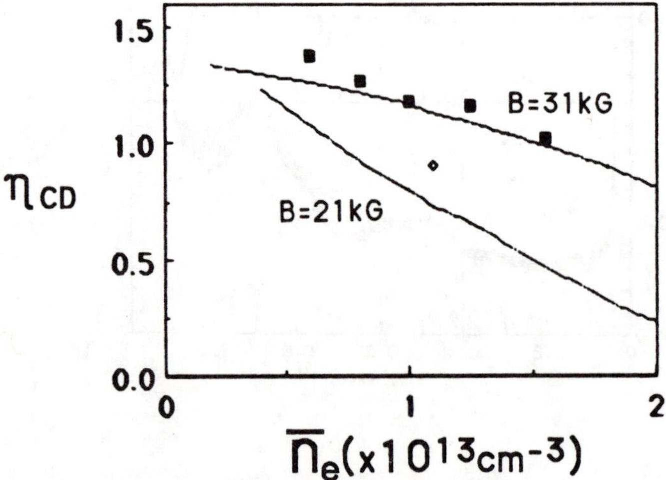

Fig.3 - Current drive figure of merit versus density with $I_p=500kA$.

STUDY OF m=2 ACTIVITY DURING LOWER-HYBRID-DRIVEN, LOW q DISCHARGES IN PLT

R.W. Motley, R. Bell, S. Bernabei, A. Cavallo, T.K. Chu,
F. Söldner, J. Stevens, and S. von Goeler
Princeton Plasma Physics Laboratory, Princeton University
P.O. Box 451, Princeton, NJ 08544

ABSTRACT

Lower hybrid driven discharges in PLT with $q < 2.5$ are unstable to m=2 instabilities, as observed on Mirnov loops. The mode structure grows within a few milliseconds to encompass most of the plasma column. After the mode slows down from a few kilohertz and locks in, a mini disruption occurs, giving rise to a drop in the population of the high energy electron tail, a rise in the impurity level, and a loss of 30-50% of the plasma energy.

It has been shown that lower hybrid waves can stabilize the m=0 (sawteeth) and m=1 instabilities in a tokamak plasma. It is believed that the mechanisms for this stabilization are a broadening of the current profile and an increase in the plasma conductivity. If current broadening is the dominant stabilizing mechanism, one might expect that the m=2 mode might become more prominent.

We have examined the MHD stability of ohmic discharges with added lower hybrid power, using (1) a set of 16 Mirnov loops arranged poloidally around the vacuum vessel, (2) Thomson scattering to determine $T_e(r)$, (3) second harmonic cyclotron radiation to determine the location of a disturbance, and (4) soft X-rays.

Typical PLT discharges with $q > 3$ show sawtooth or m=1 activity unless the lower hybrid power is high enough to stabilize the plasma. No m=2 activity is observed.

M=2 modes become observable if $q < 2.5$. A typical low q ($q > 2.2$) discharge in PLT is shown in Fig. 1. The MHD activity usually occurs in one or two bursts several tenths of a second after the RF power is applied and is observed on the Mirnov loops (Fig. 2). The signal shown on Fig. 3, the sum of the top and bottom loops minus the inside and outside loops, provides a quick identification of m=2 activity. More positive identification of the nature of the mode can be obtained by comparing the amplitudes or phases of signals on loops at each of the 16 locations around the poloidal circumference (Fig. 4).

Only m=2 (as well as m=1) modes have been observed during the mature RF-driven stages of the low q discharges. On the other hand, both m=2 and m=3 activity has been detected during the period of current rise.

The MHD activity usually leads to gross changes in the plasma column. As shown in Fig. 1, there are two pauses in the rise of current (at 0.51 and 0.64 sec), accompanied by growing m=2 modes (Fig. 5). Immediately following disturbances like this, the impurity level rises, the tail population drops sharply, as indicated by decreases in the second harmonic emission (Fig. 6), and the electron temperature declines by ≈ 30-50% (Fig. 7). The total energy content of the plasma decreases by a similar fraction.

By observing a number of channels of the second harmonic emission, it is possible to observe the growth in time and position of the m=2 modes. The mode appears to start towards the outside of the plasma column and rotates at a few kilohertz. Within 5-10 msec the mode grows almost to the center of the plasma column, increases in amplitude, and slows down to a rotation speed below 1 kilohertz. Finally, the mode appears to lock in and cease rotating. This behavior is confirmed by the Mirnov loops (note Figs. 3 and 4).

ACKNOWLEDGMENT

This work was supported by the U.S. Department of Energy Contract No. DE-AC02-76-CHO-3073.

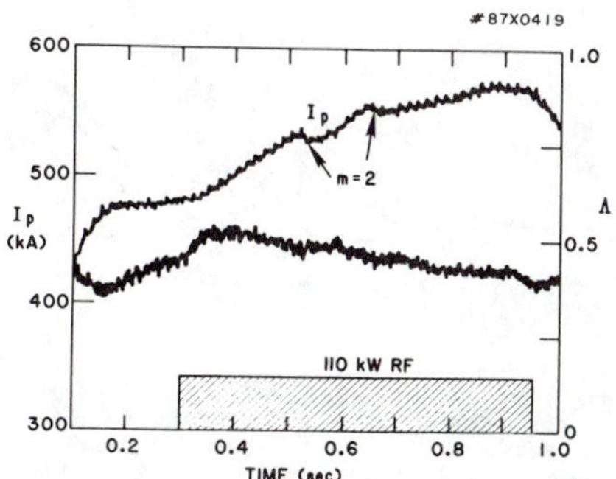

Fig. 1: Plasma current and $\Lambda = \beta_p + l_i/2$ waveforms for ohmic + 110 kW RF pulse in PLT. $\bar{n}_e = 10^{13}$ cm^{-3}, $B_t = 20$ kG, $q_L \geqslant 2.2$.

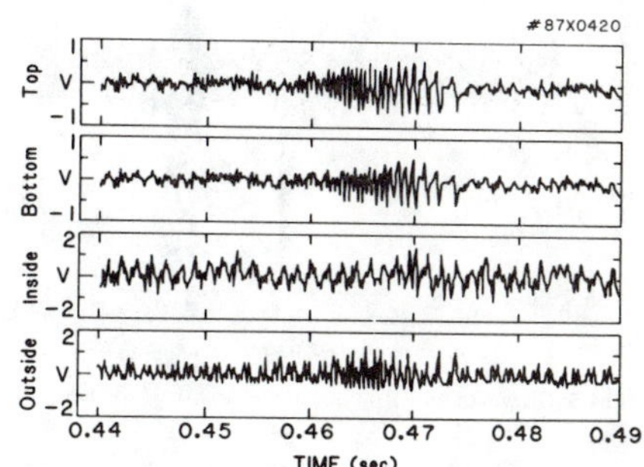

Fig. 2: Signals on four of the Mirnov coils positioned poloidally around the torus at the top, bottom, inside, and outside. An m = 2 oscillation grows from t = 0.46 to 0.47 sec.

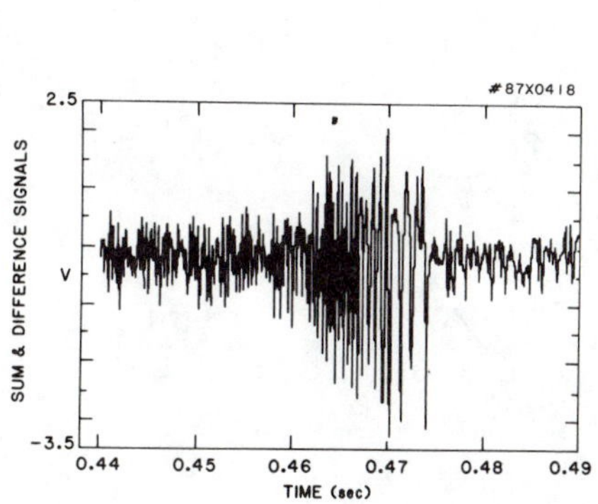

Fig. 3: Sum and difference signals from the four Mirnov coils (top + bottom − inside − outside) to monitor m = 2 activity.

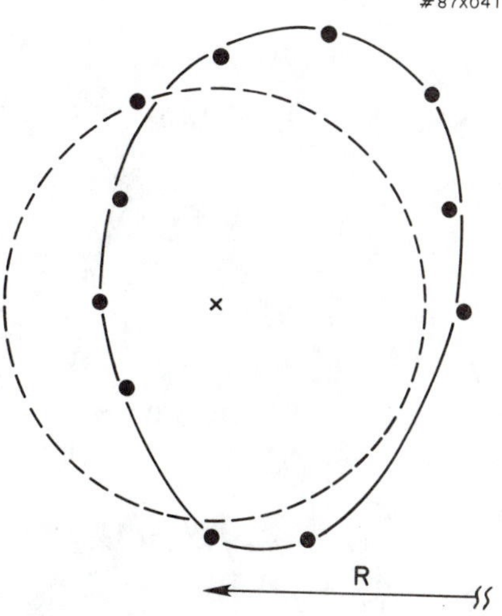

Fig. 4: Amplitude of signals on 11 of the 16 coils, showing m = 2 modes at t = 0.51.

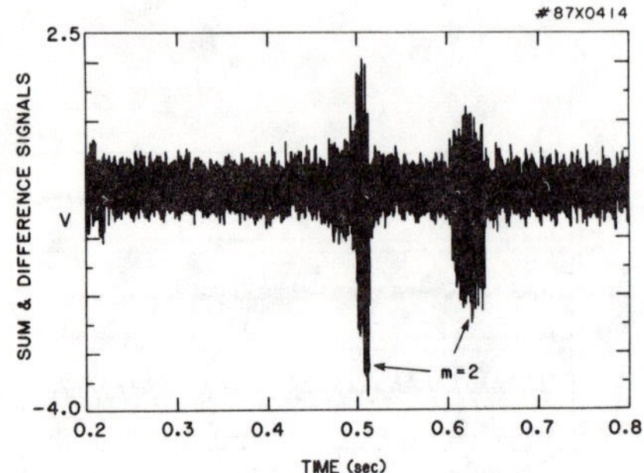

Fig. 5: Growth of m = 2 modes at t = 0.51 and 0.64 sec.

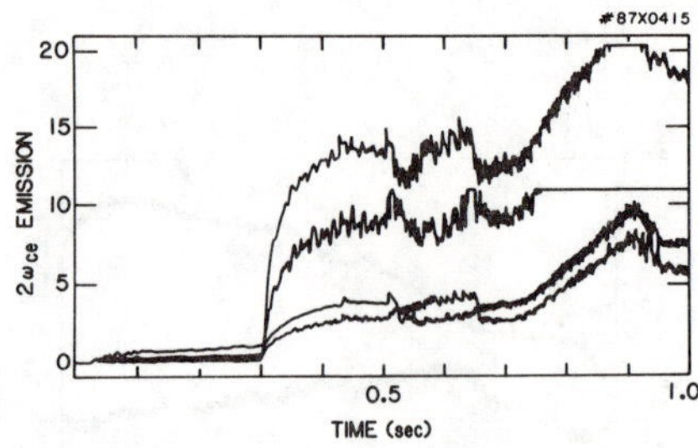

Fig. 6: Time behavior of the 2 ω_{ce} emission from PLT showing loss of high energy tail electrons at t = 0.51 and 0.64 sec as a result of m = 2 activity. The largest signal arises from electrons near the discharge center.

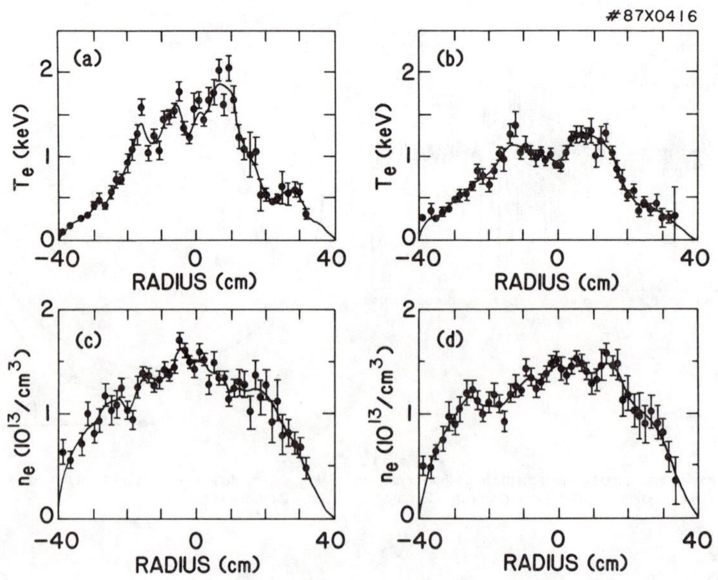

Fig. 7: Density and temperature profiles before the RF pulse (a and c) and after m = 2 activity (b and d).

CURRENT PROFILE BROADENING DURING SAWTOOTH SUPPRESSION BY LOWER HYBRID CURRENT DRIVE

T.K. Chu, R.W. Motley, R. Bell, S. Bernabei, A. Cavallo,
J. Stevens, and S. von Goeler
Princeton Plasma Physics Laboratory, Princeton University
Princeton, NJ 08544

ABSTRACT

Current-carrying channel during lower hybrid current drive is broadened from that in inductively driven discharges. For the PLT sawtooth suppression experiment ($\bar{n}_e = 1 \times 10^{13}$ cm^{-3}, I = 500 kA, B = 30 kG), the rate of decrease of internal inductance (or the rate of increase of the width of current channel) with wave power is ~0.1/MW. At $P_{rf} \simeq 450$ kW, the power level at which sawtooth oscillations are suppressed and the m = 1 magnetic island is eliminated, the radius of the current-carrying channel is estimated to increase by 1.5 cm.

Several recent experiments[1] have demonstrated that sawtooth oscillations can be suppressed by lower hybrid current drive. Since internal disruptions are associated with the existence of a q = 1 surface in the core region of the plasma, their suppression indicates that the current-carrying channel during lower hybrid current drive may be broadened from that in inductively driven discharges and that q becomes greater than unity in the entire plasma column. This paper presents indirect experimental evidence of a broadening of the current profile during lower hybrid current drive. These results include, during current drive, (1) an outward shift of the inversion surface of sawtooth oscillations, (2) an increase of the width of the m = 1 magnetic island (prior to a crash), and (3) the decrease of internal inductance. The results are based on experiments in the PLT tokamak, described in the first reference in Ref. 1. The PLT tokamak has a major radius R_o = 132 cm and a minor radius a = 40 cm. The experimental parameters are $\bar{n}_e = 1 \times 10^{13}$ cm^{-3}, I_p = 500 kA, B = 30 kG. The waves are launched from two arrayed waveguide grills at -60° phasing.

Figure 1 shows the radial distribution of the sawtooth oscillation (determined from 2 Ω_{ce} emission) in inductively driven discharges (P_{rf} = 0). Sawtooth peaks at the major radius position R ≈ 135.5 cm. The position of the inversion surface on the weak field side is $R_{inv} \approx 145.5$ cm. Figure 2(a) shows R_{inv} as a function of lower hybrid wave power P_{rf}. The inversion surface moves outward by ~2 cm when P_{rf} is increased to 245 kW. (At still higher power, sawtooth oscillations are suppressed and an inversion surface no longer exists. A steady-state m = 1 oscillation is present till $P_{rf} \simeq 450$ kW.[2])

Figure 2(b) shows the width of the m = 1 magnetic island (on the weak-field side) just prior to a sawtooth crash as a function of P_{rf}. At P_{rf} = 0, the island width is $\lesssim$ 1 cm. This width increases to $\sim$10 cm at 300 kW. An increase of island width indicates a flattening of the q profile in the vicinity of the q = 1 surface.

Another indirect indication of profile change during current drive is gained by comparing profiles of electron temperature in inductively driven discharges to profiles of hard X-rays in current-drive discharges. Figure 3 shows such a comparison. The half width at half amplitude of the X-ray profile is broader by $\sim$3 cm than that of the $T_e^{3/2}$ profile at this power level (P_{rf} = 590 kW).

Finally, the plasma internal inductance ℓ_i also decreases with wave power, indicating a profile broadening. In inductively driven discharges, $\ell_i/2 \approx 0.6$. Figure 4 shows the change of internal inductance, $\Delta(\ell_i/2)$, as a function of rf power. It decreases at a rate of $\sim$0.1/MW. A rough estimate can be obtained by assuming the current is distributed uniformly over a radius r_o and $\ell_i/2 = 1/4 + \ell n(a/r_o)$. Thus, $\Delta r_o/r_o = -\Delta(\ell_i/2)$. At P_{rf} = 450 kW when the m = 1 mode is eliminated, $\Delta r_o \approx 0.05\ r_o \approx 1.5$ cm. A corresponding percentage decrease of j(0) and increase of q(0) can also be expected.

REFERENCES

[1] T.K. Chu *et al.*, AIP Conf. Proc. 129, 131 (1985); also in Nucl. Fusion 26, 665 (1986); C. Gormezano *et al.*, AIP Conf. Proc. 129, 131 (1985); F. Soldner *et al.*, Phys. Rev. Lett. 57, 1137 (1986); K. McCormick *et al.*, Phys. Rev. Lett. 58, 491 (1987).

[2] S. Bernabei *et al.*, 11th Int. Conf. on Plasma Physics and Cont.Nucl. Fusion Res., Kyoto, Japan (1986), paper IAEA-Cn-47/F-II-1.

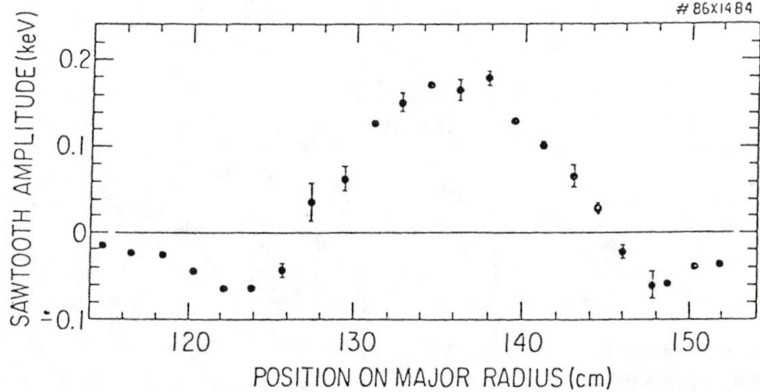

Fig. 1: Radial distribution of sawtooth amplitude in an ohmically heated discharge (P_{rf}) = 0. Position of the inversion surface on the weak field side is $R_{inv} \simeq 145.5$ cm. $I_p = 500$ kA, $\bar{n}_e = 1 \times 10^{13}$ cm^{-3}, B = 30 kG.

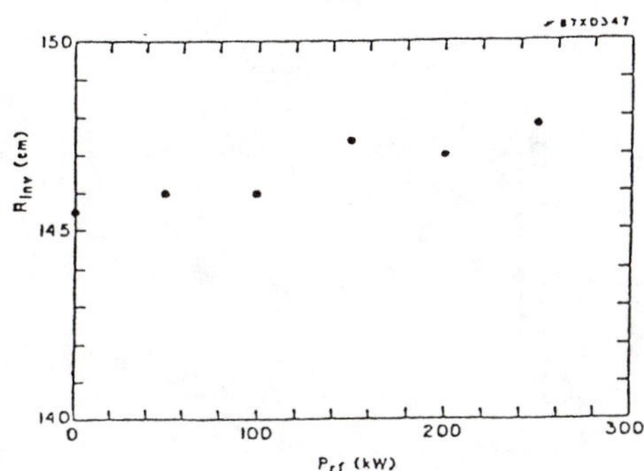

Fig. 2a: Position of inversion surface, R_{inv} vs. lower hybrid wave power. **2b:** Width of the m = 1 magnetic island (just prior to a crash) vs. lower hybrid wave power.

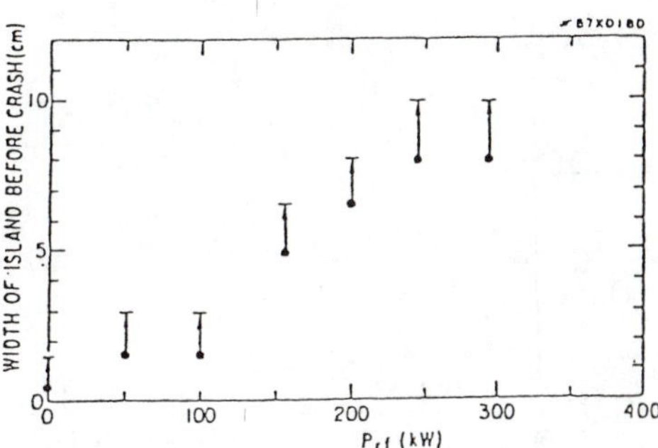

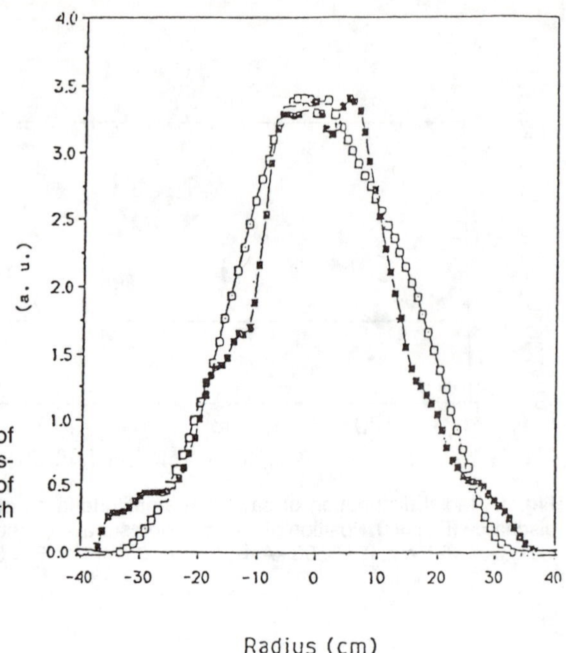

Fig. 3: A comparison of profile of $T_e^{3/2}$ in an ohmically heated discharge (solid squares) and that of hard X-ray (open squares) with $P_{rf} = 590$ kW.

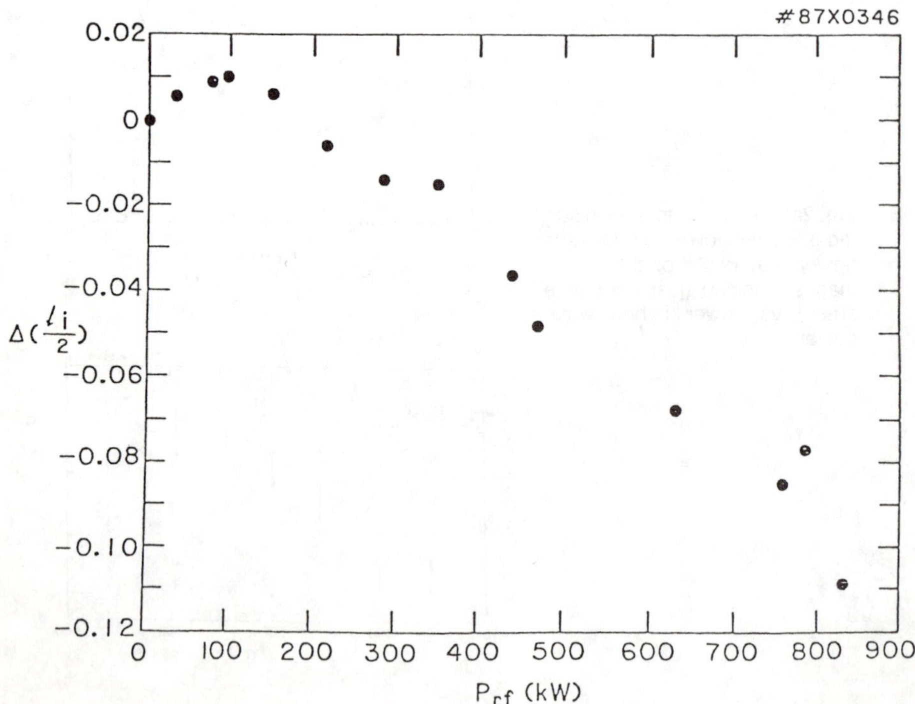

Fig. 4: Change of internal inductance, $\Delta(l_i/2)$, vs. lower hybrid wave power.

REDUCED THERMAL DIFFUSION USING LOWER HYBRID WAVES IN A TOKAMAK PLASMA

R.E.Bell, S.Bernabei, A.Cavallo, T.K.Chu, T.Luce,
M.Ono, R.Motley, J.Stevens, S.Von Goeler

Princeton University, Princeton Plasma Physics Laboratory,
P. O. Box 451, Princeton, NJ 08544

INTRODUCTION

Lower hybrid waves were launched into PLT tokamak plasmas with waves propagating with the electron drift velocity (current drive), opposing it, and symmetrically (equal power in both directions). This paper examines the dependence of electron heating on the direction of the launched LH wave. Large increases in the central electron temperature, $T_e(0)$, have been obtained in the PLT tokamak[1] using modest power levels by launching LH waves in direction of the electron drift velocity producing a current driven plasma. Waves launched opposing the current and symetrically launched waves produce much smaller temperature increases. This strong dependence of electron temperature on the direction of the launched wave with otherwise identical spectra suggests that the central confinement is affected by the suprathermal current generated during current drive. Associated with the LH current driven plasmas is a reduction in the central value of χ_e, which was calculated using measured hard x-ray, electron temperature and electron density profiles in order to estimate power deposition profiles.

EXPERIMENT

The experiment was carried out in the PLT tokamak with a major radius R = 132 cm , a minor radius a = 39 cm, and a magnetic field B_T = 3.1 T. The LH waves were launched using a 2.45 GHz 16-element grill. By adjusting the relative phasing between adjacent waveguide elements the peak value of $n_{\|}$ $(=ck_{\|}/\omega)$ was phased to propagate the wave in the direction of the electron drift velocity (CD-favoring current drive) or in the opposite direction (OC-opposing the current) with $n_{\|0}$=+1.8,-1.8 respectively. A symmetric spectrum (SYM) was produced by phasing the adjacent waveguides $00\pi\pi...00\pi\pi$ such that the launched power was evenly divided between the positive lobe ($n_{\|0}$=+1.8) and the negative lobe ($n_{\|0}$=-1.8). The target plasmas had a line average density $\bar{n}_e = 10^{13}$ cm^{-3} and the plasma current was held constant (by adjusting the current in the ohmic primary) during the application of ≈580 kW of LH power. The effective ionic charge of the plasma, Z_{eff}, was 3.4 (from Spitzer resistivity) for the LHCD and LHSYM cases and Z_{eff}6 for the LHOC plasmas. Electron temperature and density were measured along a vertical chord with Thomson scattering. Hard x-rays were measured through seven vertical chords along the major radius, giving an integrated spectrum for $\hbar\nu > 100$ keV.

Figure 1 shows the measured temperature profiles compared to the appropriate ohmic profiles before the application of rf. A peaked profile is obtained with LHCD, while only modest increases in temperature are obtained for the LHOC or LHSYM cases. In each of these cases there was a reduction in the loop voltage,for LHCD it dropped nearly to zero during the rf pulse. There was no increase in the line averaged electron density during the LH pulse and no significant change in the effective charge (Z_{eff}) of the plasma as indicated by a visible bremsstrahlung measurement.

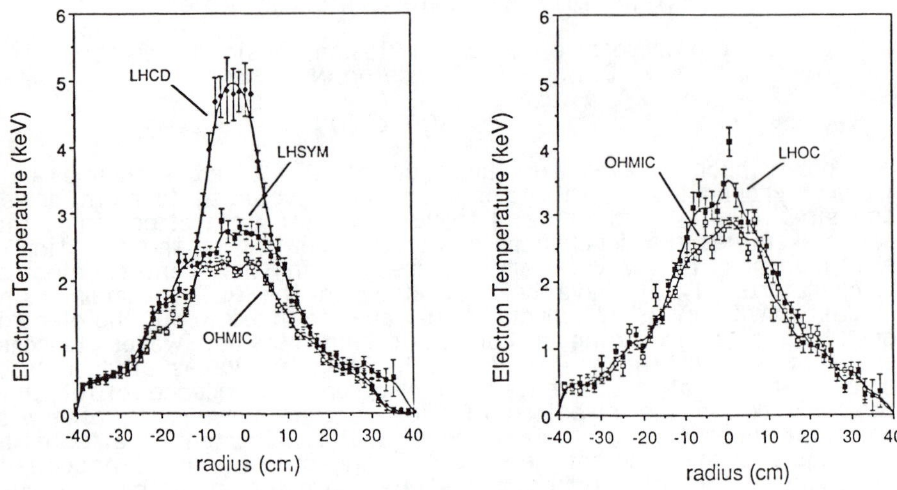

Fig. 1 Electron temperature profiles

The suprathermal current generated by the waves was estimated with the assumption that the Z_{eff} of the plasma did not change during the application of rf. The voltage on axis, V_{axis}, was calculated from the external circuits (ohmic primary and vertical field) taking into account the change in $^{1}/_{2}LI^2$. The resistivity, R_Ω, of the thermal plasma was calculated from the T_e profile during LH using the value of Z_{eff} prior to the application of rf power. In this way, the thermal current, $I_{th}=V_{axis}/R_\Omega$, was calculated, and the total plasma current, I_p, was measured, and the suprathermal current was the difference ($I_{st}=I_p-I_{th}$). For the ohmic plasmas, $I_{th} = I_p$. During LH, a suprathermal current was generated; in the non-current drive cases, suprathermal electrons are aided by the electric field to create a suprathermal tail current. Some of the plasma paramerters are given in Table I below. The global confinement time of the thermal electrons, τ_{Ee}, is the total electron energy divided by the total input power; the energy of the tail is neglected. For the ohmic plasma $\tau_{Ee}\approx 17$ ms.

Table I Summary of plasma parameters

	I_p(kA)	I_{st}(kA)	V_{axis}(v)	P_{rf}(kW)	P_{tot}(kW)	τ_{Ee}(ms)
LHCD	526	618	-0.10	585	532	19.3
LHSYM	517	414	0.13	563	630	14.9
LHOC	524	320	0.4	584	794	12.6

Sawteeth are present before the rf is applied. They are suppressed only in the LHCD case.

ANALYSIS

A one dimensional analysis was performed in order to examine the change in χ_e during LH current drive. Experimentally obtained electron temperature and density profiles were used, and the Abel-inverted hard x-ray profile, $E_{hx}(r)$, divided by $n_e(r)$ was used for the deposition profile of the rf power to the bulk electrons. It was assumed that all the rf power was eventually deposited in the bulk electrons; the energy of the suprathermal tail was not included. In this low density discharge, energy losses by electrons to ions are negligible as are radiation losses from the central region of the plasma. When sawteeth were present, a simple model was used to account for losses due to sawtooth transport. The energy change inside the q=1 surface during the rise of the sawtooth was estimated from the sawtooth amplitude, $\Delta T_{saw}(0)$. Dividing this energy by the sawtooth period, τ_{saw}, gave the average power used to increase the internal plasma kinetic energy during the sawtooth rise which was equivalent to the average power expelled from the interior at the sawtooth crash. The thermal diffusion is then given by

$$\chi_e(r) = \frac{\int_0^r \left[n_e \frac{\partial T_e}{\partial t} - \frac{2}{3} Q_{in} + \frac{2}{3} Q_{saw} + \frac{2}{3} Q_{loss} \right] r'\, dr'}{r\, n_e \frac{\partial T_e}{\partial r}} \tag{1}$$

where Q_{in} is the power supplied to the bulk electrons, Q_{saw} is the power transported by sawteeth ($Q_{saw}=0$ except at the sawtooth crash when $^2/_3 Q_{saw}(r) \approx n_e(r)\Delta T_{saw}(r)/t_{crash}$, where the duration of the crash $t_{crash} \ll \tau_{saw}$). Q_{loss} is radiated power, the power lost to ions, and the power stored in the tail as well as direct losses from the tail. Here it is assumed that these losses are negligible. Averaging over one sawtooth cycle gives $n_e \partial T_e/\partial t + {}^2/_3 Q_{saw} \approx n_e \Delta T_{saw}/\tau_{saw}$. $Q_{in}(r)$ is the sum of the inductive power, Q_{ind}, and the rf power, $Q_{rf}(r)$, applied to the plasma. $Q_{ind}(r)$ is proportional to $T_e^{3/2}(r)$, and $Q_{rf}(r)$ is proportional to $E_{hx}(r)/n_e(r)$.

The electron temperature profiles were symmetrized and fit with a smoothing spline function to get $\partial T_e/\partial r$. Figure 2 shows the computed power deposition profiles of the rf power compared to the ohmic profiles. The deposition profiles for inductive power and rf power were scaled with I_{th} and I_{st} respectively. The amplitude of the hard x-ray signal was larger for the LHOC and LHSYM cases than for the LHCD case. This is probably due to the presence of the electric field which causes the hot tail electrons to runaway[2]. A vertical ECE diagnostic[3], which measures the emission spectrum of the suprathermals below 500 keV, shows no change in the emission spectrum between the LHCD and LHSYM cases; this diagnostic is not sensitive to the sign of $v_{\parallel}$ and therefore indicates that the suprathermal distribution function $f(|v_{\parallel}|, v_{\perp})$ has not changed between LHCD and LHSYM.

Figure 3 compares the calculated χ_e profiles for the ohmic (dotted line) and LH cases shown in Fig. 1. The central value of χ_e in the LHCD case has decreased substantially from the ohmic case. The drop in χ_e for LHCD can be understood from the large gradient in T_e seen Fig. 1. The LHSYM and LHOC cases show an increase in χ_e. The change in χ_e in these cases results from the more peaked profiles of $E_{hx}(r)/n_e(r)$ used for the power deposition, since the T_e profiles are quite similar to the ohmic case. The larger signal seen by the hard x-ray measurement indicates that a high energy component of the tail is present for these cases and due to a longer slowing down time this may represent a loss term in Equation (1) which was not taken into account, so the value of χ_e shown in Fig. 3 may be too high for LHSYM and LHOC.

REFERENCES

1. R.E. Bell, et.al. Princeton Plasma Physics Report, PPPL-2452 (1987).
2. S. Von Goeler, et. al. this meeting.
3. T.C. Luce, et. al. this meeting.

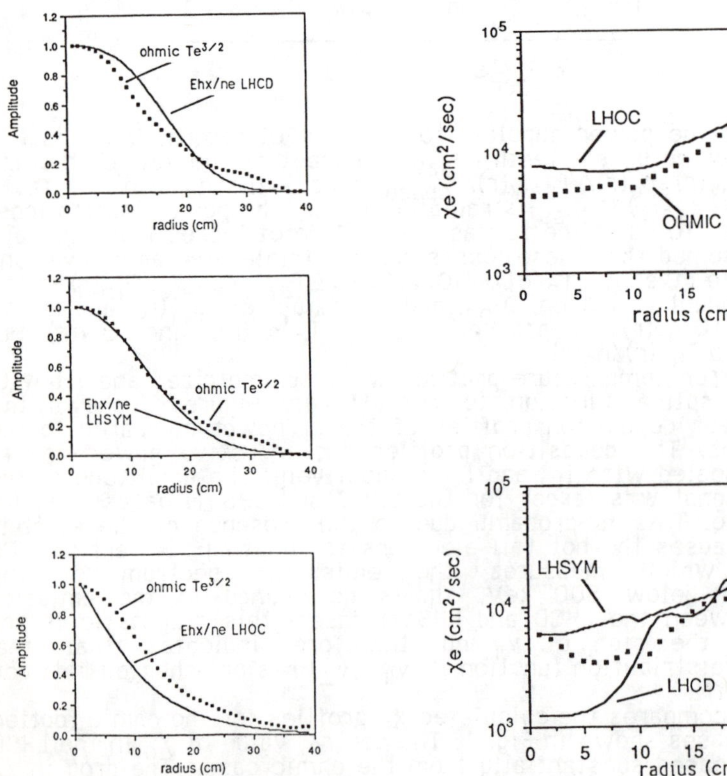

Fig. 2 Normalized power deposition profiles.

Fig 3. χ_e profiles.

SUPPRESSION OF LARGE AMPLITUDE DRIVEN SAWTEETH BY LOWER HYBRID CURRENT DRIVE

A. Cavallo, S. Bernabei, T.K. Chu, J. Stevens,W. Hooke, R. Bell, P. Colestock, R. Pinsker, J.R. Wilson, J. Hosea, G. Gammel, G. Greene, R. Motley, E.Mazzucato, S. von Goeler P. Beiersdorfer, S. Suckewer

Princeton Plasma Physics Laboratory, Princeton, NJ 08544

ABSTRACT A characteristic feature of ion cyclotron minority heating is the excitation of large amplitude sawteeth due to central power deposition and good electron-ion coupling. A similar phenomenon is expected to occur in a fusion reactor, where energetic alpha particles at the center of the plasma will also drive large amplitude sawteeth. By using ion cyclotron heating (ICH) in conjunction with lower hybrid current drive (LHCD) the suppression of large amplitude sawteeth could be studied in a controlled fashion. With P(ICH) < 0.8 MW and P(LHCD) < 0.8 MW, and $\Delta T_e/T_{e,max} = 0.2$, the lower hybrid power necessary to suppress the large amplitude sawteeth was approximately the same as that required to suppress ohmic sawteeth at the same line average electron density. In addition, it is found that LHCD does not degrade ICH discharges, nor does ICH degrade the efficiency of lower hybrid current drive.

INTRODUCTION A characteristic feature of tokamak discharges is the presence in the center of the plasma column of a sawtooth oscillation: a slow rise and abrupt collapse of the central electron density and temperature.[1,2] The drop in electron temperature is often immediately preceeded by a rapidly growing m=1 oscillation. The instability limits central temperature and plasma current as well as central energy confinement time, but does not affect global energy confinement. Thus, while of great interest theoretically, the sawtooth oscillation in ohmic discharges is a relatively benign phenomenon.

This may not remain the case in non-circular low q tokamaks with high power neutral beam injection or high power ion cyclotron heating, or in a fusion reactor. In such machines, the q=1 surface may be much closer to the plasma edge than in present tokamaks, and large amplitude sawteeth may severely limit global energy confinement time or even cause the discharge to disrupt. In fact the ability to suppress the internal disruption and in general to control the current profile could be essential for large machines.

Previous experiments[3] have shown that it is indeed possible to eliminate the sawtooth oscillation in ohmic discharges using lower hybrid current drive. An important question remains,however: can large amplitude sawteeth driven by non-ohmic heating also be suppressed, and if so at what power level?

EXPERIMENT The work to be described was done on the Princeton Large Torus (PLT), a tokamak with a major radius R = 132 cm. and a minor radius a = 40 cm. A 30 MHz, 2 MW ICH system and a 2.45 GHz, 0.8 MW LHCD system were available. As is well known, high power minority ICH is able to excite large amplitude sawteeth due to the central RF power deposition and good coupling between the electrons high temperature minority ion species.[4] [Even in PLT with q(a)= 3.2, the heat pulse from the internal disruption is seen out to the limiter.] The amplitude of the ICH sawtooth increases at first linearly with RF power but then begins to saturate at P(ICH) = 2 MW. This saturation is most likely due to the increasing importance of the m=1 precussor oscillation, which begins earlier in the sawtooth as the ICH power is increased. The m=1 island increases transport from the plasma center finally causing the sawtooth amplitude to saturate.

Another characteristic of ICH for low density [N_e = $2x10^{13}$ cm^{-3}] is a significant increase in plasma density as RF power is increased. The magnitude of the density increase is a function of antenna conditioning and machine and vacuum vessel gettering. The net result is that it is not possible to drive large amplitude sawteeth at a plasma density at which LHCD is most efficient.

Theoretically, current drive efficiency decreases inversely with increasing density[5], and experimentally this has been found to be the case. As shown in Fig. 1, the LHCD power needed to suppress the sawtooth increases strongly with increasing density. The situation is somewhat ambiguous since often the internal disruption is eliminated while a large amplitude m=1 oscillation remains near the center of the plasma. In all cases increasing the RF power still further eliminated both the sawtooth and the internal disruption.

In this experiment the LHCD power necessary to eliminate sawteeth in an ICH discharge is compared to the LHCD power needed to eliminate the m=1 and the sawtooth in an ohmic discharge <u>of the same density as the plateau density during ICH.</u>

Shown in Fig. 2 are the results of a typical experiment in which LHCD was used to suppress ICH sawteeth [$\Delta T/T$ = 0.2]. The vertical axis is the intensity of second harmonic electron cyclotron emission; for thermal plasma this is proportional to central electron temperature. The emission signal after the LHCD is applied contains a large non-thermal as well as a thermal component. The non-thermal component persists even after the LHCD power is turned off. The decrease in emission during ICH is due to a 25% rise in plasma density.

It is very clear that the sawtooth oscillation has been eliminated by LHCD; the m=1 oscillation is also suppressed.

In Fig. 3, the LHCD power necessary to suppress the ICH driven sawteeth is plotted against ICH power. The sawtooth amplitude is about 10% of the central temperature in the ohmic case and about 20% of the central temperature at maximum ICH power. Although sawtooth amplitude has more than doubled, the LHCD power necessary to suppress the sawtooth is the same.

This result is reasonable if the shapes of the normalized electron temperature profiles in ohmic and RF heated plasmas are compared. It turns out that even at the highest ICH powers when strong electron heating is observed, the profile shape is the same as that of an ohmic discharge at much lower temperature. Since the current profile is determined by the temperature profile, they should be the same in both cases. It is reasonable to expect that the same amount of LHCD power will suppress the internal disruption independent of sawtooth amplitude.

<u>CONCLUSION</u> Lower hybrid current drive can be applied to ICH driven discharges without decreasing energy confinement time or ion heating. Operating both ICH and LHCD together is more difficult than operating each alone, especially at high power levels. Edge plasma conditions tolerable for one sustem may cause the other to fail. However, it is possible for both systems to function simultaneously given some attention to plasma position and vacuum vessel conditioning.

Most importantly, it is found that the level of LHCD power necessary to eliminate sawteeth of twice the amplitude of ohmic sawteeth is the same as that necessary to suppress ohmic sawteeth at the same line average plasma density.

REFERENCES

1. VON GOELER,S., STODIEK, W., SAUTHOFF, N., Phys. Rev. Lett. 33(1974) 1201

2. KADOMTSEV, B.B., Fiz. Plazmy 1(1975) 710 [Sov. J. Plasma Phys. 1 (1975)389]

3. CHU, T.K., et. al., Nuc. Fusion 26 (1986) 666

4. MAZZUCATO, E., et. al., in Plasma Physics and Controlled Nuclear Fusion 1984, Vol 1, 433

5. Fisch, N., PPPL-2401, Theory of Current Drive in Plasmas, 1986

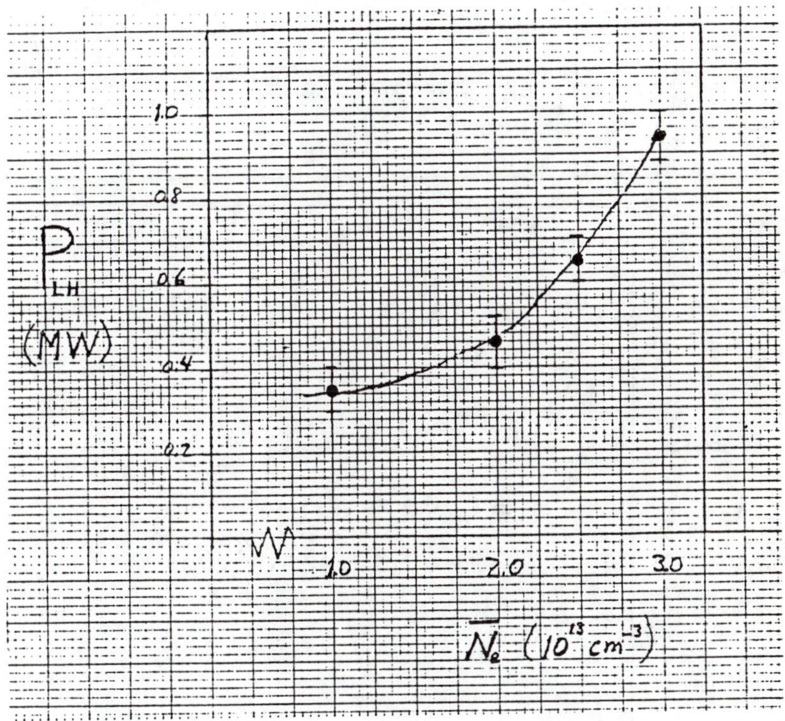

FIG. 1 LOWER HYBRID CURRENT DRIVE POWER NECESSARY TO SUPPRESS SAWTEETH IN OHMIC DISCHARGES VS. LINE AVERAGE PLASMA DENSITY

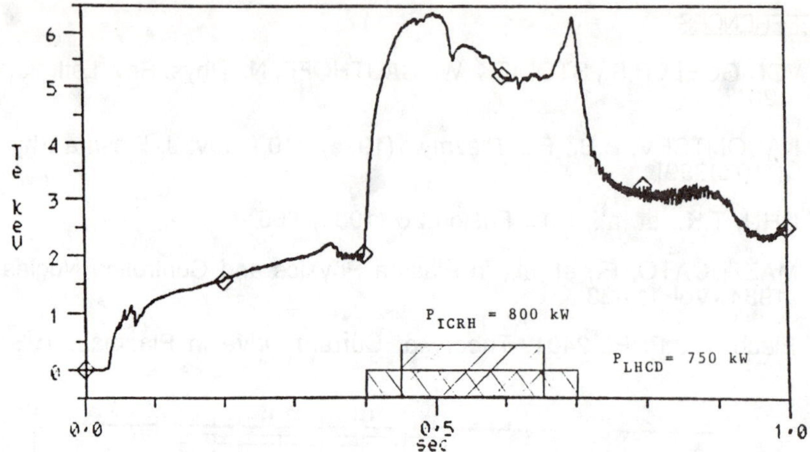

FIG. 2 ELECTRON CYCLOTRON EMISSION FROM THE PLASMA CENTER VS. TIME WITH ICRH AND LHCD APPLIED TO THE DISCHARGE

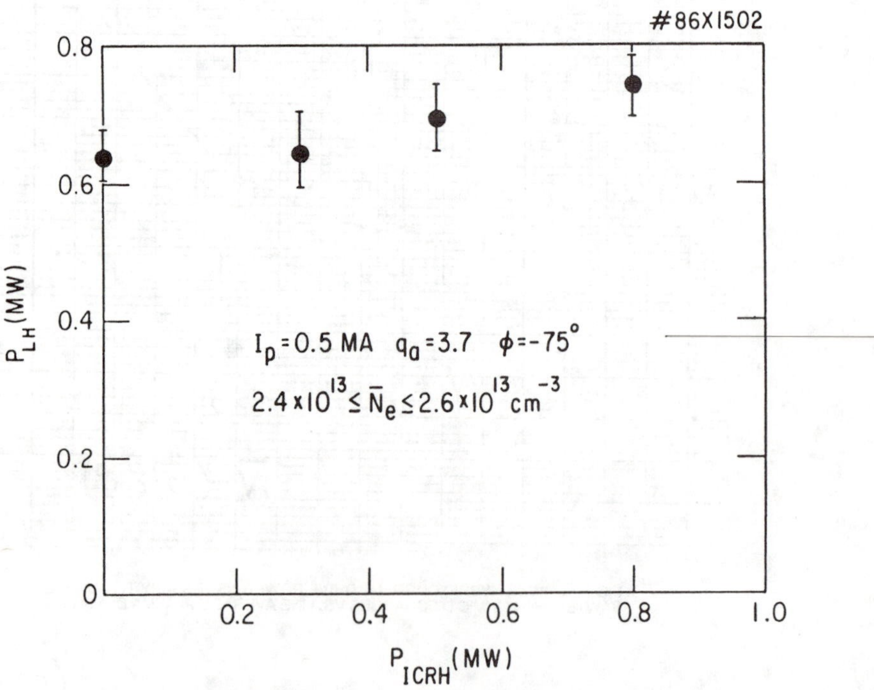

FIG. 3 LHCD POWER NECESSARY TO SUPPRESS ICH DRIVEN SAWTEETH VS. ICRH POWER. ALTHOUGH THE SAWTOOTH AMPLITUDE INCREASES BY A FACTOR OF TWO, THE LHCD POWER NEEDED TO ELIMINATE THE INTERNAL DISRUPTION REMAINS CONSTANT.

CURRENT DRIVE, ANTI-CURRENT DRIVE, AND BALANCED INJECTION.

S.von Goeler, J.Stevens, P.Beiersdorfer, R.Bell, S.Bernabei, M.Bitter, A. Cavallo, T.K.Chu, H.Fishman, K.Hill, R.Motley and R.Wilson-Princeton University-

I BIDIRECTIONAL SCENARIOS.

It is the purpose of this paper to point out (a) that severe limitations exist in ordinary lower hybrid (LH) current drive discharges in regard to the number of suprathermal electrons as well as to the kinetic energy stored in the energetic tail, and (b) that these limitations can be overcome in bidirectional scenarios like anticurrent drive, balanced injection, or strong current ramp up.

Let us first consider a limit on the number density of suprathermal electrons. It is probably safe to assume that the central safety factor q has a value close to 1. Then the central suprathermal current density j_{sth} and the number density of suprathermal current carriers n_{sth} is fixed:

(1) $q_{cyl} = 1 \rightarrow j_{sth} = e\, n_{sth}\, \overline{\beta}_z c \leq 2B_T / (\mu_o R) \rightarrow n_{sth} = (2\, B_T)/(e\mu_o\, \overline{\beta}_z c\, R)$,

where B_P and B_T are the poloidal and toroidal magnetic field, r and R the minor and the major radius of the tokomak, $\overline{\beta}_z = \overline{v}_z/c$ the average velocity of suprathermal electrons normalized to the velocity c of light, and μ_o the permeability of vacuum. For PLT, $n_{sth} \sim 10^{11} cm^{-3}$. In standard current drive discharges, the quantity $\overline{\beta}_z$ is close to 1 [3,4]. The suprathermals consequently represent less than 1% of the electrons in discharges with central plasma densities larger than 10^{13} cm^{-3}. If the mechanism for heating during LH current drive is that rf power is first transfered from waves to suprathermal electrons, and then from suprathermal to thermal electrons, this would impose a limitation on the heating power that can be applied by lower hybrid heating.

Consider now limitations in the stored energy. The stored energy in current drive discharges consists of three contributions: (a) kinetic energy E_{therm} of thermal particles, (b) kinetic energy E_{sth} of suprathermal electrons, and (c) magnetic energy associated with the plasma current I_p, $E_{magn}=(L_{int} I_p^2)/2$. In typical current drive discharges, the magnetic energy is the largest, and the suprathermal kinetic energy the smallest quantity.

(2) $$E_{magn} \quad > \quad E_{therm} \quad \gtrsim \quad E_{sth} \,.$$

The first half of Eq. 2 is valid, because the internal inductance l_i tends to be much larger than the poloidal beta, β_P. In order to see the validity of the second half of Eq. 2, we measure the suprathermal kinetic energy in form of a poloidal beta [2]:

(3) $$\beta_{Psth} = (2\mu_o / B_P^2(a))\ (1 / \pi a^2)\ {}_o\!\int^a (\overline{\gamma}\, \overline{\beta}^2\, m_o c^2\, n_{sth}\, 2\pi r)\, dr$$

Here, a is the limiter radius and $\overline{\gamma}$ the relativistic mass factor $(1-v^2/c^2)^{-1/2}$ averaged over all suprathermal electrons in a cubic centimeter. In typical current drive discharges, $\overline{\gamma}$ has a value <2.[4] In Eq.1, the number of suprathermal electrons is proportional to the suprathermal current density, and we can write Eq. 3 in a slightly different form:

(4) $$\beta_{Psth} \leq 2 < \overline{\gamma\beta^2} / \overline{\beta}_z > (\, I_A / I_p\,)$$

where the Alfven current is $I_A = (4\pi m_o c/\mu_o e) = 17kA$. The brackets <> represent suitable radial averages. Eq.4 says that β_{Psth} for unidirectional scenarios becomes of order one only for low plasma currents. In a machine like PLT with currents around 400kA, the suprathermal energy is always small compared to the magnetic energy, and it is comparable to the thermal energy only in discharges with very low density, where $\beta_{Ptherm} \ll 1$. If we apply more rf power than needed for keeping the current stationary, the result will be current ramp up, i.e. an increase of unwanted magnetic energy.

In order to enlarge the number of suprathermal electrons and their stored kinetic energy, we consider bidirectional scenarios where electrons are accelerated in both directions parallel and antiparallel to the magnetic field. These bidirectional scenarios include the following cases:

(a) Anticurrent Drive: Lower hybrid waves are launched in the direction opposite to the drift of the current carrying electrons by proper phasing of the LH grill. A tail of electrons is consequently created in this direction. In the direction of the electron drift, electron runaway takes place in the electric field resulting from the decay of the plasma current.[1] This scenario represents probably the most efficient way to create a large number of suprathermal electrons. It may not be the most desirable scenario for machine operation because of the very high energy electrons created in the runaway process.

(b) Balanced Injection: LH waves are launched in both directions parallel and antiparallel to the magnetic field. This scenario is most easily accomplished by phasing of the grill in the mode 0,π,0,π, etc.. On PLT, the phasing 0,0,π,π, etc. created a more favorable faster wave spectrum.

(c) For strong current ramp up, a sizable electric field exists because of L(dI/dt),and electron run away takes place in the reverse direction[1]. This scenario seems not efficient. However, it might have advantages because of the large electron heating associated with current drive [5].

II MEASUREMENTS OF THE SHAFRANOV PARAMETER.

It has been observed on PLT that very large increases of the modified Shafranov parameter $\tilde{\Lambda} = \beta_p + l_i/2$ occur during anticurrent drive, and it has been speculated that these increases are due to the suprathermal poloidal beta, although changes of l_i can not be ruled out. A systematic attempt on PLT to maximize these increases in $\tilde{\Lambda}$ - hoping to find $\tilde{\Lambda}$'s so large that it would be impossible to explain them by l_i - was not successful. However, the fact that we can not prove that the large $\tilde{\Lambda}$ values are due to β_{psth}, does not mean that there did not exist large β_{psth} values in the experiment. We therefore want to document these cases here.

In Fig.1 we show the time history of the plasma current, the density, and the cylindrical Shafranov parameter $\tilde{\Lambda}_{cyl}$ for a low density anticurrent drive discharge. [$\tilde{\Lambda}_{cyl} = [(4\pi R B_V/\mu_0 I_p) + (3/2) - \ln(8R/a)]$. The plasma current is raised to 220 kA with the OH transformer and then left to decay. The anticurrent drive should theoretically increase the decay of the current; in practice it inhibits the decay because of runaway electron formation (higher conductivity).The Shafranov parameter increases immediately after the rf is turned on and rises afterwards slowly to a value in excess of 2. The first rise is probably due to creation of suprathermal electrons, the second, slower rise to the decay of current. (According to Eq.3, $\beta_{psth} \sim I^{-2}$ for constant suprathermal content). It was further observed that the peak of the radial X-ray emission shifts outwards in time and it was suspected that this phenomenon is caused by an outward shift of the magnetic axis with respect to the outermost plasma surface due to high β_{psth}. However, PEST code simulations indicate that an outward slippage of the whole plasma column occurred due to an imperfection in the PLT coil system. The PEST code calculations for time t = 650 ms are shown in Fig. 2. The computations assumed that the axis coincided with the peak of the hard X-ray emission and that q=1 on axis. The calculations indicated that the measured values of $\tilde{\Lambda}_{cyl}$ were too large because of the outward shift. The true $\tilde{\Lambda}$ was 1.2, and was mostly due to $l_i/2 = .9$. Therefore it is still not fully certain that the large Shafranov parameters are caused by the suprathermal poloidal beta.

III HARD X-RAY MEASUREMENTS.

Besides the magnetic data, hard X-ray measurements indicate that the number of suprathermal electrons is larger in anticurrent drive and balanced injection discharges. The evidence is that the hard X-ray emission does not saturate with increasing rf power for anticurrent drive, whereas it saturates for current drive. Also, considerably more X-rays are emitted in high power anticurrent drive discharges than during current drive. Unfortunately, large uncertainties remain about the actual number of suprathermals in the discharge since there was no tangential hard X-ray detector available during the experiments to permit a full analysis of the distribution function of the energetic electrons [3,4].

Fig. 3 and 4 show relevant X-ray data for current drive and for anticurrent drive, respectively. The hard X-ray intensity from the central chord of the vertical hard X-ray array integrated over all photon energies is plotted versus rf power. In order to interpret the data,we discuss the nature of the X-ray emission.The slope of the hard X-ray emission falls roughly between 40 and 90 keV. For a comparison of the suprathermal emission, one consequently wants to sum over all photon energies down to ~15 keV. This is not possible for the current drive case, because appreciable heating occurs (T_e = 5 keV) and one would sample also the thermal spectrum. In Fig. 3 we consequently have used various filter foils. The case without a foil has a cut off of ~15 keV because of air absorption.The stainless steel foil cuts off at ~30 keV, the 1/8" Cu foil at ~120 keV. We believe that the stainless steel foil shows the right dependence on rf power, although the level is a little too low because it cuts off the low energy spectrum. The X-ray intensity for the current drive is initially larger than for anti current drive. For low rf power an electric field is still present, and field and waves work together to produce the maximum amount of suprathermal electrons. For large rf power the hard X-ray emission saturates for current drive. A small increase of the intensity is still observed for high powers; it may be due to electron runaways in the reverse direction during current ramp up, although $V \leq 0$ for $P_{rf} \geq$ 600keV in this case. The hard X-ray emissions for anticurrent drive and balanced injection behave similar and increase linearly with rf power.For high rf powers, the hard X-ray emission for anticurrent drive is more than twice the emission during current drive.

IV DISCUSSION.

We have provided evidence that the number of suprathermal electrons in bidirectional discharges is much larger than in normal current drive discharges for large rf powers. In bidirectional scenarios, we observe very large Shafranov parameters, and the hard X-ray emission is larger by a factor of 2 and does not saturate for high rf power. Admittedly, large uncertainties in the number of suprathermals exist. Hopefully this can be improved in experiments with more rf power. An important question is why we see less heating for bidirectional scenarios.[5,6] A possible explanation is that the electrons tend to be more energetic for anticurrent drive and that they leave the discharge instead of thermalizing in the plasma. The heating of the limiter on PLT seems to point in this direction. On the other hand, it seems to us that this is a very complex question because of the sawtooth stabilization that accompanies current drive, and because we do not know whether wave heating or improvement of confinement near the plasma center are crucial for causing the large temperature increases.

Acknowledgements:The support and many discussions with Dr. H.P.Furth and Dr. J. Hosea are gratefully acknowledged. We thank the PLT crew under Chris Mann, and J. Gorman, J. Lehner, and K.S.McGuire for excellent technical assistance.The work was supported by US DOE Contract No.DE-AC02-76-CHO-3073.

REFERENCES:

(1) J. Stevens et al.:Proc.5th Conf. RF Pl.Heating, Madison, p.164(1983).

(2)V. Mukhovatov, V.Shafranov: Nucl. Fusion 11, p. 605 (1971).

(3) S. von Goeler et al.: Nucl. Fusion 25, p. 1515 (1985).

(4) J. Stevens et al.: Nucl. Fusion 25, p. 1529 (1985).

(5) R. Bell et al.: Princeton report PPPL 2452 (1987)

(6) R. Bell et al.: Paper at this conference.

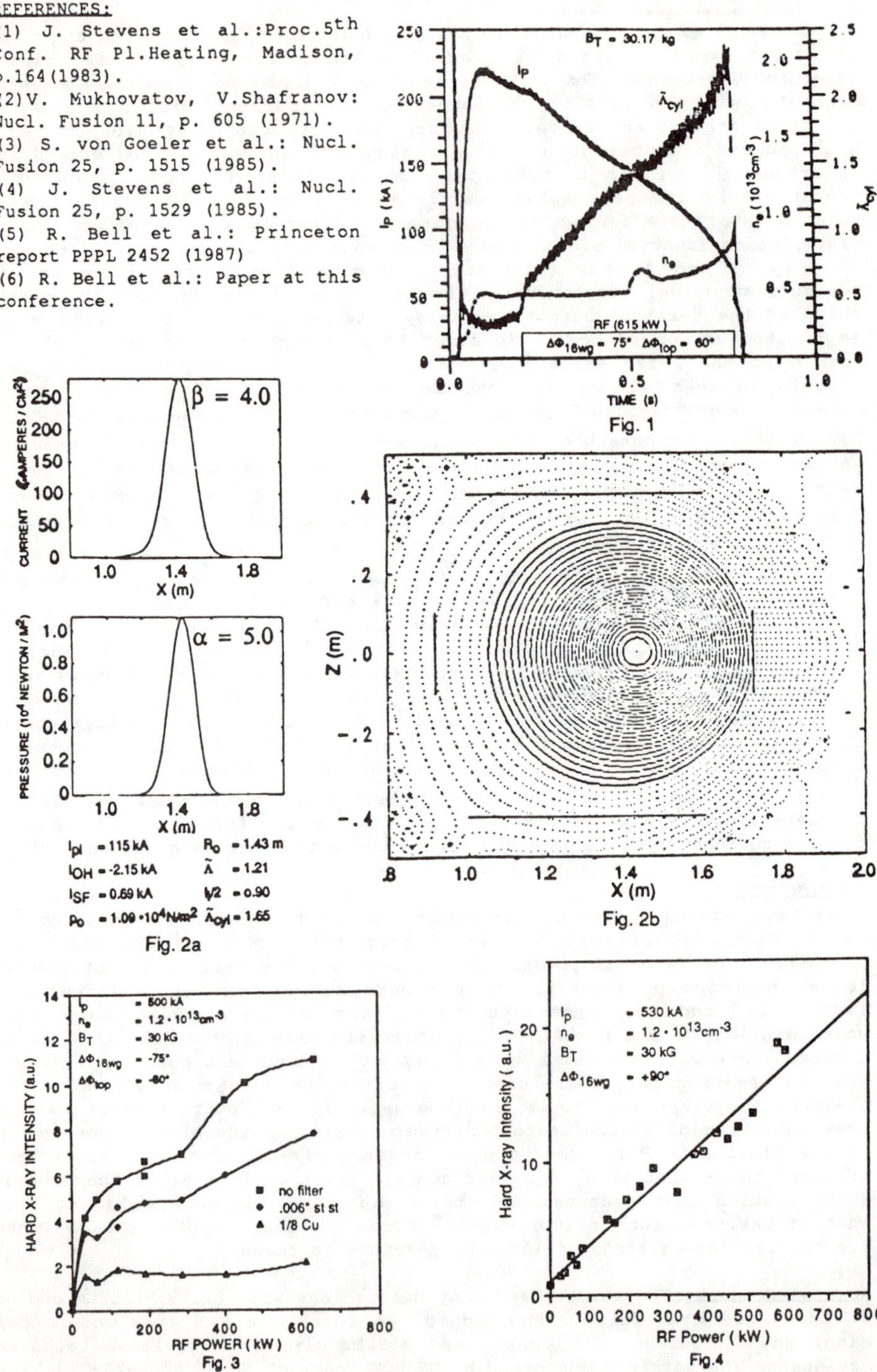

Fig. 1

Fig. 2a

Fig. 2b

Fig. 3

Fig.4

MODELING OF VERTICAL ECE DURING LOWER HYBRID CURRENT DRIVE ON PLT

T. C. Luce, P. C. Efthimion, N. J. Fisch, R. E. Bell, J. E. Stevens
Princeton University, Princeton, NJ 08544

ABSTRACT

Electron cyclotron emission of both polarizations has been measured on the PLT tokamak during 2.45 GHz lower hybrid current drive. The experimental configuration allows identification of radiation at any frequency with the kinetic energy of the emitting electron. Fitting functions and a theoretical model have been used in an attempt to reproduce the measured spectra.

EMISSION THEORY

All of the calculated spectra discussed in this paper are generated from single particle emission theory neglecting dielectric effects and absorption. While these effects could be important for quantitative comparisons to the experiment, they are not significant for the qualitative aspects of the emission spectra to be discussed here. The emission spectra are obtained by summing the emissivity for a single electron over a distribution f of electrons in momentum space. For emission perpendicular to the magnetic field direction this can be written for the two polarizations:

$$j^{(\mathrm{X})} = \frac{q^2\omega}{c}\sum_{n=1}^{\infty} p_n^3\,\mathrm{H}(p_n^2)\int_{-1}^{+1} dy\,(1-y^2)\,\mathrm{J}_n'^2\Big(\frac{\omega}{\omega_0}p_n\sqrt{1-y^2}\Big)\,f(p_n,y) \qquad (1)$$

$$j^{(\mathrm{O})} = \frac{q^2\omega}{c}\sum_{n=1}^{\infty} p_n^3\,\mathrm{H}(p_n^2)\int_{-1}^{+1} dy\,y^2\,\mathrm{J}_n^2\Big(\frac{\omega}{\omega_0}p_n\sqrt{1-y^2}\Big)\,f(p_n,y) \qquad (2)$$

where p_n^2 is the resonant momentum normalized to m_0c, y is the cosine of the momentum pitch angle, ω_0 is the rest mass cyclotron frequency, n is the harmonic number, and J_n is the Bessel function of the first kind of order n. The Heaviside function appears because there is no upshifted harmonic emission. The measurements take advantage of the electron cyclotron resonance condition by looking perpendicular to the magnetic field and along a constant field surface. While losing direct information on the direction of the parallel momentum, a correspondence between particle energy and radiated frequency is obtained through the resonance condition:

$$\omega = \frac{n\omega_0}{\gamma} \qquad (3)$$

where γ is the standard relativistic factor. The plasma is assumed to be optically thin to the radiation above the upper hybrid resonance and a viewing dump is employed to eliminate spurious radiation from wall reflections.

EXPERIMENTAL ARRANGEMENT

The millimeter-wave receiver is a swept heterodyne radiometer with an antenna system designed to view either the X or O mode polarization. The instrument sweeps in frequency between the upper hybrid resonance and the second harmonic once every five milliseconds. An absolute calibration of the antenna and receiver was performed with a liquid nitrogen (77 K) blackbody.

The data discussed here are from a series of shots where the current of 320 kA was held flat by 330 kW of lower hybrid power. The line-average density was $n_e = 1.0 \times 10^{13}$ cm^{-3}.

MODELING OF THE EMISSION

To describe the emission spectrum of this steady state, various functional forms have been used. The x-ray bremsstrahlung measurements previously made on PLT were compared with a two-temperature function:[1]

$$f = C_N \begin{cases} e^{-p_\perp^2/2T_\perp} e^{-p_\parallel^2/2T_{\parallel f}} & 0 \leq p_\parallel \leq p_{\parallel max} \\ e^{-p_\perp^2/2T_\perp} e^{-p_\parallel^2/2T_{\parallel b}} & 0 \geq p_\parallel \end{cases} \tag{4}$$

For this analysis, $T_{\parallel b}$ is set equal to $T_\perp$ since the experimental arrangement precludes obtaining direct information on the forward-backward asymmetry. The best fits to the x-ray data usually had this form so the assumption should not be too restrictive. The ECE measurements made on Alcator C were compared to a function with five parameters:[2]

$$f = f_0 e^{-E/T + C/E} e^{Ap^B \cos^2\theta} \tag{5}$$

For this work, C is set equal to zero because the third harmonic emission covers the region where C should be important. A functional form for modeling emissions during lower hybrid current drive was proposed by Giruzzi *et al.*:[3]

$$f = C_N e^{-p_\perp^2/T_\perp} \qquad P_1 \leq p_\parallel \leq P_2 \tag{6}$$

This simply places all of the electrons in the RF created plateau.

DISCUSSION

Only the distribution function proposed by Giruzzi *et al.* reproduces the qualitative characteristics seen in the data—a distinct peak for both the second and third harmonic emission (see Fig. 1). The O mode is shown since the absorption and dielectric effects should be less important. This spectrum was generated by taking $P_1 = 0.7\, m_0 c$, $P_2 = 1.0\, m_0 c$, $T_\perp = 60$ keV, and $n_s = 6.0 \times 10^{11}$ cm^{-3}. Both the density and lower bound of the plateau seem too large, and the calculated current driven is 1 MA even with an empirical correction for backward going particles.

Taking $T_\parallel = 900$ keV, $T_\perp = 150$ keV, $p_{\parallel max} = 1.0\, m_0 c$, and $n_s = 1.0 \times 10^{12}$ cm^{-3} in Eqn. 4, the relative heights of the second and third harmonics are reproduced but with no intervening minimum. This also produces a calculated plasma current of 500 kA. However, the large superthermal density—10% of the bulk density—seems unphysical.

For Eqn. 5 a "best fit" was obtained by looking at the X/0 ratios. This should minimize any errors from calibration, although it sacrifices sensitivity to the fitting parameters. For parameters $A = 7.5$, $B = 0.0$, $T = 100$ keV, and $f_0 = 1.5 \times 10^{10}$, the ratios look reasonable, but the individual harmonics do not fit at all. The calculated $T_\perp$ and $T_\parallel$ are reasonable (70 keV and 400 keV respectively), but the density and current are very large (2.0×10^{12} cm^{-3} and 1.7 MA).

From the previous discussion, it is not clear whether the unsatisfactory fits arise from the constraints imposed by the fitting functions or from poor data. To model experimental data, some convenient fitting function is employed. A fit to the data is not guaranteed, and the lack of fit does not carry any physical implication. On the other hand, if an inversion can be made which shows the data to be physically reasonable, the data can then be compared with a theoretical model. The quality of the fit then shows how well this model describes the experiment.

A simple inversion has been found which indicates the data are physically reasonable. Assume the emission at any frequency is from electrons at only one point in momentum space. The frequency determines the energy of the particles, and having both polarizations gives the pitch angle and density. The results are shown in Fig. 2. Notice that the density of particles needed to model the emission falls off with energy, and the scatter in pitch angle is small. Also, the points all lie in the region of the launched spectrum.

The time-asymptotic model of Fisch and Karney[4] for the current drive distribution has been compared to the data. Although an exhaustive study has not been completed, it appears that the model does not generate spectra which exhibit the two characteristic peaks. Two effects not included in the model may account for this. Relativistic bending of the resonance region would add more high energy particles to the distribution. This would increase the emission from the third harmonic. Particle confinement losses could alter significantly the steady-state distribution. For particles in the 100–400 keV range in PLT, the 90° collision times and the confinement times are of the same magnitude, while the model assumes perfect confinement. Refinement of the inversion scheme may give further insight into the physics necessary to model the experiment.

ACKNOWLEDGEMENTS

One of the authors (TCL) would like to acknowledge the help of Bob Cutler, Mike McCarthy, and the PLT technical staff in the construction and installation of the experimental apparatus. All work performed under U.S. DoE contract no. DE-AC02-76-CHO-3073.

1. S. von Goeler, J. Stevens, *et al.*, Rev. Sci. Instrum. **57**, 2130 (1986).
2. I. H. Hutchinson, K. Kato, S. C. Texter, Rev. Sci. Instrum. **57**, 1951 (1986).
3. G. Giruzzi, I. Fidone, G. Granata, R. L. Mayer, Phys. Fluids **27**, 1704 (1984).
4. N. J. Fisch, C. F. F. Karney, Phys. Fluids **28**, 3107 (1985).

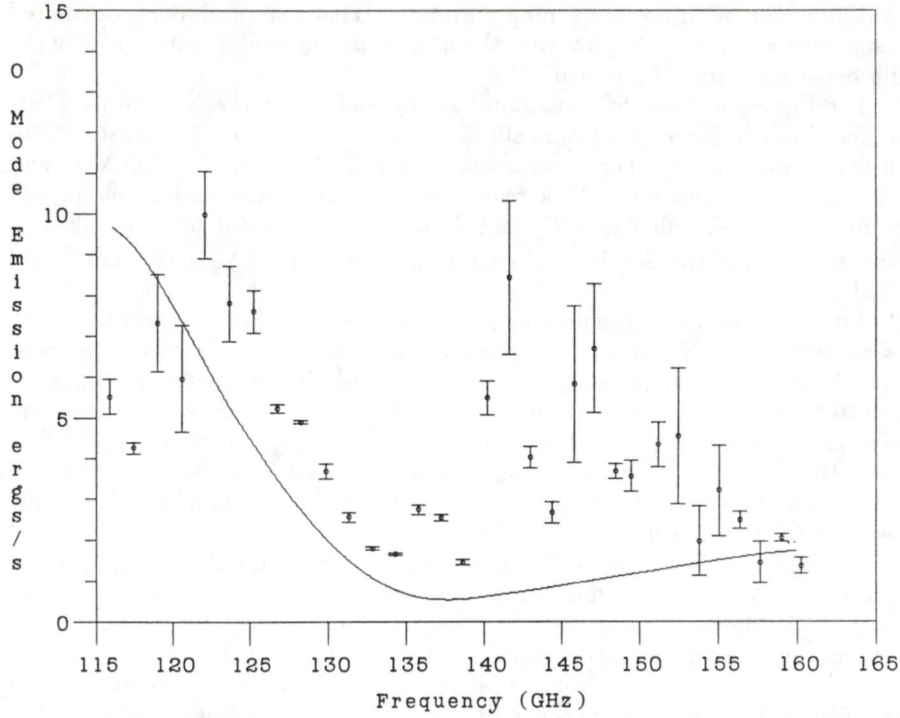

Fig. 1. The solid curve is Eqn. 3 for the parameters given in the text. The points are calibrated O mode emission from one frequency sweep during the steady state. The error bars represent statistical error from the calibration, not a complete error analysis.

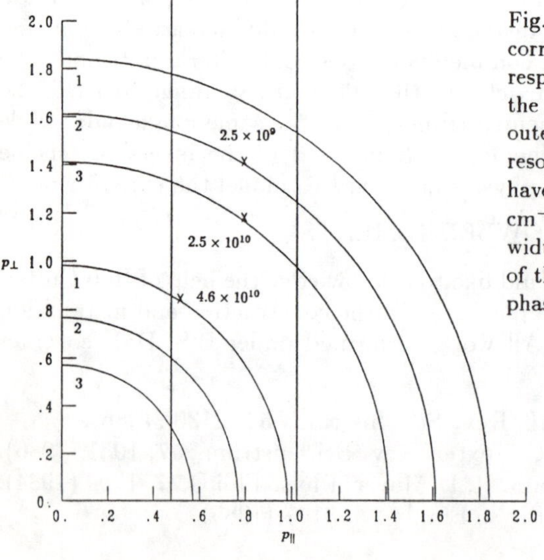

Fig. 2. The curves labeled 1, 2, and 3 correspond to ω = 122, 135, and 148 GHz respectively. The inner three curves are the second harmonic resonances and the outer three curves are the third harmonic resonances for these frequencies. The axes have units of m_0c, and the density is in cm^{-3}. The vertical lines represent the full width at half maximum of the main peak of the launched wave spectrum for $-90°$ phasing.

FAST WAVE DISPERSION, DAMPING, AND ELECTRON CURRENT DRIVE IN THE IRVINE TORUS

R. C. Platt and R. McWilliams
University of California, Irvine, CA 92717

ABSTRACT

Fast waves with frequencies near the mean gyrofrequency were excited in a toroidal magnetized plasma. Experimental measurements of wave dispersion were found to be in good agreement with predictions from cold plasma theory. Experimental measurements of wave damping lengths have been made. Measured damping lengths were found to be anomalously short when compared to predictions for electron Landau-damping, transit-time magnetic pumping and collisional damping. Unidirectional fast waves drove steady-state electron currents. Peak efficiencies up to $\eta = InR/P = 6 \times 10^{-2}$ A/W(10^{13} cm^{-3})m were observed with up to 14% of the wave energy converted to poloidal magnetic field energy.

Considerable interest has developed in the possibility of using fast waves with frequencies in the regime $\omega_{ci} \ll \omega \ll \omega_{ce}$ to drive toroidal electron currents. Current drive by fast wave toroidal eigenmode excitation[1] and unidirectional waves launched from an antenna designed to excite fast waves in a tokamak[2] has been reported. This paper presents results of a series of experiments that investigated fast wave dispersion[3], current drive by unidirectional, phased-array-launched fast waves[4], and fast wave damping.

The experiments were conducted on the Irvine Torus (see Figure 1). This device has a major radius of R = 55.6 cm, minor radius limiter set at a = 3.7 cm and can produce steady-state toroidal magnetic fields of $B_\phi \leq 2.5$ kG. There is no ohmic pulse and hence no toroidal dc electric field. The plasma was formed by impact ionization of argon by energetic electron beams thermionically emitted from filaments in the vacuum vessel. The beams circulate in both directions around the torus so there is no net current in the plasma without introduction of the fast waves.

The fast waves were excited by a sixteen-element phased array antenna[4] designed to produce primarily the E_θ required for fast wave excitation while allowing control of the toroidal wavelength as well. Each antenna element could be phased independently, and with $\pi/2$ phase shifts between adjacent elements the principal vacuum indices of refraction excited were $n_\theta = 13.6$ and $n_z = 10.0$ at a frequency of 100 MHz. Appropriate phasing allowed the wave to be launched in either toroidal direction, or both directions at once, and in either poloidal direction.

Experimental measurements of the fast wave energy trajectories, radial wavelengths and phase velocities with varying wave frequency, plasma density, and magnetic field have been made in the Irvine Torus. The results were in good agreement with predictions from cold plasma theory over the parameter range investigated. For instance, the theory, with approximations appropriate for the parameter range of the experiment, predicts a radial wavelength

$\lambda_r = \frac{2\pi c^2 k_z \omega_{ce}}{\omega\omega_{pe}^2}\left[1 - \frac{2\omega_{pe}^2}{n_z^2 \omega_{ce}^2}\right]^{1/2}$. Figure 2 shows radial wavelengths measured from radial interferograms as a function of wave frequency. The solid line is calculated from the preceeding equation for the principal vacuum wave number excited by the antenna. Similar agreement between theory and experiment was found for the wave energy trajectory and radial phase velocity.

Measured inverse toroidal damping lengths, k_{zi}, were anomalously large when compared to theoretical predictions for collisional damping, electron Landau damping (ELD) and transit-time magnetic pumping (TTMP). Cold plasma theory with the inclusion of collisions predicts

$k_{zi}(\text{collisions}) \simeq \frac{\omega_{pe}^2 \nu_{ei} k_z}{n_z^2 \omega\omega_{ce}^2}$ where ν_{ei}, the electron ion collision frequency, is the largest collision frequency in the plasmas used. The hot electron dielectric tensor must be used to calculate k_{zi} for ELD and TTMP. Calculation of the tensor elements requires a specific form for the electron distribution function. The electron beams used to produce the plasma form a high energy tail on the distribution function[5] so a simple model for $f(\underset{\sim}{v})$ might have two populations: a bulk population with density n_{eo} and temperature T_{eo} and a tail population with density n_{el} and temperature T_{el}. Bulk electron temperatures were $T_{eo} \lesssim 10$ eV while the electron beams were injected with energies typically greater than 300 eV. Since the toroidal wave phase velocity was $\omega/k_z \approx 3 \times 10^9$ cm/sec, damping by ELD and TTMP was expected to occur primarily on the beam electrons. The theory then gives

$$k_{zi}(\text{ELD}) \simeq \sqrt{\pi}\,\frac{n_{el}}{n_o}\,\frac{\omega^2\omega_{pe}^2}{c^2 k_z \omega_{ce}^2}\,\alpha^3 e^{-\alpha^2} \quad \text{and}$$

$$k_{zi}(\text{TTMP}) \simeq \sqrt{\pi}\,\frac{n_{el}}{n_o}\,\frac{(n_z^2 - \varepsilon_\perp)\,\omega\omega_{pe}^2}{2n_z^2 c^2 \omega_{ce}^2}\,(2T_{el}/m_e)^{1/2} e^{-\alpha^2}$$

$\alpha \equiv (m_e/2T_{el})^{1/2}\omega/k_z$. Nonlinear effects have been neglected since the fast wave power was under one watt for the wave damping measurements.

Figure 3 shows k_{zi} measured versus wave frequency. The results show k_{zi} decreasing with increasing wave frequency. This behavior is expected for resonant damping and is contrary to expectation for collisional damping. The measured values are, however, three to four orders of magnitude larger than those calculated from the preceeding equations for any of the three mechanisms. The anomalously large measured values of k_{zi} may be due to scattering of the fast waves by drift wave density fluctuations. Substantial wave amplitude was found in the frequency regime in which drift waves were expected to occur, i.e. low frequencies, and the measured strength of the associated density fluctuations was $\langle(\delta n/n)^2\rangle^{1/2} \simeq 6 \times 10^{-2}$. The fluctuation level was approximately constant across the plasma radius.

Antani and Kaup[6], and Andrews and Bhadra[7] have investigated the effects of density fluctuations on fast wave damping in tokamak plasmas. Although the model of reference 6 considers only the scattering of coherent fast waves into an incoherent, turbulent form, when their results are applied to this experiment, rough agreement was found. Reference 7 considers the possibility of broadening of the initial n_z spectrum excited by the antenna with resultant enhancement of resonant damping. This result may have bearing on the observed current drive by fast waves described later in this paper where observations suggest resonant damping to be the primary damping mechanism. Mode conversion of fast into slow waves stimulated by density fluctuations [8] may also play a role in the observed damping lengths. While the parameter regime of this experiment makes quantitative application of these theories uncertain, the importance of drift wave density fluctuations on the damping of fast waves is certainly indicated.

Unidirectional fast waves were observed to drive steady-state toroidal electron current in the Irvine Torus. The direction of the wave-induced electron flow was in the direction of the imposed axial phase velocity of the wave. The current was slightly, but not significantly, larger for waves launched in one toroidal direction compared to the other.

It was observed during this experiment that as the electron beam injection energy was lowered below about 300 eV, current generation ceased. This energy corresponds to that at which $\alpha = (m_e/2T_{el})^{1/2}\omega/k_z = 2.8$. It is expected that resonant damping is effective only for $\alpha \lesssim 3$, so the loss of current drive at beam energies below 300 eV is consistent with resonant wave damping.

Figure 4 shows the dependence of wave-driven current on wave power. The maximum current of 1.34 A was produced by 20.5 W of power coupled into the wave. At this maximum current, 14% of the wave energy was converted to poloidal magnetic field energy. From the graph, the central data points yield a current drive efficiency of $\eta = InR/P \simeq 6 \times 10^{-2}$ A/W (10^{13}cm^{-3})m. This value of η compares very favorably with figures for slow wave current drive in similar experiments.

In summary, this paper has described a series of experiments investigating fast wave dispersion, damping, and current drive. The dispersion and current drive results showed that fast wave propagation in the Irvine Torus was well described by cold plasma theory and that unidirectional fast waves can drive electron currents with relatively high efficiency. The experimentally measured damping lengths were anomalously short when compared to predictions for electron Landau damping, transit-time magnetic pumping, and collisional damping. It is suggested that scattering of fast waves by drift wave density fluctuations may be responsible for the observed damping lengths.

This work supported by DOE Grant #DE-FG03-86ER53231.

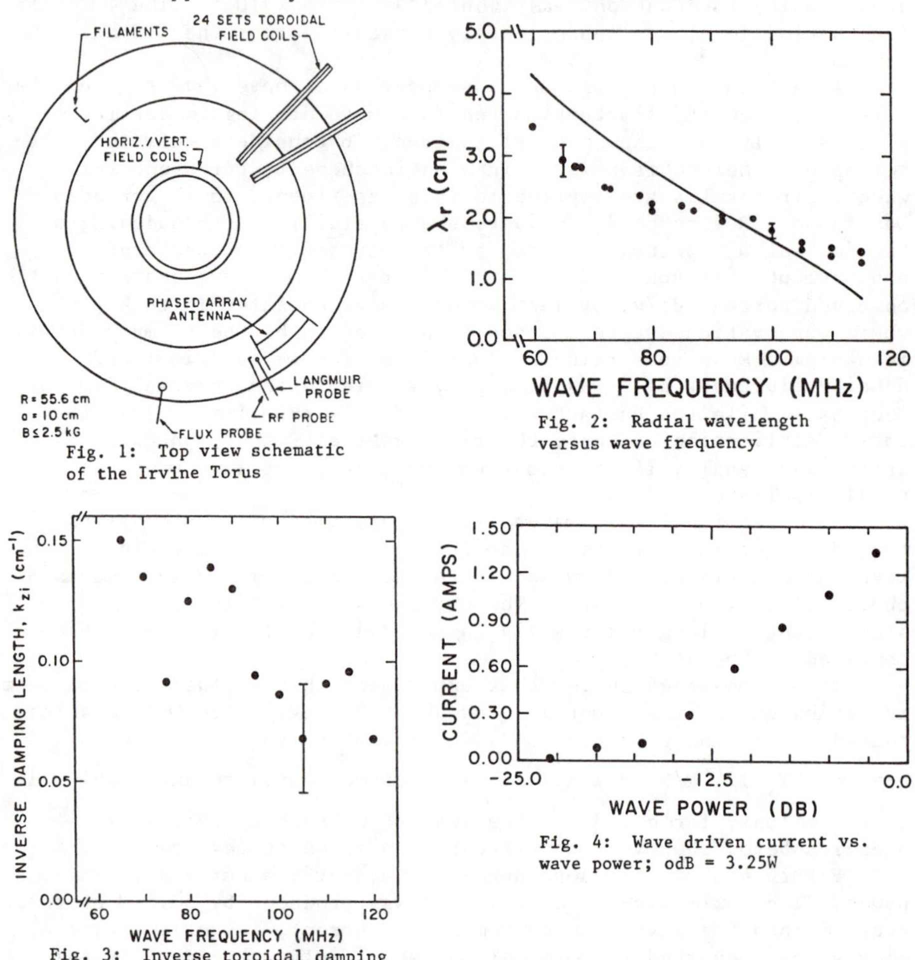

Fig. 1: Top view schematic of the Irvine Torus

Fig. 2: Radial wavelength versus wave frequency

Fig. 3: Inverse toroidal damping length versus wave frequency

Fig. 4: Wave driven current vs. wave power; odB = 3.25W

1. J. Goree, M. Ono, P. Colestock, et al., Phys. Rev. Lett. 55, 1669 (1985).
2. K. Ohkubo, Y. Hamada, Y. Obana, et al., Phys Rev. Lett 56, 2040 (1986).
3. R. C. Platt and R. McWilliams, Phys. Rev. Lett. 57, 2276 (1986).
4. R. McWilliams and R. C. Platt, Phys. Rev. Lett. 56, 835 (1986).
5. H. Okuda, R. Horton, M. Ono, et al., Phys. Fluids 28, 3365, (1985).
6. S. W. Antani and D. J. Kaup, Phys. Fluids 27, 1169 (1984).
7. P. L. Andrews, D. K. Bhadra, Nuc. Fusion 26, 897 (1986).
8. P. L. Andrews, Phys. Rev. Lett. 54, 2022 (1985).

800 MHZ FAST WAVE CURRENT DRIVE EXPERIMENTS IN PLT

R. I. Pinsker, P. L. Colestock, S. Bernabei,
A. Cavallo, G. J. Greene, R. Kaita, and J. E. Stevens

Princeton University, Princeton NJ 08544

ABSTRACT

We present results from a medium power ($P_{RF} \lesssim 200$ kW) fast wave current drive experiment at 800 MHz. Two couplers were used: a phased array of six unshielded loop antennas, and a 4 × 3 array of dielectric-loaded waveguides. Coupling measurements with the waveguide array showed dependence on phasing and $\bar{n}_e$ qualitatively in agreement with theoretical expectations of fast wave loading. With either coupler, rf-induced effects on fast electrons, such as current drive and enhanced ECE, exhibit the same density limit $\bar{n}_{crit} \simeq 1.0 \times 10^{13}$ cm^{-3} as has been observed in previous 800 MHz slow wave experiments on PLT.

INTRODUCTION

Lower hybrid waves have been found to have several important applications to tokamaks: steady-state current drive with an efficiency 70% of the theoretical value[1], start-up and ramp-up, electron heating[2], and control of MHD fluctuations[3]. However, the well-known density limit for interaction with fast electrons introduces a constraint on the use of lower hybrid waves in dense tokamak plasmas, such as are characteristic of current reactor designs. This motivates the experimental study of current drive with plasma waves other than the lower hybrid wave, which might retain the benefits of LHCD while having a higher density limit, if any. In the present work, we attempt to launch the other propagating branch of the cold plasma dispersion relation in the range of frequencies $\Omega_{ci} \ll \omega \ll \Omega_{ce}$, which is referred to as the fast wave. Previously reported tokamak experiments have been carried out on JIPP T-IIU[4] and JFT-2M[5]. Neither experiment has unequivocally demonstrated current drive above the density limit for slow waves. The experiments reported here are the first to employ a phased array of dielectric-loaded waveguides in a tokamak.

APPARATUS

The two couplers used in these experiments were an array of six unshielded resonant loops (each conductor 3.2 cm wide, and 11.4 cm of the conductor exposed to the plasma) oriented so that the rf current flows in the poloidal direction in order to couple to the fast wave, and an array of twelve dielectric-filled ($\epsilon = 8.8$) rectangular waveguides each 8.6 cm in the toroidal direction by 5.4 cm in the poloidal direction arranged in four columns of three. Fig. 1 shows the two couplers as installed in an outside midplane port on PLT. The diagnostics used to measure the effects of the rf on the plasma included $2\Omega_{ce}$ emission, perpendicular charge exchange and a high frequency rf probe in the limiter shadow to detect parametric decay. Several types of receiving antennas were used to detect relatively low frequency ($\sim n\Omega_{ci}$) emissions from the plasma.

RESULTS

Coupling measurements with the six loop array showed strong plasma loading, but the loading for the entire array did not depend on the array phasing. This fact, along with the dependence on forward power level of the optimal tuning stub positions for matching to the vacuum, suggests that an rf produced glow existed within the antenna structure, modifying the phase at the plasma. This glow did not prevent the coupling of significant power to the plasma, as is shown below. The coupling measurements made with the waveguide array show dependence on phase in qualitative agreement with theory: since the edge density and edge density gradients required for low reflection coefficients are much larger for the fast wave than for the slow wave, the curve of the reflection coefficient as a function of toroidal phase angle has the same shape in the plasma case as it does in the vacuum. The net reflection coefficient for the array as a function of toroidal phase angle is shown in Fig. 2. A series of phase scans at different line average densities was performed with a pair of waveguides, with the results shown in Fig. 3. As is clear from Fig. 3, raising the density at a fixed phase angle lowers the net reflection coefficient, while even at high densities the reflection is minimized at zero phase. The net reflection coefficient for the entire array as a function of line average density with a fixed toroidal phase angle of 180° is compared with a similar density scan with the 'wide' 800 MHz slow wave grill[1] in Fig. 4. We may conclude from these scans that the waveguide array indeed excited the fast wave in the plasma.

However, with either coupler, the effect of the rf on fast electrons, as observed by changes in $2\Omega_{ce}$ emission and at high rf power by changes in plasma current and loop voltage, had the same density limit as the slow wave at 800 MHz. Experiments with the loop array with a coupled power of 30 kW showed that the rf-induced enhancement of ECE disappeared between $\bar{n}_e = 7.0 \times 10^{12}$ cm^{-3} and $\bar{n}_e = 9.4 \times 10^{12}$ cm^{-3}; at the latter density, a sharp increase in neutron emission from the deuterium plasma, parametric decay spectra and the formation of a fast ion tail were observed. The rapid rise and fall of the fast ion signal on charge exchange and the frequencies of the parametric decay sidebands imply that the fast ion tail is formed only at the plasma surface near the coupler. At lower density and higher power ($\bar{n}_e = 2 \times 10^{12}$ cm^{-3}, $P_{net} \simeq 130$ kW) approximately 150 kA of rf-driven current was observed using the loop array, but the efficiency of the current drive did not depend significantly on the array phasing.

Though the waveguide array's coupling characteristics were much more easily understood in terms of fast wave loading than those of the loop array, and the waveguide array's power handling capabilities were much superior (maximum $P_{net} \simeq 340$ kW), the two antennas produced similar effects on the plasma. We performed a density scan in which the ohmic drive was shut off after plasma formation with a fixed toroidal phase angle of 180° on the waveguide array, and compared $\Delta \dot{I}_p \equiv \dot{I}_p|_{\mathrm{rf}} - \dot{I}_p|_{\mathrm{no\ rf}}$ with data from a similar scan performed with the 'wide' 800 MHz slow wave grill phased at 90°. While the coupling properties of the slow and fast wave arrays were quite different, as shown in Fig. 4, the current drive efficiencies were virtually indistinguishable (Fig. 5). The density limit for the current drive effect was also the same — in both experiments no significant current drive was seen above $\bar{n}_{crit} \simeq 9 \times 10^{12}$ cm^{-3}. Fig. 6 shows the broadening of the 800 MHz pump observed during this scan with an rf probe in the limiter shadow (received power integrated over a 60 MHz band centered on 800 MHz). The sudden broadening apparent above $\bar{n}_{crit}$, accompanied by the appearance of ion cyclotron quasimodes, was also seen in the slow wave experiments[6]. Current drive effects show a sharp density limit coinciding with the onset of parametric decay and the fast ion tail, but the enhancement of ECE has a much softer density limit. The rf-induced enhancement of ECE at 142 GHz and the enhancement of neutron emission are shown as a function of $\bar{n}_e$ in Fig. 7. These data were taken in a scan where I_p was held constant at 500 kA with the ohmic transformer.

An experiment was performed in which the 16-waveguide 2.45 GHz slow wave system[1] was used to create a fast electron population at $\bar{n}_e = 1.5 \times 10^{13}$ cm^{-3}, then the 800 MHz 4 × 3 waveguide array was turned on with $P_{net} \sim 150$ kW to search for coupling between the 800 MHz rf with fast electrons produced by the 2.45 GHz system. No incremental changes in the plasma current or loop voltage were produced by the 800 MHz rf, demonstrating that our current drive density limit is not due to the lack of a fast electron tail on which to damp the waves, as has already been proven for slow wave systems.

CONCLUSIONS

The dependence of reflection coefficient on plasma parameters and on phasing observed with the 4 × 3 dielectric-filled waveguide array is consistent with theoretical expectations of fast wave loading. However, all the effects of the rf on the tokamak plasma are similar to those seen in previous experiments in the same device under similar plasma conditions in which slow waves were launched at the same frequency. The fact that similar phenomena in the neighborhood of $\bar{n}_{crit}$ are observed regardless of the character of the launched waves has obvious implications for any theory attempting to explain the density limit in terms of parametric decay or stochastic ion heating, unless mode conversion processes play an important role in both slow and fast wave experiments.

REFERENCES

1. J. E. Stevens, *et al.*, Princeton Plasma Physics Laboratory Report PPPL-2426, 1987, submitted for publication.
2. R. E. Bell, *et al.*, presented at this meeting.
3. K. McCormick, F. X. Söldner, D. Eckhartt, F. Leuterer, H. Murmann, *et al.*, Phys. Rev. Lett. **58**, 491 (1987).
4. K. Ohkubo, Y. Hamada, Y. Ogawa, *et al.*, Phys. Rev. Lett. **56**, 2040 (1986).
5. T. Yamamoto, A. Funahashi, *et al.*, presented at 11th Int. Conf. Plasma Phys. and Cont. Nuc. Fusion Res., Kyoto, Nov. 1986, paper IAEA-CN-47/F-II-5.
6. F. Jobes, S. Bernabei, *et al.*, Physica Scripta **T2/2**, 418 (1982).

Fig. 1a Six-loop 800 MHz fast wave coupler in PLT

Fig. 1b 4 × 3 dielectric-filled waveguide array in PLT

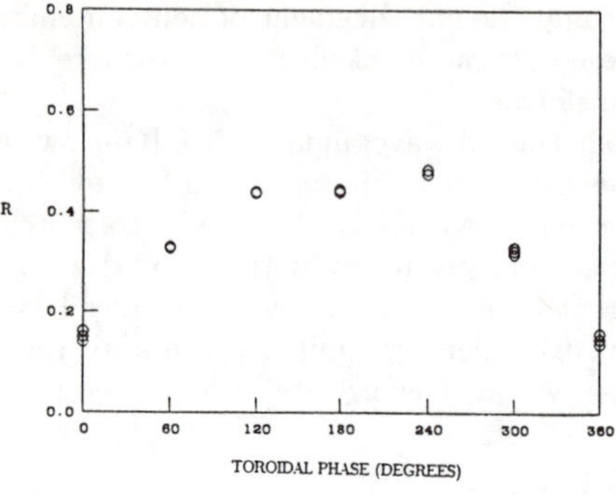

Fig. 2 Net reflection coefficient for 4 × 3 array vs. toroidal phase angle. $B_T = 31.3$ kG, $\bar{n}_e = 1.5 \times 10^{13}$ cm^{-3}.

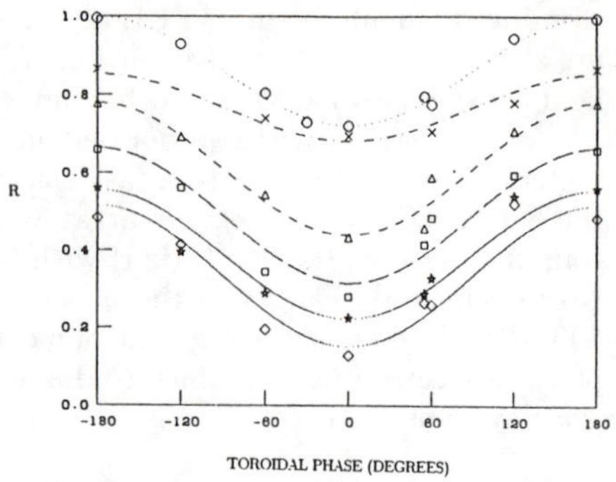

Fig. 3 Net power reflection for 2-waveguide subset of the 4 × 3 array vs. toroidal phase angle at six different densities. The curves are fits of the form $a + b\cos\phi$.
○ = vacuum × = 0.6×10^{13} cm^{-3} △ = 1.1×10^{13} cm^{-3}
□ = 1.5×10^{13} cm^{-3} ⋆ = 2.0×10^{13} cm^{-3} ◇ = 2.6×10^{13} cm^{-3}

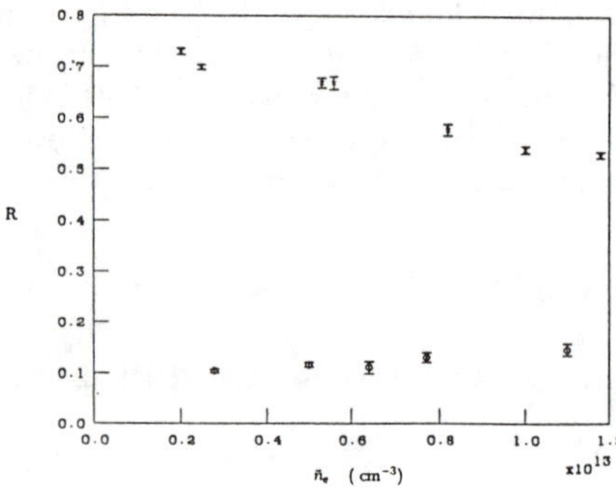

Fig. 4 Net power reflection for 4 × 3 fast wave array (crosses) and for 6 × 1 slow wave array (circles) vs. $\bar{n}_e$. Toroidal phase of 180° for fast wave array, 90° for slow wave array.

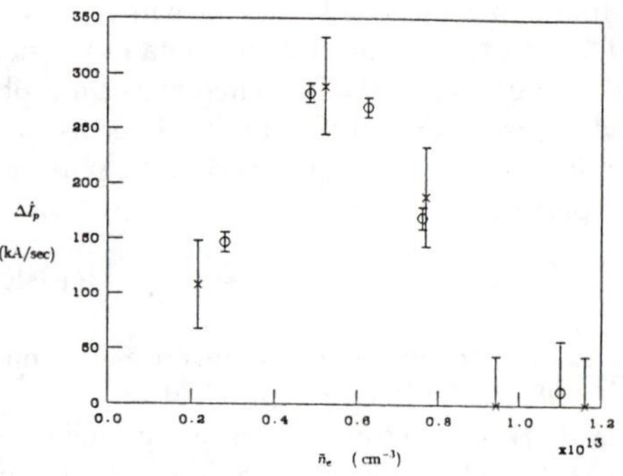

Fig. 5 Comparison of $\Delta \dot{I}_p$, normalized to net power of 160 kW, vs. $\bar{n}_e$. Slow wave array—circles, fast wave array—crosses.

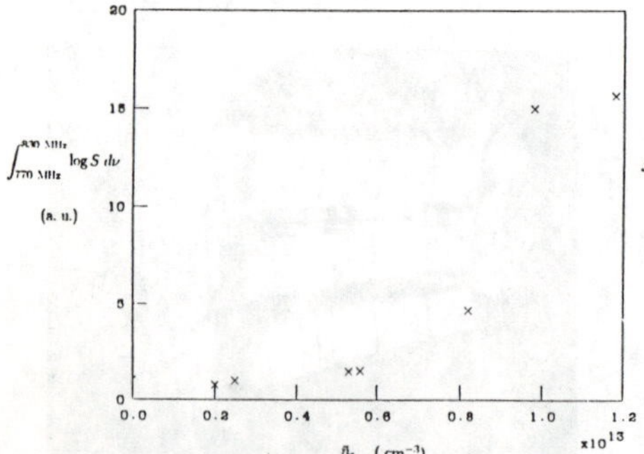

Fig. 6 Broadening of 800 MHz pump observed with rf probe, from the same $\bar{n}_e$ scan as Figs. 4 and 5. Data only from fast wave array.

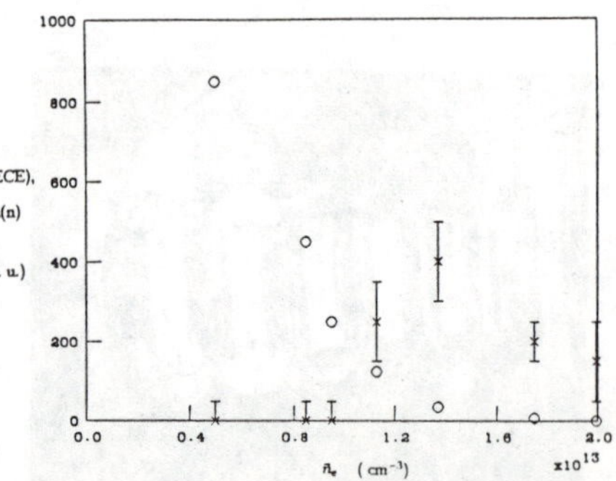

Fig. 7 Rf enhancement of ECE at 142 GHz (circles) and of neutron emission (crosses) vs. $\bar{n}_e$. $B_T = 31.3$ kG, $I_p = 500$ kA.

FAST WAVE CURRENT DRIVE EXPERIMENT ON JFT-2M

Y.Uesugi, T.Yamamoto, K.Hoshino, H.Ohtsuka, H.Kawashima, M.Hasegawa*, S.Kasai, T.Kawakami, T.Matoba, T.Matsuda, H.Matsumoto, M.Miura, M.Mori, K.Odajima, H.Ogawa, T.Ogawa, S.Sengoku, T.Shoji, N.Suzuki, T.Tamai, T.Yamauchi, K.Hasegawa, A.Honda, I.Ishibori, Y.Kashiwa, M.Kazawa, K.Kikuchi, Y.Matsuzaki, K.Ohuchi, H.Okano, E.Sato, T.Shibata, T.Shibuya, T.Shiina, T.Tani, K.Yokoyama

Japan Atomic Energy Research Institute,
Naka-Machi, Naka-Gun, Ibaraki, 319-11, Japan

ABSTRACT

Current drive experiment by 200 MHz fast waves was carried out in JFT-2M tokamak in order to investigate a capability of the fast wave current drive in hot and dense plasma. Two phased loop antennas were employed to excite fast waves, which had a loading resistance of 5 ohms. A plasma current of 30 kA was generated by an rf power of 40 kW at a density of $2x10^{12}$ cm^{-3}. It is probable that the slow waves contribute to current drive at the low density below the LHW current drive density limit. The efficiency as high as that of slow LHW at a low density region decreases rapidly with the density and no rf current was generated above $3x10^{12}$ cm^{-3}. This density limit is two times higher than that expected from LHW current drive scaling at the same frequency. It seems that the occurrence of the density limit relates with the accessibility of the launched waves to the high density region and a lack of rf power. When the same loop antennas were employed to excite fast waves at a frequency of 750 MHz, we have not observed large differences except for the electron tail temperature near the density limit between the loop antenna and grill launcher excitations.

INTRODUCTION

Lower hybrid wave current drive(LHCD) has been established successfully for steady state operation of a plasma current and for recharging of a ohmic transformer[1,2]. When applying to hot and dense reactor plasmas, however, LHW is not accessible to hot core region since the wave power is absorbed at plasma surface. This scheme can be applied to current profile modification to obtain a good confinement in tokamaks.

Fast waves in the frequency range, $\omega_{ci} \ll \omega \ll \omega_{ce}$ can penetrate into a center region without suffering strong resonances even in reactor plasmas. Fast waves have a possibility to drive the plasma current of high density and high temperature plasmas by Landau damping in the same way as LHW[3,4]. The linear theory predicts that the absorption of fast waves is weak compared with LHW because the polarization is almost perpendicular to the toroidal magnetic field. The electron

* On leave from Mitsubishi Electric Co.

temperature of several keV is required for sufficient absorption in plasma. The accessibility and coupling conditions give an optimum frequency for fast wave current drive(FWCD). In JFT-2M, a frequency between 200 to 300 MHz is suitable for effective FWCD.

EXPERIMENTAL SETUP

The FWCD experiments were carried out in the JFT-2M tokamak with a major radius of 1.31 m. The target plasma was circular or D-shaped defined by inner or outer graphite limiters located in the mid-plane. An rf power of 100 kW is divided into two lines by 3 dB hybrid coupler. The phasing between two lines is controlled by a line stretcher. The impedance matching between 50 ohm transmission line and antennas is tuned with double stub tuners. Fast waves are excited by two phased loop antennas. Each loop antenna has a single layer Faraday shield and graphite plates on both sides which protect from particle bombardment. The parallel wavenumber spectrum is broad and has a peak near $N_z = 2.5$ when $\Delta\phi = 90°$. The rf power at a frequency of 750 MHz was fed from LHW rf system.

COUPLING CHARACTERISTICS OF LOOP ANTENNAS

In order to obtain clear results of FWCD, the antenna is required to excite purely electromagnetic fast waves in plasma. In practice, it is possible to excite slow waves by the loop antenna with Faraday shield. The loading resistance is shown in Fig.1 as a function of the density. The loading resistance increases with the density and depends strongly on the distance between the antenna and the plasma boundary. This dependence is related with the edge density. These loading characteristics show the same density dependence as that of the conventional ICRF loop antenna. The present loop antennas excite the electromagnetic fast waves effectively as well as the ICRF loop antenna. The loading resistance at $\bar{n}_e = 2\times10^{13}$ cm^{-3} is about 5 ohms and the loading efficiency is 80 % at maximum.

CURRENT DRIVE BY 200 MHz FAST WAVE

The rf power was injected into the plasma with a constant plasma current. The rf driven current was deduced from the loop voltage drop in the usual manner. The density dependence of the loop voltage drop is shown in Fig. 2. The horizontal axis is the LH resonance frequency normalized by the rf frequency. The loop voltage drop shows rapid decrease with the density. The LHCD density limit occurs at $\omega_{LH}/\omega = 0.5$-0.6. The FWCD density limit is $\omega_{LH}/\omega = 1.4$ in the hydrogen plasma, which corresponds to $\bar{n}_e = 2.2\times10^{12} cm^{-3}$. This density limit is more than two times higher than that of LHCD density limit at a same frequency. In this density region, the cold LHW resonance layer exists inside the plasma column. Its location depends on the density profile. The parametric decay instabilities were observed with a Langmuir probe near and above the density limit and disappeared above $\bar{n}_e = 1 \times 10^{13} cm^{-3}$. The observed parametric decay waves seems to relate with the slow waves parasitically excited by the loop antennas or mode-converted from the launched fast waves. The parasitic slow waves have a possibility of current drive at the low density. We think that the density limit is caused by the accessibility of the launched waves with broad Nz spectrum to the high density region and a lack of rf power not by the parametric instabilities. Figure 3 shows the figure of merit of FWCD as a function of the density. The efficiency is as high as that of LHCD in the low density region and degrades with the density. The efficiency in deuterium plasma is larger than that in hydrogen. The density limits are 2.2×10^{12} cm^{-3} in hydrogen and 3×10^{12} cm^{-3} in deuterium, respectively.

FREQUENCY DEPENDENCE OF FWCD

In order to study the frequency dependence of FWCD, the same loop antennas as that of 200 MHz FWCD were used to excite fast waves at a frequency of 750 MHz. The wave power launched by two loop antennas is concentrated on $N_z < 2$. A theoretical estimation gives that the single pass damping of the fast waves with $N_z < 2$ is only several percent in JFT-2M plasma. Figures 4(a) and (b) show the density dependence of the loop voltage drop and the tail electron temperature measured with a soft X-ray PHA. The reference data of 750 MHz LHCD is also shown. Slow waves were launched by two waveguides grill. In this configuration, slow waves with $N_z < 2$ are dominant. The density dependence of $\Delta V_L / V_L$ of FWCD is almost same as that of LHCD at 750 MHz. The density limit is $(8\text{-}10) \times 10^{12} cm^{-3}$ in both cases. The observed density limit corresponds to $\omega_{LH} / \omega = 0.6$ for hydrogen. However, the tail electron temperature produced by the launched waves shows a clear difference between them. The tail temperature in FWCD is sustained at about 70 keV when $\bar{n}_e \leqq 9 \times 10^{12}$ cm^{-3}. In the case of the grill excitation, the tail electrons are not produced above $\bar{n}_e = 6 \times 10^{12} cm^{-3}$. This is consistent with the experimental observation that the loop voltage drop in FWCD is slightly larger than that of LHCD around $\bar{n}_e = 6 \times 10^{12} cm^{-3}$. In addition, the intensity of the parametric decay waves is weak even at the density near and above the density limit in the case of loop antenna excitation compared with that in grill excitation. It seems that the loop antennas excite slow waves dominantly at the low density and the small effect of FWCD appears at the high density region below the density limit. The probable reasons for slow wave excitation are the mode conversion from fast waves and insufficient shielding of the parallel electric field at a high frequency. The clear investigation of FWCD distinct from LHCD requires further experimental efforts.

ACKNOWLEDGMENT: The authors are grateful to Drs. S. Mori, K. Tomabechi, M. Tanaka, Y. Tanaka, A. Funahashi and Mr. K. Suzuki for their continuous encouragement.

REFERENCES

1. F. Jobes, et al., Phys. Rev. Lett., 52, 1005(1984).
2. F. Leuterer, et al., Phys. Rev. Lett., 55, 75(1985).
3. K. Theilhaber and A. Bers, Nucl. Fusion, 20, 547(1980).
4. K. Ohkubo, et al., Phys. Rev. Lett., 56, 2040(1986).

FIGURE CAPTIONS

Fig.1 Density dependence of the loading resistance normalized by the vacuum value, taking the distance between the loop antenna and the plasma boundary as a parameter. P_{rf}=1 kW.

Fig.2 Density dependence of the loop voltage drop. The horizontal axis is the LHW frequency normalized by the rf frequency.

Fig.3 Density dependence of the figure of merit of FWCD.

Fig.4 Density dependence of the loop voltage drop(a) and the tail electron temperature(b) in FWCD(Loop) and LHCD(Grill). at a frequency of 750 MHz.

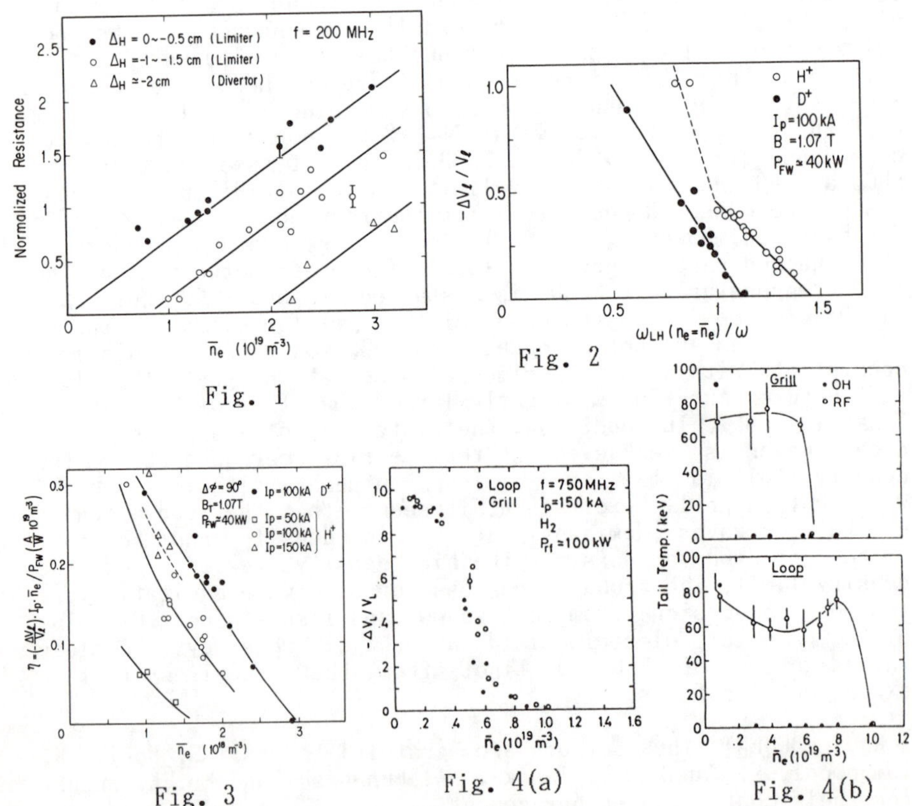

Fig. 1

Fig. 2

Fig. 3

Fig. 4(a)

Fig. 4(b)

PARTICLE TRANSPORT SIMULATION OF LOWER-HYBRID CURRENT DRIVE EXPERIMENTS ON THE VERSATOR II TOKAMAK

K-I. Chen, S.C. Luckhardt, M.J. Mayberry,* M. Porkolab

Plasma Fusion Center and Research Laboratory of Electronics
Massachusetts Institute of Technology, Cambridge, Massachusetts 02139

ABSTRACT

The one-dimensional particle transport equation has been solved numerically to simulate temporal and spatial evolutions of density behavior observed during 800MHz and 2.45GHz LHCD experiments. In order to fit the 800MHz profiles, the inward pinch velocity has to be increased several fold. However, for the 2.45GHz case, the reduction of the diffusive loss near the periphery seems to be needed.

In the earlier Versator II LHCD experiments, it has been shown that significant density increases observed in both Versator 800MHz and 2.45GHz LHCD experiments are due to the improvement of the particle confinement by a factor of two.[1,2] The purpose of this paper is to model the density profiles with particle transport code and obtain values of relevant transport coefficients.

The one-dimensional particle transport equation[3] is

$$\partial n_e(r,t)/\partial t = \frac{1}{r}\frac{\partial}{\partial r}\left(r\left[D\frac{\partial}{\partial r} + v\right]n_e(r,t)\right) + S(r,t) \quad (1)$$

where $n_e(r,t)$ is the local electron density, D is the diffusion coefficient, v is the convective velocity and S(r,t) is the source function of electron which is mainly due to ionization of hydrogen neutrals. Equation (1) can be solved numerically by the Crank-Nicolson implicit method.[4] In modeling the anomalous D and v, the simplest radial dependences have been chosen as a starting point; namely, the diffusion coefficient D is constant and the convective velocity is linearly proportional to the radius $v(r)=v_0\cdot r/a$. This simple model has been used by other authors[3,5] and appears to be adequate to explain their experimental results. However, it will be necessary to introduce additional spatial variation of D and v to accurately model some of our experimental results. Model simulated profiles obtained through different trial values of D's and v's are compared with experimentally obtained profiles. Then D and v are determined by the best fit of the profile.

In the absence of 800MHz RF power, the density profiles are relatively flat ($n_{e0}/\bar{n}_e \sim 1.2$). (See Fig. 1.) To describe the temporal and spatial profiles in the ohmic phase, D=15000cm2/sec and $v(r)=600\cdot r/a$ cm/sec are needed. The agreement between the modeling and the experimental values is good within 4% over the entire radial profile. During injection of the 800MHz RF power in the current drive mode, the density initially increases and the profile becomes more peaked ($(n_{e0}/\bar{n}_e)_{max} \sim 1.9$). Note that the source term is decreasing during the course of the RF pulse. One interesting question is whether this density increase is due to the increase of v and/or the decrease of D. For the three representative simulations we have performed, Table 1 summarizes the parameters used in these simulations. Figure 2 shows results of three simulations with approximately the same v/D value. Simulation #1 yields the best fit with several fold increase of v and the unchanging D value. Note that this simulation correctly predicts both the initial increase and the subsequent decrease of the electron density while the D and v are held constant. Simulations #2 and #3 predict continually increasing density behavior which is not consistent with measured profiles. Therefore, a factor of ∿8 increase of the inward pinch velocity appears to be mainly responsible for the reduced outward transport. After termination of the RF pulse, the particle transport deteriorates. This is correlated with an anomalous doppler relaxation mode burst which occurs after the RF pulse.[1] Using the pre-RF injection values of D and v results in profiles which are in significant error (∿75%) in the peripheral part of the discharge. However, if the transport model shown in Fig. 3(a) is used; i.e., the diffusive loss is larger near the outside, the predicted profiles are in considerable better agreement (10%-20%) with experiment.

In the initial ohmic phase during 2.45GHz LHCD experiments, the density profile is more peaked ($n_{e0}/\bar{n}_e \sim 2$). D=5000cm2/sec and v(r)= 2100·r/a cm/sec produces a good fit to the observed density profiles. During the 2.45GHz RF current drive pulse, the density increases and the profiles becomes broader (see Fig. 4). Four representative models have been used in simulating these experiments. Table 2 lists values of D and v used in these simulations. Best results were obtained in simulation #1, where the model of D shown in Fig. 3 was used. D was increased near the center, and decreased near the periphery. The code simulation is in much better agreement with the measured profile as shown in Fig. 4. When the value of D (simulation #2) or the value of v (simulation #3) was increased by a factor of two, the experimental profiles could not be reproduced. In both simulations, the density increased too quickly and was too peaked (see Fig. 5). Simulation #4 uses a model shown in Fig. 3, which decreases the value of v near the center by an order of magnitude and increases its value by a factor of two near the outside. This particle transport model does not maintain the density desired. In particular, the density decreased more rapidly than the experimental results. To model the post RF-phase, the pre-RF D and v values have been used. This model is in good agreement with the experimental profiles.

It can be shown from Eq. (1) that for source free, steady state particle transport, the ratio v/D is determined by the ratio of an n. Table 3 shows the value of v/D from the profile measurement and from the model used to yield the best fit. The values of v/D obtained from profile measurement and best fit models are in close agreement.

Particle confinement improvement has been confirmed in both 800 MHz LHCD and 2.45 GHz LHCD cases which have quite different shapes of the density profile. Thus, there does not seem to be a correlation between the peaked profile and better confinement.

In summary, the inward convective velocity must be increased by ∿8 to model the 800MHz experimental result. In the 2.45GHz case, the diffusive loss had to be reduced near the plasma periphery. The difference in the behaviors of particle transport between two LHCD experiments might be related to the difference of wave penetration and current generation in the two different density and RF frequency regimes. The physicsl reason for particle confinement improvement during lower-hybrid current drive experiments in Versator II is not understood at present.

ACKNOWLEDGEMENTS

One of the authors (K.C.) would like to acknowledge useful discussions with James L. Terry. This research was supported by US Department of Energy (Contract DE-AC02-78ET-51013).

REFERENCES

1. S.C. Luckhardt, et al, Phys. Fluids 29 , 1985 (1986).
2. M.J. Mayberry, K.I. Chen, S.C. Luckhardt, M. Porkolab (submitted to Physics of Fluids).
3. J.D. Strachan, et al, Nucl. Fusion 22, 1145 (1982).
4. A.R. Michell and D.F. Griffiths, "The Finite Difference Method in Particle Differential Equation", New York, Wiley (1980).
5. B. Richards, et al, Bull. APS 30, 1567 (1985).

* Present address: GA Technologies Inc., San Diego, CA. 92318

800 MHz LHCD

Before RF		During RF			After RF	
$D(\times10^3 cm^2/sec)$	$v(\times10^3 cm^2/sec)$		$D(\times10^3 cm^2/sec)$	$v(\times10^3 cm^2/sec)$	$D(\times10^3 cm^2/sec)$	$v(\times10^3 cm/sec)$
15	$0.6 \times \frac{r}{a}$	Model Simulation #1	15	$5.2 \times \frac{r}{a}$	$D_1 = 15$, $D_2 = 25$, $r_1 = 5cm$, $r_2 = 10cm$	$0.6 \times \frac{r}{a}$
		Model Simulation #2	6	$2 \times \frac{r}{a}$		
		Model Simulation #3	2	$0.6 \times \frac{r}{a}$		

Table 1. Transport parameters used on model simulations of 800MHz LHCD

2.45 GHz LHCD

Before RF		During RF			After RF	
$D(\times10^3 cm^2/sec)$	$v(\times10^3 cm^2/sec)$		$D(\times10^3 cm^2/sec)$	$v(\times10^3 cm^2/sec)$	$D(\times10^3 cm^2/sec)$	$v(\times10^3 cm/sec)$
5	$2.1 \times \frac{r}{a}$	Model Simulation #1	$D_1 = 8$, $D_2 = 2$, $r_1 = 5 cm$, $r_2 = 8 cm$	$2.1 \times \frac{r}{a}$	5	$2.1 \times \frac{r}{a}$
		Model Simulation #2	5	$4.2 \times \frac{r}{a}$		
		Model Simulation #3	2.5	$2.1 \times \frac{r}{a}$		
		Model Simulation #4	5	$v_1 = 0.5$, $v_2 = 4.2$, $r_1 = 3 cm$		

Table 2. Transport parameters used on model simulations of 2.45GHz LHCD

800 MHz LHCD

Time (ms)	Radius (cm)	v/D (cm^{-1}) from profile measurement	v/D (cm^{-1}) from model
17.5	5	0.025	0.016
23	5	0.14	0.135

2.45 GHz LHCD

Time (ms)	Radius (cm)	v/D (cm^{-1}) from profile measurement	v/D (cm^{-1}) from model
19.5	4	0.154	0.17
26.5	4	0.04	0.10

Table 3. The ratio of pinch velocity (v) and diffusion coefficient (D) from experimental density profiles and from model simulations

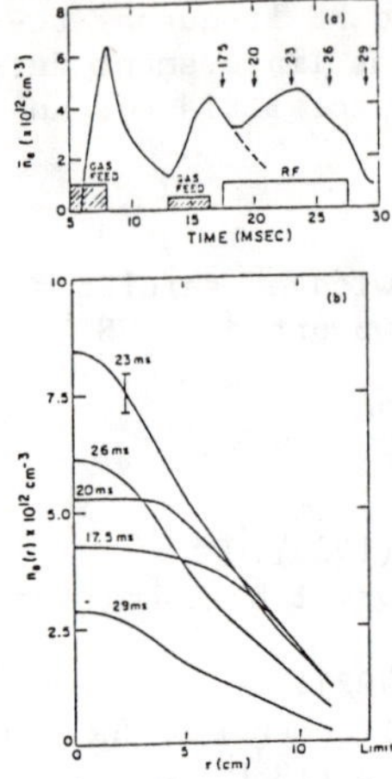

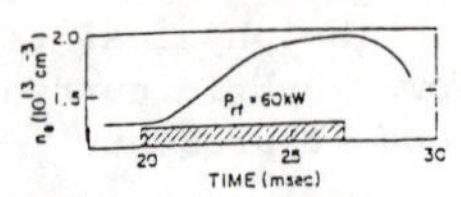

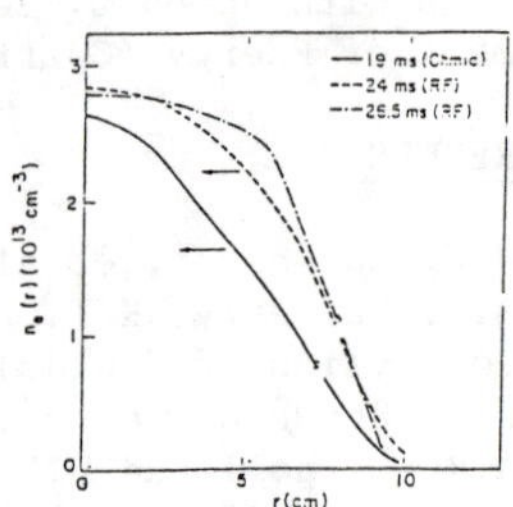

Fig. 1. The temporal evolutions of the line-averaged density and the density profile during 800MHz LHCD

Fig. 4. The temporal evolutions of the line-averaged density and the density profile during 2.45 GHz LHCD

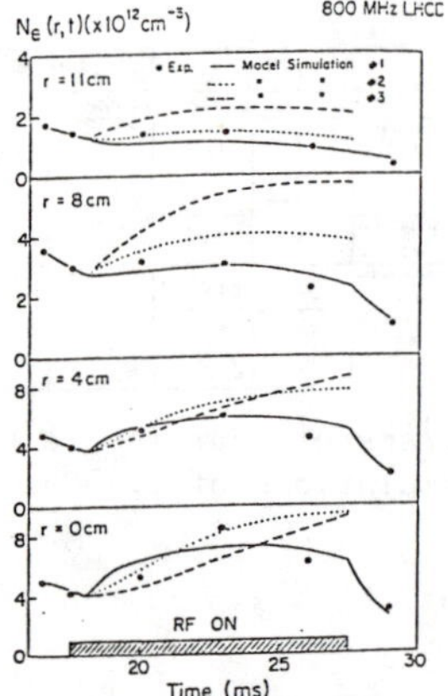

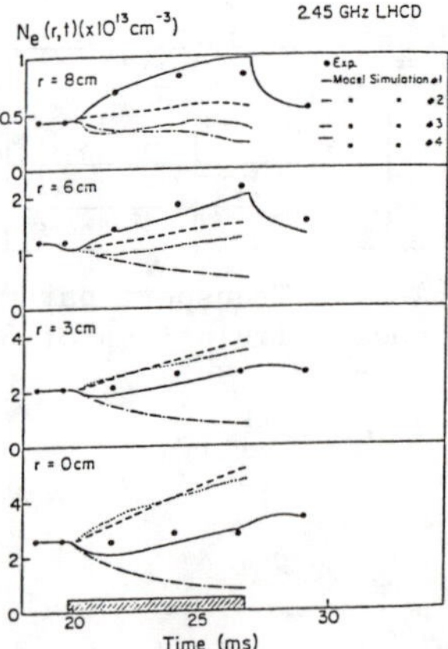

Fig. 2. Three model simulations of the density evolution on 800MHz LHCD at various radii

Fig. 5. Four model simulations of the density evolution on 2.45 GHz LHCD at various radii

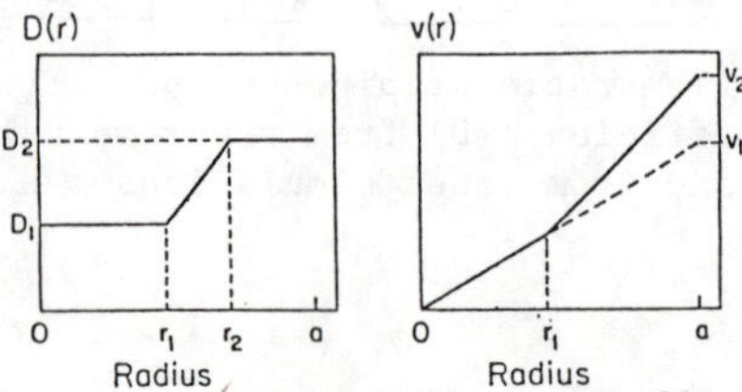

Fig. 3. Modified transport model of diffusion coefficient and pinch velocity

LOWER HYBRID CURRENT DRIVE MODEL AND ITS VERIFICATION WITH ASDEX DATA

K.Yoshioka and T.Okazaki
Energyy Research Laboratory, 1168 Moriyamacho Hitachi, 316 JAPAN

M.Sugihara and N.Fujisawa
Japan Atomic Energy Research Institute, Nakamachi Nakagun, 312-02

ABSTRACT

A simple analytical model for evaluating the lower hybrid current drive efficiency in the presence of a DC electric field is developed and its result are compared with the experimental data from the ASDEX tokamak. The model is on the basis of an analytical solution of the "adjoint equation" to the Fokker Planck equation and considers the finite spread of the rf wave spectrum. The dependency of current drive efficiency on the DC electric field predicted by the model agrees well with that of the experiment within the experimental error.

INTRODUCTION

The quasi-steady state operation using rf current drive in combination with inductive current drive is planned for the next generation tokamak fusion reactors. In this mode of operation, LHCD (lower hybrid current drive) is used for transformer recharge and current ramp-up. In these situations, the inductive DC electric field significantly influences the nature of the current drive, A reliable model for rf current drive with DC electric field is needed for reliable fusion reactor design.

Generally, the current drive efficiency J_N/P_N (driven current J_N divided by dissipated power P_N) is given by

$$\frac{J_N}{P_N} = \frac{\int du^3 \Gamma \nabla W_S}{\int du^3 \Gamma \nabla W_S (\frac{u^2}{2})} \qquad (1)$$

where Γ is the rf-induced flux in velocity space, W_S is the solution of the equation adjoint to the corresponding Fokker Planck equation[1], and $u=v/v_{th}$ is the velocity normalized by the thermal velocity. The quantities J_N and P_N are normalized with env_{th} and $v_0 nmv_{th}^2$, respectively.

Karney and Fisch have obtained the exact solution W_S by numerically integrating the adjoint equation[1] and compared with PLT experimental data[2]. In this comparison, the efficiency J_N/P_N has been approximated as $J_N/P_N=(1/u)\partial W_S/\partial u$, which means that the rf waves are localized in velocity space, and by adjusting the location of wave deposition as a fitting parameter, an excellent agreement with experimental data has been observed.

However, the actual, experimental spectrum has a broad profile and the flux Γ itself also is a function of DC electric field[3]. In such a situation, nature of current drive is expected to be different from those in previous studies. We develop a simple analytical expression for the current drive efficiency, considering the effect of finite spread of rf spectrum as well as the existence of DC field. The expression is on the basis of an approximate solution of W_S and Γ , but is found to be correct enough through the check by 2D Fokker Planck code and ASDEX experimental data.

MODEL

The equation adjoint to the 2D Fokker Planck equation under the presence of DC electric field is given by[1)]

$$-\frac{1}{u_R^2}\frac{\partial W_S}{\partial u_R}+\frac{1}{2u_R^3}\frac{\partial}{\partial \mu}(1-\mu^2)\frac{\partial W_S}{\partial \mu}+\frac{\partial W_S}{\partial u_{R\parallel}}=-u_{R\parallel} \tag{2}$$

where $u_R=u\sqrt{2E_N}$, $u_{R\parallel}$ is its toroidal component, $\mu=u_{R\parallel}/u_R$, and $E_N=E_\parallel/E_{DR}$ is the DC field normalized by the Dreicer field E_{DR}. In the case of LHCD, only electrons with $|\mu|=<1$ are pushed by rf waves. In this case, Eq.(2) can be solved analytically as

$$W_S=-\frac{8}{Z+5}\frac{\mu}{\alpha^2}\left[\frac{\alpha^*}{2}u_R^2+\ln\left|1-\frac{\alpha^*}{2}u_R^2\right|\right] \tag{3}$$

with $\alpha^*=\mathrm{sgn}(\mu)\alpha$, and $\alpha=12/(7+Z)$. If we use the rf-induced flux from 1D Fokker Planck estimation as $\Gamma=1/u^2-E_N$, we can evaluate Eq.(1) as

$$\frac{J_N}{P_N}=\frac{(J_N/P_N)_0}{1+(\frac{5+Z}{8})E_N(J_N/P_N)_0}\left[1-(1-\alpha)\frac{(J_N/P_N)_1}{(J_N/P_N)_0}\right]\frac{1}{\alpha} \tag{4}$$

where

$$\left(\frac{J_N}{P_N}\right)_0=\frac{4}{5+Z}\frac{u_2^2-u_1^2}{\ln(u_2/u_1)} \tag{5}$$

is the steady state efficiency and

$$\left(\frac{J_N}{P_N}\right)_1=\left(\frac{J_N}{P_N}\right)_0\frac{\ln\left|(1-\alpha E_N u_2^2)/(1-\alpha E_N u_1^2)\right|}{\alpha E_N(u_1^2-u_2^2)} \tag{6}$$

is the efficiency when steady state flux $\Gamma=1/u_2$ is assumed. The quantities u_1 and u_2 are the lower and upper limit in which the rf wave spectrum is contained.

CHECK OF MODEL WITH 2D FOKKER PLANCK CODE

The validity of using 1D estimates for the flux Γ is checked by the exact numerical solution of the 2D Fokker Planck equation. The code intergrates the following equation;

$$\frac{\partial}{\partial u_\parallel} D \frac{\partial f}{\partial u_\parallel} + \frac{Z+1}{4u^3} \frac{\partial}{\partial \mu} (1-\mu^2) \frac{\partial f}{\partial \mu} + \frac{1}{2u^2} \frac{\partial}{\partial u} \left(\frac{1}{u} \frac{\partial f}{\partial u} + f \right)$$

$$-E_N \frac{\partial f}{\partial u_\parallel} = 0 \qquad (7)$$

where D is the dimensionless diffusion constant proportional to the square of rf wave amplitude. The equation is finite differenced using up-wind control volume scheme and solved by BCG (bi-conjugate gradient) method. Mesh sizes are $u \times \theta = 100 \times 100$.

Fig.1 shows the comparison of the numerical results with our analytical model. The plus direction in E_N implies that the direction of rf acceleration coincides with that of DC electric field. Numerical results with $D \geqq 3 \times 10^{-2}$ coincides with the analytical model. The results of smaller D asymptotically approaches to the curve of $1/u(\partial W_S/\partial u)$ with $u \simeq 5$. this means that with weaker rf field the rf power absorption occur locally just at the lower edge of the rf-spectrum in the velocity space.

CHECK OF MODEL WITH ASDEX EXPERIMENTAL DATA

The ASDEX LHCD experimental data are compared with our model. The raw data in the form of I_{OH} (transformer current change rate) versus P_{rf} (rf power) for various electron densities transformed into the normalized current drive efficiency (J_N/P_N) as a function of E_N. Results are shown in Fig.2. The solid curve is the estimation by the present model with $u_\parallel = 7$ to 14, which corresponds to the Brambilla spectrum of ASDEX 8-grill launcher being in the range of $n_\parallel = 1.5$ to 3.0 assuming the electron temperature $T_e = 1$keV. It can be seen that the model agrees with the experimental data within the scattering of the data.

ACKNOWLEDGEMENTS

The authors would like to thank Dr. Fritz Leuterer for the valuable discussion as well as his providing the ASDEX data.

REFERENCES

1. C.F.F.Karney and N.J.Fisch, Phys. Fluids 29 (1986) 180
2. C.F.F.Karney, N.J.Fisch,and F.C.Jobes, Phys. Review A 32 (1985) 2554
3. K.Borrass and A.Nocentini, Plasma Phys. Controlled Fusion 29 (1984) 1299

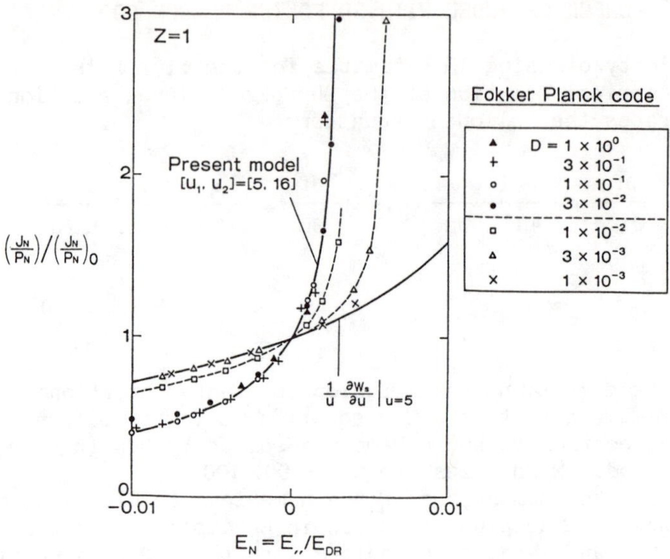

Fig.1 Comparison of results from the present model and 2D Fokker Planck Code

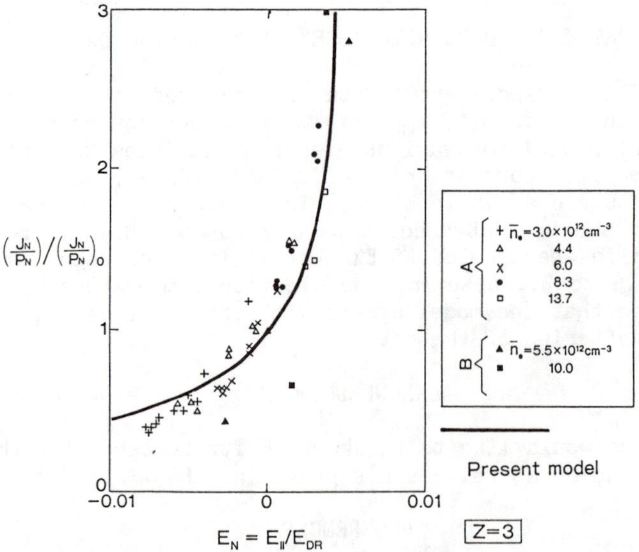

Fig.2 Comparison of results from the present model with ASDEX experimental data

THREE WAVE MODE CONVERSION IN THE LOWER HYBRID FREQUENCY RANGE

Suwon Cho and D. G. Swanson
Auburn University, AL 36849

ABSTRACT

It has been shown that propagation of lower hybrid rays sometimes involves cases where three types of waves are coupled through linear mode conversion in the same region of space, namely the cold lower hybrid wave, the warm lower hybrid wave, and an ion Bernstein mode. This three wave mode conversion problem is analyzed in terms of a sixth order differential equation which describes coupling between all branches, and analytic coupling coefficients are found for the case without the local absorption term. The formal solution of this equation shows that a non-trivial amount of the incident cold wave energy goes into an ion Bernstein mode, which is then absorbed by electrons via Landau damping, rather than converting into the warm branch totally, when they are closely coupled in space.

DISPERSION RELATION

The simplest dispersion relation for lower hybrid waves near ion cyclotron harmonics is given as

$$\frac{k_\parallel^2}{k_{Di}^2}\left(1-\frac{\omega_{pe}^2}{\omega^2}\right)-\frac{k_\perp^2}{k_{Di}^2}\left(1+\frac{\omega_{pe}^2}{\omega_{ce}^2}\right)-\frac{1}{2}\wp Z'(\sqrt{\xi})-\frac{\nu}{\delta}Fe^{-\lambda_i}I_n(\lambda_i)=0 \tag{1}$$

where n is the harmonic number being considered and $\nu=\frac{\omega}{\omega_{ci}}$, $\lambda_i=\frac{1}{2}k_\perp^2\rho_{Li}^2$, $\xi=\frac{\nu^2}{2\lambda_i}=(\omega/k_\perp v_i)^2$, $\delta=\nu-n$, $\zeta_{ni}=\frac{\omega-n\omega_{ci}}{k_\parallel v_i}$, and $F=-\zeta_{ni}Z(\zeta_{ni})$. Expanding $k_\perp^2$ about $k_{0\perp}^2$ which will be determined later, we obtain the cubic equation in $k_\perp^2$, with $\tilde{\delta}=\frac{\delta}{n}$,

$$k_\perp^6-\left(a_4+b_4\tilde{\delta}\right)k_\perp^4+\left(a_2+b_2\tilde{\delta}+c_2\tilde{\delta}^2\right)k_\perp^2-\left(a_0+b_0\tilde{\delta}\right)=0 \tag{2}$$

which describes the coupling between all three waves [1].

MODE CONVERSION EQUATION

If the magnetic field is given by $B(r)=B/(1+\tilde{r})$ and $\tilde{r}=r/R$, $\delta/n=r/R$ where R is the distance between the center of the toroidal device and the point of the harmonic resonance, and r is the horizontal distance from the resonance. The bicubic equation Eq. (2) is transformed into a sixth order differential equation

$$\psi^{vi}+(a_4+b_4\tilde{r})\psi^{iv}+(a_2+b_2\tilde{r}+c_2\tilde{r}^2)\psi''+(a_0+b_0\tilde{r})\psi=0. \tag{3}$$

Although Eq. (3) seems to be the best model to describe three wave coupling when the harmonic resonance occurs near the lower hybrid turning point, it has a critical disadvantage that one can not find solutions analytically, since it has a quadratic term in r and the usual Laplace transform technique gives a second order differential equation with an essential singularity at the origin in the transformed space. This very stiff and high order differential equation is almost impossible to solve numerically.

Without the quadratic term, the equation is analytically tractable but it does not give any coupling between the cold and the warm branch, which must be a major effect around the turning point.

A way to compromise between physics and mathematics is to neglect both c_2 and b_0 terms in order to get

$$\psi^{vi} + (a_4 + b_4\tilde{r})\psi^{iv} + (a_2 + b_2\tilde{r})\psi'' + a_0\psi = 0. \tag{4}$$

Substituting $\mu r = x - x_0$, we obtain the differential equation in a normalized form

$$\epsilon^2\psi^{vi} + (x-\sigma)\psi^{iv} + x\psi'' - \gamma^2\psi = \frac{\mu^2}{\epsilon^2}(x - x_0)\left(1 - \frac{1}{F}\right)(\psi^{iv} + \mu^2\psi'') \tag{5}$$

with $\mu = -\sqrt{\frac{b_2}{b_4}}$ $(= -k_{0\perp})$, $x_0 = \mu\frac{a_2}{b_2}R$, $\epsilon^2 = \frac{\mu^3}{b_4}R$, $\sigma = \mu\left(\frac{a_2}{b_2} - \frac{a_4}{b_4}\right)R$, $\gamma = -\frac{a_0}{b_2\mu}R$. The right hand side of Eq. (5) represents the localized absorption due to the harmonic resonance. Without the local absorption term ($F = 1$), this equation has a conserved quantity

$$P = \epsilon^2(\psi^v\psi^{iv*} + \psi^v\psi''^* - \psi^{iv}\psi'''^*) - \sigma\psi'''\psi''^* + \gamma^2(\psi'''\psi^* - \psi''\psi'^* + \psi'\psi^*) - c.c.. \tag{6}$$

ASYMPTOTIC SOLUTIONS

The homogeneous part of Eq. (5) is solved formally by using Gambier–Schmitt–Swanson theory [2,3]. For cold waves, asymptotic solutions are given as

$$\psi_{1,2} \simeq \begin{cases} -\sqrt{\pi}\gamma^{-3/2}x^{1/4}\exp\left[\pm 2\gamma x^{1/2} - iY_0 + i\left\{\begin{array}{c}\pi\\ \frac{\pi}{2}\end{array}\right\}\right] & x > 0 \\ +\sqrt{\pi}\gamma^{-3/2}|x|^{1/4}\exp\left[\pm 2i\gamma|x|^{1/2} - iY_0 \pm i\frac{\pi}{4}\right] & x < 0 \end{cases} \tag{7}$$

where $Y_0 = \epsilon^2 + \sigma - \sum_{q=1}^{2}\frac{i+\alpha_q}{k_q}$ and $k_q = (-1)^q$ and $\alpha_q = \frac{1}{2}(-1)^q(\epsilon^2 + \gamma^2 + \sigma)$. Considering the time dependence $e^{-i\omega t}$, when they propagate, we find ψ_1 has $v_\phi < 0$ and $v_g > 0$, and ψ_2 has opposite signs [4]. For warm waves, we obtain, with positive v_ϕ and v_g for ψ_3, and negative ones for ψ_4,

$$\psi_{q+2} \simeq \frac{2\pi i}{\Gamma(1 - i\alpha_q)}\exp\left[-i\left(k_q + \alpha_q\ln|x|\right) - \frac{\pi}{2}\alpha_q\mathrm{sgn}(x)\right] \tag{8}$$

$-\infty$	x	$+\infty$	$\frac{P}{2\pi i}$
$\psi_1 + \frac{1}{\tau}\psi_4 + \psi_5$	ϕ_1	$-\psi_2 + \frac{1}{\tau}\psi_5$	$-\frac{1}{\tau^2}$
$\left(1-\frac{1}{\tau}\right)\psi_2 + \psi_3$	ϕ_2	$\left(1-\frac{1}{\tau}\right)\psi_2 + \frac{1}{\tau}\psi_3$	$\frac{1}{\tau^3} - \frac{1}{\tau^2}$
$\frac{1}{\tau}\psi_4$	ϕ_3	$\psi_4 + \left(\frac{1}{\tau}-1\right)\psi_5$	$\frac{1}{\tau} - \frac{1}{\tau^2}$
$-\psi_2$	ϕ_4	$-\psi_2 + \psi_3 + \psi_6$	$\frac{1}{\tau}$
ψ_5	ϕ_5	$-\psi_1 - \psi_2 - \psi_3 - \psi_4 + \psi_5 - \psi_6$	0
ψ_6	ϕ_6	ψ_2	0

Table 1: Asymptotic behavior of physical solutions

where $(q = 1, 2)$. The Asymptotic form of the hot wave solutions are given by

$$\psi_{5,6}(x) \simeq \begin{cases} \sqrt{\pi}\epsilon^{7/2}x^{-9/4}\exp\left[\mp i\frac{2}{3\epsilon}x^{3/2} + i\left\{\begin{array}{c}\frac{\pi}{4}\\ \frac{3\pi}{4}\end{array}\right\}\right] & x > 0 \\ \sqrt{\pi}\epsilon^{7/2}\left|x\right|^{-9/4}\exp\left[\mp\frac{2}{3\epsilon}\left|x\right|^{3/2} + i\left\{\begin{array}{c}0\\ \frac{\pi}{2}\end{array}\right\}\right] & x < 0 \end{cases} \tag{9}$$

and ψ_5 has $v_\phi < 0$ and $v_g > 0$, and ψ_6 has opposite signs.

PHYSICAL SOLUTIONS

Collecting relevant contours and adjusting levels of the Riemann surfaces, we find the transfer matrix which connects asymptotic solutions from both sides of x.

$$\mathrm{M}(\Psi^+, \Psi^-) = \begin{pmatrix} 1 & 1 & 0 & 0 & 0 & -1 \\ 0 & 0 & 0 & 0 & 0 & 1 \\ 0 & \tau-1 & \tau & 0 & 0 & 1-\tau \\ \tau-1 & 0 & 0 & 1 & \tau-1 & \tau-1 \\ \tau & 0 & 0 & 1 & \tau & \tau \\ 0 & -\tau & \tau & 0 & 0 & \tau \end{pmatrix} \tag{10}$$

where $\tau = e^{2\eta} = e^{\pi(\epsilon^2+\gamma^2+\sigma)}$. Since any solution can be expressed in Ψ^+ or Ψ^-, $\tilde{\Lambda}^+\Psi^+ = \tilde{\Lambda}^-\Psi^-$ so that $\Lambda^- = \widetilde{\mathrm{M}}(\Psi^+, \Psi^-)\Lambda^+$ where Ψ's are column vectors consisting of six independent solutions and Λ's are vectors having expansion coefficients for their elements.

There must not be any incoming warm wave for $x < 0$ (the low field side) nor any incoming hot wave for $x > 0$, when the cold wave is launched. In addition, we require that there exist no components of exponentially growing solutions. These boundary conditions lead to $\lambda_2^- = 0$, $\lambda_4^- = \frac{1}{\tau}$, $\lambda_5^- = -1$, $\lambda_2^+ = -1$, $\lambda_3^+ = 0$ and $\lambda_5^+ = \frac{1}{\tau}$. It is worthwhile noting that none of the incident energy goes to the warm wave on the other side, which justifies the assumptions about the form of the equation. Assuming that we call the conversion to the wave propagating on the other side of the incident wave *tunneling*, we obtain $T = e^{-2\eta}$, $C = 1 - e^{-2\eta}$, and $R = 0$. To get T and C, asymptotic solutions are normalized with respect to P.

All six independent solutions representing physical phenomena are constructed and summarized in Table (1). The tunneling and conversion coefficients found above are universally obtained for all four propagating solutions. The other two solutions ϕ_5 and ϕ_6 are exponentially decaying or growing functions.

RESULTS AND CONCLUSIONS

When the absorption term is not included, $k_\perp^2$ vs. ν from two dispersion relations, Eq.(1)(light), and Eq.(2)(heavy) without c_2 and b_0, are plotted for $n_i = 3.5 \times 10^{19}$, $T_i = 722eV$, $n_\parallel = 2$, $\omega = 5.03 \times 10^9$, and $R = 1.3m$ in Fig.(1), where solid lines represent real parts and the dotted ones imaginary parts. The tunneling coefficient is found to be 18%. This result agrees with the value calculated from $\exp(-2\int \text{Im}k_\perp dx)$ with careful identification of the branch for $k_\perp$ satisfying Eq.(1). With the absorption term, the simple integral method is no longer comparable and the full solutions of Eq. (5) should be sought.

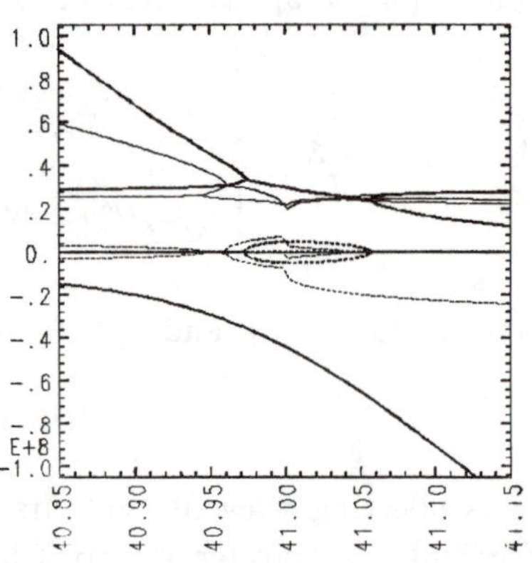

Figure 1: $k_\perp^2$ vs ν .

In conclusion, when the high harmonic resonance occurs near the lower hybrid turning point, some of the incident wave energy goes to the ion Bernstein wave resulting in reduced efficiency of ion heating.

ACKNOWLEDGMENT

This work has been supported by the U. S. DOE under contract DE-FG05-85-ER53206.

REFERENCES

1. D. G. Swanson, S. Cho, et. al., 1986 IAEA Conference, Kyoto, Japan, IAEA–CN–47/F–IV–4.
2. D. J. D. Gambier and J. P. M. Schmitt, Phys. Fluid 26, 2200, (1983).
3. D. J. Gambier and D. G. Swanson, Phys. Fluid 28, 145, (1985).
4. Suwon Cho and D. G. Swanson, Bull. Am. Phys. Soc., Vol. 31, No. 9, 1577, (1986).

EFFECT OF HOT ELECTRON TAIL IN CURRENT DRIVE ON LH WAVE ABSORPTION

I.P.Shkarofsky, MPB Technologies, Dorval H9P 1J1

V.Fuchs and M.M.Shoucri, IREQ, Varennes, J0L 2P0

ABSTRACT

Ray tracing and absorption with a hot electron tail distribution, which occurs in LH current drive, are treated with model relations for the perpendicular temperature and slope of the velocity distribution. Both non-relativistic and relativistic 2-D Fokker- Planck solutions are fitted. We find least path length for complete absorption with the relativistic model, a longer path with the non-relativistic model, and a much longer path when no fast electrons are included. Hot electron damping can dominate over collisional plus electron Landau damping. This is true when the ray can reach near the plasma center and when ion damping is negligible. If a mode conversion surface occurs, the ray is limited to the outer plasma and perpendicular ion Landau damping can dominate.

INTRODUCTION

This paper uses model solutions of both the relativistic and non-relativistic Fokker- Planck equation, with long tail electron velocity distributions as occur in lower-hybrid current drive. We compare damping of rays for an equilibrium Maxwellian distribution with either of the above two solutions. Ray tracing of slow waves in the LH frequency range has been performed by Ignat[1], Bonoli and Englade[2] and Valeo and Eder[3]. Here we include all the various mechanisms that absorb the wave, namely (1)collisional, (2)electron Landau, (3)perpendicular ion Landau damping with up to six ion species and (4)damping by hot electrons in the tail of the distribution. We find that mechanisms (1) and (2) can damp the wave completely, but often only after a very long path length involving multiple wall reflections. When mechanism (4) is added, especially with the relativistic analysis, the wave damps out more quickly in a shorter path length. The form of the electron distribution function, subject to a rf wave spectrum causing current drive, is obtained from the solution of the Fokker-Planck equation with a quasi-linear diffusion term added to it. Computer programs[2,3] combining solutions of the Fokker-Planck equation with ray tracing can become very large and complex. We use an alternative approach, combining ray tracing with fitted equations, modeling the solution of the Fokker-Planck equation as given by Fuchs et al, both for its non- relativistic[4] and relativistic[5] version. They give a model fit equation for an effective perpendicular temperature and for the slope of the electron velocity distribution function in the raised plateau region.

ANALYSIS

The ray tracing code we are using has been developed by combining Ignat's code[1] with the ray tracing subroutines in Bonoli and Englade[2]. We can use either profiles with Shafranov's expansion of the magnetic field, including the Shafranov shift(see Eqs.(21a)-(30) in Ref.2), or the profiles in Ref.1, Sec.II. In contrast to Refs.1 and 2, we use the expanded forms of all the elements of ε,the warm dielectric tensor[6], so that ε_{22} differs from ε_{11} and both ε_{13} and ε_{23} are non-zero. A low- density profiled plasma is allowed outside of the plasma boundary up to the wall, but once a ray passes outward halfway between the boundary and the wall, it is reflected by reversing the sign of the radial wavenumber. Up to six ion species can be included separately as in Ref.1. The ray integration procedure we use is that given in Ref.1, namely a predictor-corrector routine with a Runge-Kutta(Gill's method) starter. The starter is used not only to start the ray but also everytime the ray is reflected and everytime the path increment is altered. The path increment is programmed to automatically decrease when the ray moves outward towards the plasma boundary and to increase to its original value on its way back inward. Ten variables are integrated along the ray trajectory. In addition to variables for time, space (r=radius, θ=poloidal angle and ϕ=toroidal angle), wavenumber (k_r, $m_k=rk_\theta$ and $n_k=(R+r\cos\theta)k_\phi$ where R is major radius), used in Refs.1-3, we also integrate the path length, absorption coefficient (imaginary part of the perpendicular wavenumber, $k_\perp$) and phase change from free space. Electron-ion collisions are included as in Ref.2, but extended to all cold ε elements. We follow Ref.1 in calculating wavenumber derivatives analytically and spatial derivatives numerically. The ray is started at any point near the plasma boundary subject to the conditions that ε_{33} is negative and that $k_\perp$ satisfies the dispersion equation for the slow wave branch with real $k_\perp$ and k_r. Initially, the perpendicular($n_\perp$) or poloidal refractive index is equated to a small number or zero, whereas the parallel($n_{||}$) or toroidal refractive index is assigned a value determined by the wave-exciter. As noted in Refs.1 and 2, $n_\perp$ can increase to values up to 50-100 during propagation of a ray and $n_{||}$ can increase as well up to about 4-5. When $n_{||}$ becomes large especially in the vicinity of the plasma center, the effect of electron Landau damping is enhanced. When $n_\perp$ is large, perpendicular ion damping can dominate in certain situations. The equations for Landau damping of the bulk thermal electrons are obtained from the imaginary parts of plasma dispersion function in the hot ε 33, 22 and 23 elements with zero harmonic number. Shkarofsky[7] gives equations for perpendicular ion Landau damping with $k_{||}$ corrections. Let $h=(k_{||}/k_\perp)^2$

$$d=(\omega/k_\perp v_{ti})^2\ ,\ k^2=k_\perp^{\ 2}+k_{||}^{\ 2}\ ,\ A=(\omega_{pi}^{\ 2}/\omega k v_{ti})(\pi/2)^{\frac{1}{2}}\exp[-2(\omega/2kv_{ti})^2]$$

then the imaginary parts of ε are: A(d+h-2hd) for 11, A for 22, A(1-h+hd) for 33, $-A(1-d-h+2hd)k_{||}/k_\perp$ for 13 and the contributions from the 12 and 23 elements are negligible. Here, ω is the angular frequency, ω_{pi} is the angular ion plasma frequency and $v_{ti}^{\ 2}=(T_i/m_i)$ is the ion thermal velocity squared.

The main addition that we have done is to calculate the absorption due to the raised tail of the electron distribution that arises subject to current drive, again from only the $\mathrm{Im}(\varepsilon_{33})$, $\mathrm{Im}(\varepsilon_{22})$ and $\mathrm{Re}(\varepsilon_{23})$ elements with zero harmonic number. We use the following equations, valid even relativistically under the a-priori assumption that the momentum($\underset{\sim}{p}$) distribution function, f, is separable into $f(\underset{\sim}{p})=f(p_{||})[\exp(-p_\perp^{\ 2}/2mT_\perp)]/(2\pi T_\perp/mc^2)$ with $\int f(p_{||})dp_{||}=1$, where m is the electron rest mass, c is the velocity of light, and $T_\perp$ is a slowly varying function of $p_{||}$, but assumed constant below. Let v_t be a normalizing velocity, spatially dependent on the square root of the electron temperature. Define

$$\lambda=k_\perp^{\ 2}T_\perp/(m\omega_{ce}^{\ 2})\ ,\ \gamma_0^{\ 2}=(1+2T_\perp/mc^2)/(1-n_{||}^{\ -2})\ ,\ F(z)=f(zmv_t)mv_t\ ,\ z=\omega\gamma_0/(k_{||}v_t)$$

Let ω_{ce} and ω_{ph} be the angular electron cyclotron and hot electron plasma frequencies and let I_0 and I_1 be modified Bessel functions. Then

$$\mathrm{Im}(\varepsilon_{33})=-\pi\gamma_0(\omega_{ph}/k_{||}v_t)^2\ e^{-\lambda}\ I_0(\lambda)F'\quad \text{with } F'=dF/dz \text{ and } \int F(z)dz=1$$

$$\mathrm{Im}(\varepsilon_{22})=-2\pi(\omega_{ph}/\omega\ v_t)^2\ (T_\perp/m\gamma_0)\lambda\ e^{-\lambda}\ [I_0(\lambda)-I_1(\lambda)]\ F'$$

$$\mathrm{Re}(\varepsilon_{23})=-\pi(k_\perp/(k_{||})(\omega_{ph}/v_t)^2\ (T_\perp/m\omega\omega_{ce})\ e^{-\lambda}\ [I_0(\lambda)-I_1(\lambda)]\ F'$$

A quasilinear diffusion coefficient, D, is included in the linearized 2-dimensional Fokker-Planck equation. The coefficient D is assumed to be constant in the parallel direction between z_1 and z_2 in the relativistic analysis, or between v_1 and v_2, both normalized to v_t in the non-relativistic analysis. Here $z_{1,2}^{\ 2}=v_{1,2}^{\ 2}(1+\beta^2 2T_p)/(1-\beta^2 v_{1,2}^{\ 2})$, where $\beta=v_t/c$ and $T_p=T_\perp/mv_t^{\ 2}$, subject to $z_2 \geqslant p_2=z_2(T_p=0)$. The spatial dependence of D, besides through its limits, is taken proportional to v_t/n_e, where n_e is bulk electron density, since D is normalized to them. We need relations for T_p, F' and ω_{ph}. The non-relativistic relations for the first two are respectively given in Fuchs et al[4], Eqs.(37) [with $v_{\perp 0}^{\ 2}=1+T_p$] and (A29) [with the I's in Eq.(A17) evaluated in terms of the error function]. The hot electron density is $n_e(v_2-v_1)F_p$, where F_p is the distribution magnitude at the plateau. Here F_p, v_2, v_1 and the thermal density depend on spatial coordinates. Fuchs et al[5] provide the relativistic versions of these equations. F_p is obtained from Ref.4, Eq.(45) with p_1 replacing v_1, where $p_1^{\ 2}=v_1^{\ 2}/(1-\beta^2 v_1^{\ 2})$. The plateau extends from p_1 to z_2, so the hot electron density is $n_e(z_2-p_1)F_p$. Let Z be the effective ion charge, $\alpha=2(1+Z)/(2+Z)$ and $b^2=1/(1-\beta^2 v_1^{\ 2})$. Introduce the function

$$y(x)=-v_1^{\ 2}+(v_1\gamma_1/x)^{2-\alpha}\ [(1+\gamma_1)/(1+\gamma)]^{\alpha}\ (x^2+1+T_p)/b^2$$

$$\text{where } \gamma_1^{\ 2}=b^2(1+\beta^2 y)\ ,\ \gamma^2=1+\beta^2(x^2+1+T_p)\ ,\ T_p=\int y(x)dx/[2(z_2-z_1)]$$

with limits z_1 to z_2. We expand $y(x)$ to order β^4 ($\beta=0$ is non-relativistic) and solve by iteration for T_p. The value of F' in the plateau region is

$$F'(z)=X/Y \ , \ X=D_m zF_m T_p - F_p z\gamma_0 [a_z I_2^{(3)}/T_p + \gamma_0 I_0^{(3)} - \gamma_0{}^2 I_2^{(5)}/T_p]$$

$$Y=D_t T_p + \gamma_0 a_z I_2^{(3)} + \gamma_0{}^3 z^2 I_0^{(5)}, \ I_n^{(m)}(z)= \int dx[x^{n+1}\exp(-x^2/2T_p)]/(x^2+z^2)^{m/2}$$

$$a_z=(1+Z)/2 \ , \ D_t=D[\exp(-w_2)-\exp(-w_1)], \ D_m=D_t(T_p=1)$$

$$F_m=\exp(-z^2/2)/(2\pi)^{\frac{1}{2}}, \quad 2w_{1,2}T_p=z^2[1/(v_{1,2}\beta)^2-1]-1/\beta^2$$

subject to $w_2=0$ for $z\leqslant p_2$. The above completes the model relativistic analysis.

RESULTS

Figures 1 and 2 show results for the PLT tokamak, and respectively correspond to Fig.3 in Bonoli and Englade[2] and to Fig.1 in Ignat[1], except that a Fe impurity is specified in Fig.1 and that a H instead of a D plasma is used in Fig.2. Also at the plasma center, we take D=3.6, v_1=2.5 and v_2=7.5. The profiles used are the same as given in Refs.1 and 2, with some parameters specified in Figs.1b and 2b. The differences from Ref.2 in the ray trace towards the third wall reflection and afterwards are due to slightly different equations and to our forced imposition of wall reflection at a certain radius.

Using the relativistic model, the arrow in Fig.1a shows the position where the ray suffers 40dB absorption as indicated by curve 1 in Fig.1b. Using the non-relativistic model, the ray has to propagate further to the end of the trace in Fig.1a to suffer 40dB absorption as shown by curve 2 in Fig.1b. With no hot electrons and a Maxwellian distribution, the attenuation is very much less, even after further propagation, as shown by curve 3 in Fig.1b. Figure 1c based on the relativistic model shows the relative absorption coefficient. Ordinary electron damping is due to collisions, with Landau damping also operational near the peak in the curve. However, complete damping near the end of the trace is caused only by the hot electrons. There, $n_{||}$ has increased to 4.2 from 1.33, and $n_\perp$ to 127. Electron heating occurs near the plasma center. The physics is similar for the non-relativistic model, but a longer path length is required.

Figure 2 shows a situation where the ray is restricted to the outer plasma region by a mode conversion surface[1] and an inner LH surface. The absorption is due to perpendicular H ion Landau damping, as seen in Fig.2c, which causes the 40dB absorption, shown in Fig.2b, at the end of the trace in Fig.2a. Whereas $n_{||}$ decreases from 2.5 to 1.25, $n_\perp$ increases to 640. The electron damping, either with a Maxwellian or non- relativistic or relativistic analysis, has a much lesser effect, so that Figs.2a and 2b are the same for all three approaches. Consequently, ion heating occurs in the outer plasma region. Perpendicular ion Landau damping with D instead of H is much weaker and the ray would not be damped for the same path length, as shown in Ref.1, Fig.1. Similar results were obtained with other conditions and other tokamaks. As the frequency increases or as the density decreases, mode conversion surfaces disappear and the rays penetrate deeper during the first pass inward. The above may be relevant in explaining the density limit and ion heating observed in LH experiments.

REFERENCES

1. D.W.Ignat, Phys.Fluids 24, 1110 (1981).
2. P.T.Bonoli and R.C.Englade, Phys.Fluids 29, 2937 (1986).
3. E.J.Valeo and D.C.Eder, J.Computational Phys. 69, 341 (1987).
4. V.Fuchs,R.A.Cairns,M.M.Shoucri,K.Hizanidis and A.Bers, Phys.Fluids 28, 3619 (1985).
5. V.Fuchs,M.M.Shoucri,C.N.Lashmore-Davies and A.Bers, Bulletin Am.Phys.Soc. 31, 1423 (1986).
6. T.W.Johnston, Can.J.Phys.40, 1208 (1962).
7. I.P.Shkarofsky, Bulletin Am.Phys.Soc. 31, 1626 (1986).

ACKNOWLEDGMENTS

Dr.Shkarofsky wishes to thank Dr.Ignat and Dr.Bonoli for their ray tracing codes. This work is part of a joint project of NRC, Canada, and Hydro-Quebec, carried out by IREQ, INRS, Univ.of Montreal, MPBT Inc. and Canatom Inc.

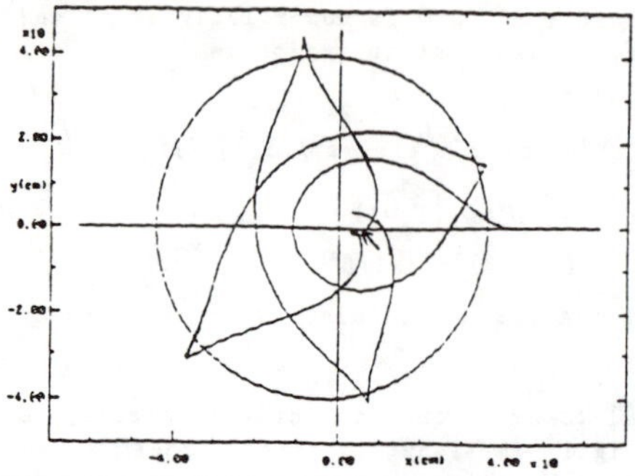

FIG.1a. Ray Trace in Plasma Cross Section. 40dB absorption occurs at end of full curve with non-relativistic analysis of hot electrons, and at arrow with relativistic analysis.

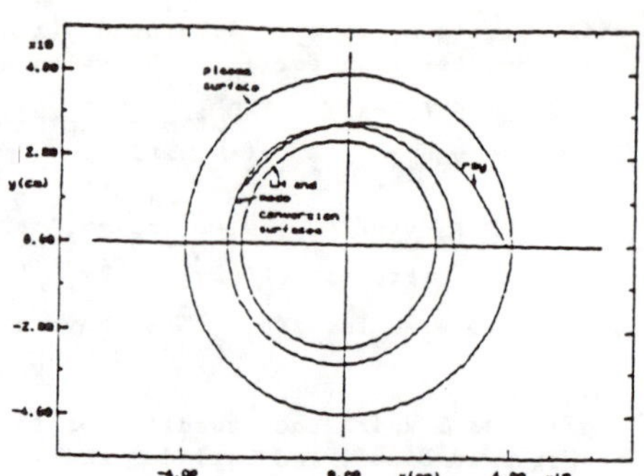

FIG.2a. Ray Trace in Plasma Cross Section. 40dB absorption occurs at end of ray curve near mode conversion surface due to damping by ions, both if hot electrons are included or not.

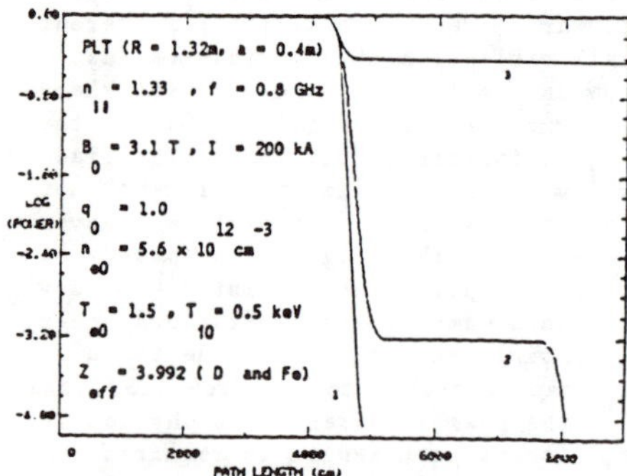

FIG.1b. Power Absorption Versus Path Length. Hot electrons are treated in curve 1 relativistically and in curve 2 non-relativistically. Curve 3 is for a Maxwellian distribution (no hot tail).

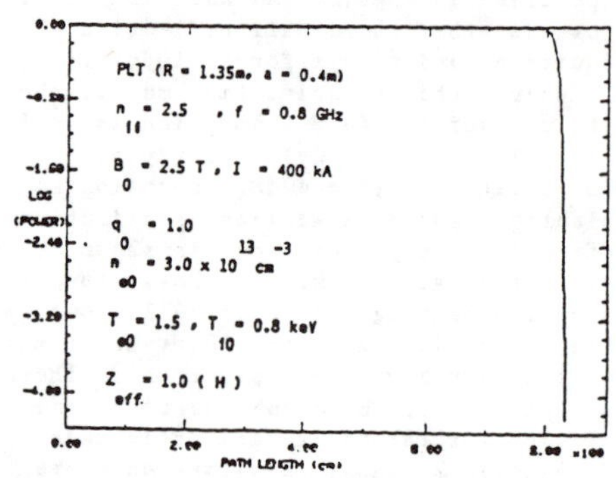

FIG.2b. Power Absorption Versus Path Length, Irrespective of Whether Hot Electrons Are Included or Not.

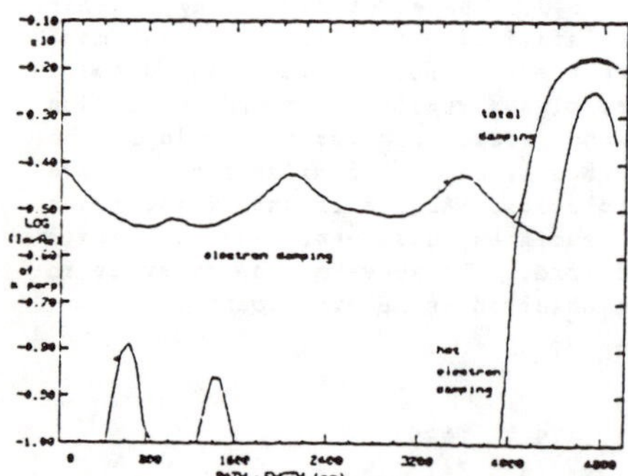

FIG.1c. LOG(Imaginary $k_{\perp}$/Real $k_{\perp}$) Versus Path Length, Showing Relative Magnitudes of Damping Mechanisms(Relativistic Analysis Used). Here near end of path, hot electron damping dominates over collisional plus Landau damping.

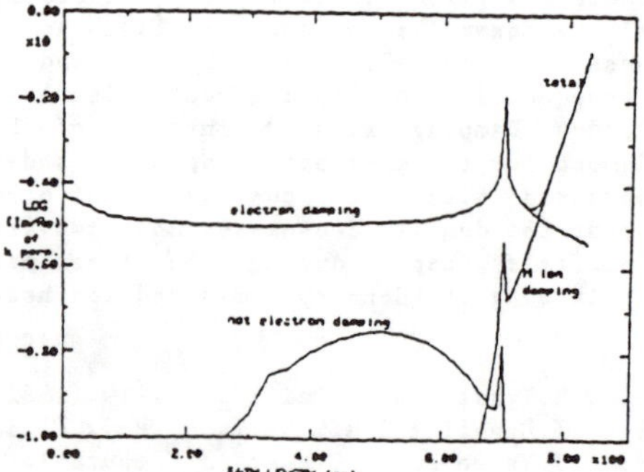

FIG.2c. LOG(Imaginary $k_{\perp}$/Real $k_{\perp}$) Versus Path Length, Showing Relative Magnitudes of Damping Mechanisms. Here, most important near end of path is perpendicular H ion Landau damping.

THE CAUSE OF THE LOWER HYBRID CURRENT DRIVE DENSITY LIMIT

L.H. Sverdrup[(a)] and P. M. Bellan
Caltech, Pasadena, CA 91125

ABSTRACT

We present a model which predicts within a factor of two the observed density limits of all major current drive experiments. This model is based upon linear mode conversion destroying upshifted $k_{//}$ modes which would ordinarily fill the spectral gap.

INTRODUCTION

Fisch[1] predicted that DC toroidal currents could be driven in tokamaks by injecting suitably phased lower hybrid waves. However, it has always been found that current drive works only up to a (rather low) "density limit" not predicted in Fisch's theory.

We present here a new and very simple model which predicts within a factor of two the observed density limits of all major current drive experiments[2-17]. This model is based upon (i) the dependence on parallel wavenumber $k_{//}$ of the coalescence of the two modes involved in linear mode conversion of a lower hybrid wave into a hot plasma wave, and (ii) the upshift in $k_{//}$ that is associated with the filling of the spectral gap in current drive. Before deriving the model, let us briefly review (i) and (ii).

$k_{//}$ DEPENDENCE OF MODE COALESCENCE

Stix[18] showed that when hot plasma effects are included, the lower hybrid dispersion relation becomes

$$k_\perp^4 \varepsilon_{th} + k_\perp^2 \varepsilon_\perp + k_{//}^2 \varepsilon_{//} = 0 \tag{1}$$

where $\varepsilon_\perp = 1-\omega_{pi}^2/\omega^2 + \omega_{pe}^2/\omega_{ce}^2$, $\varepsilon_{//} = 1-\omega_{pe}^2/\omega^2$, $\varepsilon_{th} = -(3\omega_{pe}^2 u_{Te}{}^2/4\omega_{ce}^4 + 3\omega_{pi}^2 u_{Ti}{}^2/\omega^4)$. Equation (1) is quadratic in $k_\perp$ giving two modes:

$$k_\perp^2 = \frac{-\varepsilon_\perp \pm (\varepsilon_\perp^2 - 4k_{//}^2 \varepsilon_{//}\varepsilon_{th})^{1/2}}{2\varepsilon_{th}} \quad . \tag{2}$$

The small $k_\perp$ mode is the launched lower hybrid wave (cold mode) which propagates from the plasma periphery towards the lower hybrid layer (where $\varepsilon_\perp$=0); in the vicinity of this layer the cold mode converts[18] linearly to the large $k_\perp$ hot plasma mode which propagates back towards the periphery and is strongly damped. Mode conversion occurs at the point where the two modes described by Eq.(2) coalesce, i.e., where

$$k_{//}^2 = \varepsilon_\perp^2/4\varepsilon_{th}\varepsilon_{//} \quad . \tag{3}$$

UPSHIFT OF $k_{//}$

There is a well known[20,21] problem concerning use of Fisch's theory to explain the results of current drive experiments. According to Fisch's theory, lower hybrid waves drive current by imparting momentum to electrons in the tail of the distribution function. These electrons can interact with the wave because their velocity u satisfies the resonance condition $u=\omega/k_{//}$. Yet, in all experiments[13] the parallel refractive index launched is typically $n_{//}=ck_{//}/\omega=1.5\text{–}10$ so that, in order for electrons to be resonant with the wave, they must have $3\times10^9<u<2\times10^{10}$. In the experiments the electron temperature has ranged from 50-2000 eV ($3\times10^8<u_{Te}<1.3\times10^9$). Thus, except for the hottest of these plasmas there ought to be essentially zero electrons capable of resonantly interacting with the wave. What seems to happen is that the wave creates its own tail by pulling electrons from the bulk out to high velocities via parallel wave-particle resonant interaction. In order to do this, at least some component of the launched wave $k_{//}$ spectrum must interact resonantly with electrons in the bulk, and so a spectral component must develop[19-21] which has a parallel phase velocity much lower than the launched value.

We will not attempt here to decide what mechanism causes the $k_{//}$ upshift. Instead, we will simply accept the upshift as an experimentally observed fact; i.e. experiment has shown that $k_{//}$ is upshifted for part of the launched power in such a way that

$$k_{//}=\omega/u_{Te} \tag{4}$$

OUR MODEL

We postulate that the density limit occurs when the $k_{//}$ predicted by Eq.(4) - i.e., the $k_{//}$ interacting with the bulk -- is of such a value to satisfy Eq.(3). When this occurs, then: (i) mode conversion takes place for the upshifted $k_{//}$ component which thus becomes strongly attenuated by *perpendicular*[18] damping processes, and so cannot pull electrons from the bulk to the tail by *parallel* wave-particle resonance, so that (ii) there are no tail electrons to resonantly interact with the unshifted (i.e., high phase velocity) component of the incoming wave, so that (iii) there is no current drive. We emphasize that, for the upshifted $k_{//}$ component, mode conversion takes place even though $\omega>\omega_{lh}$.

Proceeding with the mathematical derivation we equate Eq.(4) and Eq.(3) and obtain

$$\frac{\omega_{pe}^2}{\omega^2}=\left[2\left(\frac{3}{4}\frac{\omega^4}{\omega_{ce}^4}+3\frac{m_e{}^2}{m_i{}^2}\frac{T_i}{T_e}\right)^{1/2}+\frac{m_e}{m_i}-\frac{\omega^2}{\omega_{ce}^2}\right]^{-1} \tag{5}$$

All the major current drive experiments had $T_i/T_e\sim0.3$ - 1; however there was a fairly large range of frequencies, magnetic fields and observed density limits. Figure 1 plots Eq.(5) for hydrogen (solid lines) and for deuterium (dashed lines) and also shows the experimentally measured normalized density limits of a large number of experiments (for comparison, Fig.1 also shows the predictions of Wegrowe and Engelmann[24]). Figure 1 also plots Eq.(5) for Argon which was used in the Caltech Encore plasma[23]. Table I presents the same information but with non-normalized parameters; hydrogen gas is assumed, unless specified otherwise, and for cases where T_i and/or T_e were unspecified, an estimate of T_i/T_e=0.3 was used to calculate the predicted density limit.

It is clear from Fig. 1 and Table I that Eq.(5) predicts the experimental observations for a wide variety of parameters. Several[7,12,15] experiments were carefully controlled so as to determine the dependence on just one parameter. Equation (5) is consistent with the observations of these experiments: in particular, Eq.(5) is consistent with the dependence on ω observed in Versator[15] and Petula-B[10,17], the dependence on B observed in FT[7], and the dependence on m_i observed in ASDEX[12]. Possible reasons for the modest (<factor of two) error are: (i) a lack of standardization in definition of the density limit, (ii) some quoted densities are line average rather than peak, (iii) impurities could cause a change in the effective values[8] of m_e/m_i and $(m_e/m_i)^2(T_i/T_e)$, (iv) hot tail electrons could change the effective T_e, (v) variation of B with major radius. Considering all these possible sources of error, the agreement is excellent.

From Eq.(5) and Fig. 1 it is seen that there are essentially two regimes of interest (i) a low field region (slope on LHS of Fig. 1) where the density limit is independent of ω and proportional to B^2, and (ii) a high field region (flat portion of curve in Fig. 1) where the density limit is independent of B, and is instead determined by T_i/T_e, ω, and m_i. The change from regime (i) to regime (ii) occurs at $\omega \approx \omega_{gm}(2T_i/T_e)^{1/2}$, where $\omega_{gm} = (\omega_{ci}\omega_{ce})^{1/2}$ is the geometric mean frequency.

Supported by NSF Grant ECS-8414541.

(a) Present address: Western Research Corporation, San Diego CA 92121

1 N. J. Fisch, Phys. Rev. Lett., **41**, 873 (1978).

2 T. Yamamoto et.al., Phys. Rev. Lett., **45**, 716, (1980).

3 S. C. Luckhardt et al., Phys. Rev. Lett., **48**, 152, (1982).

4 S. Bernabei et.al., Phys. Rev. Lett., **49**, 1255 (1982).

5 M. Nakamura et.al., J. of Phys. Soc. of Japan,**51**, 3696, (1982); S. Tanaka et.al., in Proc. of the Tenth Int. Conf. on Plasma Physics and Controlled Nuclear Fusion Research, London (1984), Vol. 1, p.623.

6 M. Porkolab et.al., in Proc. of the Tenth Int. Conf. on Plasma Physics and Controlled Nuclear Fusion Research, London (1984), Vol 1, p.463.

7 A. Santini, IAEA Technical Com. Meeting on Non-Inductive Current Drive in Tokamaks, Culham UK, 1983; Proc. Culham Lab Report CLM-CD (1983), p.278; see also Table I in Ref. 22 .

8 F. Alladio et al., Nuclear Fusion, **24**,725,(1984).

9 V. V. Alikkaev et al., Ref. 7 , p. 313.

10 C. Gormezano et.al., in Proc. of the Tenth Int. Conf. on Plasma Physics and Controlled Nuclear Fusion Research, London (1984), Vol. 1, p.503.

11 K. Toi et.al., ibid, p.523.

12 F. Leuterer et.al., ibid, p.597.

13 M. Porkolab, IEEE Trans. on Plasma Science, **PS-12**, 107 (1984).

14 M. Porkolab et al., Phys. Rev. Lett., **53**, 450 (1984).

15 M. J. Mayberry et.al., Phys. Rev. Lett., **55**, 829, (1985).

16 M. Yoshikawa et.al., IAEA llth Int. Conf. on Plasma Physics and Controlled Nuclear Fusion Research, Kyoto (1986), paper A-I-1.

17 F. Parlange et al., ibid, paper F-II-3.

18 T. H. Stix, Phys. Rev. Lett., **15**, 878 (1965)

19 P. Bonoli, IEE Trans. on Plasma Science, **PS-12**, 95 (1984).

20 S. Succi et al. in Proc. of the Tenth Int. Conf. on Plasma Physics and Controlled Nuclear Fusion Research, London (1984), Vol. 1., p.549.

21 E. Canobbio and R. Croci, ibid, p.567.

22 J-G. Wegrowe and F. Engelmann, Comments Plasma Phys. Controlled Fusion, **8**,211 (1984).

23 L. H. Sverdrup and P. M. Bellan, Bull. Am. Phys. Soc.,**31**,1578 (1986).

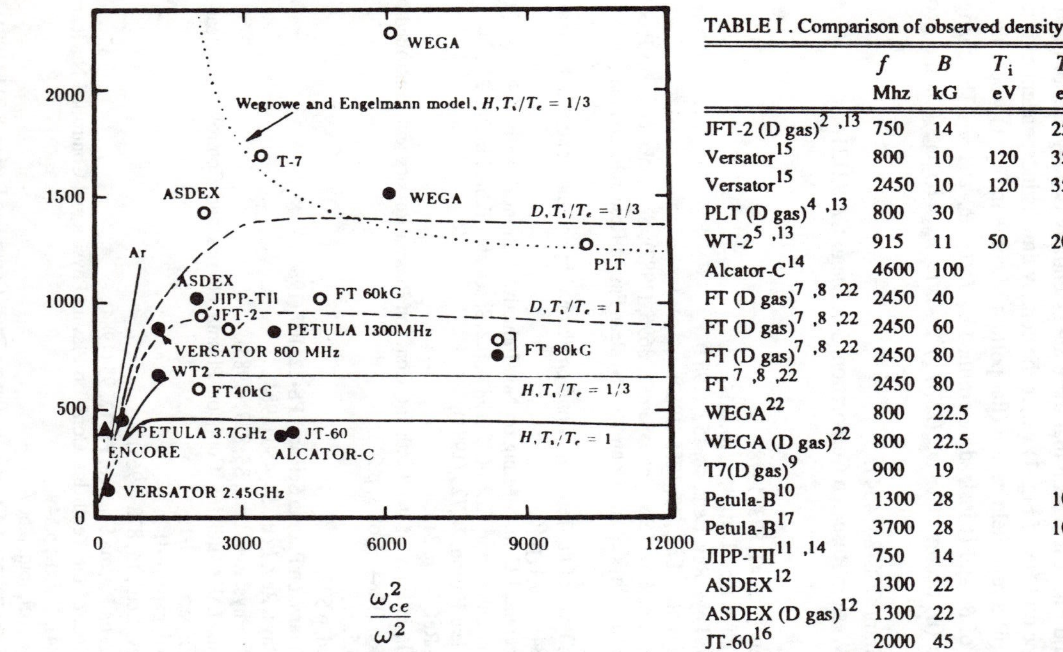

Fig. 1. Density limit predicted by Eq.(5) v. experiments: Solid line is Eq.(5) for hydrogen, dashed line for deuterium. For experiments, solid circles (●) indicate hydrogen, open circles (○) deuterium, and triangle (▲) Argon (all references in Table I). For comparison, dotted line shows Wegrowe and Engelmann model[22] for hydrogen, $T_i/T_e = 1/3$.

TABLE I. Comparison of observed density limit with Eq.(5) prediction

	f Mhz	B kG	T_i eV	T_e eV	$n_{observed}$ $10^{12}cm^{-3}$	$n_{Eq.(5)}$ $10^{12}cm^{-3}$
JFT-2 (D gas)[2,13]	750	14		250	6	9
Versator[15]	800	10	120	350	7	5
Versator[15]	2450	10	120	350	10	12
PLT (D gas)[4,13]	800	30			10	11
WT-2[5,13]	915	11	50	200	7	7
Alcator-C[14]	4600	100			100	180
FT (D gas)[7,8,22]	2450	40			45	94
FT (D gas)[7,8,22]	2450	60			75	103
FT (D gas)[7,8,22]	2450	80			60	100
FT[7,8,22]	2450	80			55	48
WEGA[22]	800	22.5			12	5
WEGA (D gas)[22]	800	22.5			18	11
T7(D gas)[9]	900	19			17	14
Petula-B[10]	1300	28		1000	18	14
Petula-B[17]	3700	28		1000	80	73
JIPP-TII[11,14]	750	14			8	6
ASDEX[12]	1300	22			20	15
ASDEX (D gas)[12]	1300	22			30	27
JT-60[16]	2000	45			20	34
Encore (Ar gas)[23]	450	1.5	5	10	1	.3

LOWER HYBRID WAVE PROPAGATION IN A TOKAMAK

D.C. Stevens and H. Weitzner

Courant Institute of Mathematical Sciences, New York University

ABSTRACT

Cold plasma model wave propagation in an axisymmetric equilibrium has been formulated and a code writen for use at lower hybrid frequencies. A resonance other than the usual lower hybrid resonance has been demonstrated and shown relevant to tokamaks such as Petula-B. A spurious resonance has appeared and work is proceeding to suppress it.

I. COLD PLASMA MODEL WAVE PROPAGATION IN AN AXISYMMETRIC EQUILIBRIUM

We start with Maxwell's equations with the cold plasma dielectric matrix $\underset{\approx}{\varepsilon}$ and cylindrical coordinates r,θ,z in which the toroidal equliibrium magnetic field in a tokamak or RFP is in the θ direction. We assume an axisymmetric equlibrium independent of θ. The ratio of the poloidal to total equilibrium magnetic field is $\sin\alpha \geq 0$. The principal elements of $\underset{\approx}{\varepsilon}$ are $\varepsilon_{\perp}$, $i\varepsilon_x$, and $\varepsilon_{||}$, each dependent on r and z and we define $\Lambda = c/\omega$ and $\underset{\sim}{D} = \underset{\approx}{\varepsilon}\cdot\underset{\sim}{E}$. We assume that the electromagnetic wave fields have time and theta dependence $\exp i(\omega t + M\theta)$, $\mathrm{Im}\,\omega < 0$. We introduce two generalized electromagnetic potentials $\lambda(r,z)$ and $\mu(r,z)$ such that $rE_\theta = \Lambda\lambda$, $rB_\theta = \Lambda\mu$, and for the four-vectors $M(E_r,E_z,B_r,B_z) - r(B_z,-B_r,-D_z,D_r) = -i\Lambda(\lambda_{,r},\lambda_{,z},\mu_{,r},\mu_{,z})$. Maxwell's equations reduce to $\Lambda(E_{r,z}-E_{z,r}) = -iB_\theta$, $\Lambda(B_{r,z}-B_{z,r}) = iD_\theta$. The representation is most easily obtained by considering $M \neq 0$ and $M = 0$ separately, see Refs. 1 and 2. We might solve the four-vector relation for (E_r,E_z,B_r,B_z) as linear combinations of λ,μ and their derivatives. When these relations are inserted into the remainder of Maxwell's equations two second order partial differential equations result. It is this system which we have finite differenced. The algebraic elimination of (E_r,E_z,B_r,B_z) is possible provided the determinant of the system does not vanish or

$$\Delta \equiv (M^2-\varepsilon_{\perp}r^2/\Lambda^2)[M^2 - (\varepsilon_{\perp}\cos^2\alpha + \varepsilon_{||}\sin^2\alpha)(r^2/\Lambda^2)] + \varepsilon_x^2(r/\Lambda)^4\cos^2\alpha \neq 0 .$$

Another quantity of some interest is $e = \varepsilon_{\perp}(\varepsilon_{\perp}\cos^2\alpha + \varepsilon_{||}\sin^2\alpha)$.

Before turning to lower hybrid waves in particular we comment generally on the system. It is easy to verify that provided $\sin\alpha \neq 0$, the two second order differential equations are <u>not</u>

singular at cyclotron resonance.[1] Thus, we conjecture that $\underset{\sim}{E},\underset{\sim}{B}$ and their derivatives are smooth across cyclotron resonance. Since Re $\underset{\sim}{E}^*\cdot\underset{\sim}{J} = 0$ off resonance, we expect it to vanish at resonance and no heating occurs at cyclotron resonance. Hence, heating occurs in the cold plasma model only when $\underset{\sim}{E},\underset{\sim}{B}$ or $\underset{\sim}{J}$ become unbounded. We treat unbounded solutions by the usual device of adding a small non-hermetian part to $\underset{\approx}{\varepsilon}$, specifically $i\eta\underset{\approx}{I}$, and we take the limit numerically as $\eta \to 0$. We comment on this limit further shortly.

In order to discuss unbounded solutions we must consider the mathematical structure of the system. Obviously, we have a fourth order system of partial differential equations which is singular at $\Delta = 0$. We discuss $\Delta = 0$ shortly. A simple calculation shows that the system always has two imaginary characteristics and the remaining two characteristics are imaginary if and only if $e > 0$. Thus, for $e > 0$ the system is elliptic, for $e < 0$ the system has both elliptic and hyperbolic properties and the surface $e = 0$ is a "sonic" transition. The "sonic" transitions occur at $\varepsilon_\perp = 0$, the usual lower hybrid resonance, and $\varepsilon_\perp \cos^2 \alpha + \varepsilon_{||} \sin^2 \alpha \equiv f = 0$. The transition $f = 0$ is a direct consequence of the toroidal geometry. In a non-periodic geometry the equivalent transition would be $\varepsilon_\perp + \varepsilon_{||} = 0$. Although $|\varepsilon_{||}| \gg |\varepsilon_\perp|$, $\sin^2 \alpha = 0$ on the magnetic axis so that $f = 0$ will occur in tokamaks in which $\varepsilon_\perp$ never vanishes. We show the curve $f = 0$ superimposed on the numerical calculations.

Much is known about linear partial differential equations for which there are surfaces of change of type. That experience suggests that in general surfaces of change of type are not usually regions of singularities of solutions. For the system that occurs here only on those curves which lie on the surface $e = 0$ <u>and</u> which are tangent to a flux surface are singularities likely to occur. Additionally, singularities may appear on the characteristic surfaces leaving these curves. These phenomena have been demonstrated numerically by a study of the model equations $\left((x-y^2+i\eta)u,_x\right),_x \pm u,_{yy} = 0$. We do not give the results here, but it is easy to find that the curve of change of type $x-y^2 = 0$ is tangent to a "flux surface" x = constant only at $x = y = 0$, so that solutions are singular at $x = y = 0$ and on some characteristics leaving this point for the + sign only. Thus, for the cold plasma model we expect singularities, and heating only on the curves of tangency of resonance $e = 0$ and flux surfaces.

We must now examine possible singularities at $\Delta = 0$. We cannot yet give definite numerical evidence or analytic arguments, but we believe no singularities appear here. First, we note that the linear sixth order system is regular at $\Delta = 0$. Second, for small η only when the solution is demonstrably numerically unreliabe do we see any effects at $\Delta = 0$. For ordinary differential euqations this problem is well-known, and the "singularity" at $D(x) = 0$ in

$[N(x)y'(x)/D(x)]' = M(x)y(x)$ is eliminated by introduction of the system $y'(x) = D(x)w(x)$, $[N(x)w(x)]' = M(x)y(x)$. Thus, we believe no singularities appear at $\Delta = 0$, although we do have numerical problems which so far limit the accuracy and reliability of our solutions.

II. NUMERICAL COMPUTATION

So far we have carried out calculations with a Soloviev equilibrium with magnetic flux function

$\psi(r,z) = \frac{1}{4}\lambda(r^2-r_o^2)^2/r_o^2 + \frac{1}{2}\mu z^2$, current flux function $I^2(\psi) =$

$1 - 2\beta_p\mu\psi/r_o^2$ and density profile $n(r,z) = n_o[(1-n_v)(1-\psi/\psi_M)^2 + n_v]$. We have adjusted the constants to approximate Petula-B with[3] r_o=72cm, a = 17cm, B_o = 2.8T, $\bar{n}_e = 6\times10^{13}\text{cm}^{-3}$, I_p = 150kA, ω = 1.3GHz,[3] and we have taken M = 32.55 in the accompanying figures. We give the contour lines of Re E_θ, Im E_θ, Re B_θ, Im B_θ, and energy deposition = Re $(\tilde{E}^*\cdot\tilde{J})$. We have assumed that there are antennas in the boundaries which are sources of B_θ, but not $E_\theta = 0$. Only the middle third of each side is a non-zero source of B_θ. In these figures the horizontal coordinate is z and the vertical coordinate is r. The inside of the tokamak is at the bottom of the figure and the outside is at the top. We have taken a grid of 80 × 80 points. The value of η is .03 and should be compared with diagonal elements of $\underset{\approx}{\varepsilon}$ which are of order one. The resonance curve $f = 0$ is superimposed on all the figures and is indicated on one. Particularly in the curves for Im E_θ there is indication of irregular behavior on an oval with diameter about one half the box size. This curve is $\Delta = 0$, and if η were made smaller more irreproducible and irregularities would appear. For this mesh, .03 is about the smallest reliable η. No resonance at $\varepsilon_\perp$ is present in the domain. Heating clearly appears at a point of tangency of $f = 0$ and a flux surface. The resonance is quite sharp to ~1% as M is varied. Clearly M should be an integer but for preliminary studies we vary M. We appear to find some plasma heating near the magnetic axis, but we consider this a numerical artifact. In order to verify this conclusion we would need smaller grids and smaller values of η. Only after we have eliminated the false singularity at $\Delta = 0$ can we perform such studies entirely satisfactorily. But such studies done to date support the conclusion that no heating occurs at the magnetic axis.

REFERNCES

1. H. Weitzner, Wave Propagation Based on the Cold Plasma Model, CIMS-NYU Report MF-103.

2. H. Weitzner, Comm. Pure Appl. Math., 33 99 (1985).

3. C. Gormezano, et al., in Radio Frequency Plasma Heating, Sixth Topical Conference, Callaway Gardens, Georgia, 1985, ed. D.G. Swanson (American Institute of Physics, N.Y., 1985), p. 111.

Acknowledgment: This work was supported by the U.S. Department of Energy, Office of Fusion Energy.

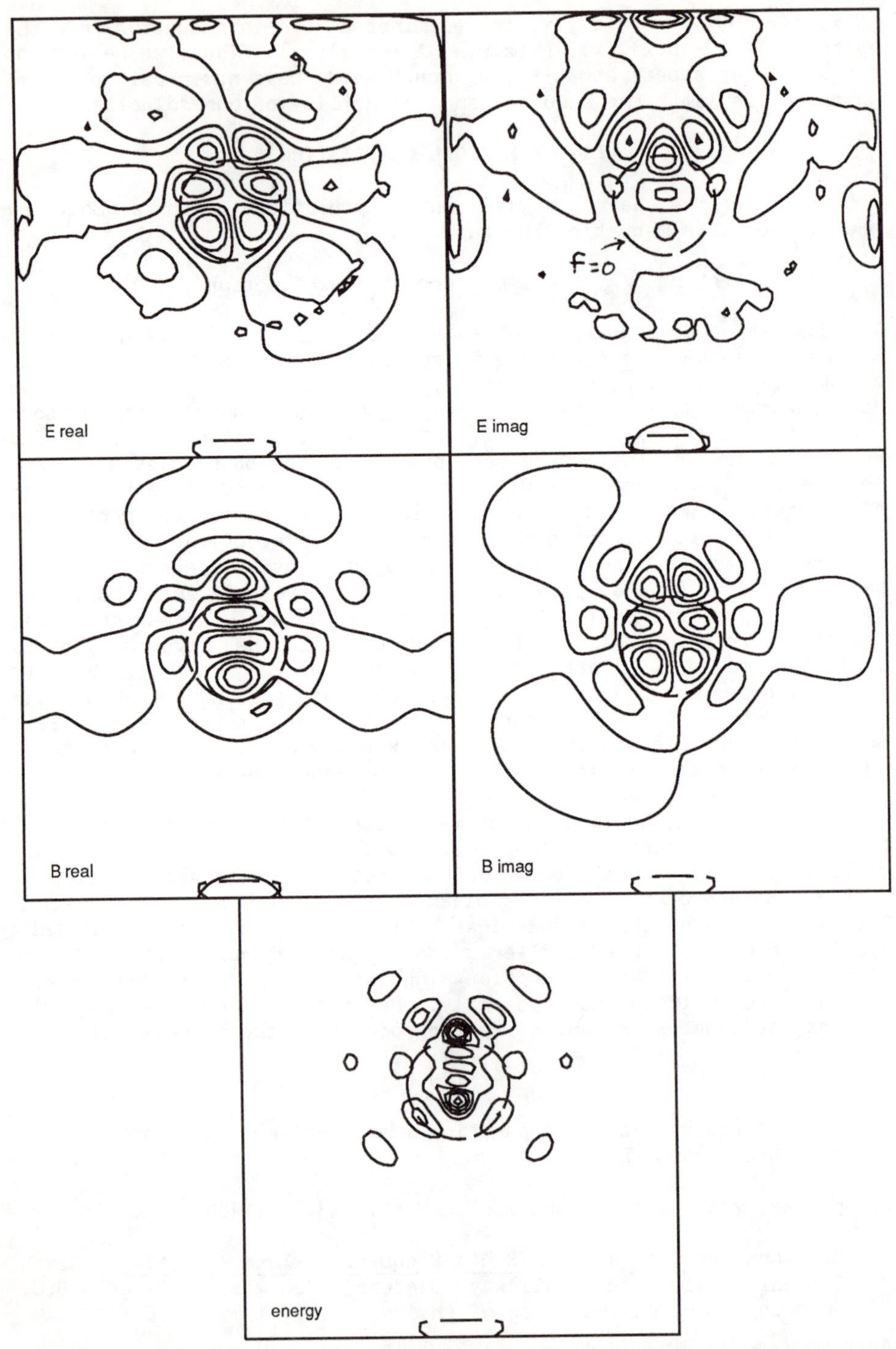
E real
E imag
f=0
B real
B imag
energy

RAY TRACING ANALYSIS OF LH FAST WAVES IN CCT

T.K. Mau, K.F. Lai, R.J. Taylor
University of California, Los Angeles, CA 90024

ABSTRACT

Lower hybrid fast wave field profiles have been measured in the CCT tokamak. These field maps exhibit characteristics of large $n_{//}$ wave components, with $n_\perp/n_{//} \approx 1$, and the wave energy appears to be heavily damped in the toroidal direction. We use a 3-D toroidal ray tracing code to study this wave behavior. The spectral filtering of the fast wave can be partially attributed to mode conversion to the slow wave for the low $n_{//}$ components. In most parameter regimes, subsequent strong absorption of the converted slow wave via electron Landau and higher harmonic ion damping results.

INTRODUCTION

Current drive was observed in the UCLA Continuous Current Tokamak (CCT) when the fast wave was launched from a toroidal array of 8 large-area loops.[1] While the current drive mechanism is being resolved, we take advantage of the steady state feature of CCT to study the detailed propagation and absorption physics of the fast wave launched into the torus. We have measured the wave field profiles of the fast wave in the lower hybrid (LH) range of frequencies, using specially constructed, movable magnetic probes.[2] The wave is generated either by a single large-area antenna module or by a small loop which can be inserted radially along the equatorial plane inside the plasma. The field maps depict characteristics of large $n_{//}$ wave components, with $n_\perp/n_{//} \approx 1$, while the wave energy appears to be heavily damped in the toroidal direction. Independent probe measurements also confirm the existence of short wavelength electrostatic modes in some parameter regimes, pointing to the possibility of <u>m</u>ode <u>c</u>onversion to the <u>s</u>low <u>w</u>ave (MCSW).

RAY TRACING MODEL

In this paper, we study the LH fast wave in CCT with a 3-D toroidal ray tracing code,[3] in order to verify the observed wave characteristics and identify the damping mechanisms. For the cases we study, $\omega \gg \omega_{LH}$, so that the cold plasma dispersion relation $D(\vec{r},\vec{k},\omega)=0$ can be used. The ray equations are solved in (R,ϕ,Z) space while the plasma profiles are described in flux-surface coordinates (ρ,θ,ϕ). Absorption along the ray is expressed in terms of the damping decrement γ, defined by $\gamma=2\,\mathrm{sign}(\partial D/\partial\omega)D_I/|\partial D/\partial\vec{k}|$ where D_I is the imaginary part of D due to a specific damping process. For electron Landau damping, $D_{Ie}=2\pi^{1/2}(\omega_{pe}^2/\omega^2)n_\perp^2 n_{//}^2 x_{oe}^3 \exp(-x_{oe}^2)$, where $x_{oe}=\omega/k_{//}v_e$, while for ion harmonic damping, $D_{Ii}=2\pi^{1/2}(\omega_{pi}^2/\omega^2)n_\perp^4 x_{oi}^3 \times\exp(-x_{oi}^2)$, with $x_{oi}=\omega/k_\perp v_i$, which is valid for $\frac{1}{2}k_\perp^2\rho_i^2 \gg 1$.[4] The last term is usually significant only for the slow mode. For Coulomb collisional absorption, D_{Ic} is obtained simply by replacing m by $m(1+i\nu_{ei}/\omega)$ in D, with $\nu_{ei}=9.2\times10^{-11}Z^2n_e(cm^{-3})T_e^{-3/2}(keV)\ln\Lambda$.

In the LH regime, cutoff and mode conversion between the fast and slow waves involve turning points in 3 dimensions, which, for all practical purposes, do not coincide along the ray. As such WKB approximation is always valid and mode conversion and reflection are adequately described by the ray equations.[5] We start each ray at a fixed distance inside the plasma, where only a window of incident n_ϕ can propagate, being bounded by coalescence with the slow wave from below and by evanescence from above. The ray is allowed to undergo specular reflection as it hits the wall.

NUMERICAL RESULTS

We investigated the LH fast wave in two regimes: (A)high field(1kG),high frequency(80MHz), and (B)low field(250G),low frequency(21MHz). The plasma is a weakly ionized He gas with $Z_i=1$, being sustained by energizing the OH primary at a frequency of 3 kHz. The plasma profiles take the form: $n_e=n_{eo}(1-(\rho/a)^2)^{\mu_N} + n_{ea}$ and $T=T_o(1-(\rho/a)^2)^{\mu_T} + T_a$. For our calculations, $n_{eo}=2\text{-}4\times10^{13}cm^{-3}$, $n_{ea}=2\times10^{10}cm^{-3}$, $\mu_N=1\text{-}1.5$, $T_{eo}=20eV$, $T_{ea}=10eV$, $T_i=0.5T_e$, $\mu_T=0.5$, $q(0)=1.$, $q(a)=40(I_p=1kA)$. The CCT plasma has the following geometry: R=150cm, a=40cm, elongation $\kappa=1$ and axis shift $\delta=0\text{-}0.5$.

Case (A) $B_0=1kG$, f=80MHz: In this case, $n_{eo}=2\times10^{11}cm^{-3}$, $\mu_N=1$ and $\delta=0$. Five rays are launched from the outboard edge along the equatorial plane and followed for about 2 toroidal transits, with initial n_ϕ of 2.5, 3.0, 3.5, 4.0 and 5.0. Their trajectories are plotted in Fig.1(a) in the form of ρ/a as a function of $L/2\pi R$, where L is the arclength. We note that for $n_\phi=2.5$ and 3.0, the incident wave undergoes MCSW which then propagates radially outward. The $n_\phi=2.5$ ray is confined to the outer periphery of the plasma while for $n_\phi=3$, MCSW actually takes place deep inside the plasma. For the other 3 rays, penetration to the core occurs in almost every radial transit, without MCSW. From Fig.1(a), we see that the higher n_ϕ is, the more readily the rays penetrate to the center. Apparently, these are also waves with high values of $V_{g\perp}/V_{g//}$. In fact, for maximum $V_{g\perp}/V_{g//}$, it can be shown that $n_{//}^2=0.77\omega_{pe}^2/(\omega\omega_{ce})$, $n_\perp/n_{//}=0.83$ and $(V_{g\perp}/V_{g//})_{max}=0.31$. In Fig.1(b),(c), we plot respectively $|n_\perp/n_{//}|$ and $|V_{g\perp}/V_{g//}|$ along the toroidal distance for initial $n_\phi=5$. $|V_{g\perp}/V_{g//}|$ reaches a maximum of 0.31 at the plasma center while on the average $|n_\perp/n_{//}|=1$. We conclude that the experimentally observed field pattern has features similar to those of the waves with maximum $V_{g\perp}/V_{g//}$. It should be noted that at higher densities, the characteristic $n_{//}$ is observed to increase as $n_e^{1/2}$, but still $n_\perp/n_{//}=1$. However, for these rays, only weak damping has been observed over the length of the trajectories.

Case (B) $B_0=250G$, f=21MHz: For this case, $n_{eo}=4\times10^{11}cm^{-3}$, $\mu_N=1.5$ and $\delta=0.3$, the outward shift being set by the vertical field. Rays are launched from three locations, namely: outboard, inboard and top. These rays all have similar trends with respect to their incident n_ϕ. At the lower end of the propagating n_ϕ window, the wave quickly undergoes MCSW, which is subsequently trapped in the plasma periphery, suffering weak collisional absorption. For intermediate values of n_ϕ, the wave usually mode converts during its initial transit, reflects from the wall with an anomalous increase in $n_{//}$ and is then damped by electrons, ions or both. An example of this is shown in Fig.2 for a ray with n_ϕ=8 incident from the inboard midplane. In this case, electrons account for 46% of the absorbed power while the rest is deposited in the ions. In Fig.2(d), at ϕ=2 rad, MCSW is clearly indicated. Much of the absorption takes place on the outer half of the plasma and in some cases, it takes only one toroidal transit for total absorption. On the high end of the n_ϕ window, waves launched from the inboard side and top still get converted, but the slow wave is only weakly damped by electrons. For those launched from the outboard edge, the fast wave is well focused in the magnetic axis, suffering moderate electron damping.

We also studied the case of a small loop being placed in the geometric center of the torus and aligned to excite mainly the fast wave.[2] Because of the location of the launcher, only high n_ϕ(>18) wave components can propagate in its vicinity. Mode conversion and subsequent damping can occur for the lower n_ϕ waves. For higher n_ϕ, the fast wave persists and is observed to focus towards the magnetic axis, on which most of the power is deposited.

CONCLUSIONS

Much of the experimentally measured LH fast wave characteristics have been demonstrated by 3-D ray tracing analyses. Mode conversion to the slow wave can take place deep inside the plasma and lead to strong damping of the wave energy by both electrons and ions. However, higher n_ϕ wave components are less susceptible to this process and thus dominate the global wave structure. Finally, wave scattering off density fluctuations may also play an important role and ray tracing studies in such a medium may have to be considered.

This work is supported by USDOE Contract DE-AM03-76SF-0010, Mod.1001.

REFERENCE

1. R.J. Taylor, J. Nucl. Mat. 145-147(1987)700.
2. K.F. Lai, et al, this conference.
3. T.K. Mau, R.W. Conn, Bull. Am. Phys. Soc. 28(1983)1091.
4. D.W. Ignat, Phys. Fluids 24(1981)1110.
5. D.G. Swanson, A.H. Glasser, Bull. Am. Phys. Soc. 27(1982)965.

FIGURE CAPTIONS

1. (a) Ray trajectories with initial n_ϕ=2.5,3.0,3.5,4.0,5.0 for case(A). (b) $|n_\perp/n_{//}|$ and (c) $|V_{g\perp}/V_{g//}|$ as a function of toroidal angle ϕ for initial n_ϕ=5.0 in case(A).
2. Example of a ray in case(B): (a) trajectory in (R,Z) space, (b) normalized ray power along R, (c) $n_{//}$ evolution and (d) $(n_\perp/n_{//})^2$ of fast(F) and slow(S) waves along ray.

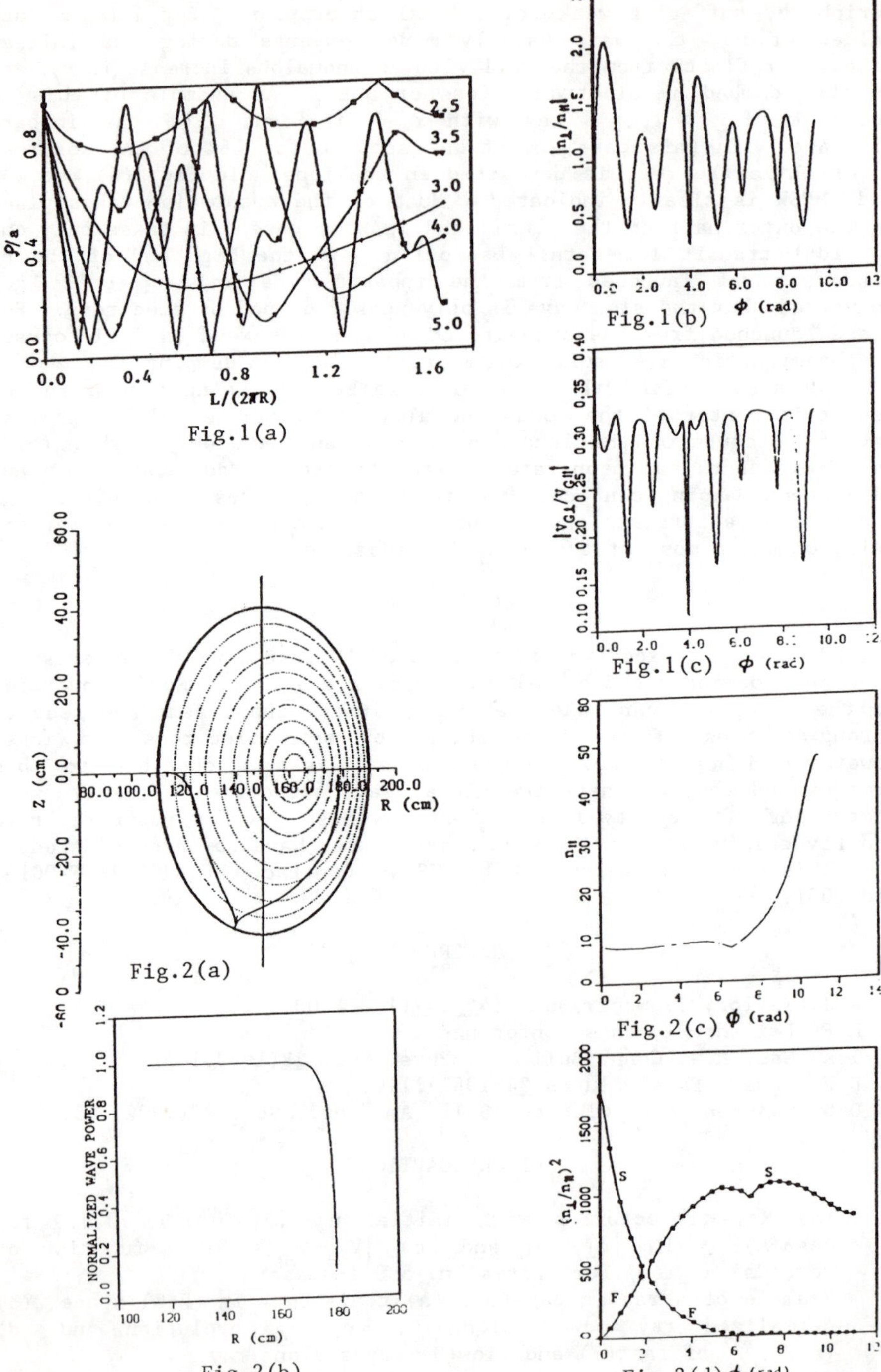

Fig.1(a)

Fig.1(b)

Fig.1(c)

Fig.2(a)

Fig.2(b)

Fig.2(c)

Fig.2(d)

ICRF HEATING IN FUSION PLASMAS

Karl Steinmetz
Max-Planck-Institut für Plasmaphysik,
EURATOM Association, D-8046 Garching, F.R.G.

Abstract
A review is given of the potential of ICRF heating in fusion plasmas, which, in recent years, has become comparable in its achievements to those of conventional neutral beam injection. It is attempted to present an overview of ICRF heating activities in various fusion devices. The main emphasis is put on ICRF in tokamaks, which apply it most intensively and at the highest power level. Some of the important technological aspects and problems relating to ICRF operation in fusion experiments are also described, this being followed by a summary of most recent experimental results and a comparison with theoretical expectations. Finally, the prospects of ICRF heating for future activities in devices for controlled nuclear fusion research are discussed.

INTRODUCTION

In order to reach the parameters for plasma ignition, ohmic heating seems to be insufficient and auxiliary heating of the plasma is required. Neutral beam injection (NBI) and heating in the ion cyclotron range of frequencies (ICRF) are used in present fusion devices /1-12/ as additional heating methods with the highest power levels to study plasma stability and confinement. In the past NBI has proved to be a reliable heating scheme not causing deleterious impurity problems. However, the application of NBI in present-day experiments is already hampered by the unavoidable refuelling rendering the plasma density control rather difficult. With respect to reactor application, NBI based on the established technique of positive ions has the disadvantage of shallow power deposition. Moreover, the injectors have to be placed close to the machine and be shielded, the problem of the nuclear shutter is not yet solved, and the particle energy requirements of the neutrals significantly lower the system efficiency.

ICRF heating may offer many advantages for auxiliary heating of future experiments operating close to reactor conditions. Heating and particle refuelling are decoupled, the rf generators can be located far away from the device and the overall system efficiency is higher. Recent investigations performed in the experiments listed in Table I have demonstrated the high heating potential of ICRF for various plasma configurations and have allowed critical comparison with NBI. Generally, ICRF experiments are performed either in the two-ion mode (minority heating, mode conversion), the second harmonic regime ($2\Omega_{ci}$) or with ion Bernstein waves (IBW). The purposes of applying ICRF to tokamak, stellarator (Heliotron E) and mirror (TMX-U) plasmas are mainly bulk plasma heating with the emphasis towards confinement studies as well as heating of passing ions to reduce their collisional trapping in the central cell thermal barriers of tandem mirror machines.

Exp	R(m)	a(m)	$\bar{n}_e$ (10^{13} cm^{-3})	f(MHz)	P_{rf} (MW)	t_{rf} (sec)	Heating scheme
Tokamaks:							
JET	3	1.2 (b=1.7)	< 4	25-55 <u>48</u>	7.2	< 8	D(H), D(^{3}He), ICRH + NI
JT-60	3	0.9	< 4	110-130	1.5	< 2	$2\Omega_{CH}$ in H
TEXTOR	1.75	0.46	< 6	25-29	2.2	< 2	MC(D(H))
ASDEX	1.67	0.40	< 9	30-115 <u>33.5/67</u>	2.7	< 2	D(H), $2\Omega_{CH}$ in H and H/D, ICRH + NI
JFT-2M	1.31	0.42 (b=0.60)	< 5	15.4/16.8	2	0.3	MC(D(H)) ICRH+NI
PLT	1.32	0.4	< 4	25/42 <u>30</u>	4.3	0.25	D(H), D(^{3}He) $2\Omega_{CH}$ in H, IBW, ICRH + NI
JIPPT -IIU	0.91	0.23	< 7	40	2.2	< 0.1	MC(D(H)), IBW
TFR	0.98	0.19	10	60	2	0.1	MC(D(H)),D(H)
ALC.C	0.64	0.13	20	180	0.5	0.07	D(H),$2\Omega_{CH}$,IBW
Stellarator:							
Hel.E	2	~ 0.2	< 2.5	26.7	1.5	< 0.2	MC(D(H)), IBW
Mirror:							
TMX-U	-	-	< 0.4	1.5.-4	0.24	0.04	slow wave

<u>Table I:</u> List of experiments (typical data achieved): R, a, $\bar{n}_e$, f, P_{rf}, t_{rf} are the major and minor radii, the line-averaged electron density, the frequency, the power launched by the antennae and the rf pulse length.

THEORETICAL BASIS

Heating in the ion cyclotron range of frequencies is related to the existence of a resonance layer in the plasma cross-section, like the ion cyclotron layers and/or the two-ion hybrid resonances /13/. ICRH can be applied at the fundamental frequency to heat a hydrogen or helium minority in a deuterium bulk plasma (D(H), D(^{3}He)) or to utilize mode conversion, MC(D(H)), or at the hydrogen second harmonic frequency $2\Omega_{CH}$ to heat pure hydrogen plasmas or hydrogen-deuterium mixtures of about equal parts. In all these schemes a fast

wave (with the electric field polarized perpendicularly to the magnetic field) is excited by the antennae, while ion Bernstein wave heating works with the slow electrostatic wave with the electric field polarized parallel to B_0. An overview of the regimes applied in various experiments is listed in Table I.
A variety of theoretical models and numerical computations have been developed to describe the rf properties of the ICRF antenna and its field pattern, the propagation of the waves and the absorption of the rf power. Many antenna calculations based on slab models can be found, e.g. in /14-16/. They are partly integrated into fast simulations of the whole tokamak /17-19/. Ray tracing computations in the full tokamak geometry, giving essentially the first transit absorption, have been applied to various machines, but they are reliable only in relatively large devices /20-22/. Recent investigations are based on global wave solutions: Semi-analytic models are applied in /23,24/. Cold plasma approximations including collisional damping only are given in /25,26/, while finite Larmor radius (FLR) effects have been added in /27/. Brambilla et al. /28/ have considered FLR effects and kinetic damping by solving the integrodifferential equations. The integration of these models into transport codes should ultimately lead to a determination of ICRF power deposition profiles and allow the evaluation of the transport properties of ICRF-heated plasmas. Good agreement between the results of ray-tracing computations, which affort reasonable confidence for large plasma sizes, and experimentally determined electron heating power density profiles have been found at JET /2/.

ICRF TECHNOLOGY

High-power ICRF technology has made significant progress in recent years. Tubes of up to output powers of 2.5 MW as well as fully coaxial, tunable high-power amplifiers in the frequency range 30 to 120 MHz are available. The coaxial transmission lines are usually pressurized (SF_6 or dry air) with voltage stand-off capabilities of about 30 to 40 kV which, however, still appear to be too low for reliable high-power, long-pulse operation. Recent test bed investigations /29/ indicate much higher voltage stand-off capabilities (> 60 kV) due to optimization of the ceramic spacer geometry.
The most critical components for launching ICRF waves are the antennae. Present designs are based on all-metal geometries with double or single-layer Faraday screens. Optically open Faraday screens as applied in JT-60 and ASDEX /3, 30/ have shown no disadvantages with respect to plasma and rf performance, but they provide much better conditions for active cooling than opaque versions. Figure 1 shows, as an example, a poloidal cross-section of ASDEX with a loop antenna at the low-field side. Low-field side excitation is particularly useful for heating in the $2\Omega_{ci}$, D(H), D(^{3}He) and IBW regimes /2,3,5,6,9,12/, while the mode conversion scheme (MC(DCH)) requires high-field side launching /4,7,8,10,11/. The location of ICRF antennae at the inner wall of future devices appears to be unacceptable from the point of view of reactor relevance.

Up to now ICRF antennae have been designed conservatively with respect to the averaged rf power densities in front of the antennae (defined as launched rf power per Faraday screen surface). The values reached in recent and present experiments are summarized in Fig. 2. The power density varies typically in the range 0.3 to 0.7 kW/cm^2, which is still too low as compared with, for example, the approximate 3 kW/cm^2 required for NET. Thus, the emphasis in antenna development has to be put on increasing the power density limit of ICRF antennae.
Another critical issue is the distance between the plasma edge (separatrix) and antenna. On the one hand, this distance has to be kept as small as possible to provide maximum loading resistance of the antenna, and, on the other, it has to be compatible with plasma requirements, such as accessibility of the H-mode with divertor plasmas. In the case of ASDEX a reasonable compromise of about 3 cm, determined by its divertor configuration and scrape-off layer, has been found /32/. Nevertheless, the response of the rf system to fast changes in coupling as observed on transition into the H-mode in ASDEX, JET, D-III D /33-35/ or in pellet refuelling is still strong. Figure 3 displays the response of the maximum rf voltage in both ASDEX vacuum transmission lines, the reflected power at both transmitters and the loading resistance of both antennae during an H-mode transition indicated by the drop of the H_α-signal in the divertor. Owing to the development of the edge electron density and temperature pedestals the loading resistance abruptly decreases after the H-transition, leading to a rising system voltage (the rf power is kept constant by feedback control) which ends up in system breakdown. To solve these problems, the voltage stand-off capability of the lines on ASDEX has been improved, whereas fast power-feedback seems to be inadequate.

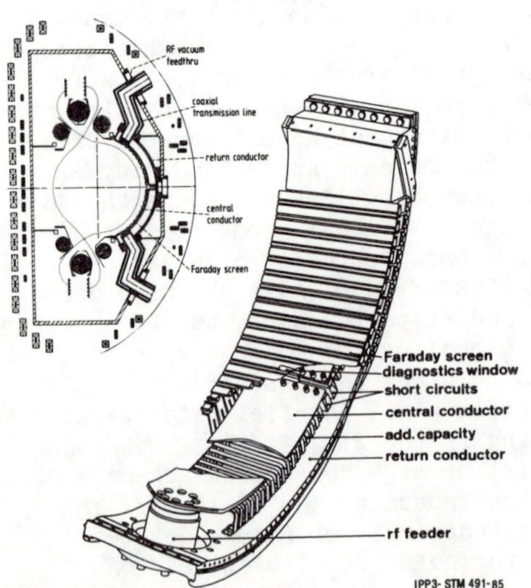

Fig. 1 Poloidal cross-section of ASDEX with an ICRF antenna.

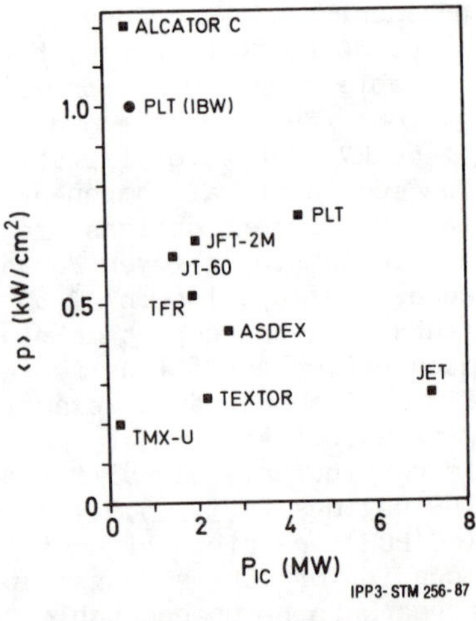

Fig. 2: Averaged rf power densities at the antennae.

EXPERIMENTAL ACHIEVEMENTS

The typical parameter ranges of electron density and plasma temperature of various ICRF experiments are displayed in Fig. 4. TMX-U works at low density and low magnetic field ($\bar{n}_e = 4\text{x}10^{12}$ cm^{-3}, $B_o = 0.5$ T). On the other hand, ICRF has been applied to high-field, high-density tokamak plasmas in TFR and ALCATOR C with $\bar{n}_e$ up to $2\text{x}10^{14}$ cm^{-3} at $B_o \leq 12$ T.

A measure of the ICRF heating capability is the plasma-volume-averaged rf power density, plotted in Fig. 5 vs the volume of the machines. The smallest devices are found to work at the highest rf power densities in the plasma.

Heating efficiency: The success of rf heating experiments has been characterized by a rough measure, the so-called heating efficiency $\eta = \bar{n}_e\ (\Delta T_{io} + \Delta T_{eo})/P_{rf}$. In order to compare different devices, it is useful to normalize the data to the plasma volumes V_p and the "target" confinement times τ_E. Figure 6 gives an (incomplete) overview of experimental results, taken from the literature, where the dimensionless heating efficiencies are

$$\eta^* = \eta \cdot V_p/\tau_E$$

Within a factor of two the normalized heating efficiencies (which do not take profile effects into account) are found to be rather independent of the configuration of the experiment, the plasma size, the rf power and the heating scheme. No significant differences of ICRH (D(H), $2\Omega_{ci}$, MC(D(H)), IBW) and NBI within the range of uncertainty of η^* are observed.

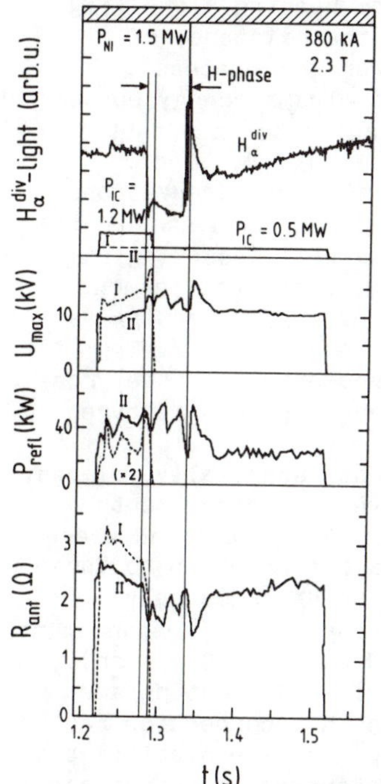

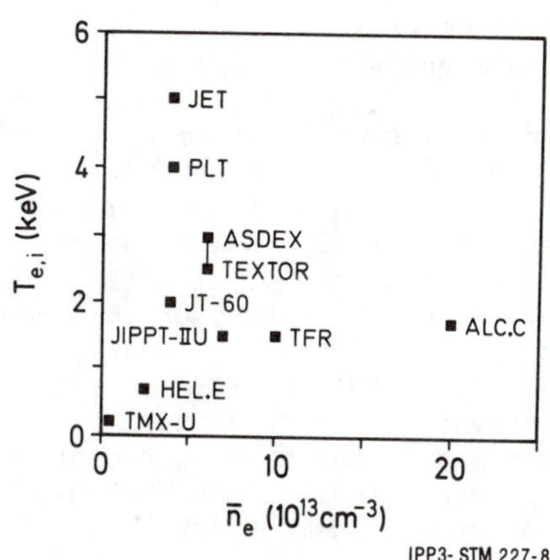

Fig. 4: Typical parameter ranges of $T_{e,i}$ and $\bar{n}_e$ of various experiments.

Fig. 3: Response of the ASDEX rf system to fast changes in coupling during an H-mode transition.

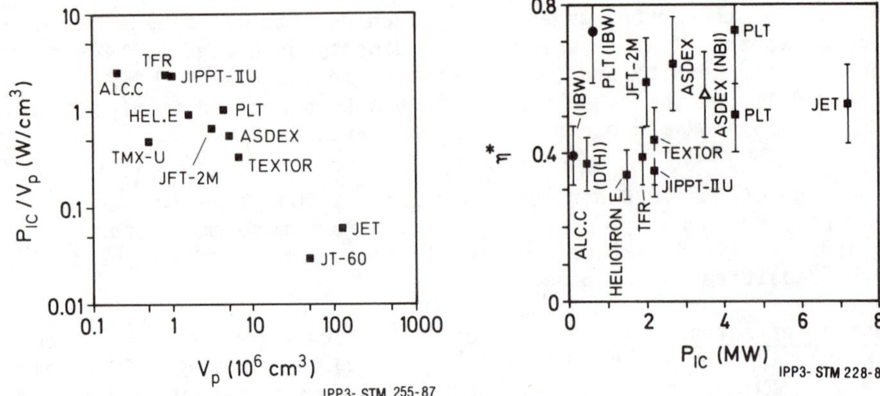

Fig. 5: Plasma-volume-averaged rf power densities of various experiments

Fig. 6: Summary of normalized heating efficiencies

Confinement: The key problem of tokamak research is the accessibility of confinement regimes with τ_E-values sufficiently high to reach ignition. Figure 7 displays the confinement of ICRF and NI-heated plasmas in ASDEX and JET /5,1/. At higher power the rf scenarios (D(H), $2\Omega_{CH}$) essentially follow an L-mode scaling, while in ASDEX the D(H) scheme appears to be slightly better than $2\Omega_{CH}$. When ICRH and NI are combined, the degradation of confinement (at the same power) seems to be less severe than with beam heating alone (Fig. 7a). In particular, ICRH + NI data indicate the existence of a substantial residual confinement at high power (τ_E^*-plateau). This is consistent with modelling the increase of plasma energy content by $W_p(t) = W_p(OH) + \Delta W_p\ (1-\exp(-t/\tau_{inc}))$ leading to an offset-linear scaling of the form $W_p = W(0) + \tau_{inc}\ P$ /1,2,4,5,7,8/. Although the offset-linear scaling agrees with many observations based on $2\Omega_{CH}$, D(H), D(^{3}He) and NBI heating, most data can be fitted with the Kaye-Goldston scaling /36/ as well (assuming a continuous decrease of τ_E with power). An important confinement aspect is the attainment of the H-mode with pure ICRF heating as first found in ASDEX /37/ and also observed in JFT-2M /7/. The comparison of NBI and ICRH L- and H-modes documents that the gross confinement structure of tokamaks is an inherent plasma feature independent of the heating technique.

As a general finding, the particle confinement essentially follows the energy confinement behaviour. In low-power IBW experiments /6,9/, however, a decoupling of τ_p and τ_E has been observed where the particle confinement times increase by a factor of up to three although energy confinement does not change or degrades only slightly. However, it is not yet evident whether the τ_p enhancement can be kept at high rf power level (compared with the OH power).

The electron temperature profiles with on-axis ICRF heating are found to be strongly peaked within the sawtooth reconnection radius compared with, for example, NBI heating (Fig. 8). The stabilization of sawteeth proposed by many groups is therefore an important aspect

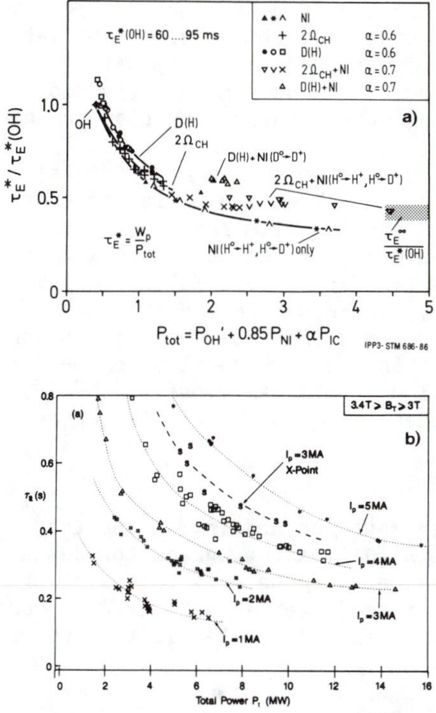

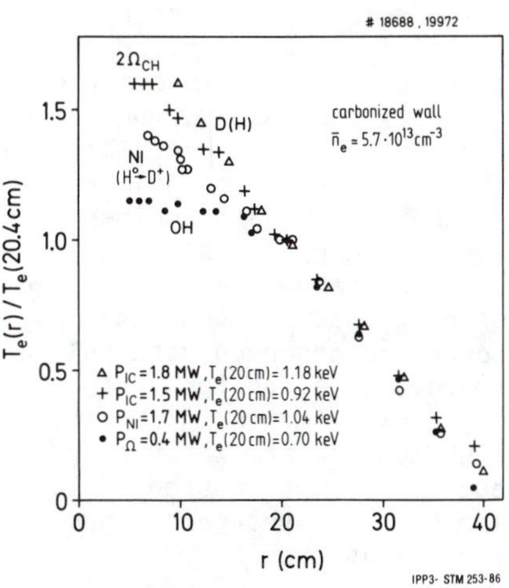

Fig. 8: Normalized electron temperature profiles of ICRF and NBI heating.

Fig. 7: Confinement of ICRF and NBI heated plasmas as observed on ASDEX (a) and JET /1/ (b).

in increasing the central confinement and in reaching high central temperatures and high fusion reaction rates.

Impurities and density limit: Although the heating efficiency and confinement are found to be comparable to or even slightly better than those of NBI-heated plasmas, some ICRF-specific problems remain to be solved. First of all, ICRF is, so far, always accompanied by a higher metal impurity concentration than observed with NBI. In some experiments the metal impurity level in fact limited the maximum launchable rf power and/or the rf pulse length owing to disruptions. The mechanisms responsible for the impurity generation are not yet fully known. In TFR, JET, JFT-2M and PLT (IBW heating) /10,38,39,40/ some evidence is found that part of the metal impurity release takes place at or close to the antennae, which might be related to particle acceleration by high rf electric fields excited close to the antennae. Erosion and sputtering of the wall (e.g. by fast ions) do also play an important role. Investigations in ASDEX /41/ showed a significant increase of the radiated power caused mainly by a higher iron concentration in the plasma. Neither the moderately improved particle confinement nor a reduction of impurity screening on account of changed plasma edge parameters is found to be responsible. Impurity sputtering by oxygen and carbon can be experimentally excluded as possible production processes. On the other hand, a clear anti-correlation between wave absorption and impurity production was observed with suprathermal ions generated in

the plasma edge region by the non-absorbed rf power /32,33/. A quantitative explanation of the measured impurity level cannot yet be given. To improve our knowledge of the processes responsible for impurity generation, more emphasis should be put on investigations clarifying the interaction between rf waves, the plasma boundary and the walls of fusion devices.
At present, ICRH experiments are concentrated on the most important topics, which are heating and confinement. The data basis for other critical issues, such as the density and β-limits are scarce. Concerning impurity problems, no increase of the density limit with rf power, as observed with NBI, has so far been found /42,43/. With respect to plans for a high-density, high-field device such as CIT /44/, experiments in the ALCATOR C tokamak indicate that high-power ICRF heating still involves problems in the high-density regime. The successful application of ICRF heating to plasmas close to the β-limit is not yet proven and is another important topic of further investigations.

CONCLUSIONS

In summary, recent and present experiments on heating in the ion cyclotron range of frequencies show significant progress towards a reliable scenario to reach ignition. Despite the high potential of ICRH to heat large, reactor-close devices, such as NET and CIT, some ICRF-specific problems (impurities, density and beta limits, antenna power density) still remain.

ACKNOWLEDGEMENTS

The author gratefully acknowledges the help of and many valuable discussions with M. Brambilla, F. Leuterer, F. Wagner and F. Wesner.

REFERENCES

/1/ P.H. Rebut et al., JET LATEST RESULTS AND FUTURE PROSPECTS, Proc. 11th Intern. Conf. on Plasma Physics and Controlled Nucl. Fus. Research, Kyoto(1986).
/2/ J. Jacquinot et al., RF HEATING ON JET, ibid.
/3/ M. Yoshikawa et al., RECENT EXPERIMENTS IN JT-60, ibid.
/4/ P.E. Vandenplas et al., CONFINEMENT STUDIES ON TEXTOR WITH HIGH POWER LONG PULSE ICRH, ibid.
/5/ K. Steinmetz et al., ION CYCLOTRON RESONANCE HEATING AND LOWER HYBRID EXPERIMENTS ON ASDEX, ibid.
/6/ M. Ono et al., ION BERNSTEIN WAVE HEATING EXPERIMENTS ON PLT, ibid.
/7/ K. Odajima et al., CONFINEMENT STUDIES OF ADDITIONALLY HEATED PLASMAS IN JFT-2M TOKAMAK, ibid.
/8/ T. Watari et al., EXPERIMENTS ON ICRF-HEATING AND FAST WAVE CURRENT DRIVE IN JIPP T-IIU, ibid.
/9/ M. Porkolab et al., RF HEATING AND CURRENT DRIVE EXPERIMENTS ON THE ALCATOR C AND THE VERSATOR II TOKAMAKS, ibid.
/10/ J. Adam et al., Plasma Phys. and Contr. Fusion 26, 165 (1984).
/11/ T. Mutoh et al., ICRF HEATING EXPERIMENT OF HELIOTRON E, Proc. Intern. Stellarator/Heliotron Workshop, Kyoto (1986).
/12/ A.W. Molvik, Proc. 13th Europ. Conf. on Contr. Fus. and Plasma Heating, EPS, Schliersee, Vol. II, 29 (1986).

/13/ T.H. Stix, THE THEORY OF PLASMA WAVES, McCraw-Hill, NY (1962).
/14/ R.R. Weynants et al., 2nd Symp. on Heating in Toroidal Plasmas, Como, Vol. 1, 480 (1980).
/15/ K. Theilhaber et al., Nucl. Fus. 24, 54 (1984).
/16/ A. Ram, A. Bers, Nucl. Fus. 24, 679 (1984).
/17/ A. Fukuyama et al., Nucl. Fus. 23, 652 (1983).
/18/ K. Appert et al., Comp. Phys. Comm. 40, 73 (1986).
/19/ M. Brambilla et al., Proc. 13th Europ. Conf. on Contr. Fusion and Plasma Heating, EPS Schliersee, Vol. I, 89 (1986).
/20/ M. Brambilla et al., Cardinali Plasma Phys. 24, 1187 (1982)
/21/ V. Bhatnagar et al., Nucl. Fus. 24, 955 (1984)
/22/ D. Hwang et al., Princeton Report PPPL-1990 (1983).
/23/ C.K. Phillips et al., Phys. Fluids 29, 1608 (1986).
/24/ P.L. Colestock, R.J. Kashuba, Nucl. Fus. 23, 763 (1983).
/25/ A. Fukuyama et al., Jap. J. Appl. Phys. 23, 613 (1984).
/26/ L. Villard et al., Comp.Phys.Repts. 4, 95 (1986).
/27/ A. Fukuyama et al., Jap.J.Appl.Phys. 23, 613 (1986).
/28/ M. Brambilla et al., to be publihed
/29/ H. Wedler et al., Proc. 14th Symp. on Fusion Technology, Avignon, Vol. I, 715, (1986).
/30/ J.-M. Noterdaeme et al., Proc. 13th Europ. Conf. on Contr. Fusion and Plasma Heating, EPS Schliersee, Vol. II, 137 (1986).
/31/ F. Engelmann et al., CONCEPT AND PARAMETERS OF NET, 11th Intern. Conf. on Plasma Phys. and Contr. Nucl. Fus. Research, Kyoto, (1986).
/32/ K. Steinmetz et al., Plasma Phys. and Contr. Fusion 28, 235 (1986).
/33/ K. Steinmetz et al., 13th Europ. Conf. on Contr. Fus. and Plasma Heating, EPS Schliersee, Vol. II, 21 (1986).
/34/ P.P. Lallia, private communication.
/35/ J. Luxon, private communication.
/36/ S.M. Kaye, R.J. Goldston, Nucl. Fus. 25, 65 (1985).
/37/ K. Steinmetz et al., Phys. Rev. Lett. 58, 124 (1987).
/38/ H. Matsumoto, private communication.
/39/ K. Behringer et al., Proc. 13th Europ. Conf. on Contr. Fus. and Plasma Heating, EPS Schliersee, Vol. I, 176 (1986).
/40/ J. Adam et al., Proc. 4th Int. Symp. on Heating in Toroidal Plasmas, Rome, Vol. I, 277 (1984).
/41/ G. Janeschitz et al., Proc. 13th Europ. Conf. on Contr. Fus.and Plasma Heating, EPS Schliersee, Vol. I, 407 (1986).
/42/ P.P. Lallia et al., Plasma Phys. and Contr. Fus. 28, 1211 (1986).
/43/ G.H. Wolf et al., ibid., 1413 (1986).
/44/ D. Post et al., PHYSICS ASPECTS OF COMPACT IGNITION TOKAMAK, Princeton report PPPL-2389 (1986).

RESULTS OF RF HEATING ON JET AND FUTURE PROSPECTS

J. Jacquinot, V.P. Bhatnagar[1], G. Bosia, M. Bures, G.A. Cottrell, Ch. David[2], M.P. Evrard[1], D. Gambier[2], C. Gormezano, T. Hellsten, A.S. Kaye, S. Knowlton[3], D. Moreau, F. Sand, D.F.H. Start, P.R. Thomas, K. Thomsen, T. Wade

JET Joint Undertaking, Abingdon, Oxon, OX14 3EA, UK.
[1]EUR-EB Association, LLP-ERM/KMS, B-1040 Brussels, Belgium
[2]EUR-CEA Association, CEN Cadarache, France
[3]Plasma Fusion Centre, MIT, Cambridge, Mass. USA

ABSTRACT

Ion Cyclotron Resonance Heating (ICRH) has been applied to various JET plasmas with net coupled powers reaching 7 MW for several seconds. Conditioning with helium discharges is shown to be effective in controlling the density and in reducing the total radiated power to about 25% of the input power. On axis heating invariably results in long sawteeth of large amplitude. In most high power cases, a sudden bifurcation to monster sawteeth as long as 1.6 s is observed. The monster sawtooth is accompanied by an improvement in energy confinement time (+20%) and by an increase in density. Central electron and ion temperatures of 7 and 5.5 keV respectively have been achieved with 6.5 MW of ICRF power. The heating rate is reduced with off-axis heating in agreement with diffusive heat transport models.

The future RF programme on JET includes upgrading the ICRF power to 20 MW and the development of current drive methods using a phased ICRF array of 8 antennae and later on a Lower Hybrid Current Drive system at 3.7 GHz with a 10 MW launcher.

1. INTRODUCTION The JET RF programme now consists of two distinct activities:

(i) Ion Cyclotron Resonance Heating (ICRH), which started operation in 1985. The first stage of this programme has recently been completed.

(ii) Lower Hybrid Current Drive (LHCD). Studies started in 1986 and approval for the construction of a first stage of the 3.7 GHz equipment has been obtained.

ICRH and NBI will ultimately (c. 1988) inject 40 MW of power in the plasma (20 MW each). LHCD is aimed at controlling the current profile in order to suppress the sawteeth and possibly to achieve higher confinement regimes.

This article, which follows several previous reports [1a, b, c,], summarizes the main experimental results obtained with ICRH. Power accountability has been deduced from the stored energy response to a step power increase [1a]. The power deposition profile has been analysed using power modulation of long pulses [1c]. As

As expected in large Tokamaks with short antennae located near the equatorial plane about 80% of the power is deposited in a narrow zone (± 10 cm for off-axis heating, ± 30 cm for on-axis heating) defined by the ionion hybrid resonance. The 20% remaining power seems to appear at the edge producing serious modifications of the scrape off layer. These edge effects can be suppressed by phasing coupling elements toroidally [3] so that most of the power is radiated at a high $k_{//}$ ($\approx$ 7 m^{-1}). The heating rate in this case was higher (+ 25%) presumably because all the power was now available for plasma core heating. These experiments were limited to low power (~ 1.5 MW) due to a low coupling resistance but suggested a redesign of the antennae to allow this mode of operation with improved coupling for the next stage of the ICRH programme.

Earlier plasma heating studies [1a,b,4] have shown that the increase of plasma stored energy is rather insensitive to the method of heating (ICRH, NBI). More extensive studies have demonstrated the interest of local heating [1b,c,5] in the plasma centre due to the diffusive nature of heat transport [5]. Another major difference between NBI and ICRF using a minority species is that NBI dominantly heats the ions and ICRF heats the electrons via a highly energetic minority. At low density NBI creates a "hot ion" mode and ICRF a "hot electron" mode.

This article is organized in the following manner. The ICRF data base is first presented with particular attention to the localisation of the heating zone and to conditions of very low plasma radiation. The resulting scaling of confinement with plasma current is discussed. We then compare the various heating scenarios using the same plasma current but different minority species or magnetic configuration (in particular X-point). Finally we discuss future prospects of RF in JET examining specific RF scenarios capable of producing a high fusion yield without significant activation.

2. EXPERIMENTAL CONDITIONS, HELIUM CONDITIONING Experiments were performed with 3 antennae, all phased (unless stated) in the toroidal monopole mode (e.g. with a radiation spectrum peaked at $k_{//}$=0). Operation just after the installation of the antennae resulted in a high fraction of plasma radiation (e.g. $p = P_{rad}/P_{tot} \approx 0.7$) of the input power. The density increase with RF was high and density disruption would set a RF power limit at low plasma current. The situation improved remarkably with high current, high power pulses which apparently carbonized the interior of the vessel. However, p stayed above 0.5 (Fig 1). Discharges in Helium were performed at the end of the operating period. The increase of density with RF was low e.g. no more than 25% for P_{RF} = 7 MW and p decreased from typically 0.55 with ohmic heating to 0.25 at maximum RF power (7.4 MW). These improved conditions remain unchanged after a return to Deuterium plasmas. Fig. 1 illustrates the effect of Helium conditioning for 2 MA plasmas both with Deuterium and Helium species.

3. SCALING OF ENERGY CONFINEMENT WITH ICRF Figs. 2 and 3 summa-

rize the available confinement data of plasmas heated by ICRF heating alone using outboard limiters (in one series of experiments the outboard limiters were removed and the plasma was limited by the side protections of the 3 antennae). The plasma current was varied from 1 MA to 5 MA. This scan was normally done with a constant ratio of the wave frequency to the vacuum toroidal field (ν/B_ϕ). Paramagnetic shift of the resonance (R_c) resulted in a variation of the resonance layer (as much as 25 cm). Fig. 2 gives the stored energy for all limiter discharges with $|R_o - R_c| < 0.55$ m. Note that on axis heating of 5 MA plasmas have only been performed with a 50 cm displacement of the resonance layer. In Fig. 3 the incremental confinement $\tau_{inc} = \Delta W/\Delta P_{tot}$ is plotted for the entire data set as a function of the resonance position. For each data subset the heating efficiency decreases as the resonance is moved out. The data is well represented by a law

$$\tau_{inc} = \tau_o \left(1 - (r_c/a)^2\right)^2 \; ; \; r_c = R_c - R_o \tag{1}$$

This result has now been obtained in the following conditions: (i) in a single shot by ramping down the toroidal magnetic field during the pulse[1b] ; (ii) by changing the frequency from shot to shot [1c], (iii) as in (ii) but in Helium discharges with Hydrogen minority heating. In this third series of experiments, the total radiated power was a factor 2 to 3 smaller than in the two other series. The fact that the τ_{inc} (r_c/a) is similar in all these series despite significant changes in radiation, suggests this data should be analyzed in terms of local heat transport models.

Rebut et al [6] have proposed an anomalous transport model in which the heat flux can be described by

$$q = n\,\chi\,(\nabla T - \nabla T_c)\;H\,(\nabla T - \nabla T_c) \tag{2}$$

Here, the anomalous heat diffusivity χ is "switched on" by the step function H when the electron temperature gradient ∇T rises above a critical value ∇T_c.

Callen et al [5] have analysed in detail the consequences of general diffusion laws and shown that Eq. (2) belongs to a functional form which best represents the following phenomenae observed with additional heating:

(i) the heat pulse propagates with a <u>velocity independent of additional heating power</u> [12];

(ii) the <u>off-set linear law</u> of $W = W_o + \tau_{inc} P$ which is a direct consequence of Eq. (2) assuming that χ does not have a strong dependence on ∇T. The 2 MA data in Fig 2 is a good illustration of the offset linear law;

(iii) <u>the decrease of heating efficiency with off-axis ICRF heating</u>. Calculations [5] shows that Eq. (1) is obtained provided that χ is assumed to be independent of ∇T but increases with radius approximately as

$$\chi(r) = \chi_o/(1-r^2/a^2) \tag{3}$$

We now use Eq. (1) and the off-set linear law to deduce, experimentally, the importance of the plasma current. We write:

$$W = W_o + \tau_o P^*_{tot} \; ; \; P^*_{tot} = 0.85 \, P_{OH} + (1-r_c^2/a^2)^2 P_{RF} \qquad (4)$$

P^*_{tot} draws its significance from the diffusive heat transport model: P^*_{tot} is the power which when deposited entirely on axis would give the same increase in W as the spatially distributed P_{OH} and the off-axis P_{RF}.

Representation of the data versus P^*_{tot} (Fig 2) shows that:

for 2 MA ; τ_o = 180 ms, W_o = 0.55 MJ

for 5 MA ; τ_o = 310 ms, W_o = 1.45 MJ

Using the diffusive heat transport model, it is also possible to correct P^*_{tot} for radiation losses. Including the full profile effects of the radiation following the prescription of [5], we find no changes of τ_o and W_o in the 2 MA data but in 5 MA it amounts to a 10% reduction of P^*_{tot} (τ_o = 340 ms).

A consequence of Eq. (2) is a relation between an averaged critical gradient $\overline{\nabla T}_c$ and W_o : $\overline{\nabla T}_c \approx W_o/2\pi^2 R_o a^3 n$. We find that $\overline{\nabla T}_c \approx 1.7$ keV m^{-1} ± .4 for both the 2 MA and 5 MA data.

The physical significance of the critical gradient ∇T_c is the expectation of a reduced heat transport in regions where $\nabla T < \nabla T_c$. Such a situation can arise with high power off-axis ICRH. Indeed the heat transport coefficient appear significantly reduced for this case [7]. The reduction concerns only the central zone located inside the heated shell and the overall effect is a loss of heating efficiency.

4. HEATING SCENARIOS, MONSTER SAWTEETH, X-POINT DISCHARGES Recent high power experiments confirm previous conclusions that there is hardly any observable differences in stored energy between scenarios using a He^3 or a H minority. In agreement with minority slowing down physics [8], He^3 minority heating, particularly at low or moderate power, is more efficient at heating ions than the H minority but at the expense of a smaller electron heating rate. Using high power on-axis heating with $n_{H,He^3}/n_D \approx 0.05$, electron heating via the minority species dominates. The ions are slowly heated by electron collisions. An example of this delayed heating is apparent on Fig. 4.

A common feature of high power on-axis heating is the sudden transition [9] from the regular sawteeth behaviour (period ≈ 300 ms) to long MHD quiescent periods of up to 1.6 s ("Monster" sawteeth). The electron temperature (Fig 4) saturates after 200 ms but the density, the stored energy (Fig 5), and the ion temperature keep increasing during the quiescent period. The monster has beneficial effects (gain of a factor of 2 to 3 on the reaction rate and of ≈ + 20% on the stored energy). However, its crash may induce a locked MHD mode which deteriorates confinement. In several cases this mode lasted for the rest of the pulse.

On the contrary when the resonance is located outside the q=1 surface the sawteeth become smaller and shorter [1b].

ICRF heating of single null X-point discharges has only been performed at 2MA with P_{RF}<6MW and 6cm separating the antenna from the plasma (Figs. 4 and 5). No clear transition to an "H mode" regime has been observed. However, the confinement is better than in limiter discharges with the same plasma current. τ_{inc} reaches 350 ms and the confinement degradation is small. The edge MHD activity is smaller than in limiter discharge. This improvement of energy confinement is also accompanied by increased impurity levels. The total radiated power fraction p increases from 0.25 (limiter) to 0.5 (X-point) and Z_{eff} from 2.5 (limiter) to 4.5 (X-point).

5. ICRF SCENARIOS PRODUCING A HIGH YIELD OF D-He³ FUSION REACTIONS

In this section, we consider ICRH scenarios capable of producing a significant amount of fusion yield from non thermal reactions between energetic Deuterium and Helium3 ions. The basic reactions are

$$D + He^3 \rightarrow H \quad (14.7\text{ MeV}) + He^4\ (3.6\text{ MeV})$$
$$D + He^3 \rightarrow L_i^5 \quad + \gamma\ (16.6\text{ Mev})$$

The second reaction only occurs with a much smaller probability and is mentioned here as a diagnostic of the process since the γ production can be monitored [10]. The cross-sections for both reactions have maxima at 440 keV. We are interested in producing a significant number of these reactions in order to study the confinement of fusion products without large neutron activation of the Tokamak.

The following ICRF scenarios have been considered:

(S1) He^3 minority heated by ICRF in a D plasma, a standard JET operating regime.

(S2) Harmonic Deuterium heating, $2\omega_{cD}$, in a He^3 plasma with NBI. In this scenario, the parameters are optimized to maximize the damping of ICRF on the Deuterium beam ions to increase their energy to the optimum of the D-He^3 cross section. Fundamental damping on a H minority is to be avoided in this scheme and therefore an H concentration smaller than 1% is required. In preliminary studies [11] up to 10% of the RF power was absorbed in this manner as deduced from observed beam acceleration above the injection energy.

(S3) $2\omega_{C\ He^3}$ scenarios in a D plasma. The 2nd harmonic damping of He^3 will become a dominant process when the ion temperature reaches 7 keV. This will be best achieved in the hot ion mode generated by NBI at low or moderate densities. Again in this scheme, fundamental damping by an H minority can occur on the high field side of the Tokamak, and thorough conditioning of the Tokamak will be required to remove Hydrogen contamination.

A Fokker-Planck calculation in these three situations has been performed [2] assuming an isotropic distribution, $T_e = T_i$ and $Z_{eff} = 1$. Figure 6 gives the fusion amplification factor $Q = P_{Fusion}/P_c$ versus P_c, the power effectively coupled to He^3 in S1 and S3 scenarios or to the beam ions in the S2 scenario.

Scenario S1 appears to give the highest Q which reaches 0.06 for JET parameters with full ICRF heating. It is also the most effective scenario from the point of view of coupling to the desired species since the damping per pass is expected to be high (≈ 70%) and no other competing damping (to H for instance) can occur. The Fokker-Planck code predicts that minority velocity distribution will become highly anisotropic with a "parallel temperature" ($T_{//}$) not exceeding 20 keV, which defines, from Döppler effects, an absorption layer of ±30 cm. However, it is likely that a velocity space instability will relax the anisotropy, increase $T_{//}$ and broaden the deposition profiles.

The production of 16 MeV γ's has already been observed in JET [10] in conditions corresponding to S1. The observed flux corresponds to about 15 kW of D-He^3 fusion reaction power or to $Q \approx 3 \times 10^{-3}$. With $T_e \sim T_i \sim 5$ keV and $Z_{eff} \approx 2$, this value corresponds to $P_c \sim 0.1$ MW m^{-3} which implies a deposition profile extending to $R_c - R_o = .6$ m. No attempt has yet been made to optimize the He^3 concentration to reach the maximum Q value.

6. THE LOWER HYBRID CURRENT DRIVE PROJECT The large temperature collapses which are observed during sawteeth with high additional power deteriorates the fusion yield and can generate a locked MHD modes with a decrease of confinement time (-25%) and, in some cases can cause a major disruption. JET has proposed a Lower Hybrid Current Drive system in order to permanently stabilize the sawteeth. The main parameters are given in Table 1.

Table 1

Frequency	3.7 GHZ	- Avoid coupling to NBI ions - Reach $n_o \approx 10^{20}$ m^{-3}
$n_{//}$	$1.2 < n_{//} < 2.4$	48 multijunction units
Coupled Power	3.2 MW (stage 1), 10 MW (stage 2)	
Current Drive	0.1 to 0.25 MA/MW, with $n_e = 5 \times 10^{19}$ m^{-3}	

The construction of Stage 1 has started and a prototype launcher is expected to be ready in mid 1988.

7. CONCLUSION The first stage of the JET ICRF heating programme has been completed and the main objectives have been achieved. Net coupled powers in excess of 7 MW, $T_{eo} = 7$ keV and $T_{io} = 5.5$ keV have been obtained. The total radiated power in less than 25% of the input power when Helium conditioning is used.

Off-axis heating suggests that the heat transport is diffusive. On-axis heating is always more efficient and slows down, rather unexpectedly, the internal sawteeth relaxation. The quiescent period lasts no more than 1.6 s but has a beneficial effect on the fusion yield. The best incremental confinement time has been obtained in X-point discharge although a transition to an H mode has not yet been observed.

The JET HF programme is being vigorously pursued with the upgrading of the ICRH to 20 MW with 8 antennae, the preparation of high fusion yield scenarios, and the development of a Lower Hybrid Current Drive system capable of driving up to 2.5 MA.

REFERENCES

[1a] J. Jacquinot et al, Plasma Phys. and Contr. Fusion 28 1 (1986)
[1b] J. Jacquinot et al, JET-P (86) 18, to be published in the Philosophical Transactions of the Royal Society
[1c] JET team, "RF Heating on JET" 11th Int. Conf. on Plasma Physics and Controlled Fusion Research, IAEA-CN-47/F-I-1 (1986)
[2] T. Hellsten, K. Appert, W. Core, H. Hamnén, S. Succi, Proc. of the 12th Eur. Conf. on Contr. Fusion and Plasma Phys. Budapest (1985)
[3] M. Bures et al, submitted for publication
[4] P.P. Lallia et al, Plasma Phys. and Conf. Fusion, 28 1211 (1986)
[5] J.D. Callen, J.P. Christiansen, J.G. Cordey, P.R. Thomas, K. Thomsen, JET P (87) 10, submitted to Nuclear Fusion
[6] P.H. Rebut, M. Brusati, M. Hugon, P.P. Lallia, 11th Int. Conf. on Plasma Physics and Cont. Fusion Research, IAEA-CN-47/E.III 4 (1986)
[7] M. Brusati, private communication
[8] V.P. Bhatnagar et al, to be presented at the EPS Conf. Madrid (1987)
[9] D.F.H. Start et al, this conference
[10] G. Sadler et al, to be presented at the EPS Conf. Madrid (1987)
[11] G.A. Cottrell et al, this conference
[12] B.J.D. Tubbing, N.J. Lopes Cardozo, JET-R(87)01.

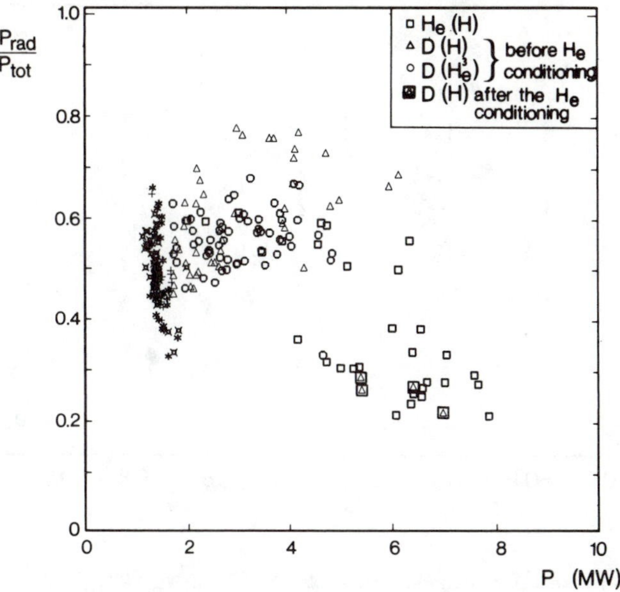

Fig. 1 Total radiated power normalised to the total input power (RF + ohmic) versus total input power. 2 MA discharges with $1 < \bar{n}_o \times 10^{-19} < 2.5$

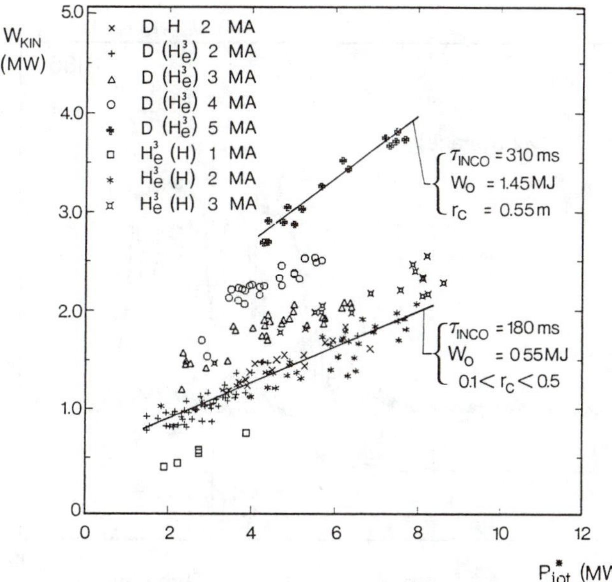

Fig. 2 Plasma stored energy versus effective input power $P^*_{tot} = 0.85\ P_{OH} + (1 - \chi^2)^2\ P_{RF}$; $\chi = (R_c - R_o)/a$. In the diffusion heat transport model with constant $\chi(\nabla T)$, the deposition of P^*_{tot} exactly on the magnetic axis produces the same increase of W than the distributed power P_{OH} and P_{RF}.

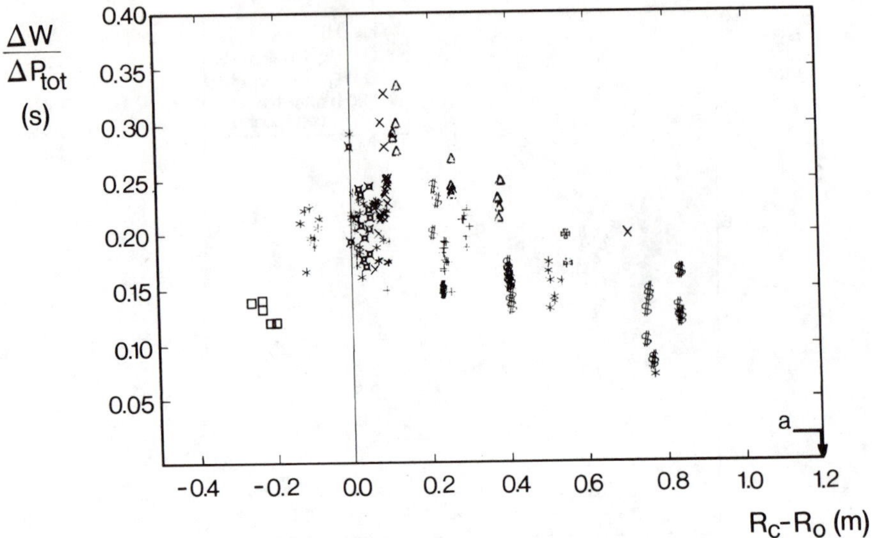

Fig. 3 Incremental confinement time $\Delta W/\Delta P_{tot}$ with $P_{tot} = P_{OH} + P_{RF}$ versus the position of the cyclotron resonance (R_c) with respect to the magnetic axis (R_o) for all JET limiter data. The symbols are explained on Fig. 2, in addition 2.5 MA data points are represented by \$. Note the general decrease of the heating efficiency with off-axis heating which can be represented by a law $(1 - \chi^2)^2$

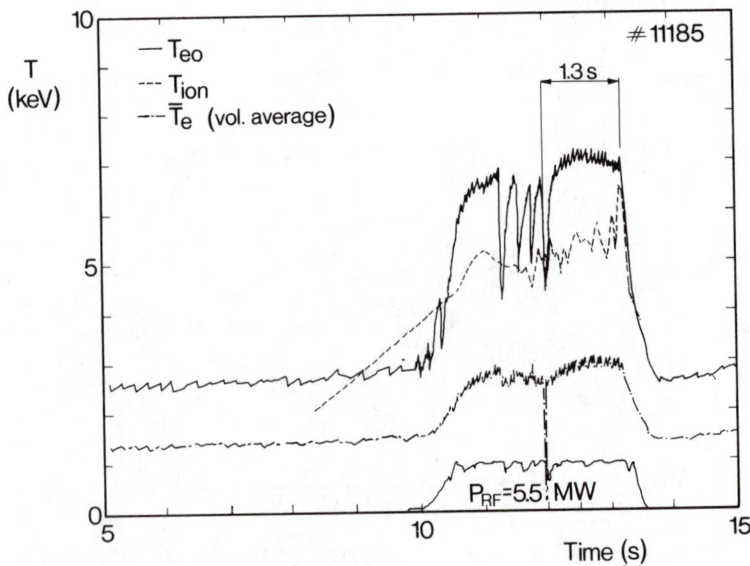

Fig. 4 Evolution of the electron and ion temperatures during 5.5 MW heating pulse (3.5 sec) in an X-point discharge (Helium plasma). 2 monster sawteeth are separated by normal sawteeth.

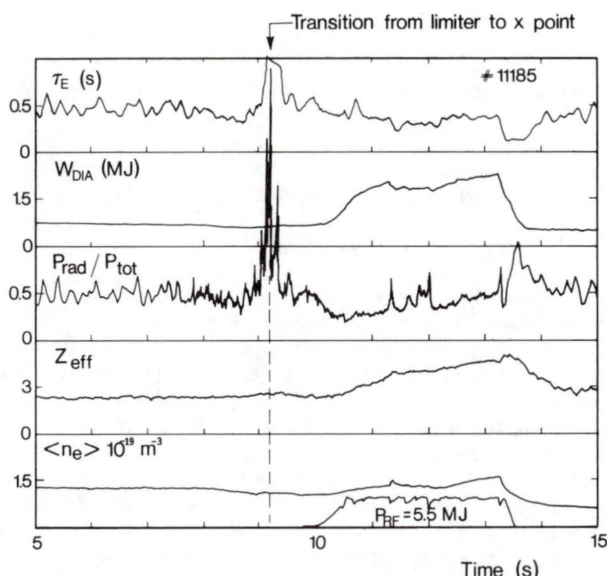

Fig. 5 Evolution of plasma parameters during a 5.5 MW heating pulse in an X-point discharge. Note the increase of density and stored energy during the monster sawtooth. The high Z_{eff} and P_{rad}/P_{tot} is typical of the high power X-point discharge.

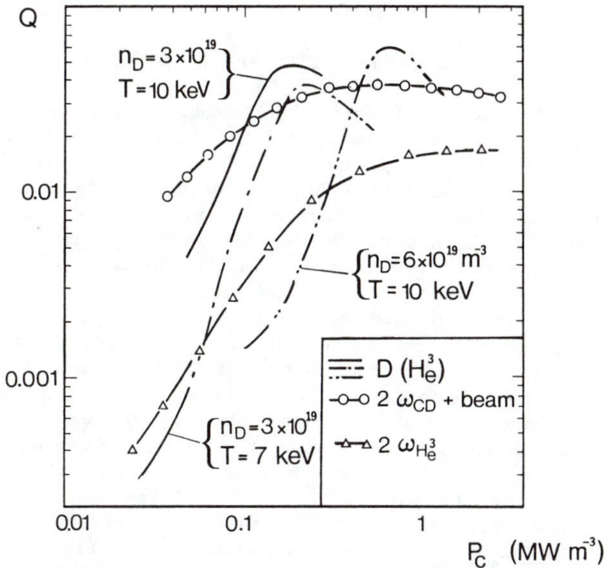

Fig. 6 Fusion amplification factor for D-He^3 reaction versus the power damped in the resonating species. 3 scenarios are considered (i) D(He^3), He^3 resonating minority with n_{He^3}/n_D = 0.02 (ii) $2\omega_{CD}$, Harmonic resonance with Deuterium beam ions, E_{inj} = 80 keV, n_{He^3}/n_D = 1 (iii) $2\omega_{C\ He^3}$, Harmonic resonance with Helium3, n_{He^3}/n_D = 0.5. In all cases Z_{eff} = 1

ION BERNSTEIN WAVE HEATING RESEARCH - A REVIEW

Masayuki Ono
Plasma Physics Laboratory, Princeton University
Princeton, New Jersey 08544

ABSTRACT

This paper summarizes research activities on ion Bernstein wave heating (IBWH) from its inception to the present day high power tokamak experiments. Basic properties and the background of IBWH are reviewed, including the basic physics experiments performed on the ACT-1 device. Then the highlights of tokamak IBWH experiments are presented, including JIPPTII-U, TNT, PLT, and Alcator C tokamak devices. In particular, unique features of IBWH heated plasmas (as compared with OH and other auxiliary heated plasmas) are being emphasized. In the theoretical area, the IBWH related research activities including linear- and non-linear wave analysis, particle simulations, and plasma transport calculations are summarized. Another important area of ion Bernstein wave research is the interaction of ion Bernstein wave with the fast wave. Coupling between the two ICRF modes can occur through a variety of processes including parametric, mode-conversion, low-frequency-fluctuation-mixing, and density-gradient-driven processes. Finally, a summary will be given for possible future IBWH research activities and remaining IBWH related issues.

BASIC PROPERTIES AND BACKGROUND

Motivations for developing ion Bernstein wave heating (IBWH)[1,2] in the late 1970's owe largely to the then on-going effort of the ion cyclotron range of frequency heating (ICRF) and lower-hybrid wave heating (LHH) research. Lower-hybrid heating investigation had successfully demonstrated efficient coupling to the plasma using a phased waveguide launcher.[3] Fast wave ICRF heating research had begun to show efficient and reliable ion heating utilizing the ion cyclotron resonances.[4] This low frequency range also offered relatively inexpensive and efficient power supplies. The natural evolution of reactor compatible heating concepts then appeared to lie in the direction of combining the advantages offered by these two rf heating concepts. From these considerations and some theoretical investigations, plasma heating by externally launched ion Bernstein waves was proposed.[1,2] It was suggested that because of lower-hybrid wavelike behavior at the antenna-plasma interface, a waveguide coupler similar to the one used for LHH can be employed. This waveguide coupler, which is tall and narrow, can fit nicely between toroidal field coils, particularly attractive for heating relatively compact ignition devices. Moreover, through ray tracing calculations in a slab geometry[2] and in a tokamak geometry,[5] it was predicted that the wave accessibility of IBW in hot-density plasmas can be excellent for a relatively wide range of launched $n_{\parallel}$. Since the wave energy is mainly carried by the bulk ion motions, the electric fields related interactions such as electron Landau damping of IBW is relatively small. The waveguide calculations similar to the one carried out by Brambilla for the LHH waveguide coupler also showed good coupling efficiencies even for a relatively simple wave guide coupler.[5] Some concerns, however, were raised for possible IBWH wave accessibility problems due to impurity ion cyclotron harmonic absorptions,[1] alpha-particle absorptions,[6] and scattering by the drift-wave turbulence.[7] These parasitic effects, although not completely negligible, were shown to be relatively weak in typical reactor parameters.[8,9,10] By combining the tokamak IBW ray-tracing code with tokamak plasma transport codes, tokamak IBWH modeling codes have also been developed.[11-13]

The finite-Larmor-radius wave formalism was established by I. Bernstein in 1968.[14] The generalized IBW dispersion relation was later given by T.H. Stix.[15] The first experimental observation of ion Bernstein wave was reported by J. Schmidt in his potassium Q-machine experiment.[16] In this experiment, a carefully aligned wire along the uniform magnetic field was used to excite the "pure" ion Bernstein wave where $k_{\parallel} \rightarrow \infty$. The first externally launched IBW was reported by M. Ono and K.L. Wong in an ACT-1 hydrogen plasma.[17] It was shown that a mode-transformation process enables the launched electron plasma wave (EPW) to change smoothly into IBW as it propagates into the plasma. This mode-transformation process is similar to the LHH mode-conversion process[18] except without wave singularities and, thus, the EPW-to-IBW conversion is nearly 100%.[2] In the ACT-1 experiments, demonstration of external launching of IBW at various harmonics,[19] absorption of IBW and resulting ion heating at various ion cyclotron harmonics,[20] mapping of the complete IBW wave dispersion relation in a hydrogen plasma which also contained deuterium and tritium-like molecular hydrogen ion species, H_2^+ and H_3^+, were carried out.[21] The CO_2 laser scattering of the externally launched ion Bernstein wave was also reported.[21] Another powerful tool for investigating the basic physics of IBWH was the computer particle simulation investigation carried out by H. Abe and H. Okada et al.[22,23] The IBWH external launching, propagation, absorption, and heating were investigated by this method. In this particle simulation investigation, the sub-harmonic heating at 3/2 Ω_i was also observed.[22] From such experimental and theoretical investigations, it was concluded that the IBWH concept looked attractive enough to move on to the tokamak heating experiments.

HEATING EXPERIMENTS

In this section IBWH heating experiments carried out on various devices including JIPPTII-U[24-27], TNT,[28-30] PLT,[13,31-34] and Alcator C[35-38] will be discussed.

IBWH Antenna - Since the size of the small to intermediate tokamak devices would not allow installation of vacuum IBWH waveguide launchers, a B_θ/E_z loop antenna was used to simulate the waveguide fields.[24] The rf currents of the IBWH antenna flow in the toroidal direction orthogonal to the fast wave antenna. As in the case of the tokamak fast wave antenna, the IBWH antenna is shielded from the plasma by the Faraday shields. This type of antenna (Nagoya Type III Coil) was used previously for ICRF slow wave excitation in mirror devices.[39] All tokamak IBWH experiments have used this type of antenna. In ACT-1, it was shown through antenna loading and wave field measurements that this type of antenna can indeed launch IBW effectively.[40,10] Recently, a CO_2 laser scattering diagnostic was used to detect IBW launched by this type of antenna in the Alcator C experiment.[35]

Ion Heating - The first tokamak IBWH heating was carried out on the JIPPTII-U device.[24] In a pure hydrogen plasma, efficient ion heating was observed when the $3/2\ \Omega_H$ layer was placed near the center of the plasma. From the measured $T_\perp(H) > T_\parallel(H)$, it was concluded that the hydrogen ions are directly heated by a non-linear heating process at the $3/2\ \Omega_H$ resonance. In this experiment, the hydrogen temperature sensitive iron line was also used to confirm the observed ion heating is central.[25] Similar heating was also observed on PLT at $3/2\ \Omega_D$ in a pure deuterium plasma.[13] The experimental observation is consistent with the processes suggested by H. Abe et al.[22] and by M. Porkolab.[41] Recently in Alcator C, the absorption at $3/2\ \Omega_H$ was directly observed using a CO_2 laser scattering system.[35]

Due to its relatively short wavelength, the ion Bernstein wave can heat bulk ions even at relatively high harmonic frequencies.[20] In the ACT-1 experiment, efficient ion heating was observed for $5\ \Omega_D$, and $5\ \Omega_T$ regimes.[20] On PLT, with a 90 MHz transmitter, $5\ \Omega_D$ heating was demonstrated at 100 kW power level.[31] No strong tail ion production was observed in these experiments.

IBW heating in the high density regime has been investigated in Alcator C.[35-38] At $\bar{n}_e = 10^{14}$ cm^{-3}, excellent ion heating of $\Delta T_i \approx$ 350 eV has been observed at $5/2\ \Omega_D$ with the IBWH power of 100 kW. In terms of usual heating quality factor $\Delta T_i \bar{n}_e / P_{rf}$, this heating represents a phenomenal 35×10^{13} eV cm^{-3}/kW. Scanning charge-exchange analyzer measurements show that the observed ion heating is central.[36] In this good ion heating regime, the particle confinement time improves by a factor of two. However, above $\bar{n}_e = 2 \times 10^{14}$ cm^{-3}, the good ion heating and improved particle confinement are no longer observed.

On PLT, ion heating with the quality factor of $6\text{-}7 \times 10^{13}$ eV cm^{-3}/kW has been observed up to the maximum power of 650 kW.[34] The central Doppler broadening ion temperature reaches 2 keV, exceeding that of the TVTS central electron temperature. Doppler broadening measurements of various ion lines show a peaked ion temperature profile indicating the heating to be central. This measured ion temperature profile is similar to the calculated profiles assuming ohmiclike ion transport, an indication of no ion transport deterioration during IBWH. These heating experiments have shown that IBWH can heat ions quite efficiently in many heating regimes without generating a large population of high energy tail ions.

Electron Heating - Due to its bulk ion heating property, a very small amount of the IBWH power is expected to flow from the heated ions. Also due to the weak wave parallel electric fields, electron Landau damping is expected to be small. However, under some circumstances, there are sufficient up-shifts of wave parallel phase velocity so that a significant electron Landau damping can occur. In JIPPTII-U, such a regime has been observed near the hydrogen minority case (Mode-II).[26,27] A global energy balance analysis[27] indicates that most of the rf power is deposited to the electrons. Since the hydrogen tail population is very small, this was attributed to a direct IBWH electron heating. A central electron heating was also reported in a smaller tokamak device, TNT.[29]

Improved Particle Confinement - During IBWH, the particle confinement shows a significant improvement.[32-34,37-38] On PLT, the plasma density increases by more than a factor of three.[34] This occurs without active gas puffing and without increasing particle recycling, as indicated by a drop in the D_α emission. From fusion neutron, spectroscopic, Bremsstrahlung, and bolometric measurements, it was concluded that the incremental density is due mainly to deuterium. A similar factor of two confinement improvement was observed for helium and selenium (high Z). Similar confinement improvement was also observed for hydrogen and silicon in the Alcator C experiment in the good ion heating regime. This Z-independent confinement improvement observed during IBWH suggests non-classical particle diffusion behavior.

Low-Frequency Turbulence - In the Alcator C IBWH experiment, the CO_2 laser scattering has detected interesting correlation between the scattered IBW signal in the plasma interior and the edge low-frequency turbulence level.[35] In Alcator C, the turbulence level appears to reach minimum near $\bar{n}_e \approx 10^{14}$ cm^{-3} where the wave signal reaches maximum. This observation is consistent with the scattering of ion Bernstein wave by low-frequency turbulence. The calculated scattering probability using Eq. (31) of Ref. 9 for T_{i0} = 1 keV, a = 12.5 cm, B_0 = 10 T, $\delta n/n$ = 0.3, yields P_s = 0.5 which is in the range of significant scattering probability.

On PLT, the scattering probability is ten times less than that of the Alcator C case. Perhaps for this reason, no obvious accessibility problem has been observed. However, in the high power regime, IBWH has actually caused a large change in the low-frequency turbulence activities.[34] The microwave scattering measurements[42] show that the scattered turbulence signal amplitude has gone down by an order of magnitude, and the turbulence frequency has up-shifted indicating an increasing poloidal plasma rotation in the electron diamagnetic drift direction. This observed poloidal drift velocity of 5×10^4 cm/sec corresponds to an electric field of 15 volt/cm in the radially inward direction. An rf-shielded probe also shows a development of large edge floating potential which is negative with respect to the chamber wall.

Plasma Energy Confinement - Plasma energy transport during high power IBWH was studied on PLT.[34] The global energy confinement time improves with the rf power mainly due to the increased plasma density. Although this confinement behavior is better than the NBI L-mode scaling,[43] there is some confinement degradation when compared with ohmic confinement. From the power balance calculation, it was concluded that the degradation was mainly due to the increased radiation loss in the electron channel. The electron thermal diffusivity, however, remained ohmiclike. The plasma current dependence on the confinement was also investigated on PLT. The ion heating efficiency and the global energy confinement are relatively independent of the plasma current.

Plasma Antenna Loading - The antenna plasma loading is not well understood at this time. There are a number of theoretical papers discussing the loading.[40,44,45] There are also a number of experiments carried out on ACT-1,[40] TNT,[28] Alcator C,[35] and PLT.[46] A considerable success in obtaining an agreement with experiment and theory was achieved in the ACT-1 experiment.[40] In the TNT experiment, a good loading with lower parallel index of refraction was reported. On Alcator C, a good plasma loading of 1 Ω has been measured. On PLT, a good plasma loading of 2 Ω has been measured resulting in 80-90% of the rf power coupled to the plasma. The increasing plasma loading with density was observed in the ACT-1, TNT, and Alcator C experiments. No strong density dependence was observed on PLT. The magnetic field dependence was relatively weak on TNT and PLT although some dependence was observed on ACT-1 and Alcator C. The experiments tend to show increased loading with decreasing launched $n_{\parallel}$ while the theory predicts decreasing loading. The recent theoretical work on the antenna loading appears to show some success in explaining the tokamak loading results.[45] However, to make meaningful comparisons, more careful parametric studies and more precise edge plasma measurements would be needed for the tokamak experiments.

Impurity Generation - During PLT IBWH, an increase in the influx of metallic impurities (predominantly iron) was observed. This high-Z impurity influx together with the longer particle confinement can result in non-negligible central radiation losses during high power (high density) IBWH. By using carbon coated Faraday shields,[47] it was possible to conclude that most of the iron influx indeed originated from the antenna Faraday shields.[34,47] This finding indicates that the Faraday shield design is one of the most important elements of future IBWH antenna development. The ion heating efficiency and the density rise, however, remained relatively independent of the impurity level.

IBW CO-GENERATION BY FAST WAVE FIELDS

Ion Bernstein wave can be excited by the ICRF fast wave fields under a variety of processes. Understanding of these processes will give a more complete picture of the ICRF plasma heating. By far the most well known case is the mode-conversion.[18,48] Since this topic falls directly under the main fast wave ICRF heating physics, it will not be covered in this paper.

Parametric Instabilities - IBW, in some circumstances, can be excited through a parametric decay process. This nonlinear process has been observed in the ACT-1 experiment.[49] Due to a relatively restrictive selection-rule condition, the parametric IBWH excitation will not likely play an important role under the normal fast wave heating conditions. A likelihood of excitation will, however, increase at the higher wave frequency regimes ($\omega > 2\ \Omega_i$) due to more available decay modes and lower threshold power level.

Density Gradient Excitation - Ion Bernstein wave can be excited through a density gradient driven process. This process was studied in the ACT-1 experiment.[50] In the context of the slab model, the E_y field of the fast wave antenna can couple to the IBW E_x fields because of the spatial derivative term, $\partial_x K_{xy}$, in the density gradient region.

Low Frequency Turbulence Mixing - Another potentially important IBW excitation process is the wave-wave mixing of fast wave with intrinsic low-frequency density fluctuations.[51] When appropriate $\vec{k},\omega$ matching conditions are satisfied, the interaction results in a beat current consistent with exciting ion Bernstein wave resonantly. This process, however, depends sensitively on the nature of the turbulence spectra existing in a given plasma. These parasitic excitations can alter the power deposition profiles of the fast wave heating experiments.

FUTURE IBWH RESEARCH TOPICS

With completion of PLT/Alcator C IBWH experiments, IBWH is perhaps entering a new phase of experimental activities and opportunities. A new major experimental initiative is the PBX-M IBWH program.[52] Together with LHCD, the IBWH is being prepared to assist PBX-M to achieve the second stability regime through pressure profile control and improved plasma transport. Due to its bulk ion heating capability, the rf power can be deposited effectively even for the case of off-axis heating. In addition, the PBX-M IBWH experiments will provide important reactor relevant information on IBWH performance in a diverted plasma, in an H-mode plasma, and in a high-beta plasma.

The H-mode like improved particle confinement regime attained in the PLT high power IBWH experiment may be utilized in high power neutral beam heating experiments operating in the L-mode regimes such as TFTR and JT-60. In fact, this combined IBWH-NBI performance can be tested relatively quickly on PBX-M.

The bulk ion heating property of IBWH may be attractive for heating devices with relatively poor high energy ion confinement such as the present day helical devices and the low plasma current tokamaks aimed to achieve second stability regime. In fact, application of IBWH is now under consideration in ATF, Heliotron E, and the Nagoya Compact Helical Device.

Another important area of research would be in the area of IBWH waveguide research. Improvements are needed in the waveguide coupling theory, particularly for the direct IBW launching case. The size of the present and future devices (such as TFTR, JET, CIT, and FT-U) are getting large enough to allow for the installation of vacuum waveguide launchers. Such a launcher would have technical advantages over the present loop antenna with Faraday shields in a reactor environment.

Finally, the stabilization of plasma instabilities by IBWH is a speculative but intriguing possibility for high power IBWH. Aside from the low frequency drift turbulence suppression observed on PLT,[34] theoretical work has been carried out for the stabilization of interchange,[53] ballooning,[54] and kink modes[55] in tokamaks. This topic of mode stabilization may be tested on PBX-M with the rf power of 2-10 MW. If such mode suppression can indeed be achieved, it will undoubtedly be one of the most important utilizations of rf for magnetic fusion research.

ACKNOWLEDGMENTS

The author acknowledges contributions from a large number of reseachers whose papers are referred to here. In particular, he thanks members of PLT, ACT-1, JIPPTII-U, Alcator C, and TNT for providing experimental data for this paper.

REFERENCES

1 S. Puri, Phys. Fluids 22, 1716 (1979).

2 M. Ono, Princeton Plasma Physics Laboratory Report PPPL-1593 (1979); M. Ono *et al.*, 8th Conf. on Plasma Physics and Controlled Nuclear Research, Brussels, T-I-2(B) (1980).

3 M. Brambilla, Nucl. Fusion 16, 47 (1976); S. Bernabei *et al.*, Nucl. Fusion 17, 929 (1977).

4 J. Hosea *et al.*, 8th Conf. on Plasma Physics and Controlled Nuclear Fusion Research, Brussels (1980), D-5-1.

5 M. Ono *et al.*, in Heating in Toroidal Plasma II, ed. E. Canobbio *et al.* (Pergamon, New York 1981) Vol. I, p. 593.

6 F.W. Perkins, in ORNL/FEDC-8311, Oak Ridge National Laboratory, Oak Ridge, TN (1983); K.L. Wong and M. Ono, Nucl. Fusion 24, 615 (1984).

7 E. Ott, Phys. Fluids 22, 1732 (1979).

8 M. Ono, Princeton Plasma Physics Laboratory Report PPPL-1900 (1982).

9 M. Ono, Phys. Fluids 25, 990 (1982).

10 M. Ono, in Course and Workshop on Application of RF Waves to Tokamak Plasmas, Varenna, Vol. I, 197 (1985).

11 M. Okamoto and M. Ono, Princeton Plasma Physics Laboratory Report PPPL-2276 (1985).

12 U.S. Cont. to INTOR Workshop, Phase II, Part IIa, INTOR Design Group, IAEA, Vienna (1985).

13 M. Ono *et al.*, AIP Conference Proc. 6th Topical Conf. on RF Plasma Heating, Callaway Gardens, GA, 1985, p. 83.

14 I.B. Bernstein, Phys. Rev. Lett. 109, 10 (1958).

15 T.H. Stix, *Theory of Plasma Waves* (McGraw-Hill, New York, 1962).

16 J.P.M. Schmitt, Phys. Rev. Lett. 31, 982 (1973).

17 M. Ono and K.L. Wong, Phys. Rev. Lett. 45, 1105 (1980).

18 T.H. Stix, Phys. Rev. Lett. 15, 878 (1965).

19 M. Ono, K.L. Wong, and G.A. Wurden, Phys. Fluids 26, 298 (1983).

20 M. Ono, G.A. Wurden, and K.L. Wong, Phys. Rev. Lett. 52, 37 (1984).

21 G.A. Wurden, M. Ono, and K.L. Wong, Phys. Rev. A 26, 2297 (1982).

22 H. Abe, H. Okada *et al.*, Phys. Rev. Lett. 53, 1153 (1984).

23 H. Okada, H. Abe et al., Phys. Fluids 29, 489 (1986).

24 M. Ono, T. Watari et al., Phys. Rev. Lett. 54, 2239 (1985).

25 K. Sato, M. Mimura, M. Otsuka, T. Watari et al., Phys. Rev. Lett. 56, 151 (1986).

26 T. Watari, in Course and Workshop on Application of RF Waves to Tokamak Plasmas, Varenna, Vol. I, 184 (1985).

27 Y. Ogawa, K. Kawahata, R. Ando, E. Kako, T. Watari et al., IPPJ-770; also in Proc. 13th Europ. Conf. Shriesee II, 169 (1986).

28 S. Shinohara, O. Naito, and K. Miyamoto, Nucl. Fusion 26, 1097 (1986).

29 S. Shinohara, O. Naito, Y. Ueda, H. Toyama, and K. Miyamoto, J. Phys. Soc. Jpn. 55, 2648 (1986).

30 S. Shinohara, O. Naito, K. Ida, and K. Miyamoto, Jpn. J. Apply. Phys. 26, 94 (1987).

31 J.R. Wilson et al., AIP Proc. 6th Topical Conf. on RF Plasma Heating, Callaway Gardens, GA (1985), p. 8.

32 J.C. Hosea et al., Plasma Phys. and Contr. Fusion 28, 1241 (1986).

33 J.R. Wilson et al., J. Nucl. Matt. 145-146 (1987).

34 M.Ono et al., IAEA-CN-47/F-I-3, IAEA Conf. Kyoto, Japan (1986).

35 Y. Takase et al., MIT PFC/JA-86-60, (Nov. 1986).

36 C.L. Fiore, et al., Bull. Am. Phys. Soc. 31, 1587 (1986).

37 M. Porkolab et al., IAEA-CN-47/F-II-2, IAEA Conf., Kyoto, Japan (1986).

38 J.D. Moody et al., in Proc. 7th Topical Conf. on Applications of RF Power to Plasmas, Kissimmee, FL (1987).

39 T. Watari et al., Phys. Fluids 21, 2076 (1978); T. Watari et al., Nucl. Fusion 22, 1359 (1983).

40 F.N. Skiff, Princeton University, Ph.D. Thesis (1984).

41 M. Porkolab, Phys. Rev. Lett. 54, 434 (1985).

42 E. Mazzucato, Phys. Fluids 21, 1063 (1978); Phys. Rev. Lett. 48, 1828 (1982).

43 S.M. Kaye, Phys. Fluids 28, 2327 (1985).

44 W.N.-C. Sy et al., Nucl. Fusion 25, 795 (1985).

45 M. Brambilla, in Proc. 7th Topical Conf. on Applications of RF Power to Plasmas, Kissimmee, FL (1987).

46 G. Greene et al., Bull. Am. Phys. Soc. 31, 1467 (1986).

47 J. Timberlake et al., Bull. Am. Phys. Soc. 31, 1467 (1986).

48 D.G. Swanson and Y.C. Nygan, Phys. Rev. Lett. 35, 517 (1975).

49 F.N. Skiff, M. Ono, and K.L. Wong, Phys. Fluids 27, 1051 (1984).

50 F.N. Skiff, M. Ono, P. Colestock, and K.L. Wong, Phys. Fluids 28, 2453 (1985).

51 G.J. Morales, S.N. Antani, and B.D. Fried, Phys. Fluids 28, 3302 (1985).

52 S. Bernabei et al., in Proc. 7th Topical Conf. on Applications of RF Power to Plasmas, Kissimmee, FL (1987).

53 V.K. Tripathi, C.S. Liu, and S.C. Chiu, Nucl. Fusion 27, 287 (1987).

54 D.A. D'Ippolito, J.R. Myra, and G.L. Francis, Bull. Am. Phys. Soc. 31, 1558 (1986).

55 D.A. D'Ippolito, Sherwood Controlled Fusion Theory Conf., San Diego (1987) 3B31.

ICRF HEATING OF CURRENTLESS PLASMA IN HELIOTRON E

T. Mutoh, H. Okada, O. Motojima, Y. Nakamura, M. Sato, H. Zushi, K. Kondo, N. Noda, H. Kaneko, T. Mizuuchi, F. Sano, S. Sudo, Y. Takeiri, K. Akaishi, T. Kawabata, Y. Ijiri, K. Itoh, S. Morimoto, A. Iiyoshi, K. Uo

Plasma Physics Laboratory, Kyoto Univ., Uji, Kyoto, Japan

ABSTRACT

Fast-wave and slow-wave heatings have been performed to study the heating properties, and to study the confinement properties of plasmas in a rare collisional regime. For the fast-wave (minority) heating, ion temperature increased from 200 eV to 650 eV at an electron density of 2.2 x 10^{19} m^{-3} at an ICRF power of 1.5 MW. With the carbonization of the vacuum chamber, the ICRF pulse became able to sustain the plasma for 100 msec without ECH pulse. For the slow-wave (Ion Cyclotron Wave) heating, T_i increased up to 1.6 keV at $\bar{n}_e$ of 0.6 x 10^{19} m^{-3}. The ion heating strongly depended on the electron density and better ion energy confinement was observed at high electron temperature region. These results were consistent with the calculation of the nonambipolar neoclassical diffusion in the rare collisional regime. This high T_i and T_e plasma condition possibly corresponds to the electron root with a positive radial electric field E_r.

INTRODUCTION

The high power and long pulse ICRF heating of currentless plasma has been investigated in Heliotron E since 1985[1]. Fast-wave (minority heating), slow-wave and ion Bernstein wave heating mode were tested[2,3] and behaviours of heating and wave excitation were investigated. In this paper, we discuss the heating characters of high power fast-wave heating modes and confinement studies of the slow wave heating mode.

FAST-WAVE HEATING EXPERIMENT

The eight antenna loops were fed by two generators and all loops were installed on the high field side inside the vacuum chamber. Normally the magnetic field strength on axis was 1.9T and the ICRF frequencies were 26.7 and 53.4 MHz. Figure 1(a) shows the typical time traces of plasma parameters without carbonization of the vacuum chamber. The target plasma was produced and heated by an ECH pulse of about 300 kW, and after that, plasma was heated by an ECH pulse (150 kW) and an ICRF pulse (1.5 MW). The electron density increased from 1.5 to 2.2 (x 10^{19} m^{-3}) during the ICRF pulse without additional gas puffing. The ion temperature of the majority deuterium increased from 200 eV to 650 eV. During the ICRF pulse, the bolometric signal and the impurity spectrum line intensities for Fe, O and Ti rapidly increased, and the electron temperature gradually decreased. The duration of effective heating was limited to less than 50 msec and the heating efficiency of the ICRF pulse was also reduced. The temperature of the high energy tail of the minority proton increased almost linearly with ICRF power, and no particular deterioration of the tail confinement was observed in the energy range of $E < 15$ keV.

After carbonization of the vacuum chamber, the bolometric loss decreased to about 30 % and the impurity line intensities were also reduced. Figure 1(b) shows the time traces of the parameters of the ICRF sustained plasma. In this case, the ECH pulse produced a seed plasma, and then the plasma was heated by the ICRF pulse only. The electron density increased from 1.5 to 3.1 ($\times 10^{19}$ m^{-3}) by enhanced hydrogen influx from the wall surface. Ion and electron temperatures were sustained at the level of 300 ∿ 350 eV. The bolometric signal slowly increased and, finally, the radiation loss terminated the plasma sustainment. Generally, the main improvement achieved by carbonization was the elongation of the duration of effective heating from 50 to 100 msec as shown in Figs. 1 (b).

The heating power dependencies of the obtained ion temperatures on two fast-wave modes are shown in Fig. 2. In this figure, two mode of fast wave heating are shown. The estimated absorbed RF power was 50 ∿ 70 % of the oscillator output power. In the minority heating mode, heating efficiency is degraded at the high heating power range above 500 kW. On the 2nd harmonic heating mode, which frequency was twice of that of the minority heating mode, the heating efficiencies that was estimated from the effective ion temperatures were relatively high. The possible explanation of the difference between these two modes is the improvement in coupling and the suppression of the parasitic effect on the boundary area due to higher frequency.

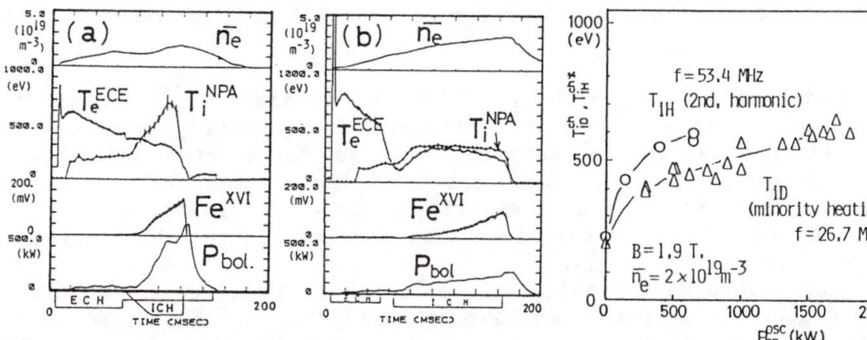

Fig.1 Temporal evolutions of the plasma parameters in minority heating mode. (freq.=26.7MHz, P=1.5(a), 0.7(b)MW)

Fig.2 Dependence of ion temperatures on ICRF oscillator power.

SLOW-WAVE HEATING IN LOW COLLISIONAL REGIME

Slow-wave heating has been an efficient ion heating method for heliotrons.[2] In this heating method, achievable ion temperature was strongly depended on the electron density, because the propagating area of the slow-wave was mainly restricted by the electron density. Therefore, the high ion temperature, $T_i > 1.0$ keV, was achieved at the electron density range of 0.4 ∿ 0.6 $\times 10^{19}$ m^{-3}. However, the obtained peak ion temperatures were also sensitive to other parameters. For example, gas puffing, ECH power and wall conditions were also sensitive factors. The obtained ion temperatures are plotted on the T_e - $\bar{n}_e$ plane in Fig. 3. In this case, the ICRF and the ECH powers were almost fixed condition. As shown in this figure, within

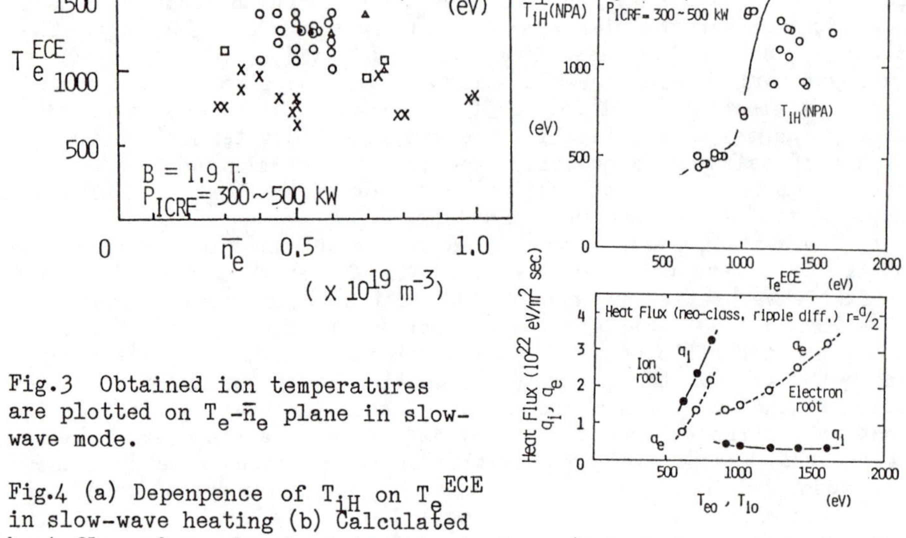

Fig.3 Obtained ion temperatures are plotted on T_e-$\bar{n}_e$ plane in slow-wave mode.

Fig.4 (a) Depenpence of T_{iH} on T_e^{ECE} in slow-wave heating (b) Calculated heat flux of neoclassical transport at r=a/2 including ambipolar E_r.

the suitable $\bar{n}_e$ range, high T_i values were obtained only when the T_e were higher than 900 eV. This tendency is more clearly shown in Fig. 4(a), which shows the ion temperature versus the electron temperature. It would appear that the critical electron temperature exists to obtain the high T_i plasma except for the constraint of an appropriate $\bar{n}_e$ value. In the experiment, we did not control the electron temperature intentionally.

There were some possible causes to explain the reason for the wide distribution of the ion temperatures obtained. The difference of the electron density profiles was one candidate, but measurements by the FIR interferometer indicated that there were no significant differences. The possibility of the charge exchange loss was also eliminated by n_o measurement using laser fluorescence spectroscopy. The effect of radiation loss power was also negligible, because the energy transfer between ions and electrons were small. The last candidate was the effect of the radial electric field on the neoclassical ripple diffusion. In helical systems, e.g. stellarator and heliotron, it is well known that the non-ambipolar ripple transport is influenced by the radial electric field which is determined by the ambipolar constraint. By using the formulation which was developed by Shain[4], the ambipolar electric field and the neoclassical heat flows were calculated. The measured $T_e(r)$, $n_e(r)$ and $T_i(0)$ were used, and $T_i(r)$ were assumed to be a parabolic profile. The results of the calculation are shown in Fig. 5. In case A, $T_i(0)$ and $T_e(0)$ are 800 eV, the radial electric field is negative which corresponds

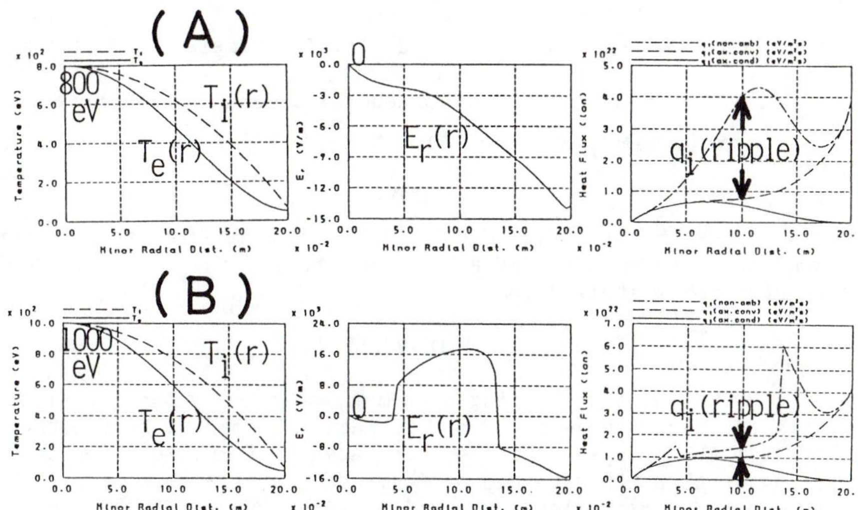

Fig.5 Calculated profiles of the ambipolar electric field and the ion heat flux. ((A); T(0)=800 (eV), (B); T(0)=1000 (eV))

to the ion root of the ambipolarity relation ship. On the other hand, when $T_i(0)$ and $T_e(0)$ are increased to 1000 eV, a positive electric field is predicted within almost half the area of radial position, which corresponds to a favorable electron root. To compare these calculations with the experimental results, the neoclassical heat fluxes at the half radius are shown in Fig. 4(b). The results of the low temperature side correspond to the ion roots and that of the high temperature side correspond to the electron roots. The change of the calculated ion heat flux is remarkably large between the two roots. The estimated heat flow at r = a/2 surface is about 50 ∿ 70 kW in the ion-root case, which could be the dominant ion loss channel. Because, the raughly estimated absorbed ICRF power inside the half radius was about 70 ∿ 100 kW. From the above considerations, it could be concluded that the characteristics of the ion heating and confinement in the slow-wave heating mode strongly suggested the substantial effects of the ambipolar electric field on the ion confinement. To confirm these neoclassical effects, the measurement of the radial electric field is now under planning.

REFERENCES

1. K. Uo, A. Iiyoshi, T. Obiki, et al. in proceedings of 11th International Conference on Plasma Physics and Controlled Nuclear Fusion Research, IAEA-CN-47/D-I-1 (1986).
2. T. Mutoh, O. Motojima, M. Sato, et al. ibid. IAEA-CN-47/D-III-2.
3. T. Mutoh, et. al. Nucl. Fusion <u>24</u> 1603 (1984).
4. K. C. Shaing, Phys. Fluids <u>27</u> 1567 (1984).

ICRF HEATING SYSTEM FOR W VII-AS

J.-M. Noterdaeme, F. Wesner, M. Söll*, F. Hofmeister, P. Grigull, A.B. Murphy
Max-Planck-Institut für Plasmaphysik, Garching bei München
* (present address: WTB, 8139 Bernried Jägerstr. 5)

ABSTRACT

The ICRF heating system for the W VII-AS modular stellarator is described, with emphasis on antenna design and the choice of heating scenarios for a stellarator.

INTRODUCTION

A 3 MW, 3 s ICRF heating system [1], powered by two generators with a frequency range of 30 to 115 MHz, was used to heat both ASDEX and W VII A. The system has been upgraded to 4 MW, 10 s for the heating of ASDEX in extended performance. It will also be used to heat the modular stellarator W VII-AS (R = 2.0 m, $\langle a \rangle$ = 0.2 m, B = 3.0 T). W VII-AS [2] has a larger plasma cross section than W VII-A, and a vacuum vessel (Fig. 1) with better access. ICRF heating scenarios have been reconsidered; improved antennas and a new antenna-line interface are to be installed.

HEATING SCENARIOS

W VII-AS uses ECRF for initial plasma formation and heating: the 70 GHz gyrotron frequency restricts the standard operating toroidal fields to 2.5 T or 1.25 T. The ICRF heating scenarios for those fields are shown in Fig. 2. We consider the possibilities for antennas positioned on the high major radius side of the plasma. Figure 3 shows representative minor cross sections of the plasma, which has toroidal period 72°. Note the different field gradients, characteristic of this stellarator. At Φ = 36° the magnetic field decreases to the outside. Possible scenarios at 2.5 T then are H minority heating, and second harmonic heating in D, He^3 or H. At Φ = 0°, the magnetic field increases with major radius. This offers in principle the possibility of mode conversion heating in an H/D plasma, which has the advantage of large single pass absorption. The relatively shallow magnetic field gradient, however, means that the minority concentration n_H/n_e must be kept within extremely restrictive bounds (between 0.325 and 0.332 for ν = 30 MHz and B_T = 2.5 T). This is not practicable in this machine, but the possibility should be seriously considered in future advanced stellarators with steeper magnetic field gradients and/or better control of species concentrations.

IBW heating, another method with good single pass absorption, could still be considered. A preliminary test of this scheme would require only small alterations to the antennas. At Φ = 36° and with a central magnetic field of 2.5 T, appropriate choice of frequency allows the 3/2 or the 5/2 harmonics of H to be positioned in the plasma centre, with integer harmonics behind the antenna.

ANTENNA SYSTEM

A preliminary conceptual design of a single antenna for W VII-AS has been presented previously [3]. In view of subsequent experimental results of JFT-2M [4] and JET (which show that there is some advantage to shaping the $k_\parallel$ wave spectrum), the present design uses an antenna pair. The feed points of each loop are placed symmetrically (Figs. 4, 5 and 6) 6.86° to each side of the $\Phi = 36°$ plane. Each antenna is fed at the bottom by its own generator. The short circuit is located outside the vessel. This arrangement allows control of both the $k_\parallel$ and k_y spectrum. This will be the first time this is tested in a stellarator. The $k_\parallel$ spectrum excited by the antennas can be shaped simply by altering the relative phasing of the generators. When the generators are operated in phase we obtain a spectrum peaked at $k_\parallel = 0$; when they are operated π out of phase, the spectrum is symmetric, depleted at $k_\parallel = 0$, and with a maximum at $k_\parallel = 5\ m^{-1}$. Intermediate phasings allow the excitation of an asymmetric spectrum. Any resultant current drive effects should be readily apparent in the otherwise currentless stellarator plasma. The k_y spectrum depends on the position of the current maximum on an antenna, which can be altered by changing the position of the external short circuit.

The present antennas were designed for maximum coupling to the plasma, consistent with two restrictions: the antenna surface is curved in only one plane, and sufficient space had to be allowed for the exploration of different plasma shapes and positions in the initial W VII-AS experimental period. Figure 4 shows the standard plasma contour at the outside edges of the antennas. A minimum distance of 4 cm between the plasma surface and the antenna screen was maintained. The estimated voltage for coupling a total of 0.5 MW in phase opposition (with 8 cm average central conductor - plasma separation) is 21 kV. The coupled power could be doubled for the same voltage by shifting the plasma up to the screen. The geometrical parameters of the antenna are given in Fig. 5. The central conductor is constructed from 1 mm copper plate; 6 mm diameter circular rods provide additional stiffness and reduce the local E fields. The central and return conductors are silverplated. The screen is made of stainless steel T shaped pieces, coated with a 20 μm layer of copper and a 6 μm layer of TiN/TiC [5]. It can easily be replaced by an optically open screen [6], which would be almost mandatory in an antenna with a surface curved in 2 planes. The antenna pair is protected on each side by movable carbon limiters, which match closely the standard plasma contour. The Faraday screen is connected to the return conductor by exchangeable side pieces. On the limiter side of the antenna a continuous stiff piece is used; the other side consists of a continuous support and an optically open series of springs (to take up the thermal expansion). Subsequent introduction of an antenna with a surface curved in two planes, closer to the plasma, and with a larger central conductor - return conductor separation, will require only the Faraday screen, the side pieces and the central conductor to be changed. This alteration will afford better coupling to the plasma, but will restrict the range of allowable plasma shapes and positions.

INTERFACE AND TRANSMISSION LINE

The interface consists of 6 1/8" vacuum insulated line elements with 25Ω impedance and incorporates newly developed ICRF components[7]. The pressurized part of the transmission line is 9" and 50Ω. The feedthroughs between the intermediate vacuum and the machine vacuum and the pressurized side consist of simple ceramic insulators, the vacuum tightness being provided by helicoflex seals [7]. Thermal expansion of the line and the antenna is compensated by bellows in the inner and outer conductors. A schematic drawing of the positions of the various components is shown in Fig.6. Between antenna and generator [8] a double stub tuner provides the matching.

RF SPECIFIC DIAGNOSTICS

Several diagnostics will be installed between the antenna and the limiter, including movable Langmuir probes to measure n_e and T_e at the boundary, and RF probes to measure components of $\tilde{B}$ and $\tilde{E}$. A video camera will be used to observe the antennas, allowing the position of the side limiters to be optimised when operating with different plasma shapes and positions.

CONCLUSION

The new ICRF heating system for W VII-AS has been designed to allow maximum experimental flexibility. Second harmonic and H minority heating, with investigation of the effect of varying the wave spectrum, and current drive experiments are all possible. The antennas can be easily modified for IBW experiments, and for optimal coupling to the eventual standard plasma configuration.

REFERENCES

1) F. Braun et al., Fusion Technology 1982 (Proceedings of the 12th Symposium, Julich), Pergamon Press, 1982, Vol. 2, 1393.
2) D. Dorst et al., Fusion Technology 1986 (Proceedings of the 14th Symposium, Avignon), Pergamon Press, 1986, Vol. 1, 139.
3) M. Söll and F. Wesner, Fusion Engineering 1983 (Proceedings of the 10th Symposium, Philadelphia), IEEE, 1983, Vol. 1, 600.
4) M. Mori et al., Plasma Phys. and Contr. Fus. 1984 (Proceedings of the 10th IAEA Conf., London), IAEA 1985, Vol. 1, 45.
5) J.-M. Noterdaeme et al., Fusion Technology 1986, (Proceedings of the 14th Symposium, Avignon), Pergamon Press, 1986, Vol. 1, 795.
6) J.-M. Noterdaeme et al., Controlled Fusion and Plasma Physics 1986, (Proceedings of the 13th Europ. Conf., Schliersee), Europ. Phys. Soc., 1986, Vol. 2, 137.
7) H. Wedler et al., Fusion Technology 1986 (Proceedings of the 14th Symposium, Avignon), Pergamon Press, 1986, Vol. 1, 715.
8) F. Wesner et al., Heating in Tor. Plasmas 1982 (Proc. of the 3rd Int.Symp., Grenoble) Com. of Eur. Com., 1982, Vol.1, 429.

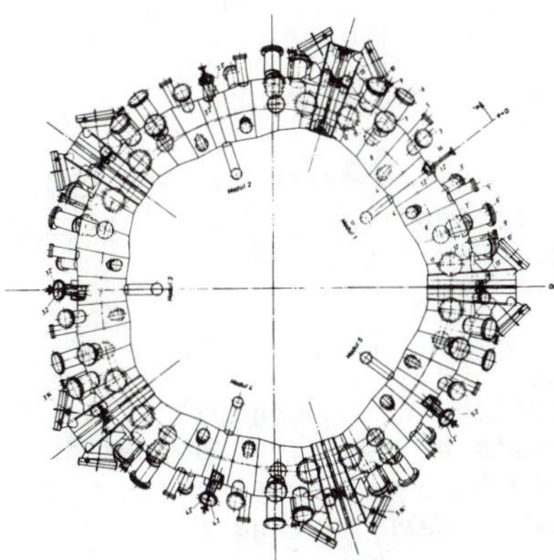

Fig.1: Vacuum Vessel of WVII-AS

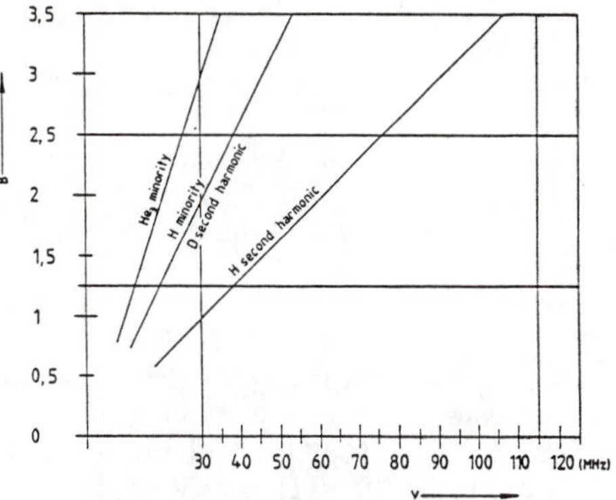

Fig.2: Heating scenarios

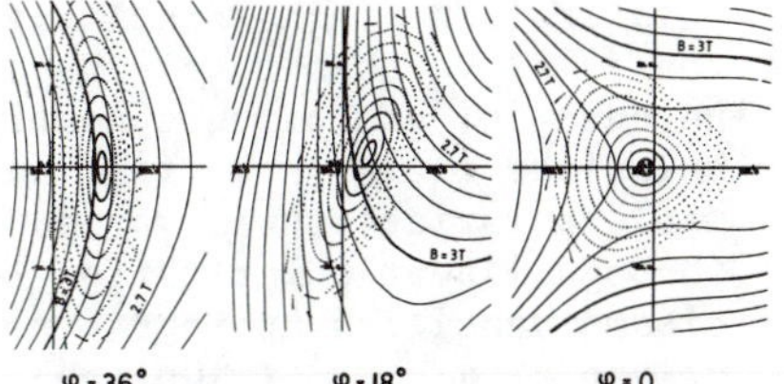

Fig.3: Constant /B/ lines and magnetic surfaces

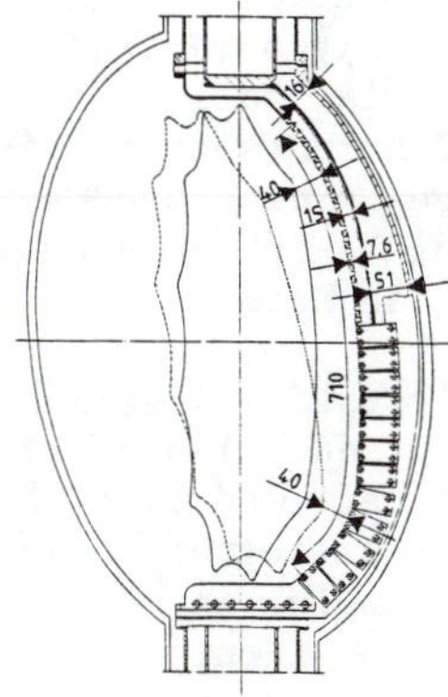

Fig.4: Poloidal cut through the antenna

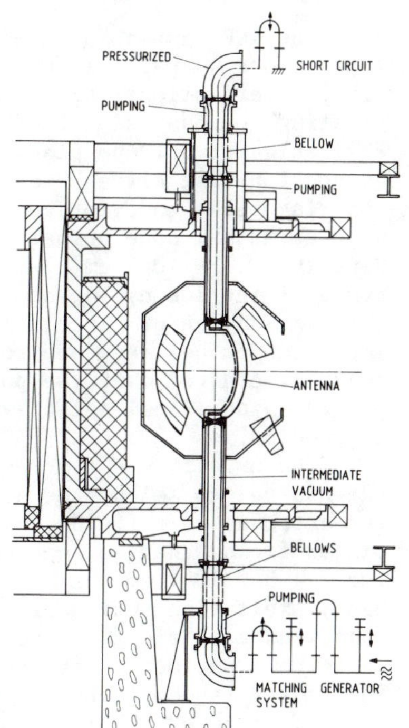

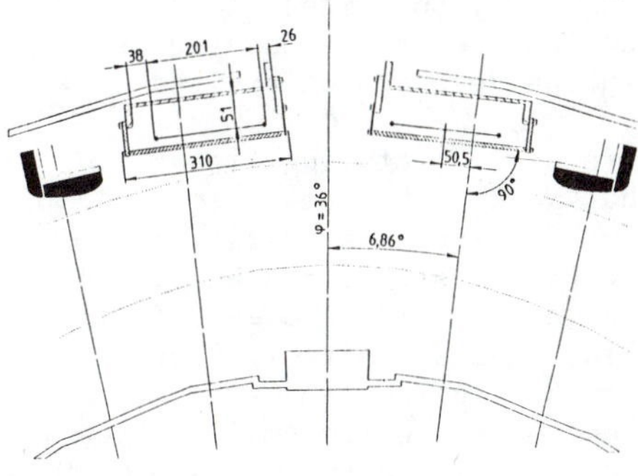

Fig. 5: Cut in the z=0 plane (all dimensions in mm)

Fig.6: Antenna and interface

ROTATING FIELD ANTENNA EXPERIMENTS IN PHAEDRUS-B

Y. Yasaka, R. Majeski, J. Browning, N. Hershkowitz, D. Roberts
University of Wisconsin-Madison
1500 Johnson Drive, Madison, Wisconsin 53706

ABSTRACT

The rotating field antenna installed in the central cell of the Phaedrus-B tandem mirror consists of two close-spaced dual half-turn ICRF antennas. The symmetry axes of the antennas are rotated 90° with respect to each other. Each antenna is driven by a separate rf amplifier, with > 200 KW power output. The polarization of the resultant antenna near fields is selected by the relative phasing of the antenna currents. In particular, the antenna set can produce nearly pure left or right circularly polarized fields. We find an increase in ion heating as the field polarization is varied from right circularly polarized thorugh linear polarization to left circular polarization, for plasma densities up to 3-4 x 10^{12} cm^{-3}, when the antenna set is driven at $\omega \sim \omega_{ci}$ (midplane). Ion temperature is diagnosed by a time of flight neutral energy analyzer. Results are compared to the predictions of the ICRF code ANTENA of Brian McVey.

The ICRF antennas which are most commonly used in heating experiments -- partial turn loop antennas -- produce linear rf field polarizations, with equal excitation of left and right-hand rotating components. Often, however, only one polarization is desirable for optimum plasma coupling. For example, in connection with Alfven wave heating, Appert[1] has identified the most favorable mode for good on-axis heating to be a surface mode with azimuthal mode number m < 0. In experiments with a moderate density (< 1 x 10^{13} cm^{-3}) single species plasma and ICRF in the range of the ion cyclotron frequency (ω_{ci}), it has been shown[2,3] that excitation of circularly polarized m = -1 fields leads to improved ion heating compared to linearly polarized fields. In general, control over the polarization of the antenna fields provides additional selectivity in the spectrum of excited eigenmodes.

The experiments described here use a "rotating field" antenna set to select the azimuthal mode number (m = +1 or -1) of the antenna near fields. We show that the polarization of the plasma fields changes accordingly. The observed ion heating in the 2 - 5 x 10^{12} cm^{-3} hydrogen plasma is best for m = -1 (left hand) excitation, which produces the largest left circularly polarized electric field component (E_+) in the plasma core.

The experiments were performed in the central cell of the Phaedrus-B tandem mirror. The hydrogen plasma (n_e ~ 2 - 5 x 10^{12} cm^{-3}, T_e ~ 40 eV, 20 msec duration) is 16 cm in radius; the mirror-mirror length of the central cell is 3.2 m. Only the central cell ICRF was active in these experiments.

The antenna set consists of two close-spaced dual half turn loop antennas, located 50 cm from mirror midplane. One antenna is rotated 90° with respect to the other (see Figure 1). Each antenna is driven at 1.31 MHz by a separate 200 kw amplifier.

When the antennas are excited with currents of equal amplitude, the azimuthal mode number spectrum of the fields produced depends on the relative phasing of the antenna currents. The amplitude of the mth azimuthal mode in the antenna spectrum is given approximately by

$$a_m = I \frac{\sin(\pi m/2)}{(\pi m/2)} \left[1-(-1)^m\right] \cos\left(\frac{\pi m}{4} - \frac{\phi}{2}\right),$$

where the antenna current $I_{W(west)}$ (see Fig. 1) = $I \cos(\omega t + \phi)$ and $I_{E(east)} = I\cos(\omega t)$. For example, for a relative antenna current phase of $\phi = -\pi/2$, 80 percent of the spectral power is in the $m = -1$ component.

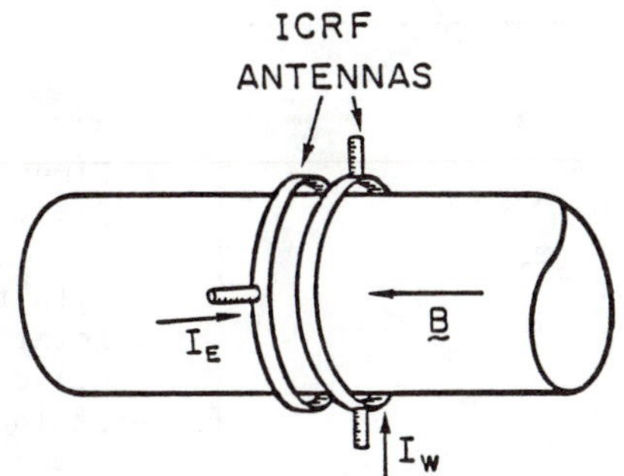

Figure 1. Schematic of the rotating field antenna set.

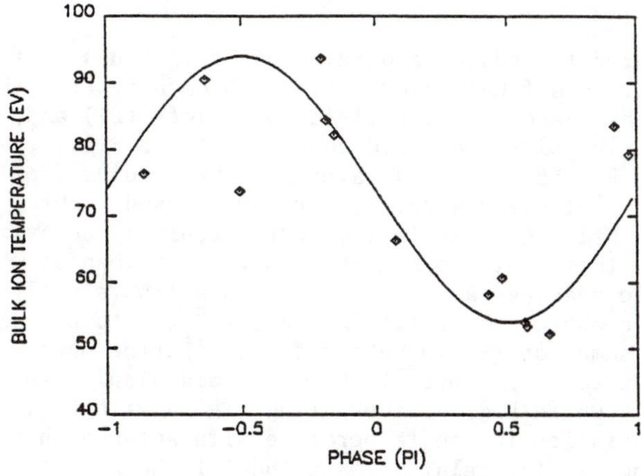

Figure 2. Variation of the measured ion temperature with the relative antenna current phase. The solid line gives the expected sinusoidal variation with phase.

Figure 2 shows the ion temperature as a function of the relative antenna phase. The perpendicular ion temperature is measured by a time-of-flight neutral energy analyzer. The antenna frequency is 1.08 ω_{ci}, where ω_{ci} is referred to central cell midplane. Again, $\phi = -0.5\pi$ corresponds to m = -1 left rotating excitation, while $\phi = +0.5\pi$ corresponds to m = +1 right-rotating excitation. The measured bulk ion temperature varies sinusoidally with ϕ, with a maximum near -0.5π.

If we separate the total antenna current power, $I_E^2 + I_W^2$, into left - (I_L^2) and right - (I_R^2) rotating components, then:

$$I_{L(R)}^2 = 0.5\,(I_E^2 + I_W^2 - (+)\,2I_E I_W \sin\phi).$$

When ϕ is maintained near $+\,0.5\pi$ to minimize I_L^2, the ion temperature remained < 60 eV although I_R^2 was varied over the range $0.3 - 0.6 \times 10^6$ A^2. On the other hand, the ion temperature increases linearly with I_L^2 as shown in Fig. 3.

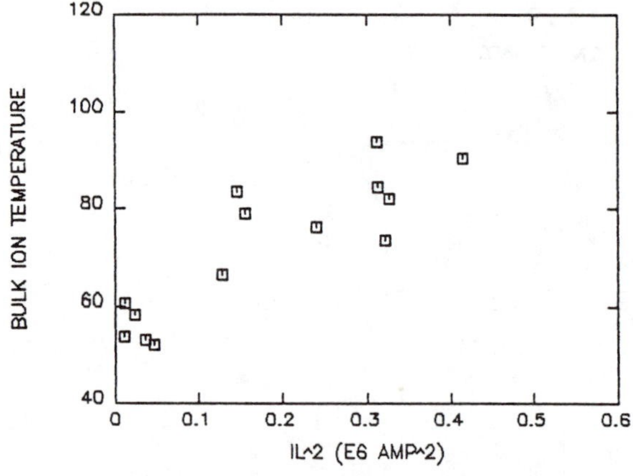

Figure 3. Scaling of the ion temperature with the left rotating (m = -1) component of the antenna power.

We have measured the radial and azimuthal components of the rf magnetic fields, as a function of radius, 25 cm away from the antenna set. We then decompose the fields onto left (B+) and right (B_-) circularly polarized field components. In Fig. 4, we display the ratio $(B_+^2/(B_+^2 + B_-^2))$, averaged over radius from 0 to 12 cm, for three antenna phases: $+0.5\pi$, -0.5π, and -0.9π. Note particularly that the power in the left circularly polarized component B_+^2 is a factor at two higher at $\phi = -0.5\pi$ than at $\phi = +0.5\pi$. Also, since phasings at $\phi = \pi$ and $\phi = 0$ are physically equivalent, we also expect the ratio $(B_+^2/B_+^2 + B_-^2))$ to be approximately the same for $\phi = -0.9\pi$ and $\phi \doteq 0$. Furthermore, since $E_z \ll E_\perp$, the ratio $(E_+^2/(E_+^2 + E_-^2))$ should display the same behavior as shown in Fig. 4. Comparing Fig. 4 with Fig. 2, we see that the variation in ion temperature with antenna phase is due to the change in the relative magnitude of the E_+

component of the rf fields. Maximum heating is obtained when the left-circularly polarized rf electric field is largest. This E_+ field is considered to be associated with the near field or m = -1 slow wave, which can propagate even above ω_{ci} due to hot ion effects.

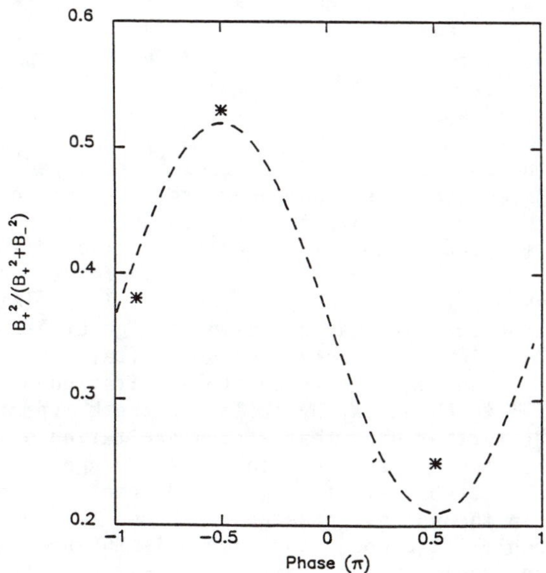

Figure 4. Amplitude of the left hand polarized rf magnetic field power, averaged over the 12 cm radius plasma core, for three values of antenna phasing. The dotted line gives the expected sinusoidal variation with phase.

In summary, we have shown that the control over the azimuthal mode number spectrum provided by phasing a pair of close-spaced dual half turn ICRF antennas can modify the polarization of the rf fields in the plasma, and hence the ion heating efficiency. Maximum heating is obtained when the antenna set is operated in m = -1 left-rotating mode. This corresponds to the largest relative magnitude of left-circularly polarized component of the rf electron field in the plasma.

ACKNOWLEDGMENTS

Supported by U.S. DOE Contract No. DE-AC02-78ET51015

REFERENCES

1. K. Appert, J. Vaclavik, and L. Villard, Phys. Fluids 27, 432 (1984).
2. Y. Yasaka and R. Itatani, Nucl. Fusion 20, 1391 (1980).
3. S. Okamura, K. Adati, T. Aoki, D.R. Baker, H. Fujita, H.R. Garner, K. Hattori, S. Hidekuma, T. Kawamoto, R. Kumazawa, Y. Okubo, and T. Sato, Nucl. Fusion 26, 1491 (1986).

CONTROL OF FAST MAGNETOSONIC MODES WITH A PHASED ANTENNA ARRAY IN THE PHAEDRUS-B TANDEM MIRROR

R. Majeski, J. J. Browning, N. Hershkowitz, S. Meassick, and J. Ferron
University of Wisconsin, Madison, WI 53706

ABSTRACT

The central cell of the Phaedrus-B tandem mirror is equipped with four dual half-turn ICRF antennas. The antennas are arranged in two close-spaced pairs, with the antenna pairs located 50 cm to either side of central cell midplane. In these experiments, two antennas (one from each pair) with an axial separation of 1 m are driven by separate RF amplifiers, each with > 200 KW power output. The relative phase of the driven antenna currents may be fixed throughout a shot, or rapidly swept (180° in < 100 μsec). We demonstrate that the relative antenna phasing can be used to control excitation of the m=+1 fast magnetosonic wave in Phaedrus, by tailoring the $k_{\parallel}$ spectrum of the antenna set. We further show that strong excitation of the m=+1 fast magnetosonic eigenmode provides stabilization against the curvature driven flute instability through ponderomotive force effects. Conversely, when the antenna phasing is chosen to suppress the m=+1 fast magnetosonic eigenmode, the plasma is MHD unstable, with density fluctuation levels on the order of $\delta n/n=1$. Results are in good agreement with predictions from the ICRF code ANTENA of Brian McVey.

The excitation of weakly damped fast magnetosonic wave eigenmodes in tandem mirror devices differs from excitation in tokamaks not only in the magnetic geometry, but in the fact that there is no tandem mirror analog to the toroidal eigenmode spectrum in a tokamak in the absence of reflecting boundary conditions at the ends of the mirror. A continuum of fast wave eigenmodes may be excited. This is also the case in a high density toroidal device, where fast magnetosonic waves are strongly damped and toroidal eigenmodes are not supported. Selective excitation of wave eigenmodes requires selectivity in the ICRF antenna spectrum.

We employ a pair of axially separated phased dual half turn ICRF antennas to control the excitation of the m=+1 fast magnetosonic wave eigenmode. The relative antenna phasing is used to tailor the antenna $k_{\parallel}$ spectrum to either strongly excite or suppress excitation of the eigenmode. The experiments were performed in the 3.2 m long central cell of the Phaedrus-B tandem mirror. The hydrogen plasma ($n \sim 2\text{-}3 \times 10^{12}$ cm^{-3}, radius ~ 16 cm, $T_e \sim T_i \sim 40$ eV, 20 msec duration) is initiated by ECH and sustained and heated solely by ICRF in the range $0.8 < \omega/\omega_{ci} < 1.1$ in the central cell. The rf systems operate at 1.31 MHz. The antenna near field is responsible for producing and heating the plasma. No endcell rf systems were active, so that without trapped plasma in the minimum-B endcells the only mechanism providing plasma MHD stability is rf ponderomotive stabilization by the central cell ICRF.[1,2] The wave fields of the m = +1 fast magnetosonic eigenmode have a profile which produces a stabilizing ponderomotive force. We find that the most obvious effect of variations in the excitation level of the mode is a major change in plasma stability. We can only produce an MHD stable plasma in the Phaedrus central cell if we phase the antennas to excite the m = +1 eigenmode.

The two dual half-turn ICRF antennas on the Phaedrus central cell used for these experiments are separated by 1 m, and driven by separate 200 KW amplifiers. For densities $< 1 \times 10^{13}$ cm^{-3} and $\omega \sim \omega_{ci}$, the only magnetosonic eigenmode which propagates in Phaedrus is the m=+1 fast wave. For our plasma parameters this mode has an axial wavelength at $\omega \sim \omega_{ci}$ of ~ 2 m, so that the ICRF antennas are spaced $\lambda/2$ apart. The k spectrum of the antenna set is a strong function of the relative phase of the antenna currents for $k_{\parallel} \sim 0.03 - 0.04$ cm^{-1}.

Figure 1 shows the antenna $k_{\parallel}$ spectrum for two values of the relative antenna phase, computed with Brian McVey's ICRF code ANTENA.[3] At 270° phasing, there is negligible power in the antenna spectrum near the wavenumber of the m=+1 fast wave, while the power near the eigenmode $k_{\parallel}$ peaks for a phase of 90°. Measurements of the plasma rf wave fields demonstrate that the variations in the antenna spectrum produce changes in m=+1 fast wave excitation. Figure 2 shows the variation in wave field amplitude as a function of antenna phase, measured near central cell midplane. Wave amplitude is a maximum for 90° phase, when the antenna $k_{\parallel}$ spectrum peaks at ~ 0.03 cm^{-1}, and a minimum when the antenna spectral power at 0.03 cm^{-1} is minimized.

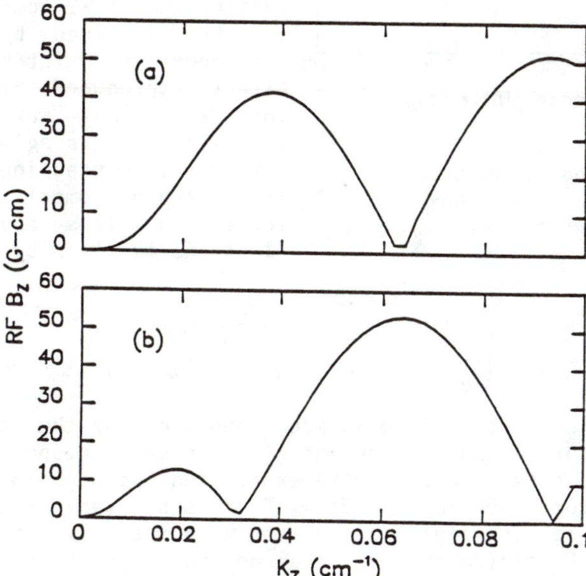

Fig. 1. $k_{\parallel}$ spectrum of the antenna array of (a) $\phi = 90^{o}$, and (b) $\phi = 270^{o}$.

We find a corresponding variation in plasma stability as the antenna phase and the excitation of the m=+1 fast wave is varied. Figure 3(a) shows the plasma density fluctuation level as the antenna phase is varied on a shot to shot basis. Note that both the ICRF power and frequency are fixed here; only the relative antenna phase

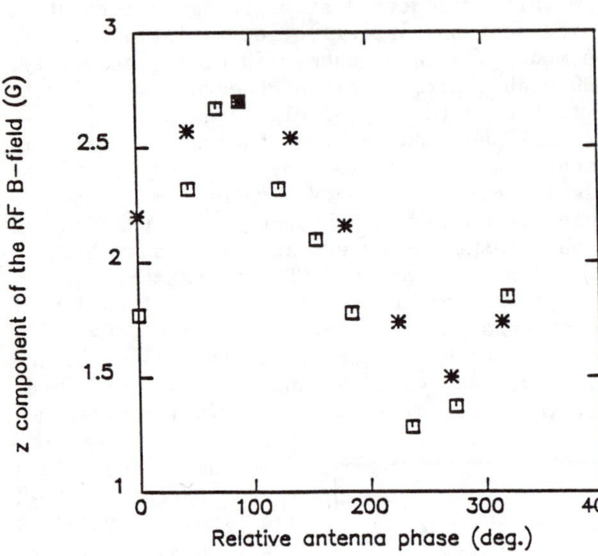

Fig. 2. Comparison of the calculated (stars) and normalized measured (boxes) z-component of the rf magnetic field.

is varied. For phases which produce strong excitation of the m=+1 fast magnetosonic wave, quiet plasmas are obtained. For phases which do not excite the m=+1 mode, Langmuir probes show large amplitude 6-10 KHz, m=+1, k ~ 0 density fluctuations, indicative of the MHD interchange instability.

The reduction in MHD activity for phases which strongly excite the m=+1 mode can be explained by ponderomotive force effects produced by the m=+1 fast wave fields. The single particle expression for the ponderomotive force on ions and electrons is given by

$$F_p(\text{ions + electrons}) = \frac{-e^2}{4m_i}\left[\frac{\nabla_r E_+^2}{\omega_{ci}(\omega - \omega_{ci})} - \frac{\nabla_r E_-^2}{\omega_{ci}(\omega + \omega_{ci})}\right] \quad (1)$$

where E_+ and E_- denote the circularly polarized components of the rf electric field which rotate in the ion and electron sense respectively. We have neglected a term in E_z in the expression for ponderomotive force since, from modeling results, E_z does not vary as a function of phase. Since the fields associated with the m=+1 mode near ω_{ci} are primarily E_- fields with a radial dependence $\sim J_o(k_r r)$, they provide a stabilizing (radially inward) contribution to the ponderomotive force.

We used ANTENA to estimate F_p as a function of antenna phase, and average throughout the plasma column. The results of this calculation are plotted in Figure 3(b). The curvature force is ~1 x 10^{-18} N. From Figures 3(a) and (b), we see that a stable plasma is produced when F_p is comparable to F_c. The largest contribution to the ponderomotive force -- and the only contribution which varies antenna phasing -- is due to the far field E_- produced by the m=+1 fast magnetosonic wave.

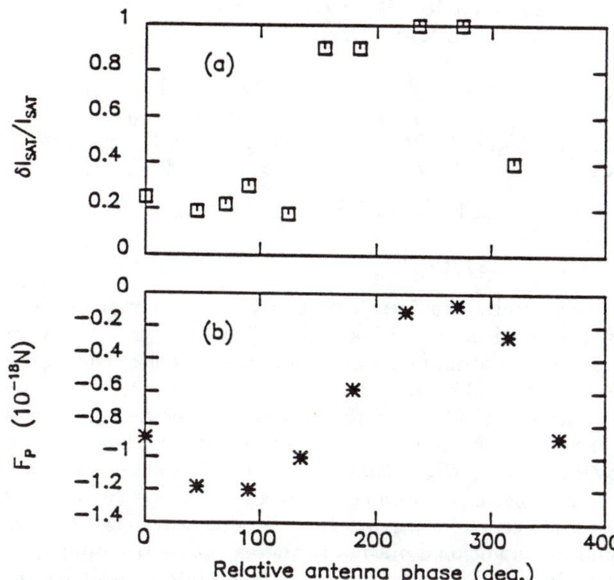

Fig. 3. (a) Density fluctuation level as measured by Langmuir probe in saturation current as a function of the relative antenna phase. (b) Calculated ponderomotive force as a function of phase.

Finally, recent theories of rf stabilization indicate that, in the absence of close fitting conducting walls surrounding the plasma column, the ponderomotive stabilizing effect of propagating wave fields should be largely canceled by sideband effects[4]. In our experiments, the presence of conducting limiters at central cell midplane[3], z = ± 0.25 m, and z = ± 1.4m may provide boundary conditions which have an effect similar to a conducting wall.

In summary, we find that a pair of axially separated phased ICRF antennas provides very good control over the excitation level of an eigenmode, in this case the m = +1 fast magnetosonic wave. Furthermore, MHD stable plasmas are only obtained in these experiments when the m = +1 eigenmode is strongly excited. The stabilization of the plasma is very well modeled by the effect of rf ponderomotive forces, which here are principally due to the fields associated with the m = +1 eigenmode.

ACKNOWLEDGMENTS

*This work is supported by U.S. Department of Energy Contract No. DE-AC02-78ET51015.

REFERENCES

1. J.R. Ferron, N. Hershkowitz, R.A. Breun, S.N. Golovato, and R. Goulding, Phys. Rev. Lett. 51, 1955 (1983).
2. J.R. Ferron, S.N. Golovato, N. Hershkowitz, and R. Goulding, "Interchange Stabilization of an Axisymmetric Single Cell Mirror Using High Frequency Electric Fields," to be published in Physics of Fluids.
3. B. McVey, "ICRF Antenna Coupling Theory for a Cylindrically Stratified Plasma," National Technical Information Service document No. DE85004960.
4. J.R. Myra, D.A. D'Ippolito, and G.L. Francis, Phys. Fluids 30, 148 (1987).

Heating and Stabilization By Slow Waves in the Tara Tandem Mirror*

S. N. Golovato, K. Brau, J. Casey, J. Coleman, M. Gerver, W. Guss, G. Hallock, S. Horne, J. Irby, J. Kesner, R. Kumazawa†, B. Lane, J. Machuzak, T. Moran, R. Myer, R. S. Post, E. Sevillano, D. K. Smith, J. Sullivan, R. Torti, Y. Yao, Y. Yasaka‡, and J. Zielinski

Plasma Fusion Center, M.I.T., Cambridge, MA 02139

ABSTRACT

The Tara tandem mirror is typically run in an effectively axisymmetric configuration by not producing β in the quadrupole anchor cells. Slow ion cyclotron wave excitation by a slot antenna located on a midplane bump in the magnetic field of the central cell provides plasma production and heating. Ion cyclotron resonances are at the minimum central cell fields on either side of the bump. Magnetic probe measurements have identified the slow wave excited by the antenna. Effective ion and electron heating is observed with a peak β of over 2%. The ICRF also provides stabilization of an $m = 1$ flute mode which is otherwise unstable in axisymmetric mirror geometry. In conjunction with a magnetic divertor at the central cell midplane, the slow wave can stabilize not only the central cell plasma but additional plasma in the axisymmetric plug cells. Destabilizing β can be produced in the plug cells by neutral beam injection, ICRF, or ECRH. By maintaining the stability of the plugs and central cell, end plugging using either ICRF or ECRH in the plug cells is obtained.

DESCRIPTION OF THE EXPERIMENT

The Tara Tandem Mirror is made up of a solenoidal central cell bounded on each side by axisymmetric mirror cells. Outboard of the axisymmetric cells are transition cells which map the flux through quadrupole mirror cells on each end. By not producing β in the quadrupole cells, Tara may be run in an effectively axisymmetric configuration where the plasma production, heating, MHD stabilization are provided in the central cell by ICRF excited by a slot or aperture antenna.[1] The magnetic field is raised by a factor of 2 in a 3 meter section at the central cell midplane, creating a magnetic "bump" with mirror "wells" on either side. The slot antenna and a gas box for localized fueling are positioned on the bump with a magnetic divertor between them. The magnetic divertor provides additional MHD stabilization, increasing the range of stable operation, particularly during end plugging experiments when additional drive is added in the axisymmetric end cells.[2]

WAVE PROPAGATION AND PLASMA HEATING

Slow wave excitation and propagation into a beach resonance is achieved by locating the slot antenna on the magnetic bump and placing the resonance in the wells on either side. Using a set of RF magnetic probes positioned along the gradient between the bump and the wells, properties of the wave fields have been measured. The axial wave number $k_{\parallel}$ equals 0.063 cm^{-1} at all radii. This corresponds reasonably well to the expected slow wave and to the 40 cm slot antenna aperture, which has an excitation spectrum peaked at about 0.078 cm^{-1}. That a short wavelength is measured even at the surface

* Supported by U.S. DOE Contract DE-AC02-78ET51013

† Institute of Plasma Physics, Nagoya University, Nagoya, Japan

‡ Department of Electronics, Kyoto University, Kyoto, Japan

indicates that the $m = 1$ fast wave is not efficiently excited. The polarization of the electromagnetic component of the wave fields has been measured with B_+ peaking on axis and B_- peaking at the edge, consistent with slow wave excitation. If it is assumed that the azimuthal modes of the wave are principally $m = \pm 1$, measurements of the azimuthal variation of the wave fields indicate that the $m = -1$ mode is strongest in the core and $m = 1$ on the outside.

Typical operation with 200-400 kW of power results in plasma densities of $2 - 4 \times 10^{12}$ cm^{-3}, $T_{i\perp} = 0.6 - 1$ keV, $T_{i\|} = 150 - 200$ eV, $T_e = 70 - 100$ eV. An average β of 1% has been achieved with a peak β of 2% . These parameters apply to the south half of the central cell, to the side of the bump where the slot antenna is located. The north central cell typically has a $T_{i\perp}$ lower by one half to one third due principally to the gas box location on the north side of the bump. There is also some indication that the divertor affects wave propagation to the north, reducing the wave amplitude there. The heating is best with the resonance at or near the minimum field of the wells. Thomson scattering measurements of the electron temperature profile show a 70 eV central temperature rising to 100-120 eV at r=10-15 cm. A rough power balance is done using the diamagnetism decay at ICRF turn-off to estimate the hot ion losses, measuring electron and ion end loss power, and estimating ionization and excitation losses for electrons and losses due to charge exchange of passing ions in the gas box. For an input power of 300 kW, about 50 kW goes to hot ions, 50-100 kW to passing ions, and 40 kW goes directly to electrons, accounting for 45-65% of the power. The rest may heat edge plasma which is lost to limiters.

The plasma density profile is best fit by a Gaussian of 10-15 cm scale length, with the limiter at 22 cm. As the power is increased the plasma radius decreases. The peak density increases but the total number of particles decreases slowly and the diamagnetism saturates at highest power. It appears that the peaking of the profile reduces the coupling of the ICRF to the plasma, with the radiation resistance of the antenna decreasing as the edge density falls. If the gas fueling rate is increased, the density profile broadens and coupling to the core plasma is recovered, with the diamagnetism no longer saturating with power. Figure 1 shows the peak density, Gaussian radius, and average β of the south central cell well vs power with sufficient gas fueling to avoid saturation of the diamagnetism. The peaking of the profile with increasing power can be seen. The profile change and the saturation of particle and energy content with increasing power may indicate some ICRF-driven enhancement of radial losses at the edge rather than a change in the power deposition profile.

The slow wave heating produces an anisotropic plasma with the diamagnetism down by a factor of two at a mirror ratio of 1.1. The magnetic probes observe a mode at $0.9\omega_{ci}$ in the wells which can be large compared to the applied ICRF. The anisotropy and β of the plasma may be sufficiently high that it is unstable to the Alfven-Ion Cyclotron instablility. The $k_\|$ of the mode is similar to the applied ICRF, indicating that it is also a slow wave. Lowering the well field by 5% reduces the anisotropy by moving the resonance to higher mirror ratio. The mode frequency drops by 5% and is lower in amplitude, as would be expected for this instability.

MHD STABILIZATION

It has been documented that when the plasma is MHD unstable, a rigid $m = 1$ flute mode is oberved in a large amplitude saturated state.[3] Stability is sensitive to a number of experimental parameters including gas fueling rate, resonance position, ω/ω_{ci} under the antenna, slot power, and divertor null radius. Stability is enhanced by high ICRF power, positioning the divertor null deeper in the plasma, keeping the ω_{ci} resonance near the well minima, keeping $\omega/\omega_{ci} > 0.55$ under the antenna, and maintaining gas fueling in a window of about 10-30 Torr-liters/sec. There is a sharp threshold for instability when any of the above parameters are varied. Figure 2 shows the effect of doubling the ICRF power during an unstable shot at low gas fueling rate. During the high power pulse the fluctuations decrease and the density builds up. There is sufficient stabilization under optimum conditions to maintain stability when drive is added in the axisymmetric end cells by ECRH, ICRF, or neutral beams. It is the stabilization provided by the slot excited fields that allows stable end plugging experiments to be done in the axisymmetric Tara configuration.

The sensitivity of stability to the resonance position is due mainly to increased drive, not to an abrupt change in the ICRF fields providing stabilization. Since the ICRF produces a very anisotropic plasma, moving the resonance onto the gradient places more β in bad curvature. Increasing the ICRF power under these conditions improves stability, indicating that the ICRF is still stabilizing. The sensitivity to ω/ω_{ci} at the antenna indicates that the $k_{\|}$ spectrum excited by the antenna is important, with lower $k_{\|}$ less stabilizing.

The gas fueling rate has a strong effect on the edge profile and thus on the ICRF coupling to the plasma. This may be why stability is lost at low gas fueling. At high gas fueling, a relaxation oscillation driven by the $m = 1$ flute mode is observed rather than the steady-state instability that is observed at other stability thresholds. This effect is not understood.

These results are in contrast to other ICRF stabilization experiments[4,5] where MHD stabilization has been attributed to ponderomotive effects of fast waves with stability resulting when $\omega > \omega_{ci}$. Assuming that the stabilizing effect of the slow waves in Tara is due to ponderomotive effects, model calculations of the wave fields indicate that the E_+ and E_- profiles are destabilizing for $\omega < \omega_{ci}$. Only the E_z component is predicted to be stabilizing for the regime of this experiment.

When drive is added in an axisymmetric end cell, the stabilizing influence of the ICRF and divertor are additive. Fundamental resonance ECRH in one axisymmetric end cell can produce instability. Figure 3 shows the stable trajectory in null radius-ICRF power space for two ECRH powers, with smaller null radius or higher ICRF power stabilizing the plasma. Since the plasma radius decreases while the stabilization improves at higher power, it is likely that line tying is not the principal effect. The divertor effect should also be reduced with a smaller plasma radius since there will be less plasma pressure at the null. This points to the key role played by the slow wave ICRF in maintaining stability in this experiment.

REFERENCES

1. R. S. Post, K. Brau, S. Golovato, E. Sevillano et al., Nucl. Fusion **27**, 217(1987).
2. J. Casey, B. Lane, J. Irby et al., (in preparation).
3. J. Irby, B. Lane, J. Casey et al.,(submitted to Phys. Fluids).
4. J. R. Ferron, S. N. Golovato, et al., Phys.Fluids **30**, (1987) (to be published).
5. Y. Yasaka and R. Itatani, Phys. Rev. Lett. **56**, 2811(1986).

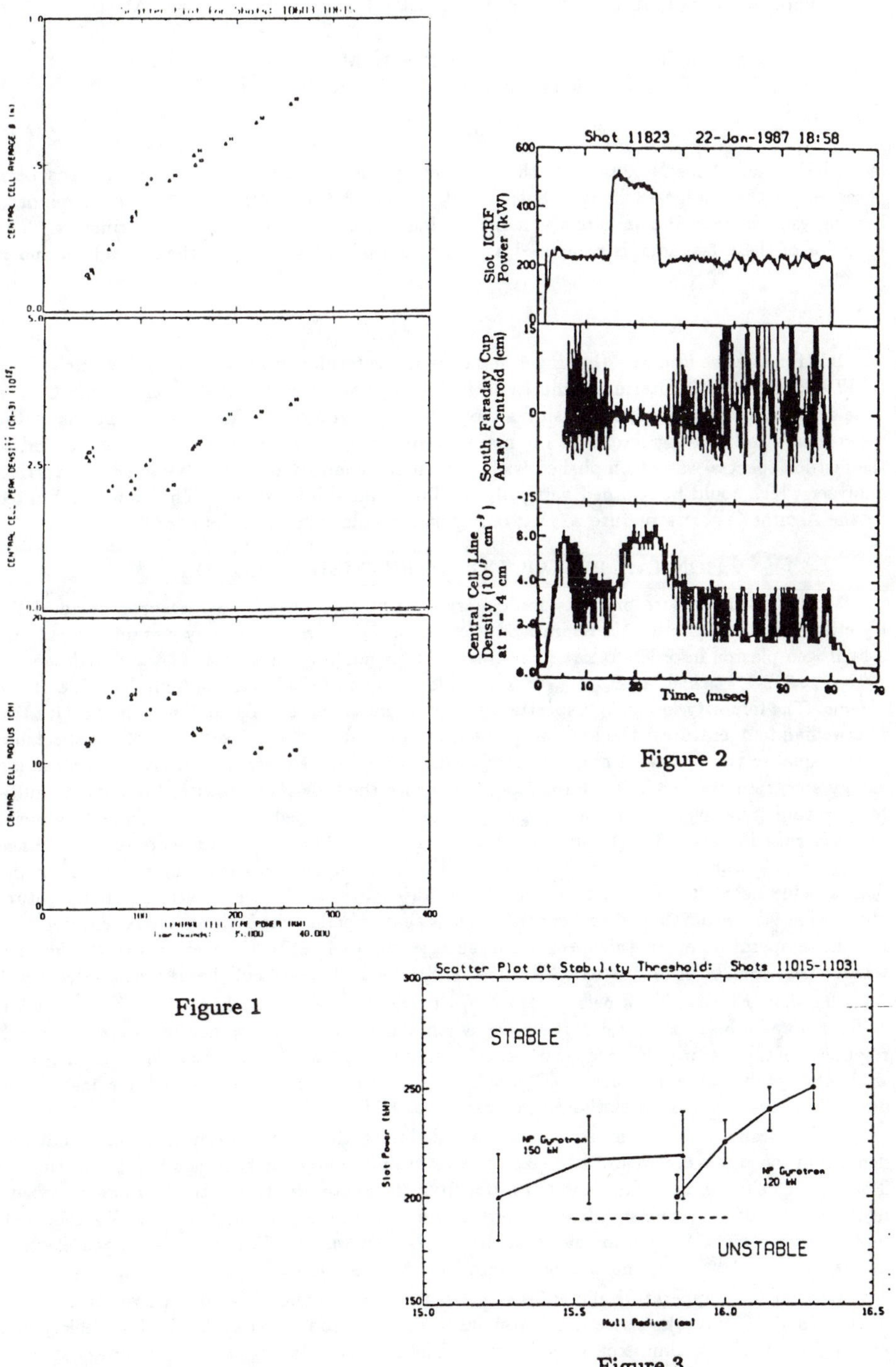

Figure 1

Figure 2

Figure 3

Minority Species Fast Ion Behavior During ICRF Heating Experiments on Alcator C.

C.L. Fiore, F. S. McDermott, J. D. Moody, M. Porkolab,T. D. Shepard
Plasma Fusion Center, MIT*, Cambridge, MA 02139

ABSTRACT

Production of fast ion tails in the minority species of the Alcator C plasma has been observed on the charge exchange neutral particle spectra for both ICRF fast wave minority heating experiments and in directly launched ion Bernstein wave (IBW) experiments. The behavior of these fast ions is compared as a function of density, concentration, radius, and rf power.

INTRODUCTION

ICRF fast wave minority (hydrogen minority in deuterium) heating and ion Bernstein wave (IBW) experiments (deuterium minority in hydrogen) have been conducted on Alcator C for a wide range of plasma conditions using a frequency of $\sim$ 180 Mhz. Further descriptions of the experiments and heating results in the majority species can be found in [1,2]. Measurement of the fast ion spectra was accomplished with a 10 energy channel, mass resolving neutral particle analyzer which could be scanned vertically to obtain radial information. The view was limited by the Alcator C port structure so that only perpendicular ions were detected.

ION TAIL PRODUCTION DURING IBW INJECTION

Multiple temperature minority ion energy distributions have been measured during IBW injection for two magnetic field regimes on Alcator C. A deuterium minority was introduced into a hydrogen plasma in order to lower the threshold for nonlinear absorption by the hydrogen at $\omega/\Omega_{cH} = 3/2$[3,4], which at $B_{toroidal} = 7.6T$ is located at $r/a = .2$ on the high field side of the plasma. The minority ion energy spectra which were measured at this field must be fit with two Maxwellian temperatures: the low energy fit is obtained from the 2-4 keV part of the spectrum and is equal to the hydrogen majority temperature; the high energy fit is derived from the ion energy spectrum above 4 keV, and is 1.3-2 keV, twice the bulk temperature. Similarly a multi-temperature minority deuterium energy spectrum was obtained when the toroidal magnetic field was raised above 9.3 T (Figure 1). At this magnetic field a non-linear deuterium resonance $\omega/\Omega_{cD} = 5/2$ occurs at the plasma center. The linear $\omega/\Omega_{cD} = 3$ resonance remains in the plasma edge near the IBW antenna at this field, at $r/a = .9$. The deuterium temperature obtained at 9.3 T from the 2-4 keV energy range is again equal to the hydrogen bulk temperature, but the temperature of the tail spectrum at energies above 4 keV is as much as 4 times the bulk temperature. For both magnetic field values, the tail density is 1/3 of the deuterium density.

The deuterium tails formed at 9.3 T are much stronger than than those formed at 7.6 T. This may be because at 7.6 T the plasma can absorb power at the nonlinear $\omega/\Omega_{cH} = 3/2$ resonance of the majority hydrogen species as well as through the linear resonance at $\omega/\Omega_{cD} = 3$, while absorption at the nonlinear $\omega/\Omega_{cD} = 5/2$ resonance is the only mechanism for heating the deuterium minority species at the plasma center at 9.3T.

Radial scans of the minority ions obtained during the IBW experiments show that the deuterium ion tail temperature is peaked at the plasma center for both the 9.3 T and the 7.6 T magnetic field regimes. The power transfer from the deuterium tail to the hydrogen majority in the inner half of the plasma for the case shown in Figure 2b is found to be 20-40 kW, with input power of 70 kW. The magnetic field for this scan was 9.5 T, the line averaged electron density was $1 \times 10^{20} m^{-3}$, and the deuterium fraction was 10%.

The ion tail temperature increases monatonically with increasing IBW power at both 7.6 T and 9.3 T. (Figure 3). There is a threshold in the tail production at 7.6 T of 30-50 kW and at 9.3 T of 15 kW, within experimental error. The nonlinear theory predicts a threshold of 10 kW for the 9.3 T ($\omega/\Omega_{cD} = 5/2$ at $r/a = .1$) experiment for a deuterium concentration of 6%, consistent with the 9.3 T data. The threshold at 7.6 T ($\omega/\Omega_{cD} = 3$ at $r/a = -.2$) is in apparent contradiction with the linear theory which predicts no threshold. It should be noted, however that the $\omega/\Omega_{cd} = 7/2$ resonance layer exists halfway in radially, so that nonlinear absorption of the IBW by the deuterium ions could also occur.

The minority ion tail temperature normalized to rf power is 110 eV/kW at $n_e = 1\times10^{20}m^{-3}$ for IBW injection at 9.3 T. The highest value for the 7.6 T IBW experiment is 38 eV/kW at $n_e = .5 \times 10^{20}m^{-3}$. The ion tail strength decreases rapidly with increasing density at both magnetic fields, and no ion tails were observed above $2 \times 10^{20}m^{-3}$ (Figure 4a). The temperature derived from the high energy part of the spectra decreases with increasing deuterium concentration at 9.3 T, decreasing from 4 keV at a concentration of 1% to 1.5 keV at 20%, at a density of $1 \times 10^{20}m^{-3}$.

HYDROGEN TEMPERATURE DURING FAST WAVE MINORITY HEATING

The fast wave minority regime produces a strong hydrogen tail via the fundamental hydrogen resonance located at the plasma center at a magnetic field of 12 T. The typical hydrogen ion energy spectrum (Figure 1) shows an increasing flux, "negative temperature" region from 2-4 keV combined with a high temperature (3-15 keV) distribution at energies > 4 keV. The "negative temperature" disappears with hydrogen concentrations > 2%. No minority heating is observed when the hydrogen concentration is above 8%.

Radial scans during fast wave injection show that the hydrogen temperature obtained from the part of energy spectrum above 4 keV is peaked at the plasma center. Figure 2a shows a minority hydrogen temperature radial profile from a plasma with a line averaged density of $1.1\times10^{20}m^{-3}$ with a hydrogen fraction of .5%. The tail is highly peaked at the center, resulting in a power transfer from the hydrogen to the deuterium of only 10-50 kW in the central half of the plasma. The input power was 160 kW in this case.

The "negative temperature" part of the spectra derived from fast wave minority heating appeared unchanged in shape or magnitude as the instrument was scanned radially. This suggests that it arises from a radius greater than the limits of the scan, ~ 7 cm and therefore does not contribute to the central plasma heating.

The hydrogen temperature is highest ($\sim 15keV$) when the $\omega/\Omega_{cH} = 1$ resonance is near the plasma center ($B_{toroidal} = 12T$) and decreases as the field is lowered and the resonance is moved toward smaller major radii. The hydrogen temperature falls to 2 keV when the $\omega/\Omega_{cH} = 1$ resonance reaches $r/a = .5$ on the high field side and continues at that temperature until the resonance is nearly at the limiter when it becomes equal to the deuterium majority temperature.

The hydrogen temperature normalized to input power as a function of line averaged electron density is shown in Figure 4b. Values of 40 ev/kW at $n_e = 1 \times 10^{20}m^{-3}$ are typical, but the hydrogen temperature decreases with increasing target density until it reaches the majority deuterium temperature at a density of $3 \times 10^{20}m^{-3}$. No minority heating was observed at higher densities.

CONCLUSIONS

Plasma heating has been effected through production of energetic minority ions for both IBW injection and fast wave ICRF on Alcator C. Minority temperatures produced by the fast wave injection were higher than those obtained during the IBW experiments, but this was accomplished with 2-4 times more rf power than was used during IBW injection. Radial profiles of the ion energy spectra showed that the power transfer from the high temperature minority species to the bulk plasma is equivalent for the IBW and the fast wave experiments, but the fast wave heated ions required twice as much input power to achieve this heating. This is in agreement with the majority heating results reported in [1,2].

Minority ion energy tails could be produced at higher minority concentrations (up to 20%) with IBW injection than in the fast wave experiments, which did not yield hydrogen heating above $n_H/(n_H + n_D) = 8\%$. Minority heating was observed at higher densities with fast wave

ICRF ($n_e = 3 \times 10^{20} m^{-3}$) than with IBW ($n_e = 2 \times 10^{20} m^{-3}$), but the available power in the fast wave experiment was 2-4 times greater than that used for the IBW experiments.

REFERENCES

1. J. Moody, et. al., this meeting.
2. T. Shepard, et. al., this meeting.
3. M. Porkolab et. al., 11th Int. Conf. on Plasma Phys. and Contr. Fus., Kyoto, Japan, 1986, IAEA-CN-47/F-11-2.
4. M. Porkolab, Phys. Rev. Lett. **54**, 434 (1985).

*Work supported by US DOE Contract DE/AC02-78ET51013

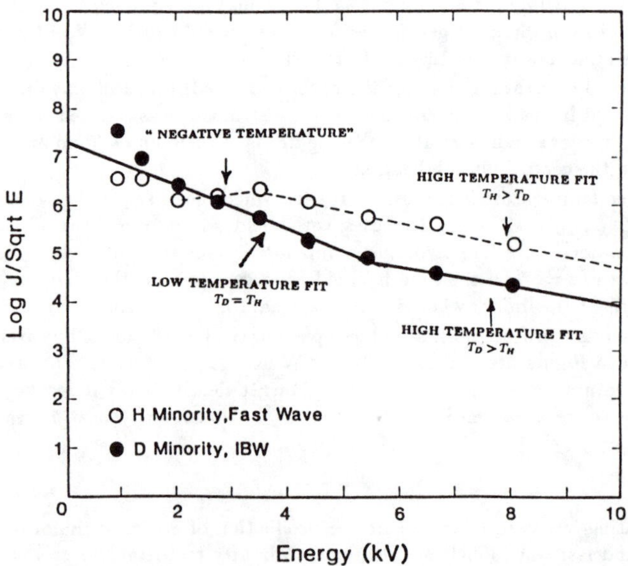

Fig. 1: Minority ion energy spectra for hydrogen (fast wave) at 12T showing "negative temperature" and high energy regions. Also shown is the deuterium (IBW) spectra at 9.3T, with the low energy fit ($T_D = T_H$) and the high energy fit ($T_D = 3T_H$).

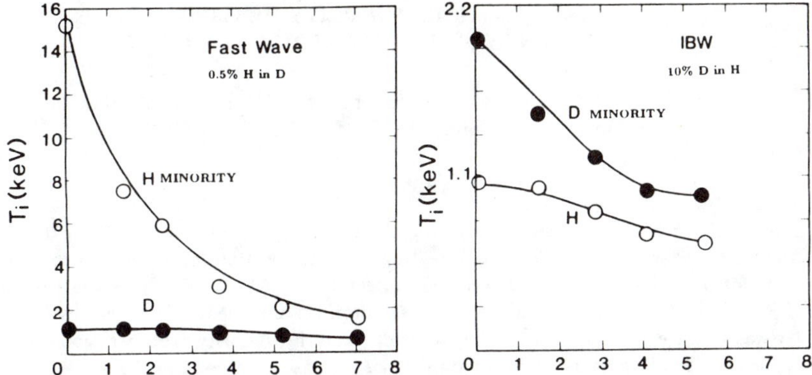

Fig. 2: Ion tail temperature as a function of radius for **a.** hydrogen fast wave, 12T $\bar{n}_e = 1.1 \times 10^{20}\,m^{-3}$, $n_H/(n_H + n_D) = .5\%$, $P_rf = 160$ kW and **b.** deuterium IBW, 9.5T, $\bar{n}_e = 1 \times 10^{20}\,m^{-3}$, $n_D/(n_D + n_H) = 10\%$, $P_rf = 70$ kW.

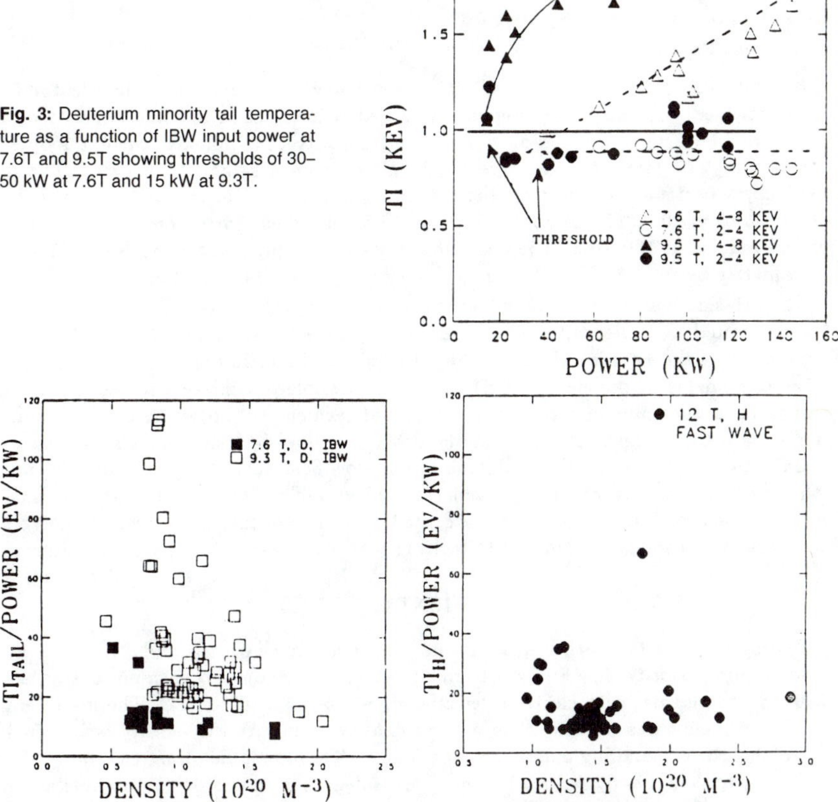

Fig. 3: Deuterium minority tail temperature as a function of IBW input power at 7.6T and 9.5T showing thresholds of 30–50 kW at 7.6T and 15 kW at 9.3T.

Fig. 4: Minority ion tail temperature normalized to input power as a function of line averaged electron density for **a.** deuterium (IBW) at 7.6T and 9.3T and **b.** hydrogen (fast wave) at 12T. No minority heating is seen above $\bar{n}_e = 2 \times 10^{20}\,m^{-3}$ for IBW or above $\bar{n}_e = 3 \times 10^{20}\,m^{-3}$ for fast wave injection.

FAST WAVE ICRF MINORITY HEATING EXPERIMENTS ON THE ALCATOR C TOKAMAK

T. D. Shepard, C. L. Fiore, F. S. McDermott, R. R. Parker, M. Porkolab
Plasma Fusion Center, M.I.T., Cambridge, MA 02139

ABSTRACT

Ion cyclotron heating experiments using the fast magnetosonic wave were carried out in the Alcator C tokamak. Significant ion heating was observed as RF power up to 450 kW at 180 MHz for pulse lengths up to 200 ms was injected into Alcator C discharges with toroidal field $B_T = 12$ T. The majority ion species was deuterium and the minority was hydrogen. The density was scanned from $\bar{n}_e = 0.5$–5.5×10^{14} cm^{-3} and the minority concentration was scanned from 0.25–8%. The largest deuterium temperature increase observed was $\Delta T_D = 412$ eV, which occurred at $\bar{n}_e = 1.4 \times 10^{14}$ cm^{-3}, $P_{\rm rf} = 300$ kW, and $\Delta t_{\rm rf} = 175$ ms. While both the central ion temperature increment and the net injected RF power decreased with density, the heating parameter $\bar{n}_e \Delta T_D / P_{\rm rf}$ remained nearly independent of density. Experiments were conducted using two different antennas. One antenna launched power only from the low-field side of the plasma, while the other launched power from both low and high-field sides. No significant difference was observed in the heating efficiencies of the two antennas.

INTRODUCTION

Minority ICRF fast wave heating experiments have been conducted in Alcator C using Faraday-shielded poloidal loop antennas described below. Each antenna consists of a two-element toroidal array. Each radiating element consists of a poloidal current loop shorted at each end. The two elements were fed in phase from a single RF source. The parallel wavenumber spectrum launched by these structures is peaked at $n_\parallel = 0$ with a half-power width of $\Delta n_\parallel = 20$. (The minor radius was 12.5 cm, which corresponds to 0.47 free space electrical radians at 180 Mhz. The condition for fast wave propagation at $B_T = 12$ T is given approximately by $n_\parallel \leq 8\sqrt{n_{14}}$ where n_{14} is the density in units of 10^{14} cm^{-3}.)

The two-side launch antenna extends poloidally from 20° above the midplane on the low-field side around the bottom to 20° above the midplane on the high-field side. By adjusting the geometry of the Faraday shields during design, and by including a dielectric sleeve below the Faraday shields at the electrical center, it was possible to achieve self-resonance with this antenna. Thus, the antenna current is maximum at each end of this antenna with a null at the center—hence the designation "two-side launch". The low-field side launch antenna extends poloidally from 135° above the midplane on the low-field side to 135° below the midplane on the low-field side. It was not possible to achieve self-resonance with this antenna. The Faraday shields on both antennas were coated with an 18 mil thick layer of molybdenum, which eliminated serious erosion due to melting that was observed in initial experiments.

EXPERIMENTAL RESULTS

Hydrogen minority heating experiments were performed with toroidal field $B_T = 12$ T, and deuterium majority. ICRF power from 0 to 450 kW was injected with pulse lengths from 0 to 200 ms into plasmas with electron densities $\bar{n}_e \sim 0.5$–6.5×10^{14} cm^{-3}. The major emphasis of the experiments was on obtaining a good density scan. As the density was increased, a limit to the power handling capability was found. The power handling capability degraded significantly at densities $\bar{n}_e > 3 \times 10^{14}$ cm^{-3}. This degradation was first noticeable at densities $\bar{n}_e > 2.5 \times 10^{14}$ cm^{-3} and gradually became more severe as the density was raised. The maximum ICRF power dropped below 100 kW at densities $\bar{n}_e > 5 \times 10^{14}$ cm^{-3}. This effect was not observed in previous experiments which were carried out at densities $\bar{n}_e < 3.3 \times 10^{14}$ cm^{-3}.

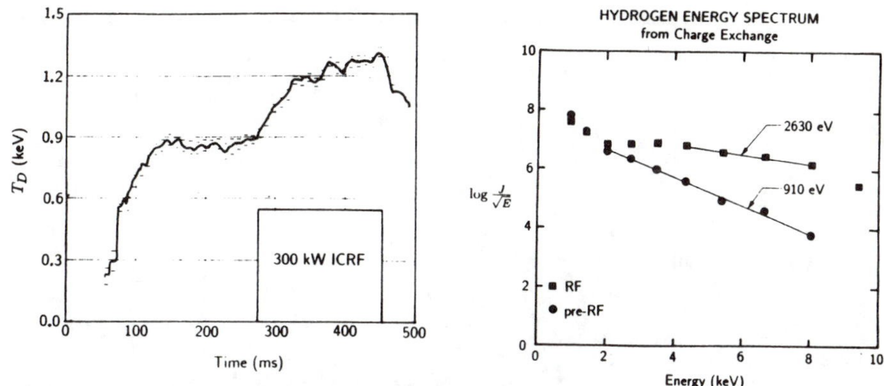

Figure 1: Deuterium temperature and hydrogen energy spectrum from charge exchange.

Parameter	Pre-RF	RF
T_D	900 eV	1312 eV
I_T	154 kA	154 kA
V_R	1.5 V	2.15 V
P_{OH}	231 kW	331 kW
P_{RF}	0 kW	300 kW
τ_E	7.7 ms	5.2 ms*
τ_{inc}	—	3.8 ms*
$\bar{n}_e$	0.84×10^{14} cm^{-3}	1.35×10^{14} cm^{-3}
n_{peak}	1.0×10^{14} cm^{-3}	2.0×10^{14} cm^{-3}
$n_H/(n_H + n_D)$	0.5%	—
Z_{eff}	2.2	3.5
q_{lim}	9.5	9.5

* Assumes 100% RF absorption.

Table 1: Parameters for the shot in Fig. 1

The power handling capability and the density dependence of its variation were the same for both antennas. The reason for the decreasing power handling capability of the antennas with density is not understood at present.

The largest deuterium ion temperature increase observed was 412 eV, which occurred at $\bar{n}_e = 1.4 \times 10^{14}$ cm^{-3}, with 300 kW of ICRF power for 175 ms (Fig. 1). Significant hydrogen tail formation was typically observed, while deuterium was always found to be thermal. Some other parameters for the shot in Fig. 1 are summarized in Table 1.

Density scans indicated that the deuterium temperature increment decreased as density increased (Fig. 2) but the efficiency, defined as $\bar{n}_e \Delta T_D / P_{rf}$, did not degrade. Efficiencies ranging from 0–20 eV $\times 10^{13}$ cm^{-3}/kW were observed at all densities and at RF power levels up to 300 kW. Heating efficiency degraded at higher powers, ranging from 0–10 eV $\times 10^{13}$ cm^{-3}/kW at 450 kW. The behavior of the deuterium temperature was the same for both antennas.

The incremental confinement time† is directly proportional to the heating efficiency for

† Incremental confinement time is defined as the change in total thermal energy (from all species) divided by the change in total heating power (RF plus change in ohmic heating power).

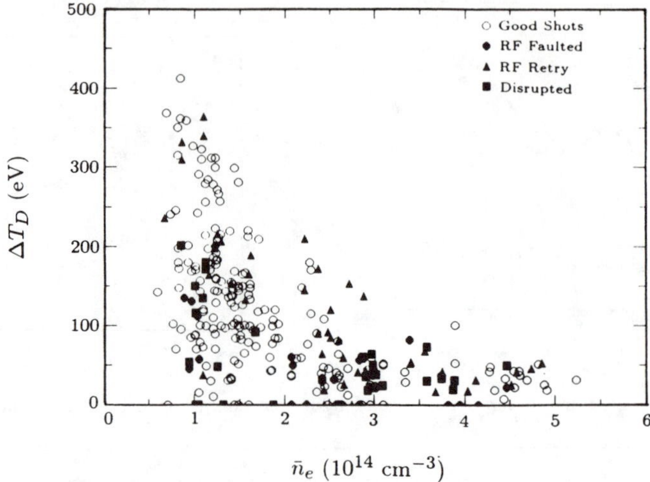

Figure 2: Deuterium heating vs. density. The behavior of the deuterium temperature is consistent with constant heating efficiency ($\bar{n}_e \Delta T_D / P_{\rm RF}$). This graph includes all data taken during November and December, 1986.

shots where the density remained constant during the RF. Incremental confinement times in these cases were typically between 0 and 3 ms, independent of density. This contrasts sharply with the ohmic (pre-RF) confinement time which increased almost linearly with density from $\tau_B \sim 6$ ms at $\bar{n}_e \sim 0.6 \times 10^{14}$ cm^{-3} up to $\tau_B \sim 30$ ms at $\bar{n}_e \sim 5 \times 10^{14}$ cm^{-3} (Fig. 3). This behavior is consistent with the empirical "JFT-2M" scaling law of Shimomura and Odajima,[1,2] which states that $\tau_{\rm inc}$ is in general primarily dependent on the minor radius of the device, and at most weakly dependent on all plasma parameter variations for a given device. Their formula is $\tau_{\rm inc} = 0.12a^2$, where a is the minor radius in meters and $\tau_{\rm inc}$ is in seconds.

It also seems possible that ripple-trapped particle losses† may be responsible for limiting $\tau_{\rm inc}$ in Alcator C. The collisional exchange time between hydrogen and deuterium for these shots is typically 1–2 ms. Simple preliminary calculations indicate that ripple trapping followed by $\mathbf{B} \times \nabla \mathbf{B}$ loss of hydrogen tail particles takes place on this same time scale. Further theoretical work is planned which will investigate this process. It is not clear whether or not there is any connection between ripple-trapping and the JFT-2M scaling law.

Significant electron heating was rarely observed during these experiments. However, slowing down of high-energy hydrogen ions on electrons is expected to be significant compared to slowing down on deuterium only for hydrogen energies greater than $\sim$ 20 keV. Such high-energy ions are very poorly confined in these discharges.

SUMMARY

Significant ion heating has been observed during minority heating experiments in Alcator C. Strong hydrogen tail formation was observed, while deuterium always remained thermal. Significant electron heating was not observed. While the ohmic confinement time was observed

† Caused by ripple in the toroidal field where the windings are interrupted by a port. For a banana-trapped particle, the banana orbit precesses toroidally until the banana tip reaches a port. At that point, the particle becomes toroidally mirror-trapped in the ripple field, and drifts vertically out of the plasma.

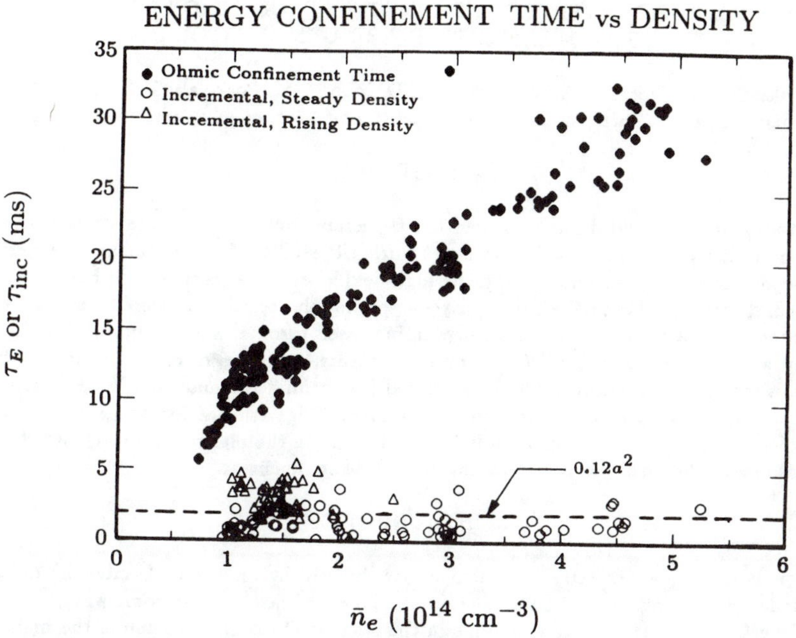

ENERGY CONFINEMENT TIME vs DENSITY

● Ohmic Confinement Time
○ Incremental, Steady Density
△ Incremental, Rising Density

$0.12a^2$

τ_E or τ_{inc} (ms)

$\bar{n}_e$ (10^{14} cm^{-3})

Figure 3: Comparison of ohmic and incremental confinement times. The incremental confinement time for shots with no density change during the RF agrees with the "JFT-2M" scaling $\tau_{inc} = 0.12a^2$.

to scale favorably with density, the incremental confinement time was independent of density. This behavior is consistent with the empirical JFT-2M scaling law. Further theoretical study is planned which will attempt to explain these results from an analysis of ripple-trapped ion loss. Simple preliminary calculations indicate that ripple-loss of energetic hydrogen ions takes place on a time scale on the order of a few milliseconds. This is comparable to the time scale for slowing down of high energy hydrogen tail ions on a deuterium background, and also agrees with the observed incremental confinement times. Thus, the transfer of energy to the deuterium may be limited by this mechanism.

ACKNOWLEGEMENT

This work was supported by the U.S. Department of Energy, contract number DE-AC02-78ET51013.

REFERENCES

1. Y. Shimomura, K. Odajima, Japan Atomic Energy Research Institute lab report JAERI-M 86-128 (1986)

2. K. Odajima, *et al.*, *Phys. Rev. Lett.* **57**:22, 2814 (1986)

STUDY OF MODE-CONVERTED AND DIRECTLY-EXCITED ION BERNSTEIN WAVES BY CO_2 LASER SCATTERING IN ALCATOR C

Y. Takase, C. L. Fiore, F. S. McDermott, J. D. Moody, M. Porkolab, T. Shepard, J. Squire
Plasma Fusion Center, Massachusetts Institute of Technology, Cambridge, MA 02139

ABSTRACT

Mode-converted and directly excited ion Bernstein waves (IBW) were studied using CO_2 laser scattering in the Alcator C tokamak. During the ICRF fast wave heating experiments, mode-converted IBW was observed on the high-field side of the resonance in both second harmonic and minority heating regimes. By comparing the relative scattered powers from the two antennas separated by 180° toroidally, an increased toroidal wave damping with increasing density was inferred. In the IBW heating experiments, optimum direct excitation is obtained when an ion-cyclotron harmonic layer is located just behind the antenna. Wave absorption at the $\omega = 3\Omega_D = 1.5\Omega_H$ layer was directly observed. Edge ion heating was inferred from the IBW dispersion when this absorption layer was located in the plasma periphery, which may be responsible for the observed improvement in particle confinement.

INTRODUCTION

Ion Bernstein waves (IBW), excited either directly by an external antenna[1] or through the mode-conversion process from the externally launched fast magnetosonic wave,[2] are studied using the CO_2 laser scattering technique in the Alcator C tokamak. Some of the main results from the IBW experiments have been reported earlier.[3] A more detailed discussion as well as results from the fast wave experiment are presented in this paper. During these experiments, molybdenum limiters of $a = 12.5\,\text{cm}$, $12.0\,\text{cm}$, or $11.5\,\text{cm}$ defined the plasma minor radius while the major radius was $R_0 = 64\,\text{cm}$. The transmitter frequency was $\omega/(2\pi) \simeq 180$–$200\,\text{MHz}$. The scattering arrangement was described in Ref. 3. The horizontal resolution (defined as $1/e^2$ in power) ranged from $\pm 8\,\text{mm}$ to $\pm 3\,\text{mm}$ depending on the optical geometry. The measurements were essentially chord averaged in the vertical direction. The location of the scattering volume is denoted by the horizontal distance from the plasma center, $x \equiv R - R_0$. The range of wavenumbers accessible by the present scattering arrangement was $10 < k < 140\,\text{cm}^{-1}$ with resolutions of $\pm 5\,\text{cm}^{-1}$ to $\pm 7\,\text{cm}^{-1}$.

FAST WAVE EXPERIMENT

Several different kinds of Faraday shielded poloidal loop antennas were used. The heating experiments[2] were carried out in two different regimes: heating at $\omega = 2\Omega_H$ in a pure hydrogen plasma ($B = 6\,\text{T}$), and heating at $\omega = \Omega_H$ in a deuterium plasma with minority hydrogen ($B = 12\,\text{T}$). Here, ω is the transmitter frequency and Ω_H is the hydrogen ion-cyclotron frequency. In both cases IBW is expected to be generated through mode conversion near the resonance layer and propagate on the high field side of the resonance layer.

The wavenumber spectrum obtained on the high field side exhibits a well defined peak and the wavenumber at the peak of the spectrum corresponds to that predicted for the IBW. On the other hand, the spectrum obtained on the low field side is much smaller in amplitude and has a monotonic fall-off, and is interpreted to be due to the coupling of the fast wave with the low-frequency density fluctuations.[4] There is a strong decrease of the scattered signal away from the mode conversion layer for both the second harmonic and the minority heating cases, which confirms that these IBW were generated by mode conversion.

The presence of two antennas permitted a comparison of mode-converted IBW excited using the two antennas. The low-field side launch (LFSL) antenna was located at the same toroidal location as the scattering diagnostic, whereas the two-sided launch (TSL) antenna was

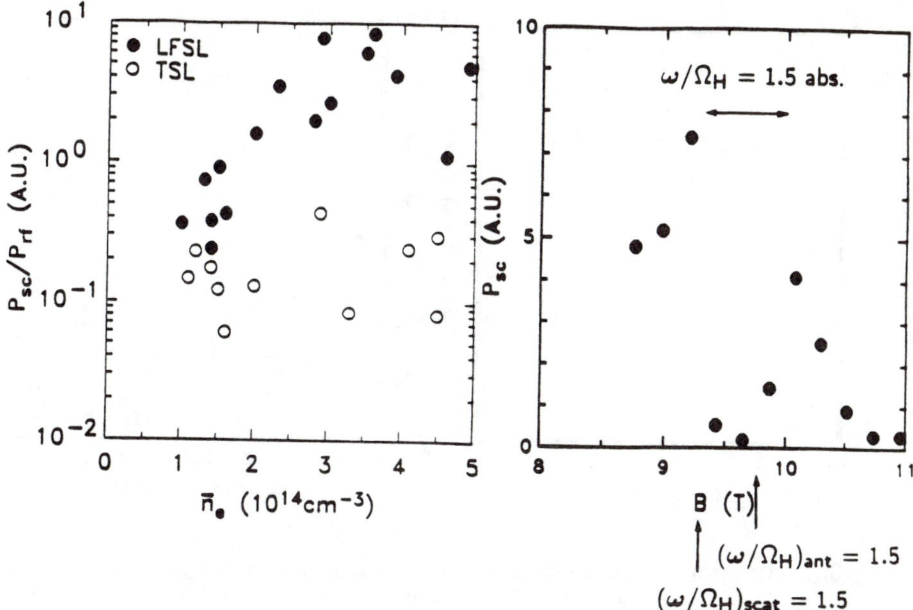

FIG. 1. P_{sc} vs. $\overline{n}_e$ for the IBW excited by the two fast wave antennas. H minority in D, 12T, $x/a = -0.5$.

FIG. 2. P_{sc} vs. B for the directly excited IBW. D minority in H, $x/a = +0.85$, $\overline{n}_e = 0.65 \times 10^{14}\,\mathrm{cm}^{-3}$ for the target plasma.

located 180° away toroidally. The result is shown in Fig. 1. The hydrogen concentration was low ($\lesssim 0.5\%$). At low density the scattered powers from the IBW excited by the two antennas are comparable indicating weak single-pass absorption and weak toroidal attenuation. At high density, however, the IBW excited by the LFSL antenna (located closer to the scattering volume) is much larger, indicating increased toroidal attenuation. Whether this is caused by damping across the resonance in the plasma interior or by damping in the plasma periphery cannot be determined from the scattering data alone. The density dependence for the LFSL antenna is stronger than that predicted by the mode conversion calculations.[5] A similar sharp dependence of P_{sc} on $\overline{n}_e$ was also observed when the TSL antenna was located at the same toroidal location as the scattering diagnostic.[6] This may be partially due to the fact that the evanescence region becomes larger as the density is decreased. The scattered power from the IBW excited by the LFSL antenna was increased by an order of magnitude at a density of $\overline{n}_e = 1 \times 10^{14}\,\mathrm{cm}^{-3}$ when the hydrogen concentration was raised to 2.5%, as expected from the mode conversion theory.

ION BERNSTEIN WAVE EXPERIMENT

The IBW was excited directly by a Faraday shielded toroidal loop antenna located on the low-field side of the torus.[1] Hydrogen majority plasma was used with varying concentrations (0–20%) of minority deuterium. The IBW can damp either linearly at an integer harmonic ion-cyclotron layer or nonlinearly[7] at a half-odd harmonic layer.

As reported earlier,[3] optimum direct excitation is obtained when an ion-cyclotron harmonic layer (e.g., $\omega = 2\Omega_H$, $3\Omega_H$) is located just behind the antenna. However, in addition to the resonant loading of $0.5\,\Omega$, there is a field-independent loading of $0.65\,\Omega$ which is not detected by scattering (for comparison, the vacuum loading was $0.35\,\Omega$), and most efficient ion heating was not necessarily obtained when largest wave signals were observed in the plasma interior

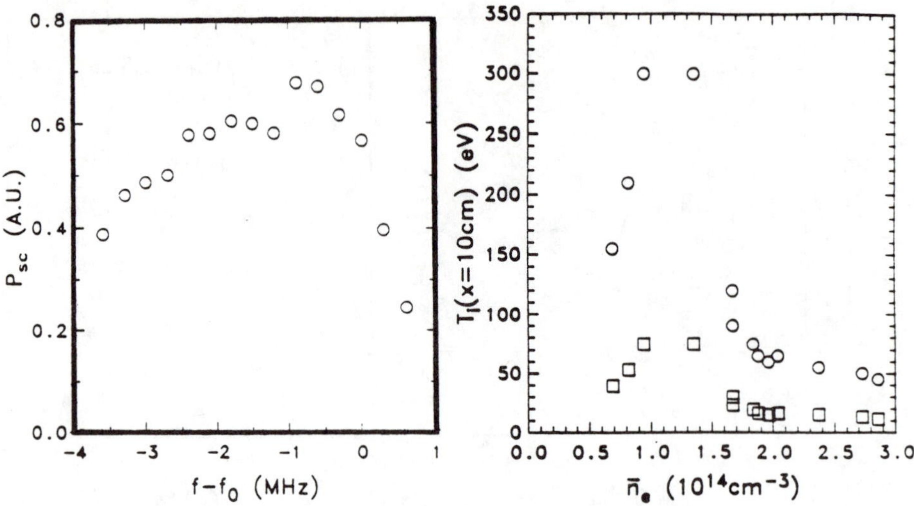

FIG. 3. An example of the downshifted frequency spectrum. Obtained at $B = 10.9\,\mathrm{T}$ and $k = 89\,\mathrm{cm}^{-1}$ during the field scan shown in Fig. 2.

FIG. 4. T_i(x/a=0.87) vs. $\bar{n}_e$. IBW antenna, D minority in H, 9.3T. The circles and the triangles represent upper and lower limits.

by scattering. In fact, the best ion heating was observed at $B = 9.3\,\mathrm{T}$ when the $\omega = 3\Omega_D$ absorption layer was located near the plasma edge and no scattered signal was detected in the plasma interior. Almost complete absorption of the IBW power is expected across this absorption layer. This is confirmed by the scattering measurements, as shown in Fig. 2 A large attenuation of the scattered signal was observed when the $\omega = 3\Omega_D$ layer was placed between the antenna and the scattering volume by varying the magnetic field. The large attenuation across this layer was also observed when the scattering volume was moved across the absorption layer at a fixed magnetic field. At high magnetic fields very broad and downshifted frequency spectra were observed, indicating the existence of a nonlinear process (such as parametric decay). As an extreme example, a spectrum obtained at $B = 10.9\,\mathrm{T}$ is shown in Fig. 3. The frequency broadening and downshift are more pronounced at higher k and higher B. A typical frequency spectrum at lower fields (e.g., $B = 7.6\,\mathrm{T}$) is symmetric about the pump frequency and the half width is less than the 300kHz resolution of the filter bank.

The scattering data can also be used as an ion temperature diagnostic.[8] The ion temperature profile obtained with this technique agreed well with that obtained by a vertically scanning charge exchange diagnostic in the inner half minor radius (the charge exchange measurement becomes unreliable in the outer half minor radius). The edge ion temperature obtained using this technique at $x/a = 0.87$, just on the high-field side of the absorption layer, is shown in Fig. 4. Because the scattering data were obtained very close to the $\omega = 3\Omega_D$ layer and the minority deuterium concentration was relatively high (estimated to be in the range 10–20% based on the charge exchange measurements), the exact value of $k_\perp$ is depedent not only on the local T_i, but also on the exact location of the scattering volume with respect to the resonance layer and also on the extent that the $\omega = 3\Omega_D$ resonance is smeared out by various mechanisms. The circles and the squares represent the estimated upper and lower bounds on T_i. Although the systematic uncertainty in the absolute value of T_i is quite large, the relative uncertainties are of the order of 30%. The edge ion temperature is higher and increases with time during the rf pulse (by up to 50%) at low densities where good ion heating is observed.[1] This observation points to the possibility that by heating ions near the plasma edge, ion transport in the plasma core is improved.

SUMMARY AND CONCLUSIONS

Mode-converted and directly excited ion Bernstein waves (IBW) were studied using CO_2 laser scattering. By comparing the relative scattered powers from the two ICRF fast wave antennas separated by 180° toroidally, an increased toroidal wave damping with increasing density was inferred. In the IBW heating experiments, wave absorption at the $\omega = 3\Omega_D = 1.5\Omega_H$ layer was directly observed. Broadened and downshifted frequency spectra were observed at high magnetic fields and at high wavenumbers. The scattering data were also used as a diagnostic for the local ion temperature. Edge ion heating was inferred from the IBW dispersion when the $\omega = 3\Omega_D = 1.5\Omega_H$ absorption layer was located in the plasma periphery, which may be responsible for the observed improvement in particle confinement.

We would like to thank the members of the Alcator group for their contributions. This project was supported by the U.S. Department of Energy, Contract No. DE–AC02–78ET51013.

REFERENCES

[1]J. Moody, et al., this conference; M. Porkolab, et al., in Plasma Phys. and Controlled Fusion (Proc. 11th Int. Conf., Kyoto, Japan, 1986) IAEA–CN–47/F–II-2.

[2]T. Shepard, et al., this conference.

[3]Y. Takase, et al., MIT Plasma Fusion Center Report PFC/JA–86–60 (1986); to be published.

[4]TFR group, A. Truc, D. Gresillon, Nucl. Fusion **22**, 1577 (1982).

[5]P. L. Colestock, R. J. Kashuba, Nucl. Fusion **23**, 763 (1983).

[6]Y. Takase, B. Blackwell, M. Porkolab, P. Colestock, Bull. Am. Phys. Soc. **30**, 1495 (1985).

[7]M. Porkolab, Phys. Rev. Lett. **54**, 434 (1985).

[8]G. A. Wurden, M. Ono, K. L. Wong, Phys. Rev. A **26**, 2297 (1982).

VISIBLE SPECTROSCOPY ON RF HEATED DISCHARGES IN THE PRINCETON LARGE TORUS TOKAMAK

D.H. McNeill, I.S. Lehrman,* G.J. Greene, M. Ono,
V.S. Vojtsenya,** and J.R. Wilson

Plasma Physics Laboratory, Princeton University, Princeton, NJ 08544

ABSTRACT

The lineshape and intensity of the H alpha line are used in calculations of the radial dependences of the neutral particle densities, H alpha emissivity, and electron production rate which are then related to particle confinement in the discharge. Stark shift of Balmer series lines is proposed for detecting rectified RF fields near ICRH antennas, provided the electron temperature varies little as the RF power is raised. The intensity ratio of Cr I lines at a Bernstein wave antenna was found to vary by less than 30% as the power was raised to 100 kW.

INTRODUCTION

Visible spectroscopy has been widely used on tokamak plasmas.[1] Here we discuss some data from the PLT tokamak, emphasizing observations that can be used as a starting point for studies of hydrogen recycling and antenna behavior in these discharges.

HYDROGEN IN THE PLASMA EDGE

A critical factor in understanding the behavior of hydrogen in a large plasma is knowledge of the energy distribution of the atoms entering the discharge. In PLT, as in other discharges, the low energy (<10 eV) spectrum is found by taking high resolution spectra of the Balmer alpha line ("H alpha" at 656 nm).[2] Figure 1 shows the H alpha lineshape for an ohmically heated PLT plasma obtained with a Fabry Perot interferometer. Here the circles denote the observed signal, the dashed curve is a computed spectrum for D alpha with a slow (0.3 eV)-fast (3-5 eV) dissociation product ratio[2] of 1:2 and a 5% residual concentration of H, and the solid curve is the convolution of this spectrum with the instrument profile of the FPI (free spectral range 0.7 nm and finesse 40). The broad, low-level portion of this spectrum is mostly light that was not blocked by a polarizer at the input optics. For this reason, the computed spectrum includes Zeeman shifted components whose intensity is 15%

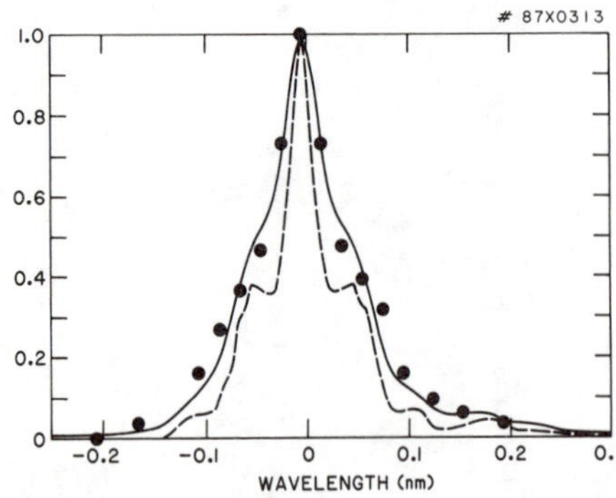

Fig. 1

of the central unshifted spectrum (B=32 kG), but should also include a contribution from first generation charge exchange atoms whose peak intensity would correspond to ≈2% of the peak for a typical ohmic discharge. Data such as Fig. 1 confirm the presence of molecules at the plasma edge (The kinetic energy of the emitting atoms originates in dissociation.) and establish the velocity distribution of atoms for use in calculations of the neutral particle behavior at the plasma edge.

Figures 2 and 3 show 1-D calculations of the local densities of D_2 molecules and D atoms (sum of slow and fast dissociation product and charge exchange atoms, with respective fixed energies of 0.3, 4, and 50 eV) and rates of production of H alpha photons

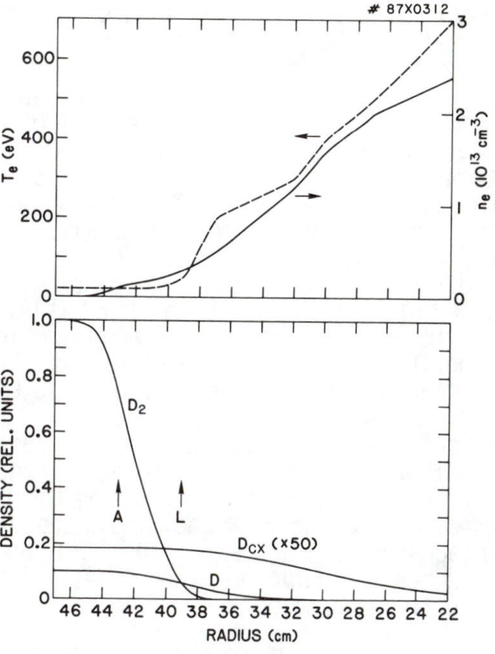

Fig. 2.

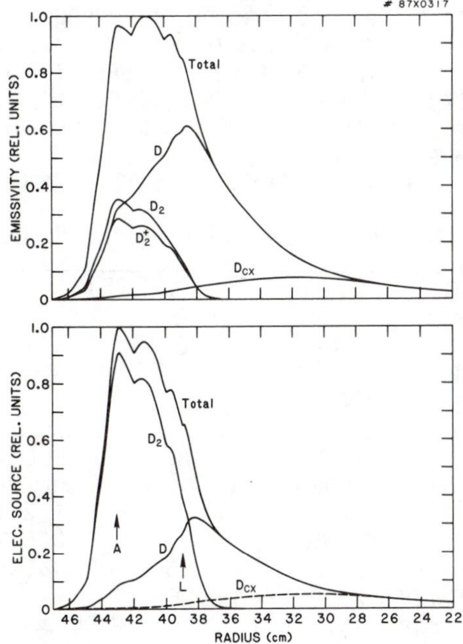

Fig. 3.

and electrons from these groups for the electron temperature T_e and density n_e profiles (measured with probes and Thomson scattering) shown in Fig. 2. Most of the rate coefficients used in these calculations can be found in ref. 3. The radii of the ICRH antenna and limiter are denoted by A and L, respectively. The plots are normalized to the following values for a single room-temperature D_2 molecule $cm^{-2}s^{-1}$ entering from the plasma edge: 8.1×10^{-6} for the densities (cm^{-3}), 3.7×10^{-3} for the emissivities ($cm^{-3}s^{-1}$), and 0.14 for the electron source rate ($cm^{-3}s^{-1}$). The measured absolute intensity of the D alpha light in discharges similar to this yields an incident D_2 flux of $\approx 3\times10^{14}$ $cm^{-2}s^{-1}$ and correspondingly larger plot scales. Thus, the scaled densities at R=47 in Fig. 2 are 2.4×10^9 cm^{-3} D_2, 2.3×10^8 cm^{-3} total D, and 9×10^6 cm^{-3} first-generation charge exchange D. The maximum total H alpha emissivity at R=43-40 is 1×10^{12} $cm^{-3}s^{-1}$ and the electron source rate is 4×10^{13} $cm^{-3}s^{-1}$.

The total number of H alpha photons produced per entering molecule is not strongly dependent on the T_e and n_e profiles. Thus, for example, a sequence of OH and 800 kW ICRH discharges have the same temperature profile and a roughly 30% increase in the density across the discharge (about 50% at the edge). This causes a 1-2% change in the computed electron and photon yield per molecule. The greatest effect of the RF power is to shift the electron and photon source profiles (cf. Fig. 3 for an OH discharge) outward. The maxima are about 11% higher with the RF on. Most (≈60 and 70%, respectively, in Fig. 3) of the H alpha and electrons from incoming molecules (and product atoms) are produced outside the limiter and, therefore, in a low confinement region. In this example the H alpha intensity increases by 50% when the RF power is applied. Scaling this to the model calculations, we find that for the observed density change, the electron and H alpha yields increase by 59 and 65%, respectively, outside the limiter, but by 31 and 28% inside the limiter. These relative increases are close to the 50% and 30% increases observed in the edge and line integrated densities. This indicates that the hydrogen particle confinement time of the core plasma is unchanged (provided the effective charge of the plasma is constant).

ELECTRIC FIELDS NEAR ANTENNA SURFACES

One possible application of the spectroscopy of hydrogen is searching for rectified RF fields near the surfaces of RF antennas in plasmas. A 10-15% increase in the width of a Balmer series line owing to the transverse Stark shift[4] resulting from the appearance of a steady-state field when the RF power is turned on could be observed. The higher Balmer series lines have smaller Doppler widths but are more sensitive to the Stark effect, so lower fields can be detected with them.

Shown here are the Doppler widths ($\Delta\lambda$, fwhm in nm) of the

Line of D	Wavelength(nm)	$\Delta\lambda(0)$	$\Delta\lambda(1)$	$\Delta\lambda(.25)$
α	656	0.38	0.44	0.39
β	486	0.27	0.48	0.28
γ	434	0.24	0.64	0.27
δ	410	0.23	0.79	0.31

first four Balmer series lines of deuterium for a thermal distribution of emitters with T=1 eV and the widths when fields of 1 kV/cm and 0.25 kV/cm are applied. These estimates show that a field of 250 V/cm should easily be detectable with the H_δ line. The actual conditions in a plasma are probably better than this, as a simulation with a realistic (nonthermal) H alpha profile of the shape shown in Fig. 1, for example, showed that the H alpha linewidth increased by about 35% (more than twice the estimate in the table) when an electric field of 1 kV/cm was applied and a polarizer used to isolate the split components. In setting up a measurement of this type on a test stand, it must be kept in mind that the high field region may be of limited extent compared to the total emitting volume and special viewing arrangements may be necessary. In low density plasmas, as at the edge of a tokamak, the shape of the Balmer lines depends on the amounts of fast and slow dissociation products present in the plasma. These amounts vary with the electron temperature, so it is important that the temperature either be high (above 5 eV), or constant. Evidence that the electron temperature is approximately constant near some RF antennas has been provided by the following spectroscopic measurement.

ELECTRON TEMPERATURE AT AN RF ANTENNA

We have used the intensity ratio of two chromium lines to investigate the electron temperature variation in the vicinity of an ion Bernstein wave antenna during RF heating experiments. The excitation cross sections for the 359.3 nm singlet and 5204, 5206, and 5208 triplet of Cr I have been measured.[5] The approximate temperature dependence of the rate coefficients for excitation of these lines (the triplet taken as a whole) and their ratio are plotted in Fig. 4. In a sequence of ion Bernstein wave heating experiments it was found that the ratio I(359)/I(521) fell by a fixed amount (≤30%) when IBW power was applied to the plasma at levels of 35-100 kW. At 20 kW, there was no observable change from the ohmic level. Assuming the initial temperature lies between 15 and 20 eV (from probe data), the variation in the temperature was less than 30% for all the RF powers, in agreement with probe data taken in the limiter shadow.

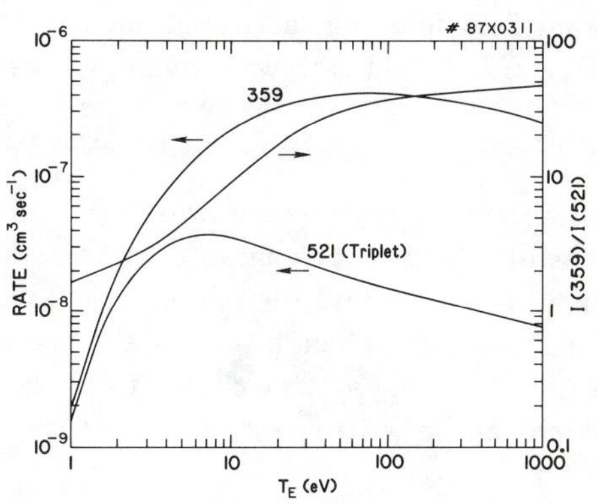

Fig. 4.

This work has been supported by the U.S. Department of Energy, Contract No. DE-AC02-76-CHO3073.

* Permanent address: Grumman Space Systems, Plainsboro, N.J.
** Permanent address: Physico-Technical Institute, Kharkov, U.S.S.R.

[1] E. Hintz and P. Bogen, J. Nucl. Mater. 128/129, 229 (1984).
[2] D.H. McNeill, et al., J. Vac. Sci. Technol. A 2, 689 (1984).
[3] R.K. Janev, et al., Atomic and Molecular Processes in Hydrogen-Helium Plasmas, to be published by Springer Verlag, Berlin (1987).
[4] E.U. Condon and G.H. Shortley, The Theory of Atomic Spectra, Cambridge Univ. Press (1963).
[5] V.V. Melnikov and Yu. M. Smirnov, Opt. Spektrosk. 52, 605 (1982).

EDGE MEASUREMENTS DURING ICRF HEATING IN PLT

I. S. Lehrman†, P. L. Colestock, G. J. Greene, D. H. McNeill, M. Ono, J. R. Wilson, D. M. Manos, S. Bernabei, and J. L. Shohet‡

Princeton Plasma Physics Lab. Princeton, NJ 08544

INTRODUCTION

Measurements of plasma parameters in the edge region of PLT during ICRF heating experiments have been performed in an attempt to ascertain the relevant processes at work in coupling RF power to a hot, dense plasma and to understand how these processes limit the power handling capability of ICRF antennas. Langmuir douple probe measurements of edge electron temperature and density were made along with a calorimeter probe to measure particle heat flux along field lines. In addition, spectroscopic observations of neutral deuterium (D_α), $CIII$, and CrI were made at various points around the torus.

MEASUREMENTS

A Langmuir double probe was used to measure edge density and electron temperature. The probe characteristic was found to be free of any RF pickup which allowed density and temperature measurements to be made before and during ICRF heating. Though more elaborate probe characteristic theories exist, such as one by Stangeby[1], we choose to fit the data to the simple double probe theory which can be written as:

$$I = I_{sat} \tanh\left(\frac{eV}{2kT_e}\right)$$

We assume that $T_i \simeq T_e$ so that the ion sound speed, C_s can be evaluated. The electron density is then:

$$n_e = \frac{2AI_{sat}}{C_s},$$

where A = effective area of the probe. Of most interest in this measurement are the relative changes in edge parameters. A calorimeter probe made of a tantalum element with a thermocouple bonded to it was used to measure heat flux along field lines in the edge region. The calorimeter is useful in estimating the edge ion temperature. The power flowing to the calorimeter can be written in terms of the plasma edge parameters:[2]

$$P = I_{sat} kT_e \left[2\frac{T_i}{T_e} + 2(1-\gamma)^{-1} - eV_f\right],$$

where γ is the secondary electron coefficient. I_{sat} and T_e are supplied from the double probe and V_f is measured at the calorimeter. The error in deriving T_i can be large, but relative changes can be obtained.

Visible spectroscopic measurements of D_α, $CIII$, and CrI were made at serveral toroidal locations A view into an optical dump along a horizontal cord at the torus midplane provided information about global behavior. A view towards an ICRF antenna provided information about the effect of the ICRF fields on the edge plasma. The antenna view provided evidence that the energized antenna is the major particle source during ICRF heating.

RESULTS

Radial profiles of density and temperature in the edge region for a typical PLT discharge are shown in Fig. 1. These discharges had a line averaged density of $2.9 \times 10^{13} cm^{-3}$ which increased to $3.4 \times 10^{13} cm^{-3}$ with the application of 300 kW of ICRF (an increase of 50% above the ohmic power). The average edge density increase in the edge region was much larger, on the order of 50%. In these discharges, and ones in which as much as $800kW$ of RF was applied, there was no appreciable change in the edge electron temperature. This is in contrast with the results of Manning et al[3] for ICRF experiments on Alcator C in which edge heating is typically observed. The calorimeter probe temperature and floating potential during a discharge is shown in Fig. 2 along with the calculated absorbed power. During RF no increase in absorbed power is seen, suggesting that no measurable change in either edge ion or electron temperature is occuring during ICRF heating.

Emmisivity calculations indicate that both before and during RF D_α emmision is localized to the edge region and is proportional to the edge density.[4] This is consistent with global D_α measurements which show a 50% increase in emission during ICRF in these discharges. Viewing towards the antenna though an increase of D_α by 157% and of $CIII$ by 155% is seen. Limiter position was changed to study its effect on edge parameters. No significant change in electron temperature or global D_α was observed, but as Fig. 3. shows, D_α, and $CIII$ emission increased at the antenna. The percentage change in D_α emission with RF was similiar for both limiter positions, but the percentage change in $CIII$ was substantially increased with the limiters moved out.

DISCUSSION

Edge measurements indicate that there is a substantial density rise in the edge region which is accompained by an increase in impurity concentration. These moderate power experiments ($300kW \leq$ RF Power $\leq 1MW$) are not expected to produce many fast ions, so the effects that are observed are attributed to edge processes. There appears to be no significant change in either edge electron or ion temperature which implies that physical sputtering is not the cause of the density rise and impurity generation. The energy that need be deposited in the edge to account for the additional ionization of edge neutrals has been calculated and shown to be small (on the order of 50 Joules). We interpret these results to suggest that the antenna near fields are perturbing the trajectories of particles in the vicinity of the antenna, causing particles to strike the Faraday shield. On striking the Faraday shield these particles liberate deuterium and impurity atoms. The RF antenna fields in the edge region and their effect on particle transport need to be determined. Direct heating of edge particles from either the evanescent slow wave or propagating fast wave is believed to be negligible.

ACKNOWLEDGEMENTS

This work supported by Grumman Space Systems and DOE Contract #DE–AC02–76–CHO–3073.

[1]P. C. Stangeby, J. Phys. B 15, 1007 (1982)
[2]D. M. Manos, J. Vac. Tech. A 3(3), 1059 (1985)
[3]H. L. Manning etal. Nucl. Fusion 26, 1665 (1986)
[4]D. H. McNeill, I. S. Lehrman, G. J. Greene, M. Ono, V. Vojtsenya, and J. R. Wilson, this conference.

†Grumman Space Systems, Plainsboro, NJ 08536
‡TSL Laboratory, University of Wisconsin, Madison, WI 53706

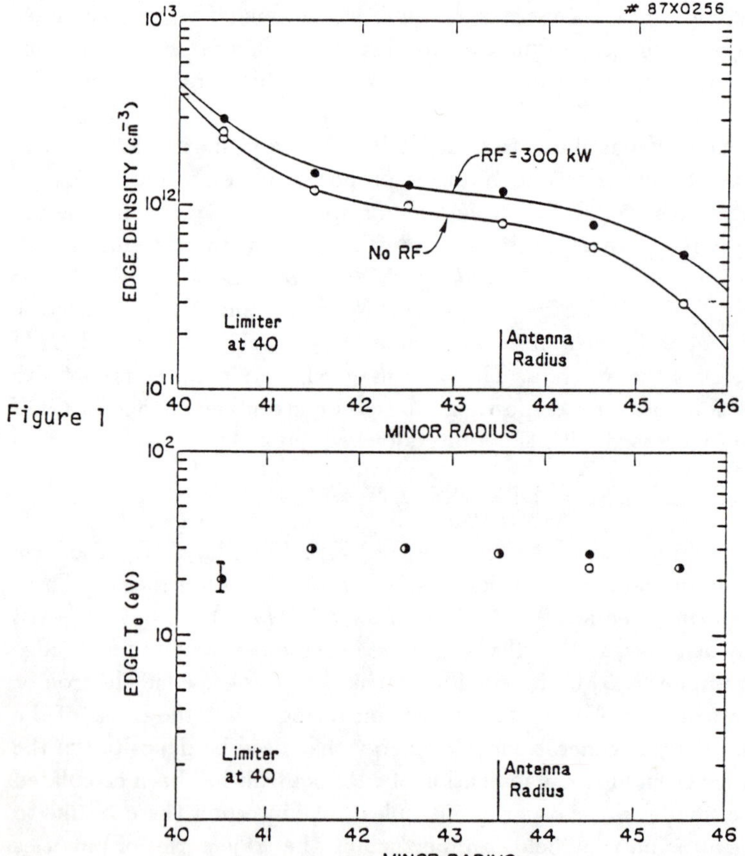

Figure 1

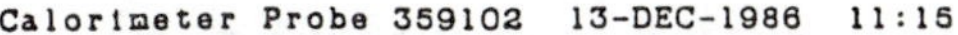

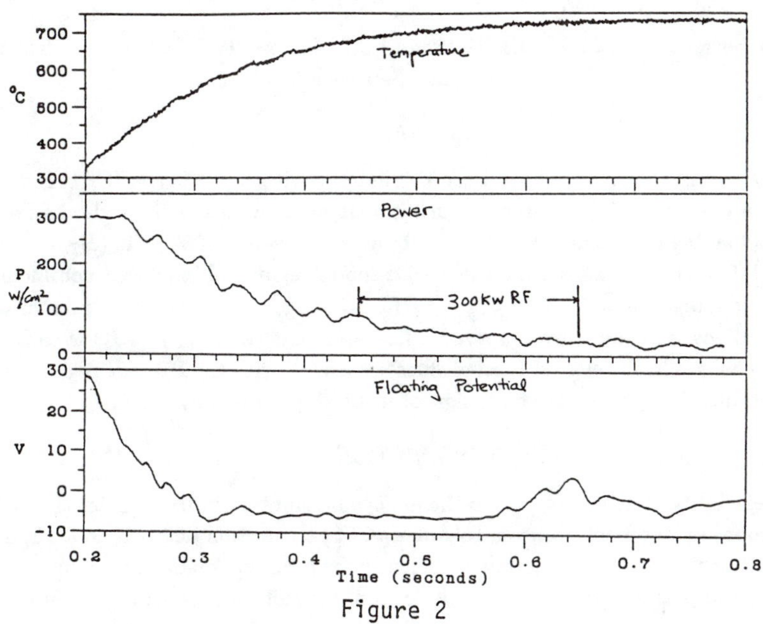

Figure 2

Global D

D Antenna

CIII Antenna

Intensity (AU)

35%   38%

93%   90%

95%   130%

No RF

225 KW

% Change

40   42

Limiter Position

Figure 3

COUPLING OF AN ICRF COMPACT LOOP ANTENNA TO H-MODE PLASMAS IN DIII-D*

M.J. Mayberry, F.W. Baity,† D.J. Hoffman,† J.L. Luxon, T.L. Owens,† R. Prater
G.A. Technologies, Inc., San Diego, CA 92138

ABSTRACT

Low power coupling tests have been carried out with a prototype ICRF compact loop antenna on the DIII-D tokamak. During neutral-beam-heated L-mode discharges the antenna loading is typically $R \simeq 1-2\,\Omega$ for an rf frequency of 32 MHz ($B_T = 21\,\text{kG}$, $\omega = 2\Omega_D(0)$). When a transition into the H-mode regime of improved confinement occurs, the loading drops to $R \simeq 0.5-1.0\,\Omega$. During ELMs, transient increases in loading up to several Ohms are observed. The apparent sensitivity of ICRF antenna coupling to changes in the edge plasma profiles associated with the H-mode regime could have important implications for the design of future high power systems.

INTRODUCTION

The feasibility of using rf waves in the ion cyclotron range of frequencies (ICRF) to heat the proposed Compact Ignition Tokamak (CIT) to ignition depends to a large extent on whether rf power can be coupled efficiently from compact launching structures to shaped, diverted plasmas with good (H-mode) confinement. An attractive antenna concept for this purpose is the compact loop antenna[1] because: 1) it can be inserted through a single port, 2) it can be tuned over a wide frequency range, and 3) it can be impedance matched internally, thus reducing standing wave voltages in the vacuum feedthrough and external transmission lines. In order to determine the coupling of the compact loop antenna to dee-shaped plasmas with limiter and divertor edge configurations, a prototype compact loop antenna[2] has been installed on DIII-D for low power tests. In this paper we emphasize coupling results obtained during neutral beam injection (NBI) experiments in which the H-mode regime of improved confinement was obtained with divertor plasmas ($I_p = 1\,\text{MA}$, $B_T = 21\,\text{kG}$, $\kappa \simeq 1.8$, $\bar{n}_e = 2-7\times10^{13}\,\text{cm}^{-3}$).

ANTENNA DESCRIPTION

A diagram of the compact loop antenna is shown in Fig. 1. The radiating element consists of a vertical current strap approximately 35 cm long, grounded at the bottom, connected to a vacuum variable capacitor at the top, and fed by a length of 50 Ω coaxial line. The entire antenna structure fits through a single DIII-D midplane port (50 cm high, 35 cm wide) and is moveable over a 5 cm range in order to facilitate impedance matching. The Faraday shield consists of two tiers of horizontal Inconel tubes which are covered with graphite on the plasma side in order to reduce metallic impurity influx into the plasma.

An equivalent electrical circuit of the antenna is shown in Fig. 2. The resonant frequency is given by $f = 1/2\pi\sqrt{LC}$, where $L \simeq 120\,\text{nH}$ is the inductance of the current strap, and $C = 38-474\,\text{pF}$ is the value of the vacuum variable capacitor. By

* Work supported by U.S. DOE Contract DE-AC03-84ER53158.
† Oak Ridge National Laboratory, Oak Ridge, TN 37831

varying C, the antenna can be tuned over the frequency range $f = 20 - 74\,\mathrm{MHz}$. At resonance, the input impedance of the antenna is given approximately by $Z \simeq \alpha^2 L/RC$ where R is the total loading resistance and $\alpha \simeq 0.22$ is determined by the position of the tap point on the current strap. The antenna losses are given approximately by $R_{loss} \simeq 0.12\sqrt{f/20\,\mathrm{MHz}}\,\Omega$ and are due mainly to the graphite in the Faraday shield.

RESULTS

Up to 6 MW of H° NBI heating power has been applied to single-null divertor discharges in deuterium. At sufficient densities ($\bar{n}_e > 2.0 \times 10^{13}\,\mathrm{cm}^{-3}$) and NBI power levels ($P_{NBI} > 2.8\,\mathrm{MW}$), a transition into the H-mode regime has been achieved in which the plasma energy confinement time becomes comparable to the pre-NBI level.[3] The time history of the antenna loading during a typical H-mode discharge is shown in Fig. 3. After an initial L-mode phase during the first 70 ms of beam injection, the H-mode transition occurs, signaled by the drop in D_α emission from the divertor region. The antenna loading resistance R exhibits a slight increase above the Ohmic level during the L-mode phase, followed by a sudden drop by a factor of two at the H-mode transition. Despite the subsequent large increase in line-average density and the decrease in the limiter/separatrix gap, the antenna loading remains nearly constant during the quiescent portion of the beam pulse (between ELMs). This is also shown in Fig. 4 where the antenna loading is plotted as a function of line-average density. While the antenna coupling to L-mode and Ohmic discharges increases linearly with line-average density, H-mode coupling is nearly independent of density. According to ICRF coupling theory,[4] the weak antenna coupling to H-mode plasmas could be due to a steepening of the edge density gradient near the separatrix accompanied by a density reduction in front of the antenna. Experimental measurements of the edge density profile at the midplane of DIII-D will be carried out in the near future using a moveable Langmuir probe.

As shown in Fig. 3, sudden drops in the density (and plasma energy), accompanied by bursts of D_α emission, can occur periodically during H-mode discharges. The phenomena appear similar to the edge localized modes (ELMs) observed in ASDEX.[5] When these occur the antenna loading increases suddenly to levels which can exceed 3 Ω. The transient increase in loading associated with ELMs is attributed to a rapid expulsion of particles from the main plasma which temporarily increases the edge plasma density in front of the antenna.

The antenna loading as a function of the antenna-plasma separation distance is plotted in Fig. 5. For all of this data, the antenna was positioned flush with the surrounding graphite tile surface, approximately 2 cm behind the face of the limiter located 75° away toroidally. For both L-mode and H-mode divertor discharges, the antenna loading increases as the gap between the plasma separatrix and the limiter is decreased. For a given antenna/plasma separation, the L-mode loading exceeds the H-mode loading by a factor of two. Under the most favorable conditions for H-mode coupling, where the limiter-separatrix gap was reduced to less than 1 cm, the antenna loading exceeded 1 Ω. Future experiments will be carried out to determine whether further improvements in H-mode coupling can be obtained by inserting the antenna closer to the plasma.

DISCUSSION

The results described above raise two main concerns about coupling ICRF fast waves at high power levels to H-mode diverted plasmas. First, because of the reduced antenna coupling to H-mode plasmas, the maximum antenna voltage and current levels required for a given output power level will be high. For example, a 2 MW antenna rating on DIII-D ($R_p = 1.0\,\Omega$ at $f = 32\,\text{MHz}$) would require a 42 kV peak voltage and 1300 A rms current rating of the vacuum capacitor. For CIT ($f = 95\,\text{MHz}$) the voltage requirement becomes more severe, namely 60 kV peak. The second problem concerns the sudden variations in antenna loading which occur at the H-mode transition and during ELMs. Such loading excursions make it difficult to maintain a constant impedance match with the transmitter in order to minimize reflected power levels. We note that such diffulties have already been encountered on ASDEX where purely ICRF-generated H-mode plasmas have been achieved.[6] Possible improvements to the antenna design which could increase the coupling to H-mode plasmas are being explored. These include: 1) deepening the antenna cavity in order to reduce the cancelling effect of image currents in the backplane, and 2) moving the current strap closer to the Faraday shield and possibly eliminating the inner tier of the Faraday shield in order to reduce the antenna/plasma separation distance.

REFERENCES

[1] S.C. Chiu et al., in *Proc. Fifth Topical Conf. on RF Plasma Heating* (Madison), 1983; F.W. Perkins and R. F. Kluge, IEEE Trans. on Plasma Sci. **PS-12**, 161 (1984).

[2] D.J. Hoffman et al., in *Proc. Thirteenth Eur. Conf. on Controlled Fusion and Plasma Heating* (Schliersee, 1986), Vol. 10C, Part II, p. 141.

[3] J. Luxon et al., in *Proc. Eleventh Int. Conf. on Plasma Physics and Controlled Fusion Research* (Kyoto, Japan, 1986), Paper No. IAEA-CN-47/A-III-3.

[4] T.K. Mau, S.C. Chiu, and D.R. Baker, GA Technologies Report No. GA-A17776 (1986) (to be published in *IEEE Trans. on Plasma Sci.*).

[5] F. Wagner et al., *Phys. Rev. Lett.* **49**, 1408 (1982).

[6] K. Steinmetz et al., *Phys. Rev. Lett.* **58**, 124 (1987).

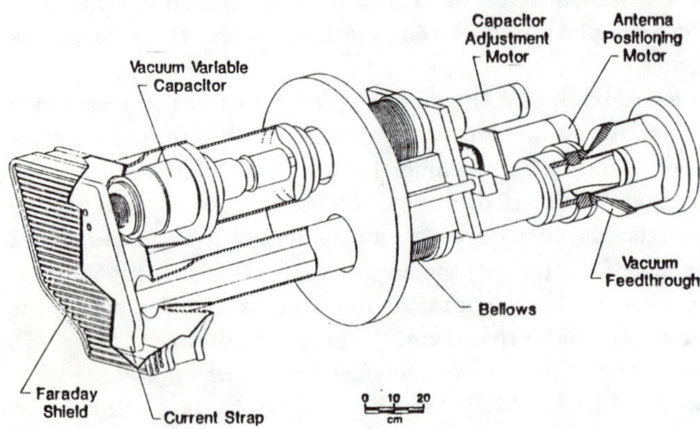

Fig. 1. Diagram of the compact loop antenna.

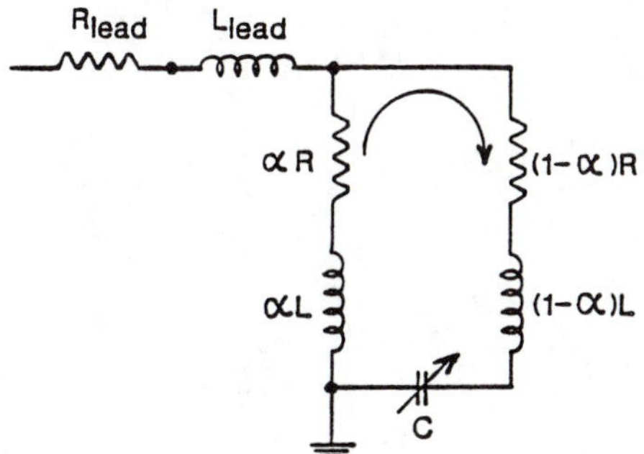

Fig. 2. Equivalent circuit of the compact loop antenna.

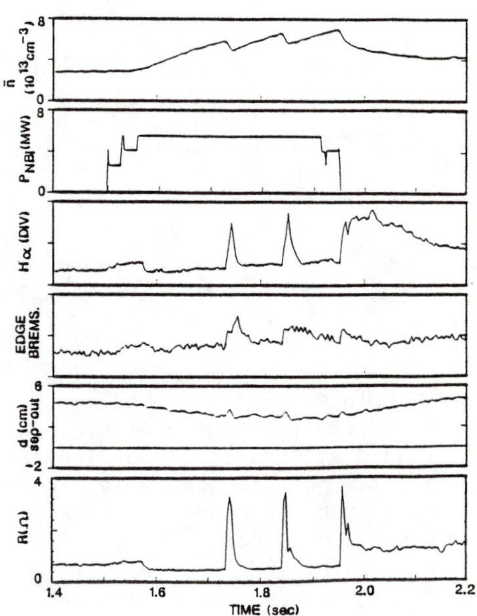

Fig. 3. Divertor discharge in which an H-mode transition takes place at $t = 1.575$ sec. Signals from top to bottom are the line-average density $\overline{n}$, neutral beam injection power P_{NBI}, the $\text{H}_\alpha/\text{D}_\alpha$ emission from the divertor region, bremsstrahlung emission from the plasma edge, gap between the limiter and separatrix $d_{\text{sep/out}}$, and the antenna loading resistance R.

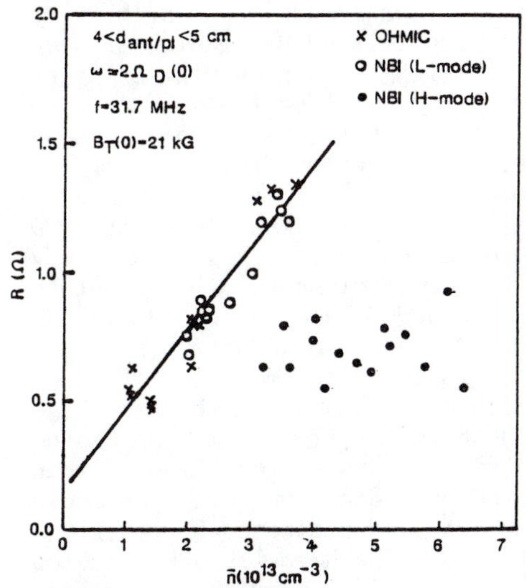

Fig. 4. Antenna loading vs. line-average density for a fixed antenna/plasma separation distance ($4 < d_{ant/pl} < 5$ cm). H-mode loading is measured between ELMs.

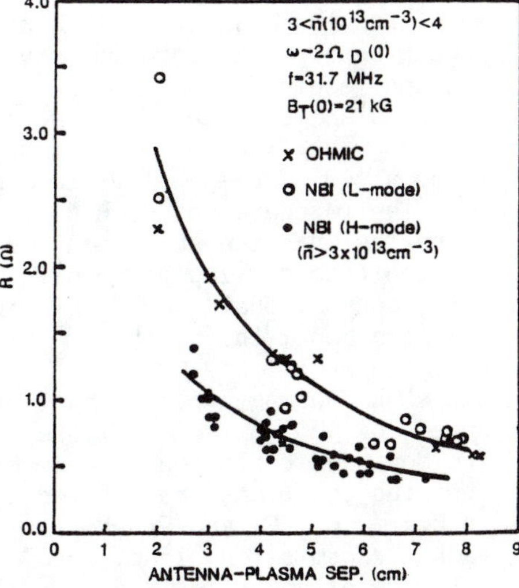

Fig. 5. Antenna loading as a function of the distance between the antenna Faraday shield and the plasma separatrix.

COUPLING PROPERTY OF ICRF 2X2 LOOP ANTENNA IN JT-60

M. Saigusa, N. Kobayashi*, H. Kimura, T. Fujii,
K. Hamamatsu, K. Annoh, K. Kiyono, M. Seki,
M. Terakado, T. Nagashima and JT-60 team

Japan Atomic Energy Research Institute,
Naka-machi, Naka-gun, Ibaraki, 311-02 Japan
*Toshiba Corporation, Hibiya, Tokyo, Japan

1. INTRODUCTION

This paper presents the first ICRF experimental results of wave coupling to the plasma and impedance matching of the plasma loaded antenna in JT-60. The generator power and frequency are 6 MW and 120 MHz, respectively, the frequency is the second harmonic ion cyclotron resonance frequency of a proton at a toroidal magnetic field of 4T. The range of experimental parameters; $\bar{n}_e$=1.3~6.0×$10^{19}m^{-3}$, Bt(0)=3~4 T, Ip=1.3~1.5 MA, R~3 m, a~0.9 m and the maximum launched power is 2.1MW.

2. COUPLING SYSTEM

The structure of JT-60 ICRF launcher[1] is sketched in Fig.1(a). The launcher is composed of a 2x2 phased loop antenna array, which can be changed the toroidal and poloidal phase differences between each two loops in order to control parallel and perpendicular wave number spectra of radiation power. The length and width of each antenna element are 23 cm and 7.5 cm, respectively. Four loop antennas are enclosed in a metal casing with open-type Faraday shield. The launcher is installed on the low field side of the torus at 40 degrees below the midplane.

The impedance of each antenna element is matched with 50Ω transmission line by each double stub tuner. It would be very difficult to get impedance matching of four elements by eight stubs with conventional stub matching method, because of the mutual coupling among the antennas. In our system, two counterplans are prepared for this problem. The one is the programmed operation of the stubs during a plasma shot, which can be driven with the high speed of 50 mm/sec. Figure 2(a) shows the time variation of the incident and reflected RF powers of the transmission lines, when the stub positions are changed by the programmed operation as shown in Fig. 2(b). P_i and P_r are the incident and reflected RF power of each transmission line, respectively. The RF pulse was injected into the plasma from 4.5 to 7.5 sec. "l_1" and "l_2" are the positions of stub1 and stub2, respectively, where stub1 is located on a antenna side and stub2 is located on the generator side. The other counterplan is the prediction of the stub matching positions calculated from the experimental data of the antenna input impedance. Figure 2(c) shows the time evolutions of the antenna input impedance;R_a+iX_a and the antenna coupling resistance;R_c, where R_a and X_a are the real and imaginary parts of antanna input impedance, respectively. In this shot, as the plasma separatrix surface is gradually shifted away from the antennas as shown in Fig.2(d), where d is the distance from the Faraday shield to the separatrix, R_c and R_a decrease and X_a slightly increases. Using both of the methods mentioned above, we can get a good impedance match within a few shots, as long as the plasma is stable.

3. COUPLING PROPERTY

The coupling properties were investigated with respect to the four phasing modes, which are (0,0),(π,0), (0,*),(π,*) as shown in Fig.1(b). The two symbols in the parenthesises are the toroidal and poloidal phase difference, respectively, and "*" means only upper 2 loop (II,III line) operation. Figure 3 shows experimental and theoretical loading resistances in the four phasing modes versus line averaged electron density. Supposing a parabolic density profile of the main plasma and an e-folding length of 3 cm in the scrape off region, the cold plasma theory including the effect of the antenna feeder current[2,3] can explain the experimental results. The coupling efficiency;η ($\eta=(R_c-R_v)/R_c$, R_v;Vacuum loading) in (0,0), (π,0),(0,*) and (π,*) phasing are 94%, 92%, 85% and 70%, respectively. The maximum launched powers are 2.1MW and 1.5MW in (0,0) and (π,0) phasing , respectively.

In (0,0) and (π,0) phasing modes, the plasma loading resistances of the upper 2 loops;(II,III) are higher than those of the lower 2 loops;(I,IV). The same tendency is also observed in the vacuum condition. One reason for this difference is the asymmetry of the resonance mode in a metal casing due to a difference between antenna feeder lengths and the gaps to the casing wall of the upper and lower loops.

Figure 4 shows the loading resistances versus line averaged electron density in (0,0),(π,0) phasing. The upper traces are theoretical values estimated by a 1-D kinetic wave theory,[4] and the lower traces are experimental results. The oscillation of loading resistance means existence of the cavity resonances. The peak radiation occurs for parallel wave numbers of 0 m^{-1} and $\sim$ 10 m^{-1} in (0,0) and (π,0) phasing , respectively. In (0,0) phasing, the cavity resonances of large amplitude were observed over a wide density regime. On the other hand, in (π,0) phasing, the cavity resonances of small amplitude were observed in only the low density regime. This difference results from the shape of toroidal wave number spectra. The width of the cyclotron resonance layer is evaluated as $k_{\parallel}R\rho_i$ (R; major radius, ρ_i; ion Lamor radius)[4]. In (0,0) phasing, since the small $k_{\parallel}$ has the major part of the radiation spectra, the resonance width is very thin and the damping of RF wave is very weak, and the cavity resonance arises. In (π,0) phasing, however, the large $k_{\parallel}$ has the major part of the radiation power, and the resonance width is thick, so the cavity resonance is suppressed.

Figure 5 shows the loading resistances of II and III loop antennas versus toroidal phase difference, where the poloidal phase difference is set at 0 degree. The behavior of experimental loading resistance due to the radiation and interchanged power between right and left antennas as shown in Fig.5(b),(c) agrees with the cold plasma theory as shown in Fig.5(a).

4. CONCLUSIONS

The coupling properties of the four phasing modes (0,0),(π,0),(0,*),(π,*) for the 2x2 phased loop antenna array can be explained by the cold plasma coupling theory. The fact that amplitudes of cavity resonance in (0,0) phasing are larger than that in (π,0) phasing can be explained by 1-D kinetic wave theory. A programmed operation of high speed stub system and the prediction of stub matching positions based on the experimental results are useful for the phased antenna array to get good impedance matching.

ACKOWLEDGMENTS

The authors wish to express appreciation to Drs. H. Shirakata, K. Uehara and staffs of Heating System Development Division and Plasma Heating Laboratory II for fruitful discussions. The continuous encouragement of Drs. K. Tomabechi, M. Tanaka, M. Yoshikawa, T. Iijima is gratefully ackowledged.

REFERENCES

1. T. Nagashima, et al., to be published in Nucl.Engineering and Design/Fusion.
2. V.P. Bhatnagar, et al., Plasma Physics and Controlled Nuclear Fusion Reseach, IAEA, Vienna (1983)
3. Y. Ikeda, et al., Japan Atomic Energy Reseach Institute Report, JAERI-M 84-191 (1984)
4. K. Hamamatsu, et al., Research Report HIFT-128 Hiroshima Univ., (1986)

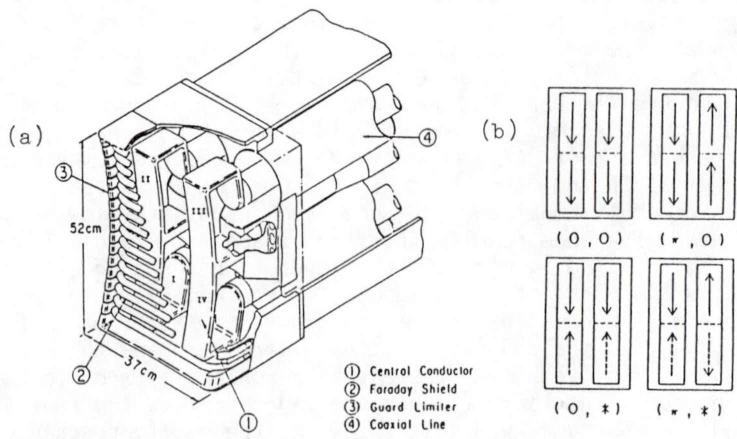

Fig.1(a) Structure of the JT-60 ICRF 2x2 phased loop antenna array.
(b) Four types of phasing modes. Broken lines are induced currents.

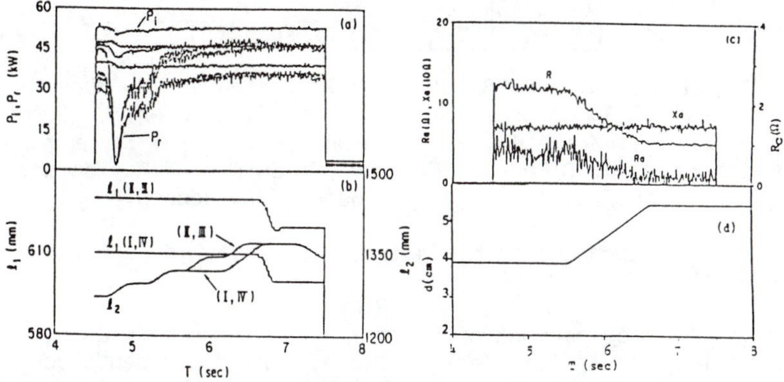

Fig.2 (a) Incident and reflected power during programmed stub operation.
(b) Time evolution of stub positions.
(c) Time evolution of antenna impedance and loading resistance.
(d) Change of distance from plasma surface to Faraday shield.

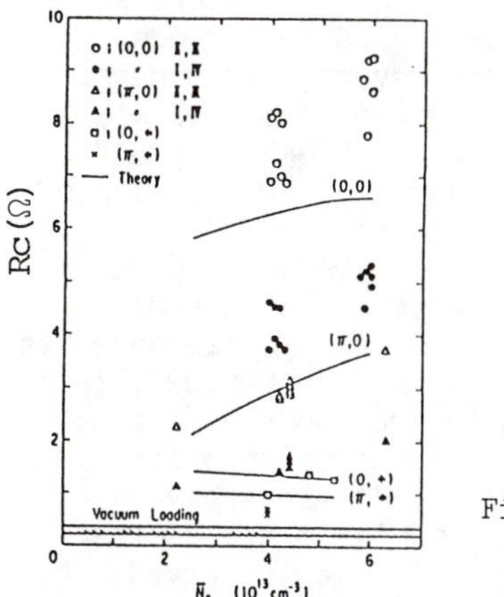

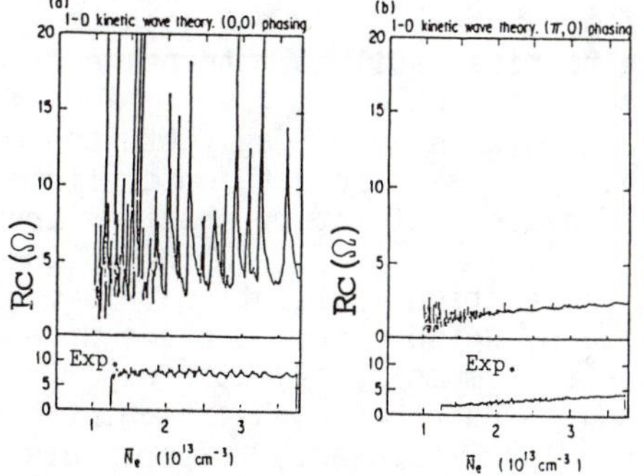

Fig.4 Loading resistance vs line averaged electron density.

Fig.3 Loading resistance of four phasing modes.

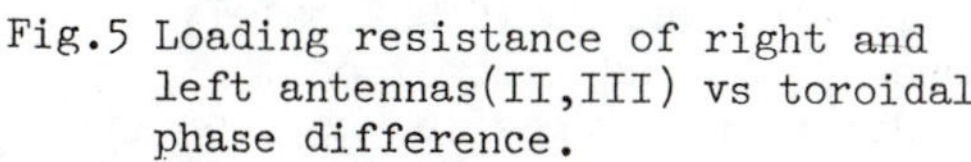

Fig.5 Loading resistance of right and left antennas(II,III) vs toroidal phase difference.
(a) Cold plasma theory.
(b),(c) Experimental results.

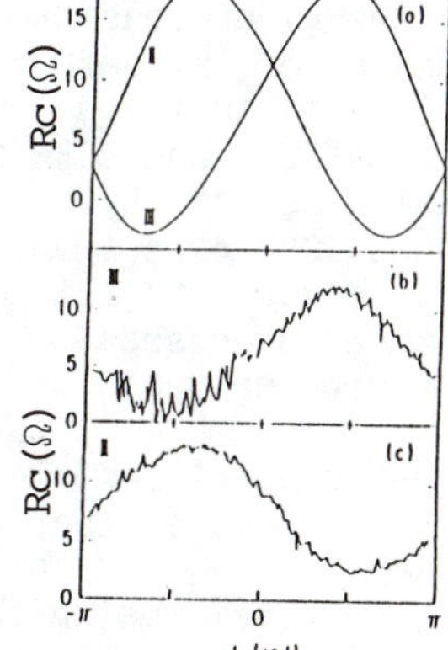

PROLONGED SAWTOOTH-FREE PERIODS IN JET DURING CENTRAL ICRH EXPERIMENTS

D.F.H. Start, D.V. Bartlett, V.P. Bhatnagar, M. Bures, D.J. Campbell, G.A. Cottrell, M.P. Evrard, D. Gambier, C. Gormezano, H. Hamnen, T. Hellsten, J. Jacquinot, A.S. Kaye, E. Lazzaro, P. Morgan, J. O'Rourke, F. Sand, J. Snipes, P.R. Thomas, B.J.D. Tubbing, T.J. Wade, J.A. Wesson,

JET Joint Undertaking, Abingdon, Oxon, OX14 3EA, UK.

ABSTRACT

Sawtooth-free periods of up to 1.6 sec have been attained in JET using high power ICRF heating alone. The experiments used hydrogen minority heating at f = 32 MHz for which the fundamental cyclotron resonance passed through the magnetic axis. A maximum of 8 MW of ICRF power was coupled to the plasma and produced central electron temperatures up to 7.2 keV. During the sawtooth-free periods, the long wavelength, coherent MHD activity is much less than that during sawtooth activity, the electron temperature saturates after about 0.3 sec but both the density and energy content continue to rise until an internal disruption eventually occurs. In the absence of sawteeth τ_E is, typically, 0.35 sec compared with τ_E = 0.29 sec when normal sawteeth are present. The sawteeth-free regime has been seen in both outer limiter and X-point discharges. Possible theoretical interpretations are discussed.

INTRODUCTION

The relaxation oscillations, commonly known as sawteeth oscillations, which occur in the centre of tokamak plasmas have been the subject of much study, both experimental and theoretical, (see Refs 1 and 2 and other work cited therein). Even so they are not completely understood and, in particular, the mechanism which triggers the collapse has yet to be identified[2)]. They have a deleterious effect on the central confinement, limit the central electron and ion temperatures and have recently been observed[3)] to expel high energy protons from D-^{3}He fusion reactions. Consequently the suppression of sawteeth oscillations would enable more effective use to be made of additional heating, especially localized heating as produced by ICRH and ECRH for example, and would ultimately benefit fusion reactivity. In recent combined neutral beam injection and ICRH experiments on JET[4)], a new regime was found in which sawteeth were stabilized for up to 1.2 sec. Subsequently this regime has been accessed for up to 1.6 sec by ICRH alone and these experiments are the subject of this paper.

EXPERIMENTS

The experiments were performed in JET limiter type plasmas having a major radius of 3 m, a minor radius of 1.2 m on the median plane and an elongation of 1.4. Up to 8 MW of ICRF power at 32 MHz was coupled to the plasma using three antennae. Two of these were operated as dipoles and the other as a monopole which could couple up to 4 MW of power. The experiments were carried out mostly using hydrogen minority heating in a ^{3}He plasma although similar results were obtained with deuterium as the majority ion species. Typically the H^+ fraction was 2% of the majority ion concentration but varied somewhat with both plasma density and RF power due to recycling effects. The toroidal field at R = 3 m was 2.1 T so that the hydrogen fundamental cyclotron resonance surface passed close to the plasma centre. Ray tracing calculations predict an absorbed power density profile which falls to 10% of its central (maximum) value at a minor radius of 0.4 m. The plasma current was 2 MA.

The electron temperature $T_e(o)$, was measured using electron cyclotron emission (ECE) diagnostics. An ECE grating polychromator provided data with sub-millisecond time resolution and a Michelson interferometer gave spatial scans on a few millisecond time scale. Density profiles were measured using a far infrared multichannel interferometer and in the present experiments the volume average density $\langle n_e \rangle$ was varied between 1.6 x 10^{19} m^{-3} and 3.7 x 10^{19} m^{-3}. The value of Z_{eff} was measured using visible bremsstrahlung and the total radiated power was monitored with a bolometric array. A wide range of magnetic diagnostics gave information on plasma current, loop voltage, plasma position, energy content and MHD activity.

The effect on the central electron temperature $T_e(o)$ of applying 7.2 MW of ICRF to a ^3He plasma with $\langle n_e \rangle$ = 2.5 x 10^{19} m^{-3} is shown in Fig. 1. The temperature rises from 2.6 keV during ohmic heating (OH) to about 6 keV during ICRF which is on for 3.5 sec at maximum power. During this heating only three temperature collapses occur with the longest crash-free period being 1.6 sec and ending at the RF switch-off. The temperature saturates during RF but the density continues to rise as does the total energy content (from diamagnetism), W, except during the crashes, as shown in Fig. 1.

One of the most noticeable characteristics of the crash-free periods is the quietness in terms of MHD activity. However, during the crash a strong n=2, m=3 oscillation is excited and may last for some time. The amplitude of this mode is shown in Fig. 1. At the highest values of $P_{RF}/\langle n_e \rangle$ an mhd mode-lock sometimes occurs whereupon $T_e(o)$ typically decreases from 6.5 keV to 4.5 keV and both the stored energy and the global energy confinement time τ_E are reduced by up to 30%. During ICRH the value of Z_{eff} rises to 2.8 from its value during OH of 2.4 which is close to Z_{eff} = 2 for a pure helium plasma.

During the crash-free periods the value of τ_E is found to be greater than that when normal sawteeth are present. This is perhaps best illustrated by the data in Fig. 2 which shows a transition from normal sawteeth to a sawtooth-free period occurring within a single pulse at a constant 4 MW RF power input. The transition is dramatically seen on the D-D reaction neutron yield from background deuterium ions showing that the ion temperature is affected in a similar way to $T_e(o)$.

The stored energy is also shown in Fig. 2 and is observed to increase at a rate $\dot{W}$ = 0.25 MW after the transition. During this time the total input power $P_T = P_{OH} + P_{ICRF}$ is less than that during the normal sawtooth activity by ~ 0.25 MW due to the fall in resistivity as T_e increases. Thus the global energy confinement time defined by $\tau_E = W/(P_T-\dot{W})$ increases from 0.29 ± 0.1 sec when normal sawteeth are present to 0.35 ± 0.1 sec when they are suppressed. Radiation loss has not been included in the definition of τ_E and a decrease in the radiation could conceivably account for the improvement in τ_E. However, this appears not to be the case since bolometer measurements show that the total radiation (~ 1.6 MW during ICRF) decreases by only 0.06 MW after the transition.

The radial profiles of T_e are shown in Fig 3 for the OH, normal sawtooth and sawtooth-free phases. The profile during the sawtooth-free phase is almost identical with that at the top of the normal sawteeth and both are more centrally peaked than those during OH. This implies that q(o) may be reduced in the sawtooth-free period although magnetic field diffusion calculations show that only a few percent reduction in q(o) can be expected in this period. Equilibrium calculations from magnetic measurements tend to support this conclusion. On the other hand the plasma inductance, ℓ_i, remains constant, or even decreases, during the sawtooth-free periods implying that the current profile j(r) is not peaking significantly. At the collapse ℓ_i decreases by ~4% indicating a broadening of j(r).

Thus it is clear that any explanation of the sawtooth suppression in terms of current profile modification will need to be sensitive to small changes in j(r). One such mechanism is the ideal mhd instability model of the sawtooth collapse in JET[2] although the source of the necessary current profile broadening is difficult to find. The fact that ICRH alone can suppress sawteeth appears to rule out non-inductive current drive and calculations indicate that the bootstrap current is too small. Other explanations involve reducing the value of q(o) at which the ideal mode becomes unstable either by trapped particle[5] or finite Larmor radius effects[6].

CONCLUSIONS

Sawtooth-free periods of up to 1.6 sec have been produced in JET plasmas by ICRH alone in both ^{3}He and D plasmas. Such periods are characterized by a reduction in MHD activity, the appearance of n=2 modes after the crash, saturation in T_e(o) and increases in density, energy content and fusion yield. A definitive explanation of the sawtooth suppression has yet to be found.

ACKNOWLEDGEMENT

The authors acknowledge with pleasure the assistance of all our colleagues in the JET team. Particular thanks go to the tokamak operating team and to the members of all the diagnostic groups involved in the measurements reported here.

REFERENCES

1. J.A. Snipes and K.W. Gentle, Nucl. Fusion 26 (1986) 1507
2. J.A. Wesson, Plasma Physics and Controlled Fusion 28 (1986) 243
3. G. Sadler, et al Proc 13th European Conf. on Controlled Fusion and Plasma Heating, Schliersee FRG 1986 Vol. I p 105
4. D.J. Campbell et al, Proc. of 11th Int. Conf. on Plasma Physics and Controlled Nuclear Fusion Research, Kyoto, Japan 1986 (to be published)
5. J. Hastie, private communication
6. G.B. Crew et al, Nucl. Fusion 22 (1982) 41

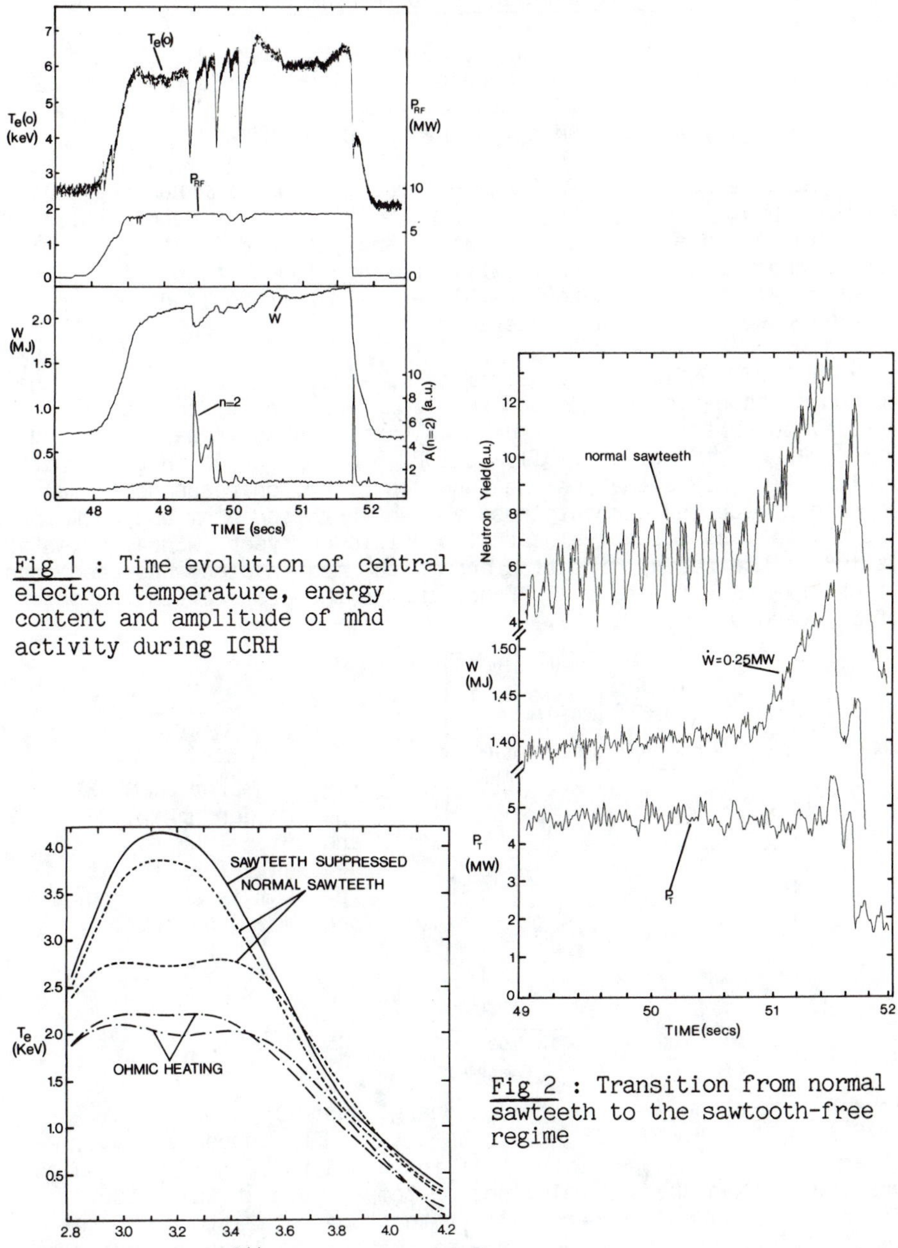

Fig 1 : Time evolution of central electron temperature, energy content and amplitude of mhd activity during ICRH

Fig 2 : Transition from normal sawteeth to the sawtooth-free regime

Fig 3 : Electron temperature profiles during the pulse shown in Fig 2

ICRF ACCELERATION OF BEAM IONS IN JET

Cottrell G.A., Bhatnagar V.P., Cordey J.G., Core W., Corti S., Hamnén H., Hellsten T., Jacquinot J., Sand F., Start D.F.H., Watkins M.,
JET Joint Undertaking, Abingdon, Oxon, OX14 3EA, UK.

ABSTRACT Combined ICRF and Neutral Beam heating has been studied in JET for $P_{RF} + P_{NBI} \leq 16$ MW. Evidence for acceleration of deuterium has been found from neutral analysis and from a study of the enhancements in DD reactivity. Fokker-Planck calculations show reactivity enhancements for the higher beam energies and heating powers expected in future experiments.

1. EXPERIMENTS The possibility of absorbing ICRF power at the harmonic resonance of injected fast ions has been discussed by several authors [1-5]. In recent JET experiments we have combined $D^o \rightarrow D^+$ neutral injection (NBI) and ICRF heating in the (H)D minority regime. In this situation the fundamental proton resonance (ω_{cH}) coincides with the harmonic resonance of both the beam and bulk ions ($2\omega_{cD}$). A mass selective neutral particle analyser (MSNPA) oriented at 15° to the radial direction and in the median plane has been used to observe the effects of ICRF heating on the distribution function of injected D^+ions (Fig 1).

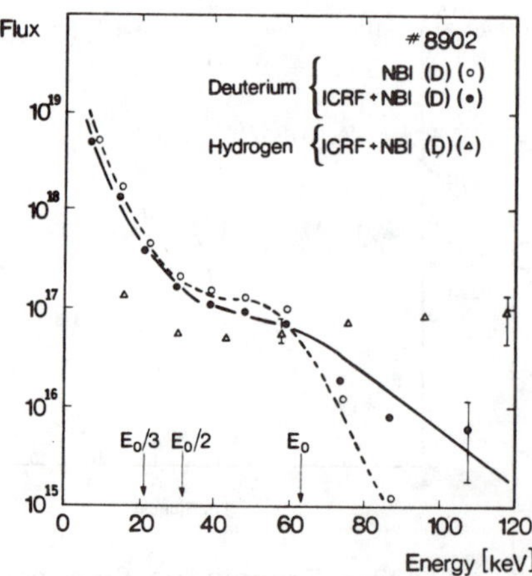

Fig 1 Deuterium MSNPA spectra during NBI (3.6 MW) and NBI + ICRF (6 MW) heating. The minority H-spectrum is also shown for the same discharge.

In the beam-alone phase of the discharge, the slope of the tail ($E>E_o$) was $T_t \approx 4$ keV ($\approx T_e(o)$). When the ICRH, tuned to $\omega=\omega_{cH}$, was later added, the slope of the tail increased to $T_t \approx 18$ keV, which was larger than the central electron temperature at this time ($T_e(o) \approx 6$ keV). In order to study the D^+ acceleration and the scaling of reactivity with plasma parameters, we have analysed the complete set of relevant combined heating discharges (up to # 10363), and compared H and ^{3}He minority heating (Fig 2). In the latter case, the $2\omega_{cD}$ resonance was located outside the plasma column. The (H)D example shows a factor of ~ 2-3 increase in DD yield when the ICRF is applied whereas in the (^{3}He)D case, the increase is small.

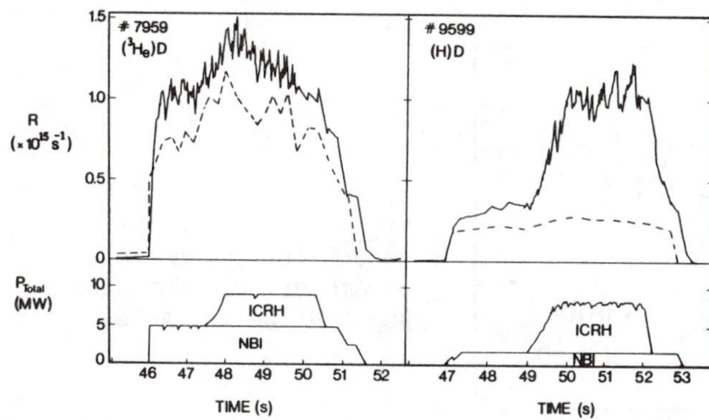

Fig 2. Comparison of fundamental ^{3}He and H heating during NBI. Upper solid line: measured DD reaction rate; Dashed line: predicted DD reaction rate.

To assess reactivity effects due to changes in plasma conditions we have used the following analysis. First, a prediction of the total DD reaction rate, R^*, was made by summing the beam-plasma and thermal rates. The beam-plasma rate was calculated using the NBI Fokker-Planck code PENCIL which does not include an ICRF operator. Absolute measured reaction rates exceed predicted rates during the NBI-only time windows, possibly because of uncertainties in knowing n_D. This uncertainty was removed by studying the behaviour of the ratios of predicted and measured reactivity $A = R^*(2)/R^*(1)$, $B = R(2)/R(1)$ where 1,2 refer to the NBI and NBI + ICRF phases (Fig 3).

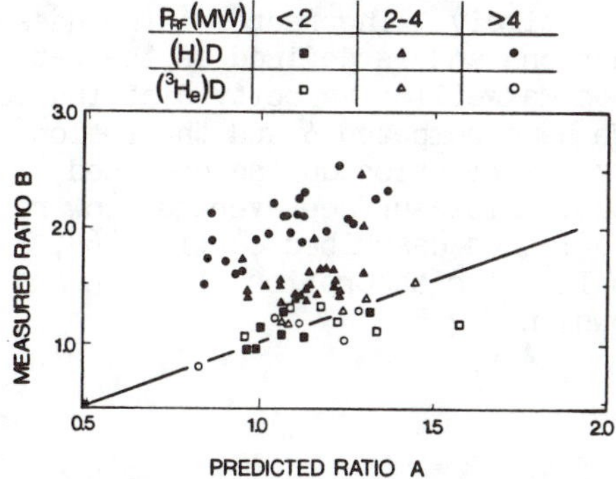

Fig 3. Measured versus predicted DD reaction rate ratios for both ICRF schemes with $D^0 \rightarrow D^+$ NBI.

The enhancement increases with RF power for fundamental cyclotron heating of H on D, whereas for fundamental heating of ^{3}He on D, the points lie close to the line A = B. We have defined an experimental reactivity enhancement parameter $\varepsilon_{exp}=B/A$. Assuming a linear scaling law for the enhancement when beam-plasma collisions are dominant, we find $\varepsilon = (1 + P_D/P_B)$, where $P_D = P_H\beta_D/\eta_H$ is the RF power coupled to the beam, P_B is the beam power, and β_D and η_H are the perpendicular fast ion beta and hydrogen-to-deuterium ratio respectively. Fig 4 shows that the scaling describes the experimental data reasonably well. The slope of the line is proportional to η_H^{-1} giving η_H=0.02±0.01 which is within experimental errors of the NPA value (derived from the low energy flux ratios) of η_H = 0.03±0.01.

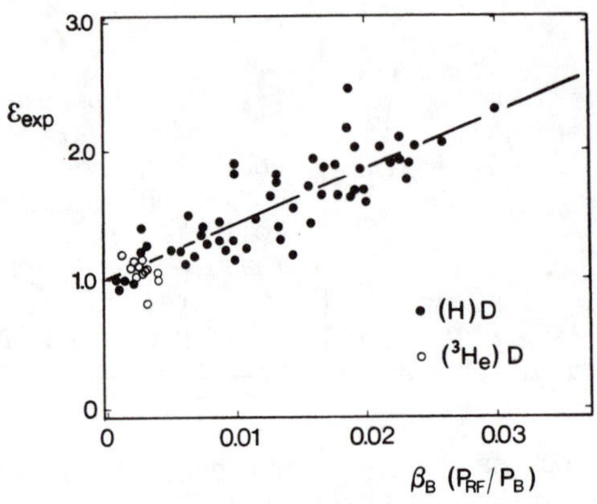

Fig 4. Experimental reactivity enhancement parameter as a function of the scaling variable.

2. THEORETICAL ASPECTS Since it is difficult experimentally to reduce the hydrogen content to a level where pure $2\omega_{cD}$ absorption can take place, the major problem with modelling cyclotron absorption at $\omega = 2\omega_{cD}$ is degeneracy with ω_{cH}. In the limit k//=0, the ratio of RF power absorbed on D and H can be written

$$P_D/P_H = n_D\, k_\perp^2\, \langle v_{\perp D}^2 \rangle\, \delta / 2 n_H\, \omega_{cD}^2 \qquad (1)$$

where $\langle v_{\perp D}^2 \rangle$, $k_\perp$ are the mean square perpendicular deuteron velocity and perpendicular wavelength respectively. The factor δ includes F.L.R. and non-Maxwellian corrections and is defined as the ratio between the RF absorption of a non-Maxwellian velocity distribution and that of a Maxwellian [5]. We have computed δ and the fusion yield for a steady-state velocity distribution during combined heating using a Fokker-Planck code. Flux-surface averaged power densities were calculated using a method described earlier [6], where the central RF power was split into absorption at $\omega=\omega_{cH}$ and $\omega=2\omega_{cD}$ using eq. (1) and are shown in Fig 5.

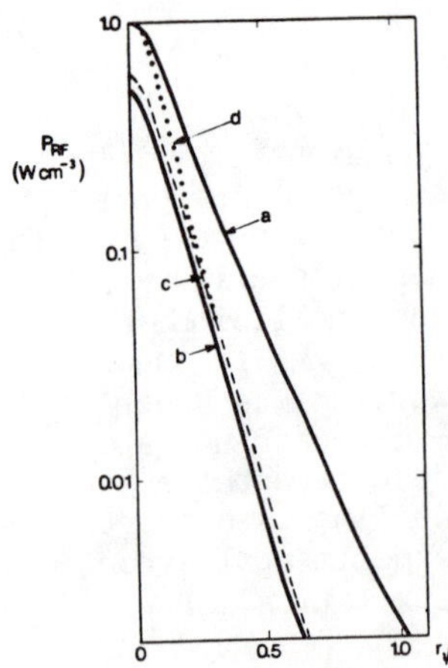

Fig 5. ICRF deposition profile for typical JET data (E_o = 80 keV, P_{RF} = 6.3 MW). Curves (a) and (b) give H and D absorption for a Maxwellian, (c) the D absorption for NBI in steady-state and (d) ICRF + NBI in steady-state.

The ratios $\int P_D dV/\int P_H\, dV$ are, for a Maxwellian 0.21, for a NBI heated plasma 0.25 and for NBI and RF combined heating 0.36. The incremental fusion energy gain, $\Delta Q = n_D^2\,(\langle\sigma v\rangle - \langle\sigma v\rangle_{th})/(P_D+P_{NBI})$ where $n_D^2\,\langle\sigma v\rangle$ and $n_D^2\,\langle\sigma v\rangle_{th}$ are the steady-state fusion yields obtained for RF + NBI and for a Maxwellian respectively is shown in Fig 6.

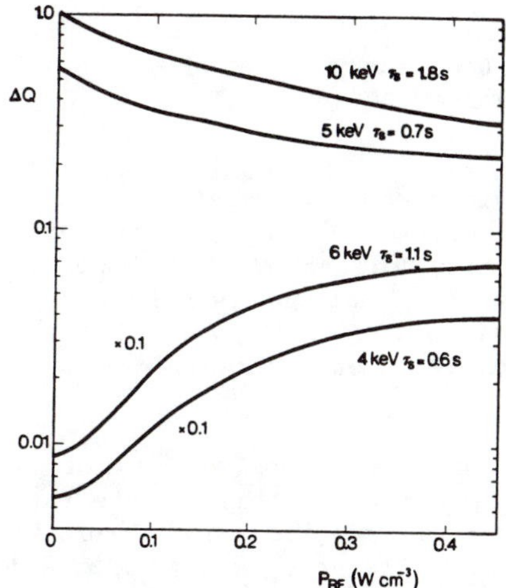

Fig 6. Incremental ΔQ versus RF power density for $T_e = T_i$. Lower curves: DD reactivity (for #9599) ; Upper curves: DT reactivity (see text). Steady-state conditions are reached after roughly one Spitzer time, τ_s.

To represent # 9599 we have used the following parameters: $n_e = 3\times10^{19}\ m^{-3}$, $n_D/n_e = 0.5$, $E_o = 80$ keV, $T_e = T_D = 4$ and 6 keV. Overall, the DD calculations show that, for low η_H, a substantial fraction of the central RF power couples at $2\omega_{cD}$ to the bulk. As the D tail develops (in a time $\sim\tau_s$) more power couples to the tail. In the outer part of the plasma, β_D is lower and therefore more RF power couples to the protons. To investigate the reactivity enhancement for a DT plasma we show ΔQ assuming T = 5 and 10 keV, $n_D = n_T = 2\times10^{19}\ m^{-3}$, $Z_{eff} = 1$, $E_o = 140$ keV. Here ΔQ is still positive but relatively smaller than for the DD case. This is because the peak in the DT fusion cross-section occurs at lower energy.

REFERENCES

1. STIX, T.H. N. Fus. 15, (1975) 737
2. ITOH, S-I, FUKUYAMA, A. and ITOH, K, N. Fus. 24 (2), (1984) 224
3. HARVEY, R.W., McCOY, M.G., KERBEL, G.D., and CHIU, S.C. N. Fus. 26 (1), (1986) 43
4. CORE, W.G.F., Proc. 13th Europ. Conf. on Cont. Fusion and Plasma Heating, Schliersee, F.R.G. (1986)
5. HELLSTEN, T., APPERT, K., CORE, W., HAMNEN, H., SUCCI, S., 12th Conf. on Contr. Fusion and Plasma Physics, Budapest, (1985)
6. HELLSTEN, T., and VILLARD, L., Proc. 14th Europ. Conf. on Contr. Fusion and Plasma Physics. Madrid. Spain (1987)

TFTR ICRF Antenna Design*

J.R. Wilson, J. Bialek, P. Bonanos, A. Brookes,
P.L. Colestock, J.C. Hosea, I. Lehrman[+], R. Ritter

Princeton University
Plasma Physics Laboratory
Princeton, New Jersey 08544

Abstract

An antenna for ICRF heating via the fast magnetosonic wave has been designed for the TFTR tokamak. The antenna will operate at 47 MHz and is designed to couple ~5 MW of rf power to the plasma. Detailed mechanical and thermal analyses have been performed to verify that the design is compatible with the heat fluxes and mechanical stresses of the TFTR environment.

Introduction

The goal of the TFTR ICRF project is to apply ~9 MW of rf power to the TFTR plasma for central plasma heating via the fast magnetosonic wave. This power is provided by four rf generators, two operating at a fixed frequency of 47 MHz and capable of 3MW each and two continuously tunable from 40-80 MHz and capable of 2 MW each. In order to couple this power to the TFTR plasma and to take advantage of the generator characteristics, two separate antennas have been designed for use on TFTR. The fixed frequency generators will be hooked to the antenna designed by PPPL and described herein. The two variable frequency generators will be hooked to an antenna designed by ORNL that is described in another paper in this volume.

Mechanical Description of the Antenna

The antenna will be inserted in a port 73.5 cm high by 89 cm wide on TFTR. The antenna consists of an outer box, two central conductor radiating elements and an array of rods that act as a Faraday shield covering the mouth of the box (fig. 1). The Faraday shield structure is composed of two rows of rods that are broken into groups of five that are attached to individual clips and bolted to the sides of the box. The sides of the box and an internal septum which separates the conductors are slit horizontally for a distance of ~10 cm back from the front in order to achieve better coupling to the plasma. This will be discussed further below. The front edges of the box have an array of carbon tiles that protect the structure from the high heat flux in TFTR. The entire antenna structure, including the tiles, is insertable through the port cover and requires no entry into the vacuum vessel. In order to find the optimal major radial position for the coupler, the entire

[+]Permanent Address: Grumman Aerospace Corp.

structure is able to translate 10 cm. This permits a compromise between obtaining the highest loading resistance without having an excessive thermal load to the box. This radial motion is accomplished through the use of a spline drive mechanism centrally mounted to the back of the antenna box. To remove the plasma and rf heat loads, the center conductors, the box walls and the central septum are all water cooled. The entire structure is modular in design allowing for parallel construction and ease of assembly. The box, drive and Faraday shield structure are bolted together.

RF design

The antenna contains two center grounded loops with a toroidal separation of 33 cm. The loops are 10 cm wide and 1.25 cm thick. Four rf cone feedthroughs are used to apply power to the structure. The loops can be driven either in phase or out of phase yielding a spectrum such as that in figure 2. By slitting the box and septum, the spectrum is influenced (fig. 3). The presence of the slits allows the spectrum to be dominated by the toroidal separation of the conductors and not the box dimensions yielding peaks at either 9 m^{-1} or 16 m^{-1}. The center conductor is kept close to the Faraday screen and the top wall of the structure so that the characteristic impedance is reduced and the voltage for a given power is minimized (~50 Ω). It is calculated that for 5 MW operation the maximum rf voltage on the antenna will be ~50 kV. The entire structure is resonated by means of external lengths of transmission line which allow for even power division and phasing.

Thermal Design

The antenna structure is subject to thermal loads from both the plasma and from internal rf losses. The rf losses are minimized by plating the inside of the box and the center conductors with silver and by having a copper plating on the Faraday shield which is subsequently covered with TiC-TiN to prevent sputtering. Rf losses to the center conductor are offset with water cooling since both conduction and radiation are ineffective for the silver plated stainless steel structure. Rf losses to the Faraday shield should be limited to ~3% of the rf power with this coating. Plasma heat load to the Faraday shield is a function of the major radial location of the antenna. At the nominal location 2 cm outboard of the limiters, a loading of ~35 W/cm^2 is expected. The Faraday rods are cooled by radiation and by conduction to the water-cooled side walls of the box. A detailed analysis of the temperature evolution (fig. 4) shows that in worst case operation, 2 second full power pulses every 5 minutes, the equilibrium temperature of the rods is ~325° C with temperature peaks of almost 400° C during a pulse.

The sides, top and bottom of the structure are protected by graphite tiles (fig. 5). These tiles are brazed to molybdenum plates which are bolted to the antenna box. Since the heat flux falls exponentially with radius, the tile is shaped to spread the heat out as uniform as possible over the tile surface. Figure 6 shows the time evolution of the tile temperature at several points as calculated using a 2-D code. For typical operation a maximum tile surface temperature of ~1300° C is reached. The ratched background temperature of the tiles is ~430° C (fig. 6).

Mechanical Loads

The mechanical structure of the antenna is subject to mechanical loads both due to disruption induced eddy currents and thermal stresses. Disruption forces on the center conductor introduce a torque of ~730 in-lbs about the center short and deflections of 1 mm at the ends. The box structure is subject to a torque around the centrally attached drive mechanism. This spline drive has a maximum sheer stress of 13,000 psi on its shield rod attachments. The forces here arise from both the disruption forces and the thermal growth of the rods. This induces a maximum stress of 18,000 psi in the front row rod welds.

Summary

An ICRF heating antenna has been designed for use on TFTR. The antenna operates at 47 MHz and is designed to handle 5 MW of rf power. The antenna operates with a hot (~350° C) Faraday screen and cooled box and center conductors. Slits in the side walls of the structure allow improved coupling and a better antenna wave spectrum. Protective tile design allows insertion close to the plasma edge. The design also allows radial motion to achieve the best balance between antenna plasma coupling and thermal loading

*Work supported by U.S. Dept. of Energy Contract No. DE-AC02-76-CHO-3073.

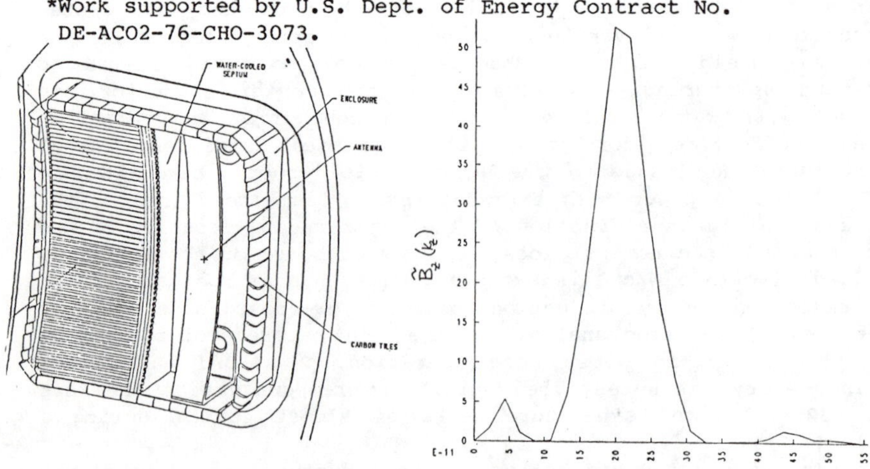

Fig. 1 Antenna Structure

Fig. 2 Fourier Spectrum of Antenna

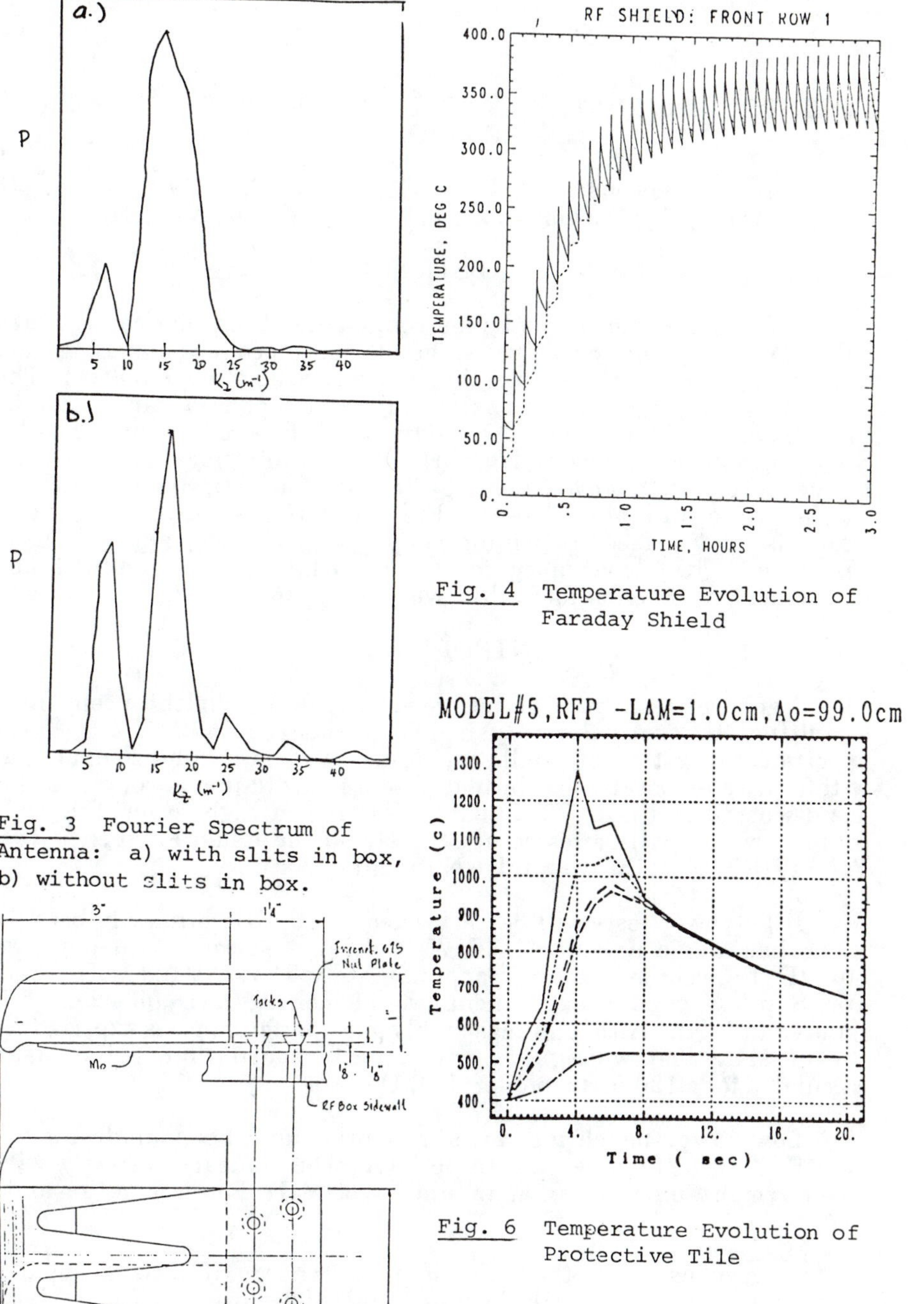

Fig. 3 Fourier Spectrum of Antenna: a) with slits in box, b) without slits in box.

Fig. 4 Temperature Evolution of Faraday Shield

Fig. 6 Temperature Evolution of Protective Tile

Fig. 5 Carbon Protective Tile and Mount For Side of Antenna Structure

TESTS OF A HIGH-POWER FOLDED WAVEGUIDE COUPLER FOR ICRF HEATING*

T. L. Owens,† G. L. Chen, G. R. Haste, P. M. Ryan
Oak Ridge National Laboratory, Oak Ridge, TN 37831

ABSTRACT

A full-scale folded waveguide coupler has been fabricated which will provide information on power handling, impedance matching, and multipactor effects. The coupler dimensions are 30 × 60 × 300 cm. The cross section of the coupler is small enough that a phased pair of couplers could be placed in a single Tore Supra or TFTR port. A single coupler could be placed in a CIT-size port. A movable back plate allows frequency adjustment over the range 78–140 MHz. Impedance matching at the waveguide has been achieved using a movable coaxial transmission line feed. Bench-top comparisons with loop antennas have been made. The folded waveguide coupler will be mounted on the Radio Frequency Test Facility for high-power tests up to 1.5 MW.

INTRODUCTION

A general description of the folded waveguide coupler concept and a simplified theoretical analysis are presented in Ref. 1. Early experimental tests on a small-scale folded waveguide and comparisons with a more elaborate 3-D finite difference calculation of coupler fields are discussed in Ref. 2. This paper describes a high-power, full-scale coupler, which will be tested at 80 MHz on the Radio Frequency Test Facility (RFTF) at the Oak Ridge National Laboratory.

High-power tests will be performed at 80 MHz primarily because RFTF has ample power (~1.5 MW) available at this frequency. At 80 MHz, the coupler length is approximately 3 m. At 120 MHz where Tore Supra waveguide experiments would operate, the coupler length is reduced to 1.4 m. Although reasonably good performance is expected for the 80-MHz test waveguide, much better performance has been calculated for a 120-MHz coupler (Ref. 1).

Low-power bench-top measurements have been made on the 80-MHz test waveguide, and these have been compared directly with measurements on a loop antenna tuned to 65 MHz. Power handling and

---

*Research sponsored by the Office of Fusion Energy, U.S. Department of Energy, under contract DE-AC05-84OR21400 with Martin Marietta Energy Systems, Inc.

†McDonnell Douglas Astronautics Company, Consultant to Oak Ridge National Laboratory.

coupling efficiency in the presence of plasma can be roughly estimated for the folded waveguide based on this loop/waveguide comparison and on knowledge of loop performance gained through experiments. Power handling and coupling efficiency estimates will be compared to calculations.

RESULTS

Figure 1 shows the poloidal distribution of folded waveguide, toroidal wave magnetic fields taken along the vertical midplane of the coupler. The poloidal distribution is reasonably smooth at a distance of 10 cm in front of the coupler. Little excitation of high-order poloidal modes is expected for this spatial distribution. Somewhat greater excitation of high-order modes occurs closer to the coupler (Fig. 1b), but most of the power is still in the lowest-order modes.

A plot of the radial decay of the toroidal wave magnetic field away from the coupler is shown in Fig. 2a. Figure 2b shows the radial decay of the toroidal wave magnetic field for a 65-MHz loop antenna having the same cross-sectional dimensions as the folded waveguide coupler. The comparison shows almost the same decay length for loop and waveguide fields. Other measurements have shown that the directions of the wave magnetic field vectors for loop and waveguide couplers are nearly the same in front of the couplers. The addition of a Faraday shield in the folded waveguide apertures does not change the field patterns but only lowers the unloaded quality factor of the waveguide about 15%.

Peak electric field for the folded waveguide coupler can be related roughly to the peak electric field between a loop antenna and its Faraday shield through the relation,

$$\frac{E_w^2}{E_\ell^2} \simeq \left(\frac{16d^2}{\omega_o \varepsilon_o V \omega L}\right)\left(\frac{f}{g}\right)^2 \frac{Q_w}{Q_\ell} h^2, \quad (1)$$

where d is the distance between the loop and Faraday shield, L is the loop inductance, V is the waveguide volume, $Q_{\ell,\,w}$ are unloaded quality factors, ω_0 and ω are waveguide and loop frequencies, f and g are field enhancement factors due to surface shapes, and h is the unloaded average wave magnetic field amplitude of the loop coupler at its load surface divided by the unloaded average wave magnetic field of the folded waveguide coupler at the same load surface. In Eq. (1), electric fields are calculated for equal power coupled to the load surface by loop and waveguide couplers. The loop antenna is assumed to have a voltage minimum at its center. The loop antenna used in the experimental comparison has a Faraday shield 1.5 cm from the current strap. The shield consists of a single row of round tubes, 1.2 cm in diameter, that yields a field enhancement factor of 1.45. From measurements of waveguide and loop coupling efficiencies to a resistive surface, the quantity, h, can be inferred using Eq. (2). It is found to be ~1.8 for resistive surfaces placed between one and ten centimeters from the couplers. For the test folded waveguide, a field enhancement factor of 2.6

is calculated. Other measured quantities are: $\omega_0 = 2\pi \times 80$ MHz, $\omega = 2\pi \times 65$ MHz, $V = 0.468\ m^3$, $L = 2 \times 10^{-7}$H, $Q_w = 2544$, and $Q_\ell = 590$. Using these values, $E_w/E_\ell \simeq 1$.

Note that the preceeding comparison is for a specific loop coupler and a specific folded waveguide coupler. Reductions in electric field for the folded waveguide coupler occur at higher frequencies and for increased vane tip radii. Table I shows the calculated improvements over the test waveguide as operating frequency and waveguide size are increased. Relative vane spacing can also be increased near the center of the coupler, where most of the Poynting flux occurs, to reduce field levels. Somewhat fewer options are available for reducing fields for the loop coupler used in this comparison. In addition, fields for the loop coupler may be considerably higher in the components that attach to the antenna (e.g., radial feed lines, feedthroughs, supports, tuning elements, etc.). Fields for the folded waveguide coupler described in the preceeding comparison are already the largest fields anywhere in the system. Finally, the position of maximum field for the folded waveguide coupler is 1.5 m back from the mouth of the coupler, well away from the plasma surface. For the loop coupler, high electric fields exist very close to the plasma surface. Tests with plasma on RFTF may provide information on the importance of this and other differences between loop and folded waveguide couplers. Another unknown is the effect of multipactor within the waveguide. Techniques to suppress multipactor may be required to maximize power handling of the folded waveguide.

Folded waveguide coupling efficiency, η_w, can be related to loop coupling efficiency, η_ℓ, by the following relation:

$$\eta_w \simeq \frac{1}{1 + h^2\left(\frac{1}{\eta_\ell} - 1\right)} . \tag{2}$$

As discussed earlier, $h \simeq 1.8$ so that, for a common loop coupling efficiency of 90% in the presence of plasma, the test waveguide coupler would have an efficiency of approximately 74%. The calculated efficiency for the test waveguide is 68%, as shown in Table I. This rough agreement between theory and measurement provides some degree of confidence in the calculated high efficiencies (94–99%) and low electric fields at 120 MHz (Table I).

REFERENCES

[1]T. L. Owens, "A Folded Waveguide Coupler for Plasma Heating in the Ion Cyclotron Range of Frequencies," IEEE Trans. Plasma Sci., **PS-14** (6): 934-46 (December 1986).

[2]T. L. Owens and G. L. Chen, "A Folded Waveguide Coupler for Ion Cyclotron Heating," Fusion Technol. **10** (3)2A: 1024 (November 1986).

Table I

Comparison of a pair of 80 MHz test couplers to a single 4 section coupler and a pair of 8 section couplers designed for 120 MHz operation in a single Tore Supra port (60 × 70 cm). Total power is 4 MW with a 10 cm plasma/coupler separation.

	4-section coupler	8-section coupler pair	10-section test coupler pair
Electric field in coupling apertures (kV/cm)	1.45	1.75	1.62
Peak electric field (kV/cm)	12.65	25.6	49.3
Plasma loaded quality factor	213	834	2609
Unloaded quality factor	23,440	13,171	5500
Coupling efficiency (%)	99	94	68

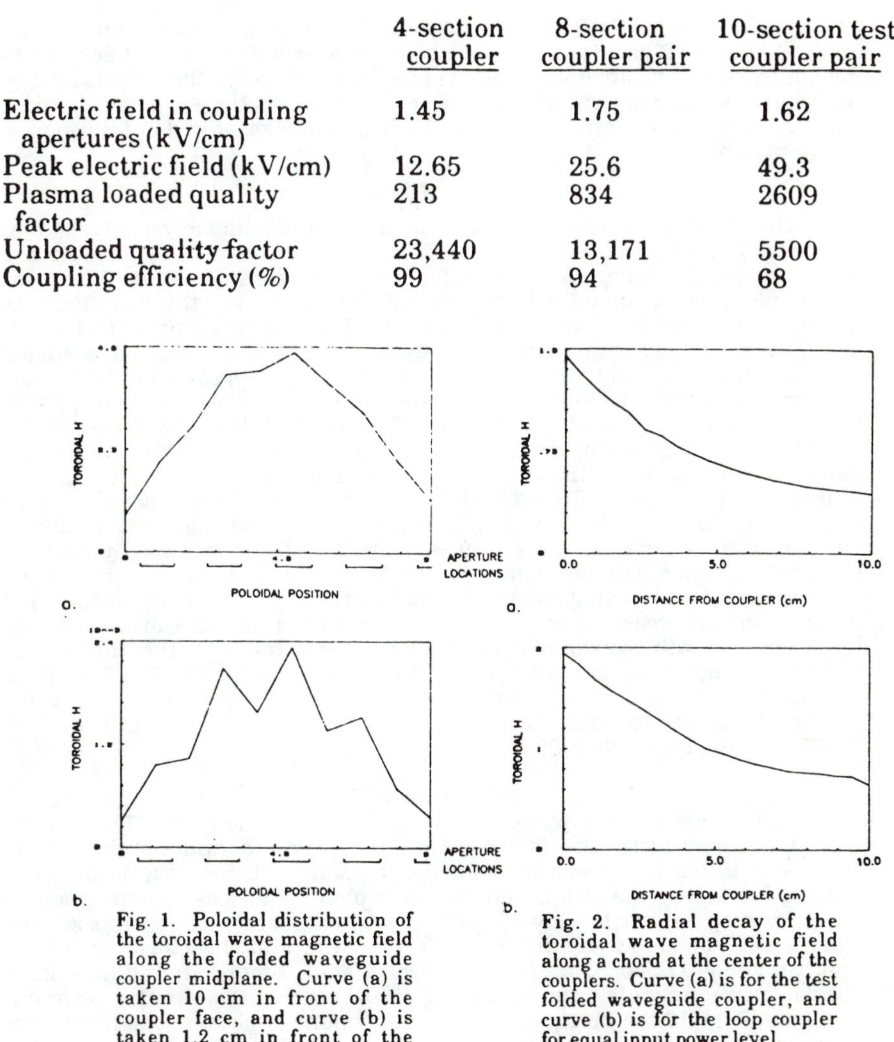

Fig. 1. Poloidal distribution of the toroidal wave magnetic field along the folded waveguide coupler midplane. Curve (a) is taken 10 cm in front of the coupler face, and curve (b) is taken 1.2 cm in front of the coupler face.

Fig. 2. Radial decay of the toroidal wave magnetic field along a chord at the center of the couplers. Curve (a) is for the test folded waveguide coupler, and curve (b) is for the loop coupler for equal input power level.

THE DESIGN OF HIGH-POWER ICRF ANTENNAS FOR TFTR AND TORE SUPRA*

D. J. Hoffman, F. W. Baity, W. E. Bryan, G. L. Chen, K. H. Luk, T.L. Owens, J. M. Ray, P. M. Ryan, D. W. Swain, and J. C. Walls
Oak Ridge National Laboratory, P. O. Box Y, Oak Ridge, TN 37831

ABSTRACT

The Oak Ridge National Laboratory is designing, fabricating, and testing antennas for TFTR and Tore Supra. The antennas will deliver 4 MW per port to the hot plasma. The antennas' designs are nearly complete; they are backed by the results of extensive modeling, analyses, and tests of the electrical, thermal, and structural characteristics of the antennas that were carried out during the design process.

INTRODUCTION

The designs for these two antennas are approximately equivalent in principle; Tore Supra's design is more complicated because of its longer pulse length, wider frequency range, and higher power density. Table I lists the design parameters for both antennas. Figure 1 shows the layout of the antennas (specifically, of the TFTR antenna). Each antenna consists of a pair of resonant double loops, separated by a solid dividing wall, in a single, movable housing designed to fit through a horizontal midplane port (approximately 60 x 70–90 cm). Each loop is designed to launch up to 2 MW of power for a total of 4 MW per port. The TFTR antenna will operate at 40–80 MHz in two bands; the Tore Supra antenna, at 35–80 MHz in a single band. Both are tuned to the frequency of operation and matched to the correct impedance (50 Ω for TFTR and 30 Ω for Tore Supra) by means of pairs of capacitors. The capacitors for TFTR were designed at ORNL to minimize electric fields at the expense of tuning range, and those for Tore Supra were designed by Comet, Ltd., for higher electric fields and broad tuning capabilities. On both antennas, a single Faraday shield structure protects the loops. The Faraday shield consists of two tiers of actively cooled Inconel tubes; the front tier is covered with semicircular graphite sleeves to minimize the introduction of high-Z impurities into the plasma from the Faraday shield. The TFTR antenna is designed for 2-s pulses; the Tore Supra antenna must sustain 210-s pulses. Plans are to install the TFTR antenna in August 1987 and the Tore Supra antenna in March 1988.

ANTENNA DESIGN

The electrical circuitry of the two antennas is identical. The current strap is separated from the Faraday shield by 1.5 cm; the width of the current strap is half that of the enclosing cavity; the side walls are solid to minimize plasma flowing by the strap; and the back plane is at least 15 cm from the current strap. For this configuration, the characteristic (measured and calculated) impedance for the strap is 65–70 Ω and the phase velocity is approximately 0.65–0.70 times the speed of light. From a distributed, lossy coaxial transmission line model (Fig. 2), the voltages and currents on various elements (Fig. 3) have been computed as functions of plasma load for a given antenna power. The load needed to achieve 2 MW (due to voltage or current limits) scales roughly as the square of the frequency. At 6 Ω/m, 47 MHz, and 2 MW per strap, the capacitor voltages are at the design limit of 50 kV and the electric field between the Faraday shield and current strap is 23 kV/cm. This

*Research sponsored by the Office of Fusion Energy, U.S. Department of Energy, under contract DE-AC05-84OR21400 with Martin Marietta Energy Systems, Inc.

Table I. Antenna design parameters

Description	TFTR	Tore Supra
Power into the port (MW)	4	4
Power per current strap (MW)	2	2
Port dimension (cm x cm)	60 x 90	60 x 70
Total frequency range (MHz)	40–80	35–80
Frequency range, first band (MHz)	40–60	35–80
RF pulse length (s)	2	210
Capacitor designer	ORNL	Comet
Capacitor voltage (kV peak)	50	50
Capacitor current (A rms)	800	750
Capacitor electric field (kV/cm)	43	100
Antenna electric field (kV/cm)	23	23
Faraday shield losses (W/cm^2)	100	100
Antenna power density (W/cm^2)	1160	1520
Antenna motion range (cm)	11	30

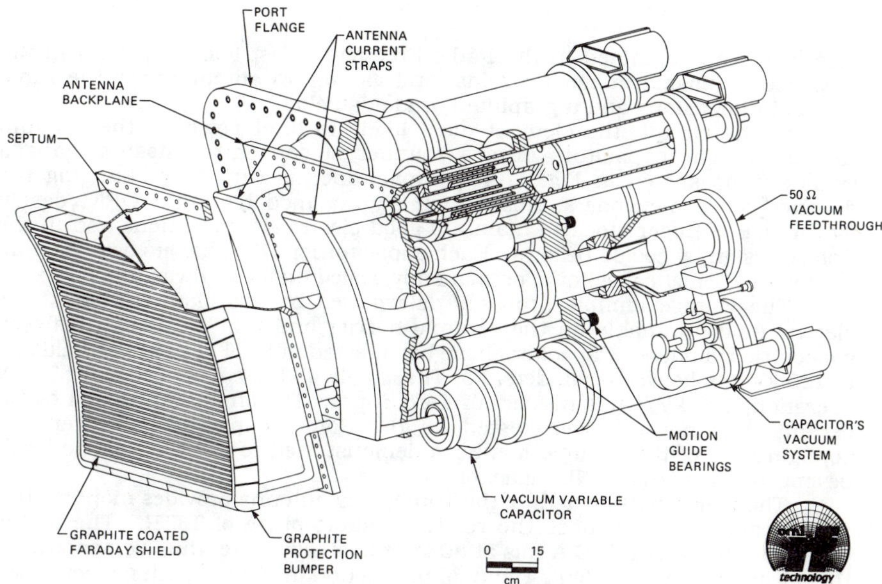

Fig. 1. The antenna for TFTR.

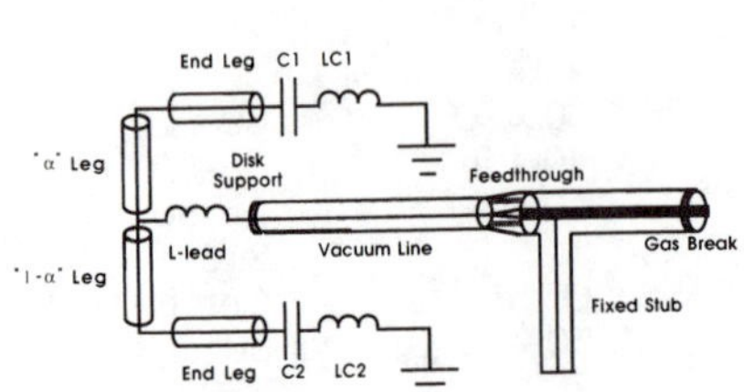

Fig. 2. The electrical circuit of the antennas. Because TFTR's current strap is not cooled, its circuit does not have the fixed stub.

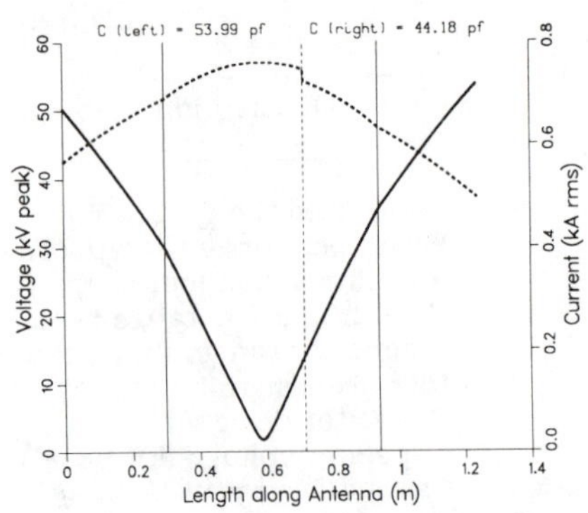

Fig. 3. Voltages and currents along the current strap. The C1 capacitor in Fig. 2 is at 0 m. The two vertical lines represent the beginning and end of the current strap, and the dashed vertical line is at the site of the feed point. This calculation is for the TFTR antenna with 2 MW at 47 MHz and a load of 6 Ω/m.

field has been sustained in the Radio-Frequency Test Facility (RFTF) in the presence of plasma, magnetic fields, and gas on an antenna with the same nominal dimensions and a graphite Faraday shield.

In the circuit, the element most likely to limit power is the vacuum capacitor. The capacitors form a number of concentric, nested, ganged cylinders, as shown in Fig. 1. The capacitance is varied by changing the distance between cylinders. This motion is accommodated in the ORNL design by the use of finger stock to maintain good electrical connections and in the Comet design by cooled bellows. Each capacitor for TFTR has its own vacuum pump; the Comet capacitor is permanently sealed under high vacuum.

The voltage limit is probably determined by two characteristics: the electric field attainable on the ceramic barrier and that attainable between concentric cylinders. Capacitors have been tested in RFTF for electric fields up to 120 kV/cm between cylinders. For TFTR, only 43 kV/cm is needed for 50-kV operation; 100 kV/cm is needed for Tore Supra. The limitation caused by the ceramic has not yet been tested on the Comet capacitor; however, the configuration of the ceramic has been demonstrated to work at voltage levels beyond 70 kV on the ORNL capacitor.

The longer pulse length of the Tore Supra antenna provides more heating in the capacitor than does the relatively short pulse of TFTR. The Comet capacitor has carried 750 A rms at 80 MHz in steady state, thus demonstrating the adequacy of the design. The final production models of each capacitor will be tested to prove their acceptability.

The Faraday shields for these antennas must survive in the plasma edge environment. Although Faraday shield losses were measured to be ≈0.15 Ω/m at 47 MHz, the shield is designed to handle losses of 0.5 Ω/m. Additional thermal loads on the antenna include plasma radiation and disruption loads. Average design heat fluxes amount to ≈150 W/cm^2. Both shields are therefore actively cooled. The design for Tore Supra specifies cooling water at 40 bar and 170°C; that for TFTR, 8 bar at 20°C. The tubes for Tore Supra will conform to the toroidal curvature of the plasma. The TFTR antenna has uncurved tubes, as shown in Fig. 1. High stresses in the Faraday shields, resulting from thermal loads and disruption mechanical forces, require that the shields be made of a very strong material. Inconel 600 was chosen for TFTR, and Inconel 625 was used for Tore Supra. The welds that attach the tubes to the frame in the TFTR antenna are designed to a stress level that is ~35% of the yield stress of Inconel 600. The tubes themselves are at stresses of up to ~70% of the yield stress. Thermally induced stresses are ~60% of the maximum principal stress. Because this is a secondary stress, the tubes can be operated closer to yield. Use of Inconel 625 holds all stresses below 50% of yield.

Surrounding the Faraday shield frame is a set of graphite bumper limiters that protect the antenna structure. The toroidal energy flow in the tokamak's plasma edge requires these bumper limiters to withstand heat fluxes up to 2 kW/cm^2 with a radial decay length of 1.5–2.5 cm. The shorter pulse length of the TFTR plasma permits some inertial cooling; the long pulse length of Tore Supra does not. The design for TFTR has ≈1-cm-thick graphite tiles bolted to the Faraday shield cooling manifold. Tests have shown that these tiles can withstand 150% of the total energy deposit at 7.5 times the tokamak repetition rate. The design for Tore Supra calls for brazing thin (3-mm) pieces of graphite to ganged cooling tubes.

The principles for antenna motion and disruption torque handling are the same for both antennas, but the implementations are different, because of differences in tokamak vacuum vessel geometry. Both antennas sustain forces of 124 kN at four motion bearings. In TFTR, the bearings are Inconel 718 rods displaced off center of the port. A 2-in. buttress-threaded rod is used to move the antenna relative to the plasma. This rod also supports radial disruption loads (equivalent to ≈2–3 bar of pressure) and vacuum loads. The Tore Supra design has two port-mounted guide bearings and two externally mounted rods. The external rods will be used to move the antenna. A single large bellows isolates the moving parts from the fixed parts in Tore Supra. Several smaller bellows around the capacitors and feedthroughs perform this function in TFTR.

Although all components have been or will be tested separately, the completed antennas will be tested on the RFTF before the final application. Primarily, the tests are intended to prove that the antennas can sustain full current and full voltage in the presence of gas, magnetic field, and plasma. Thermal cycling of the critical Faraday shield elements is required to ensure structural integrity. Finally, operational techniques for handling these specific antennas will be developed.

CONCLUSIONS

The 4-MW antennas for TFTR and Tore Supra are based on designs that have balanced the electrical requirements, thermal loads, disruption forces, and mechanical constraints. Many tests and analyses have been made to ensure that these antennas will work and be reliable. The design and testing have yielded important benefits: (1) development of a graphite plasma interface to minimize impurities, (2) development of tunable, matched antennas, (3) elimination of the requirement that plasma boundaries be configured around antenna needs, and (4) the confidence that testing affords.

ICH ANTENNA DEVELOPMENT ON THE ORNL RF TEST FACILITY *

W. L. Gardner, T. S. Bigelow, G. R. Haste, D. J. Hoffman, and R. L. Livesey
Oak Ridge National Laboratory, Oak Ridge, TN 37831

ABSTRACT

A compact resonant loop antenna is installed on the ORNL Radio Frequency Test Facility (RFTF).[1] Facility characteristics include a steady-state magnetic field of ≈ 0.5 T at the antenna, microwave-generated plasmas with $n_e \approx 10^{12}$ cm^{-3} and $T_e \approx 8$ eV, and 100 kW of 25-MHz rf power. The antenna is tunable from ~22-75 MHz, is designed to handle ≥1 MW of rf power, and can be moved 5 cm with respect to the port flange. Antenna characteristics reported and discussed include the effect of magnetic field on rf voltage breakdown at the capacitor, the effects of magnetic field and plasma on rf voltage breakdown between the radiating element and the Faraday shield, the effects of graphite on Faraday shield losses, and the efficiency of coupling to the plasma.

ANTENNA DESCRIPTION

A compact resonant loop antenna was made especially for component development and testing on the RFTF. A sectional view of this antenna is shown in Fig. 1. Radio Frequency power is fed to the antenna through a 50-Ω coaxial feedthrough, which serves as a low-loss, low-VSWR, vacuum-to-pressure interface. The antenna's resonant structure consists of an inductive "current strap" or radiating element and an adjustable commercial vacuum capacitor. Between the current strap and the plasma is a two-tiered Faraday shield. Two types of Faraday shield tubes were tested. One type consisted of copper plated over stainless steel tubing; the other consisted of graphite tile armor facing the plasma. Each tile is 2.54 cm long by 0.188 cm thick and is brazed to a copper-coated Inconel tube. These tiles cover nearly half of the tube surface. Lossy elements of the antenna (including the capacitor, current strap, and Faraday shield) are water cooled. By means of capacitor adjustment, the antenna is tunable from ~22-75 MHz. Water cooling should allow the antenna to couple ~1 MW to the plasma under steady-state conditions.

ANTENNA TESTING

At the time of this writing, 100 kW of 25-MHz rf power was available for studying antenna characteristics and limits. This is sufficient for observing high voltage and current behavior under no-plasma load conditions. The only loading is then produced by the Faraday shield (~15 mΩ for copper-plated tubes and 42 mΩ for graphite-tiled tubes) and the rest of the antenna circuit (30 mΩ). Maximum currents and voltages in the antenna under these conditions reach 1400 A and 35 kV, respectively, for the copper shield and 1200 A and 31 kV for the graphite shield. The antenna was initially tested to these levels for 100-ms pulses every 4 s with no magnetic field and pressures less than 10^{-6} torr. Interestingly, capacitor ratings are 650 A and 30 kV. No change was observed in voltage holdoff as a

*Research sponsored by the Office of Fusion Energy, U. S. Department of Energy, under contract DE-AC05-84OR21400 with Martin Marietta Energy Systems, Inc.

function of pressure (hydrogen) up to at least 10^{-3} torr. Variation of only magnetic field at first showed breakdown occurring at relatively low fields with improvement at high fields. Examination of the capacitor revealed arc tracks in the vicinity of the metal-to-ceramic seal. By adding a field grading or corona ring at this point, breakdown was eliminated over the full range of field values as indicated in Fig. 2. Initial voltage holdoff results were taken with a current-strap-to-Faraday-shield gap of 2.5 cm. This gap was reduced to 0.7 cm without affecting holdoff (i.e., breakdown strength across this gap is greater than 50 kV/cm). Figure 3 shows the effect of pressure on voltage breakdown at full field. Voltage holdoff decreases dramatically with pressure above $1\text{-}6 \times 10^{-4}$ torr. Fortunately, this pressure regime is nearly an order of magnitude above that expected in a tokamak under any credible circumstances.

Plasma loading as a function of rf power for various microwave input powers is indicated in Fig. 4. Generally, the plasma loading is small (~400 mΩ/m) but sufficient (~75% efficiency.) This indicates that 400 kW would be needed to test to equivalent conditions without loading. However, we did operate at a full 100 kW cw with plasma. The small loading is the result of low plasma density ($2\text{-}5 \times 10^{11}$ cm^{-3}) and low frequency (25 MHz). Loading observed on the DIII-D antenna, which is similar to this antenna, operating at 55 MHz and higher plasma densities, is significantly higher (~2 Ω).[2]

Of importance to present antenna designs (e.g., TFTR and Tore Supra), is the fact that the use of graphite as a Faraday shield material means no significant change in antenna operating characteristics. This was confirmed in all of the following combinations of scenarios: pulsed or cw operations in the presence of gas, magnetic field, or plasma. At 100 kW, even with plasma, the sustained electric field between the current strap and Faraday shield was greater than the design specifications for TFTR and Tore Supra. The only significant difference this Faraday shield makes to antenna performance is to increase the unloaded rf resistive losses by 42 mΩ. This increase in loss is considerably less than was expected based on published resistivity losses for graphite, and may be caused by the specific type of graphite used. Because losses are much less than expected, thermal stresses and heat loads will be much easier to deal with in antenna design.

CONCLUSIONS

The RFTF was used to test important principles being used in antenna design. Specifically, we have demonstrated that capacitive structures can be operated at or beyond voltage ratings in the presence of magnetic fields and plasmas. It was shown that graphite Faraday shields do not compromise antenna performance; thus, antennas can have a graphite-to-plasma interface to minimize contamination. In addition to these concepts, which are incorporated in the DIII-D, TFTR, ATF, and Tore Supra antenna designs, we have also refined structural cooling techniques, field grading structures, and capacitor attachment techniques. Finally, we are beginning to document breakdown characteristics as a function of power, magnetic field, gas, and plasma. Although we do not yet have enough power to push the limits with plasma, the currently documented limits serve as a guide to what is credible for design purposes. These factors (proven principles, demonstrated details of design, and characteristics of breakdown) are required to ensure that our ICRF antennas work.

REFERENCES

1. W. L. Gardner et al., Proc. 11th Symp. Fus. Eng. **2,** 1328 (1985).

2. M. J. Mayberry et al., Bull. Am. Phys. Soc. **31**, 1418 (1986).

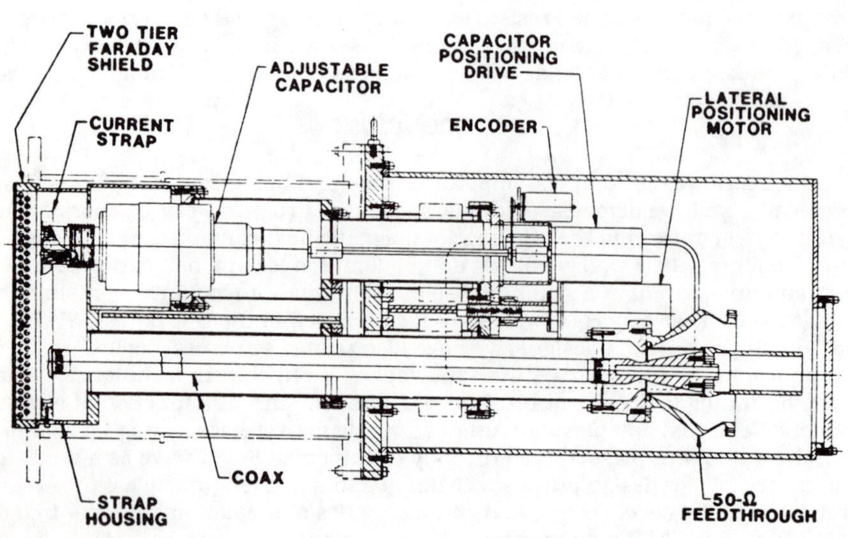

Fig. 1. Sectional view of the ORNL compact resonant loop antenna.

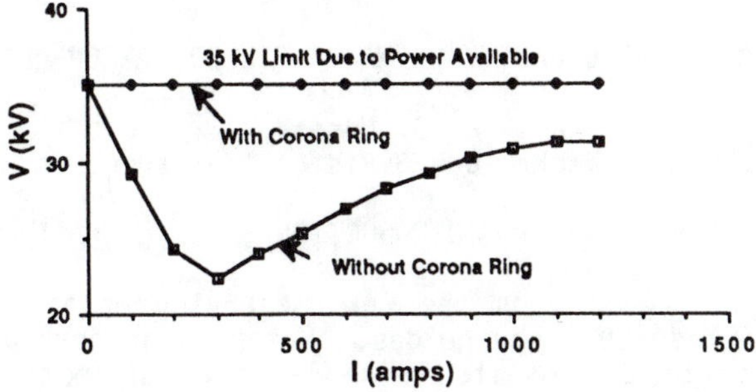

Fig. 2. Voltage standoff vs magnetic field current.

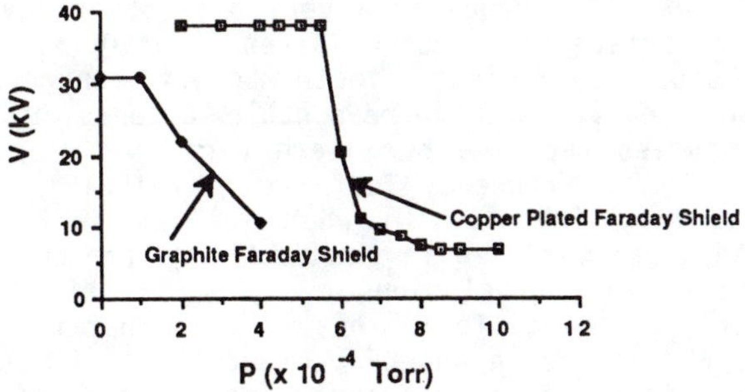

Fig. 3. RF voltage holdoff vs pressure.

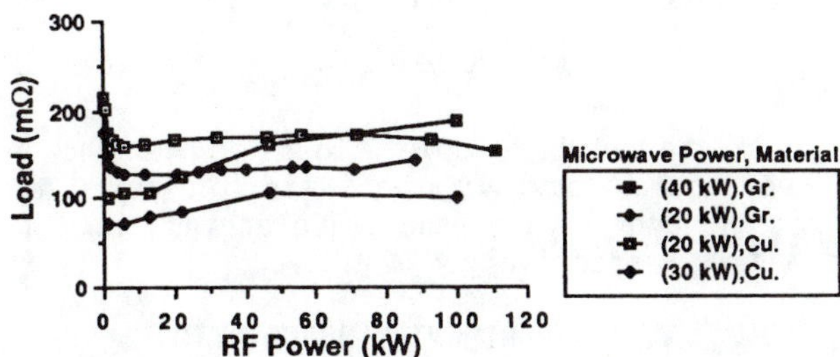

Fig. 4. Antenna load vs rf power as a function of microwave power and faraday shield material.

ICRF POWER CAPACITY OF WATER-FILLED WAVEGUIDES*

K. G. Moses
JAYCOR, Plasma Technology Division, Torrance, CA 90503

INTRODUCTION

This paper reports on the power-carrying capacity of liquid dielectric-filled (LDF) waveguides. Water is an obvious candidate liquid dielectric due to its large relative dielectric constant, high heat capacity, low cost, availability, and well-known chemistry and technology. The maximum power-carrying capacity of a transmission line is usually limited by the dielectric strength of the medium carrying the rf fields. Despite the order of magnitude reduction of waveguide dimensions, the dielectric strength of water still requires a very high power level to cause breakdown. To achieve the required stress, an LDF resonant cavity is used to store sufficient electromagnetic energy to permit testing power capacities with reasonable amounts of applied rf power. The studies reported here were performed at the Magnetic Fusion Energy, Radio Frequency Test Facility (RFTF) at Oak Ridge National Laboratory (ORNL). In what follows, we describe the techniques and apparatus used in the tests and present the results of those tests with our conclusions.

The underlying principle in this work is that the fundamental resonant mode E-M fields in an LDF rectangular cavity are directly related to the fields of waves propagating in the fundamental mode in an LDF waveguide with an equivalent cross section. The scaling relation between power P, carried by a wave propagating in a dielectric-filled waveguide, and the average power dissipated in a rectangular cavity filled with the same dielectric, resonant at the same frequency, is given by[1]

$$P = \langle P_L \rangle / (\alpha d) \quad , \tag{1}$$

where the attenuation factor α is given as[2]

$$\alpha = (\lambda_g / \lambda^2) \pi \tan \delta \quad . \tag{2}$$

Here $\langle P_L \rangle$ is the average power loss; d is the length of the resonant rectangular cavity; $\tan \delta$ is the loss tangent of the dielectric; λ_g and λ are the wavelengths in the loaded and unbounded media, respectively.

EXPERIMENTAL DESCRIPTION

The tunable rectangular cavity, (0.635 cm copper walls) used in these studies, resonated in the TE_{10} mode in a ± 10 MHz

*This work performed under U.S. DOE Contract DE-AC03-84ER52109.

frequency range centered about 80 MHz when filled with water. The 80 MHz center frequency of the cavity was selected to conform to the RFTF/ORNL 80 MHz high-power rf source. Additionally, 80 MHz is representative of the frequency range of interest for fundamental and/or second harmonic ICRF heating in fusion experiments. The rf power is coupled to the cavity through an inductive loop. A Dielectric Communications 3-1/8 in EIA coaxial reflectometer provides measurements of the incident and reflected power. The difference between the incident and reflected power at this position is equal to the average power dissipated in the cavity.

Water acts as liquid-dielectric filler and coolant. No special water treatment beyond normal laboratory procedures was used; however, to maintain low electrical conductance and prevent contamination, all conduits in contact with the water are composed of plastic or stainless steel, the only exception being the copper walls of the cavity itself. A flow meter at the input manifold monitors the volume of water/min passing through the cavity. A differential thermopile, Delta-T, model 50, measures the temperature difference between the water flowing into the cavity and the water leaving the cavity. Under equilibrium conditions, the difference between the input and output water temperatures is proportional to the power dissipated by dielectric heating of the water. A single type-K thermocouple, immersed in the output manifold, monitors the temperature of the water leaving the wave cavity. In the tests reported here, the rate at which water passes through the cavity is adjusted to maintain the water temperature in the cavity at a relatively constant level when rf power is applied to the cavity.

The dependence of ε_r' and tan δ on frequency for water[3] shows that the dielectric constant decreases as the temperature increases, but that the magnitude remains relatively constant as a function of frequency at a given temperature. The loss tangent of water exhibits a minimum in the frequency range of 10 - 100 MHz. The minimum shifts toward higher frequencies as the water temperature increases. On the other hand, the shape of the loss tangent versus frequency shows very little change as the temperature of water is varied. The frequency range used in the experiments with the cavity at the RFTF lies to the right of the minimum of the loss tangent. The log of the loss tangent exhibits a linear dependence on the log of the frequency on both sides of the minimum. Von Hippel's data shows that the tan δ data at 25.0 °C, to the right of the minimum, can be described by

$$\log(\tan\delta) = 1.06 \log f - 10.77 \quad , \tag{3}$$

where the frequency f is in Hz. We use Eq. (3) to calculate the loss tangent of water and compare the results to our experimentally measured values.

RESULTS

Water temperature in the cavity is limited to small variations by the rate of water flow to insure that the dielectric characteristics remain relatively constant during high power tests and to prevent the dimensions of the resonant (copper) cavity from changing. The rate at which the resonant frequency of the cavity shifts with variations of water temperature was measured as a function of outlet water temperature. During these tests, the position of the tuning plunger was fixed. The measurements of the resonant frequency were made by sweeping the frequency applied to the cavity under conditions of thermal equilibrium with a steady flow of water through the cavity at the rate of 6 gal/min. We found that the resonant frequency of the cavity has nearly a linear dependence on the water temperature and that the data closely fits the linear equation,

$$f_o = 0.176\,T + 76.187 \quad , \tag{4}$$

as determined by a least squares fit of the data to a straight line. Here f_o is the resonant frequency of the cavity in MHz, and T is the temperature of the outlet water in °C. The correlation factor between the data and the linear relation, Eq. (4) is 0.997, showing a very good fit. In addition, we see that a change of 1 °C changes the resonant frequency of the cavity by only ~ 176 kHz. The shift in resonant frequency with temperature is small enough to permit manual retuning of the cavity with the shorting plunger while the high power tests are in progress.

The relative dielectric constant of the low conductivity water at the RFTF was determined by measuring the wavelength in the cavity under thermal equilibrium conditions, with resonant rf frequency of 81.785 MHz applied to the LDF cavity. We found that the wavelength in the guide was 59.58 cm, giving $\varepsilon_r' = 77.3$ for the RFTF water, which compares well with von Hippel's value of 78.2 at 25.0 °C,[4] and qualitatively agrees with the trend of ε_r' with temperature.

The Q of the LDF cavity was measured in low-power tests at the RFTF using a scalar network analyzer (H-P model 8756A) driven by a sweep oscillator (H-P model 8350B). The - 3 dB bandwidth displayed by the measurement is 0.307 MHz. Our results at 81.785 MHz yield a cavity Q = 266.4 and an experimental value of the loss tangent, $\tan\delta = 0.00375$, at a water temperature of 26.3 °C. Evaluating Eq. (3), von Hippel's loss tangent data, for water at 25.0 °C, gives a value of 0.00404 at 81.785 MHz in fairly good agreement with the experimental value of loss tangent and with a proper qualitative trend with temperature.

Tests were conducted with the coupling loop immersed in the water dielectric with up to 6 kW of rf power applied to the cavity. The cavity was tuned to resonance by adjusting the shorting plunger for minimum reflected power. The reflected power from the cavity was barely detectable at its minimum value with 6 kW applied to the input.

The attenuation factor α of the LDF waveguide was found to be 4.0×10^{-4} cm^{-1} after the measured values were substituted in Eq. (2). Evaluating Eq. (1) gives

$$P = 83 \langle P_L \rangle \quad . \tag{5}$$

Thus, for 6 kW of rf power at 81.785 MHz applied to the LDF cavity, the scaling relation, Eq. (5), shows that $\lesssim$ 500 kW could be transmitted through a LDF WR1150 waveguide, terminated in its characteristic impedance. The corresponding power density is $\lesssim 1.17$ kW/cm^2. Note that this value does not represent an upper limit to the power density which could be transmitted by LDF WR1150; it is only the maximum used in the tests at RFTF/ORNL. The high power tests of the LDF cavity were terminated when the epoxy seal of the coupling loop structure began leaking. No further tests of the power-handling capacity of LDF waveguides have been conducted.

CONCLUSIONS

The results of these tests show that LDF WR1150 waveguide components, loaded with low conductivity water commonly found in fusion laboratories, can handle $\gtrsim$ 0.5 MW of rf power corresponding to density levels $\gtrsim$ 1 kW/cm^2. This achievement indicates that LDF waveguides can be competitive with vacuum or air filled transmission lines. Further, the maximum power handling capacity of LDF WR1150 lies somewhere above 1 kW/cm^2. The rf power density levels demonstrated in this test program are sufficient for application in arrays of LDF wave launchers to heat magnetically-confined plasma in fusion experiments and compact tori. Nearly an order of magnitude reduction in the size of external waveguide launchers by water loading provides significant design advantages to the reactor systems engineer. Some additional advantages of LDF launcher systems for reactor designs are: simpler shielding requirements, smaller vacuum wall penetrations, flexible rf wave launcher patterns, coupling modes, and programmable spectral content.

REFERENCES

1. K. G. Moses, "Test of the Power-carrying Capability of Water-dielectric-loaded ICH Waveguide Launchers," (to be published in Fusion Technology).
2. C. G. Montgomery, R. H. Dicke, and E. M. Purcell, "Principles of Microwave Circuits," MIT Radiation Laboratory Series, p. 36 (Boston Tech. Pub., Inc., Mass., 1964).
3. A. von Hippel, "Tables of Dielectric Materials," Report V, NDRC 14-237, Lab. for Insulation Research, MIT (1957).
4. A. von Hippel, editor, "Dielectric Materials and Applications," p. 361 (MIT Press, Cambridge, Mass., 1954).

ANALYSIS AND EXPERIMENTS FOR A WAVEGUIDE LAUNCHER IN THE ICRF FOR TOKAMAKS WITH DIVERTORS

N. T. Lam, J. L. Lee, O. C. Eldridge
and J .E. Scharer
University of Wisconsin, Madison

ABSTRACT

We present an analysis of a rectangular dielectric-filled waveguide, suitable for ICRF heating. To simulate the H-mode, we take the edge plasma to consist of a pedestal of variable length followed by a region with either a parabolic or a gaussian variation. We present numerical results for the waveguide reflection coefficient, the equivalent surface plasma impedance and the electric field profile at the aperture, for a wide range of edge plasma conditions. A dielectric-filled rectangular waveguide launcher has been designed, fabricated and tested for the ICRF wave coupling. Our theoretical analyses and measurements over the 60-130 MHz range indicate that a very high power coupling efficiency ($\geq$ 90%) can be obtained for a matched launcher with appropriate tuning of the noncontacting sliding short. An input reflection coefficient model has been developed for a matched or plasma-loaded waveguide launcher and compared with measurements.

INTRODUCTION

For coupling ICRF power into tokamak plasmas, coil antennas have succeeded in producing up to 1.5 MW/antenna and fluxes of 7 kW/cm^2. In a fusion reactor environment, a waveguide launcher may be more advantageous because of its structural rigidity, compactness and power handling capability. We present analytical and experimental studies on a dielectric-filled rectangular waveguide suitable for near-term coupling tests of ICRF heating.

PLASMA IMPEDANCE

Let x = radial direction; y = poloidal direction; z = toroidal direction. In the waveguide, the transverse fields can be written as linear combinations of TE and TM modes, i. e.

$$\vec{E}_T^w = \sum_l \vec{E}_l(y,z)\left[A_l \exp(i\beta_l x) + B_l \exp(-i\beta_l x)\right] \tag{1}$$

$$\vec{H}_T^w = \sum_l D_l^w \vec{H}_l(y,z)\left[A_l \exp(i\beta_l x) - B_l \exp(-i\beta_l x)\right] \tag{2}$$

where β_l = axial propagation constant of the l mode; $\vec{E}_l(y,z)$ and $\vec{H}_l(y,z)$ = mode transverse fields; D_l^w = mode admittance. The TE_{10} mode corresponds to $l = 1$, and it is the only mode which can propagate for the waveguide dimensions of interest.

In the slab model, the plasma fields are Fourier-analysed as, e. g.

$$\vec{E}_T^p = \frac{1}{4\pi^2} \int dk_y dk_z \vec{E}_T^p(k_y, k_z, x) \exp[i(k_y y + k_z z] \tag{3}$$

Continuities of $\vec{E}_T$ and $\vec{H}_T$ at the interface ($x = 0$) and mode orthogonality yield a formula for the reflection coefficient $\Gamma = B_1/A_1$ in terms of the surface admittance tensor $\vec{\vec{Y}}^p$ defined as

$$\vec{H}_T^p(k_y, k_z, 0) = \vec{\vec{Y}}^p \cdot \vec{E}_T^p(k_y, k_z, 0) \tag{4}$$

Assume a cold plasma. To estimate $\vec{\vec{Y}}^p$, we use the method of Bers and Theilhaber[1] specialized to the case $E_z = 0$ (this approximation is good for the plasma parameters we are considering, since the slow wave is strongly evanescent). Assuming a radiation condition at half the minor radius, the differential equations determining $\vec{\vec{Y}}^p$ have been solved numerically by a Runge-Kutta algorithm. Given the reflection coefficient Γ, one can define an equivalent surface plasma impedance by the formula $Z_p = Z_w(1+\Gamma)/(1-\Gamma)$ with Z_w = waveguide impedance for the TE_{10} mode.

In table I, we list the power reflection coefficient $|\Gamma|^2$ and plasma impedance for a reactor-like plasma with the following parameters : edge density/center density = 1% ; major (minor) radius = 480 (130) cm ; B_0 = 36 kG; 50 %-50 % D-T plasma with $n_0 = 1.5 \times 10^{14}$ cm^{-3}. To simulate the H-mode, we take the plasma density to consist of a pedestal followed by a region of gaussian variation. The waveguide has a width of 40 cm and is filled with a dielectric of $\epsilon_r = 81$. The heating is at the second harmonic of deuterium f = 55 MHz ($Z_w = 64.2\,\Omega$). Up to 10 TE and 5 TM modes have been included in the calculation.

Table I $|\Gamma|^2$ and Z_p for various waveguide heigths

	H(cm)	Power reflection coefficient	Plasma impedance (Ω)
pedestal	40.0	0.08	57.1 + j35.5
length	20.0	0.22	62.1 + j66.0
=0.0 cm	10.0	0.42	60.9 + j106.3
pedestal	40.0	0.13	69.1 + j50.7
length	20.0	0.32	62.8 + j87.2
=5.0 cm	10.0	0.54	53.0 + j125.0

Note the high reactance values. As expected, the power reflection coefficient increases as the waveguide height decreases. Also the variation in density gradient affects the reactance more significantly.

LABORATORY EXPERIMENTS

We use a 24 cm × 12 cm rectangular waveguide as a laboratory version of the launcher. Deionized water (ϵ_r = 78 in the ICRF) is chosen as the dielectric filling the waveguide. We have adopted a shorted probe excitation scheme and analyzed it theoretically in reference[2]. For a matched waveguide ($Z_p = Z_w$), we vary l_S = distance between probe and shorting plate and measure $| S_{11} |$ = magnitude of reflection coefficient at the probe for various probe radii, over a 60 - 130 MHz range. An HP8510 network analyser has been used in the measurements. Figure 1 shows the measured and theoretical $| S_{11} |$ vs. l_S for the indicated waveguide and probe parameters. The two curves agree fairly well, especially near the optimally tuned region where the power coupling efficiency can reach 90%. A small interface offset has been included in the theoretical model. Our experiments also show that deionized water gives an average Q of 185 at 100 MHz and an attenuation coefficient of .13 dB/ft, which are considered acceptable for laboratory tests.

For the unmatched waveguide coupler ($Z_p \neq Z_w$), we carry out a sensitivity analysis of S'_{11} = reflection coefficient at the probe with respect to the plasma impedance Z_p. Figure 2 shows a three- dimensional plot of $| S'_{11} |$ vs. R and X = the plasma resistance and reactance. Following the theoretical calculations, we allow R to vary from 50 to 90 Ω and X from 0 to 105 Ω. The parameter l_S has been chosen as the optimal value for $Z_p = 70 + j74\Omega$. Specifically, l_S =38.7 cm and probe radius = 2.15 cm. Notice that $| S'_{11} |$ does not vary drastically even over such a relatively wide range of Z_p. This suggests that it may be possible to design tuning schemes for the plasma loading range of interest.

ACKNOWLEDGMENTS

This work is supported by DOE grants DE-FG02-86ER52133 and DE-FG02-86ER53218. The laboratory assistance of B. Jost is also acknowledged.

REFERENCES

1. A. Bers and K. Theilhaber., Nucl. Fusion **23**, 41 (1983).

2. N. T. Lam, J. L. Lee, J. Scharer and R. J. Vernon, IEEE Trans. Plasma Science **PS-14**, 271 (1986).

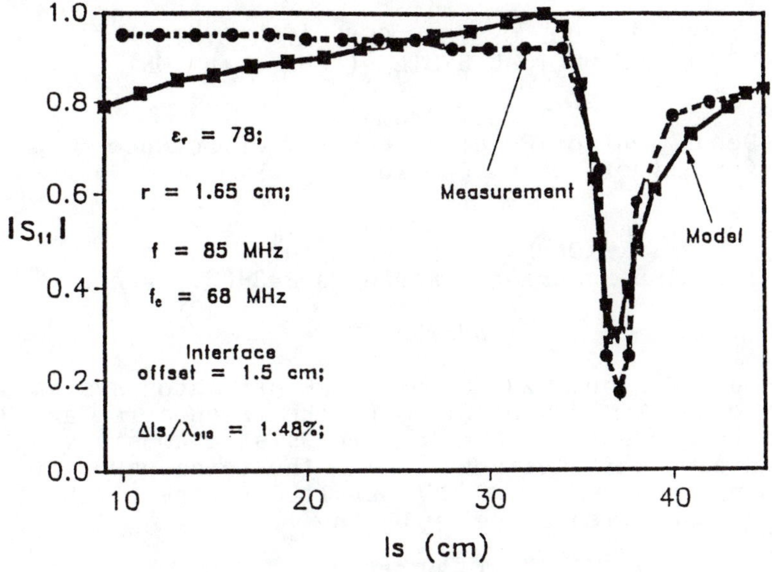

Figure 1: $| S_{11} |$ vs. l_S

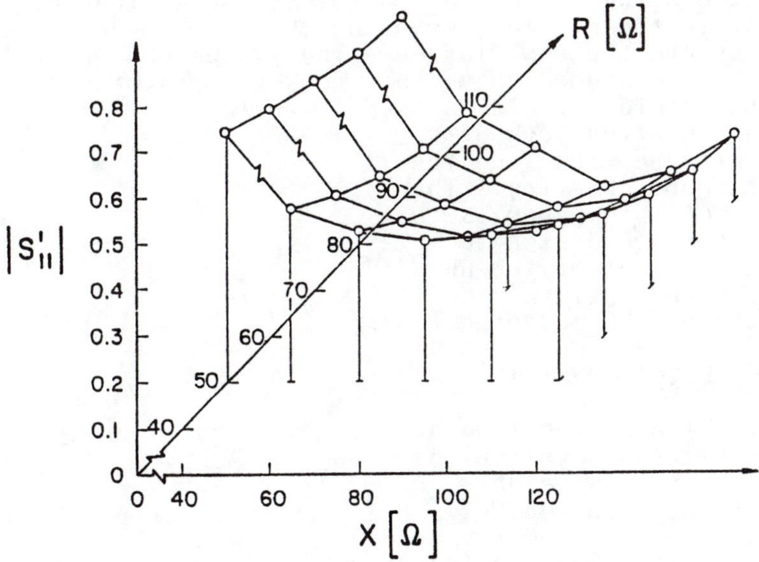

Figure 2: $| S'_{11} |$ vs. Z_p

DESIGN STUDY OF ICRF GENERATOR SYSTEM OF LARGE HELICAL SYSTEM (LHS) in TOKI

H. Toyama
Department of Physics, Faculty of Science,
University of Tokyo, Bunkyo-ku, Tokyo 113, Japan

K. Harada
Denki Kogyo Co.,Ltd.
Nakatsu-Sakuradai, Kanagawa 243-03, Japan

ABSTRACT

We have designed 20 MW, 5 sec RF generator system. The duty cycle is 1/20 minutes and the frequencies are 30 - 71 MHz in 9 pre-set channels. The system consists of 12 units and each unit delivers 2 MW. The total power of the power supply system is 60 MVA supplied by the flywheel and 900 kVA supplied directly by the line.

INTRODUCTION

The project of the Large Helical System (LHS) is now proposed to start a new National Laboratory in Toki,Japan. The electric field in the plasma is crucial in the helical system to suppress the ripple diffusion. A variety of the auxiliary heating is considered to get the desirable potential in the plasma. Even though the exact machine parameters and the auxiliary heating system are up in the air, we have designed 20 MW, 5 sec RF generator system, supposing toroidal field is 4T and the heating method is the minority heating and the second harmonic heating. The objective of this paper is to make a reference design to estimate the cost of the RF system for the optimization of the whole LHS budget.

CONCEPTUAL DESIGN

The JET has developed an ICRF system with a very high power[1]. It is straightforward now to design RF generator systems from the technical side. Frequencies of the RF generator system are determined as 30 - 71 MHz in 9 pre-set channels (30, 35, 40, 45, 51, 56, 61, 66,

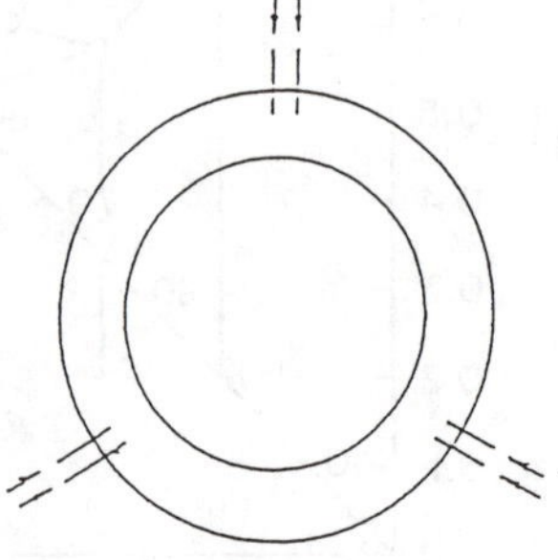

Fig. 1. Top view of LHS and rf feeders

71 MHz), supposing Hydrogen minority heating (61.1 MHz at 4 T), Deuterium second harmonic heating, and ^{3}He minority heating (40 MHz at 4 T). The RF system has 3 sets, 120 degrees apart toroidally as shown in Fig. 1.Each set consists of 2 pairs, and each pair has 2 units as shown in the poloidal cross section (Fig. 2). So, each unit is required to feed rf power of 1.7 MW to the plasma to deliver 20 MW,totally.

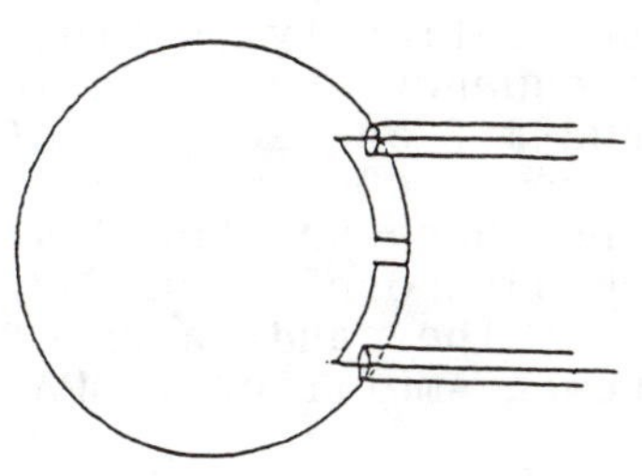

Fig. 2. Poloidal cross section and antennae

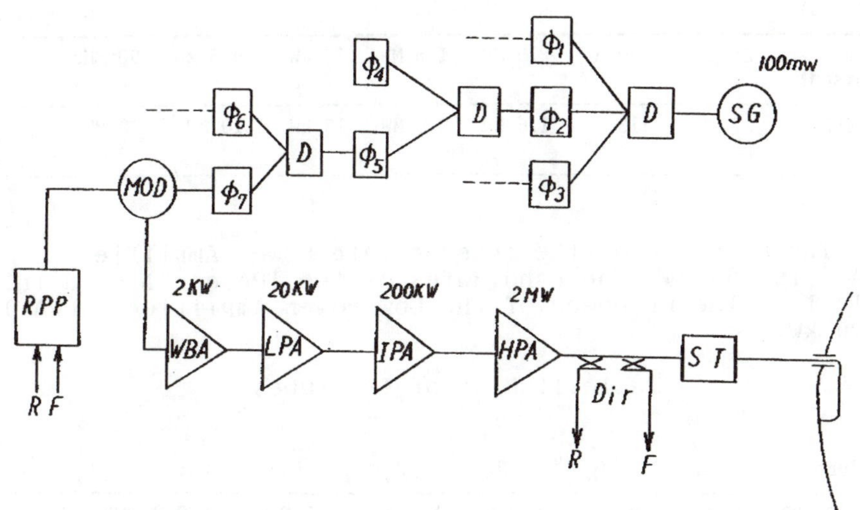

SG: Signal Generator D: Divider
Φ_i: Phase Shifter MOD: Modulator
WBA: Wide Band Amplifier LPA: Low Power Amplifier
IPA: Intermediate Power Amplifier
HPA: High Power Amplifier Dir: Directional Coupler
ST: Stub Tuner RPP: Reflection Power Protector

Fig.3 Block Diagram of an rf Generator system

Each rf generator is designed as to be able to give 2 MW. The whole system of the RF generator is shown schematically in Fig. 3. Units are operated at the same frequency and phase locked. When the phase shifter # 6 and # 7 are set out of phase, antennae are fed as "monopole". In-phase feeds, they work as "dipole", when the phase shifter # 4 and # 5 are set in-phase,and as "quadrupole" when they are set out of phase.

The candidates of the tube of the final stage, High Power Amplifier (HPA) (2 MW) are listed in Table 1.

Table I List of HPA tubes

Tube	E_b	I_b	P_o	P_p	P_{g2}	P_{g1}	f
X-2242 (Eimac)	24 kV	144 A	2.6 MW	1.4 MW	12 kW	3 kW	80 MHz
RS-2086 SK (Siemens)	24.5 kV	137 A	2.5 MW	1.3 MW	12 kW	4 kW	80 MHz
TH525 (Thomson)	24 kV	150 A	2.5 MW	1.5 MW	12 kW	4.5 kW	90 MHz
CQK-650-2 (BBC)	20 kV	113 A	1.6 MW	0.8 MW	15 kW	10 kW	80 MHz

The rf power of the Intermediate Power Amplifier (IPA) is 200 kW. The candidates of the IPA are listed in Table II. The rf power of the Low Power Amplifier (LPA) is 20 kW.

Table II List of IPA tubes

Tube	E_b	I_b	P_o	P_p	P_{g2}	P_{g1}	f
4CW150,000E (Eimac)	18 kV	16 A	200 kW	150 kW	1,750 W	500 W	108 MHz
RS-2054K (Siemens)	15 kV	17 A	180 kW	120 kW	3,000 W	1,000 W	80 MHz
TH555 (Thomson)	13 kV	22 A	200 kW	250 kW	400 W	1,500 W	80 MHz
CQK-200-3 (BBC)	15 kV	19 A	200 kW	200 kW	6,000 W	2,000 W	80 MHz

The total power required for the power supply system is given in Table III.

Table III Power supply required

(1) AC Power Source (Anode DC power are not included)

SYSTEM		ITEM	CAPACITY
Driving Amplifier System		Oscillator - modulator	1 kVA x 12 = 12 kVA
		Wide band amplifier	8 kVA x 12 = 96 kVA
		Air cooling system	1 kVA x 12 = 12 kVA
High Power Amplifier System	Low Power Amplifier	Filament power source	2 kVA x 12 = 24 kVA
		Control grid power source	0.25 x 12 = 3 kVA
		Screen grid power source	0.4 x 12 = 4.8 kVA
		Forced air cooling system	2 kVA x 12 = 24 kVA
	Intermediate Power Amplifier	Filament power source	7 kVA x 12 = 84 kVA
		Control grid power source	1 kVA x 12 = 12 kVA
		Screen grid power source	2 kVA x 12 = 24 kVA
		Forced air cooling system	2 kVA x 12 = 24 kVA
	High Power Amplifier	Filament power source	25 kVA x12 = 300 kVA
		Control grid power source	2 kVA x 12 = 24 kVA
		Screen grid power source	8 kVA x 12 = 96 kVA
		Forced air cooling system	2 kVA x 12 = 24 kVA
Transmission system		Motor control	2 kVA x 12 = 24 kVA
Control system			0.5 x 12 = 6 kVA
Linkage system			2 kVA x 12 = 24 kVA
Cooling water system		Water pump	20 kVA
		TOTAL	837.8 kVA

(2) Anode DC Power Suuply by Fly Wheel Generator

Low Power Amplifier	40 kVA x 12	480 kVA
Intermediate Power Amplifier	350 kVA x 12	4,200 kVA
High Power Amplifier	4,500 kVA x 12	54,000 kVA
	TOTAL	58,680 kVA

nearly 60 MVA

REFERENCE

[1] J. Jacquinot et al. Plasma Physics and Controlled Fusion, 28(1A), 15 (1986).

ICRF COUPLING WITH A RIDGED WAVEGUIDE ON PLT

G.J. Greene, J.R. Wilson, P.L. Colestock
C.M. Fortgang,[†] J.C. Hosea, D.Q. Hwang,[‡] and A. Nagy
Princeton Plasma Physics Laboratory
Princeton, New Jersey 08544

An ICRF ridged waveguide coupler has been installed on PLT for measurements of plasma loading. The coupler was partially filled with TiO_2 dielectric in order to sufficiently lower the cutoff frequency and utilized a tapered ridge for improved matching. Vacuum field measurements indicated a single propagating mode in the coupler and emphasized the importance of considering the fringing fields at the mouth of the waveguide. Low power experiments were carried out at 72.6 and 95.0 MHz without any external impedance matching network. Plasma loading increased rapidly as the face of the coupler approached the plasma and, at fixed position, increased with line-averaged plasma density. At the lower frequency, the reflection coefficient exhibited a minimum ($<8\%$) at a particular coupler position. At both frequencies, measurements indicated efficient power coupling to the plasma. Magnetic probe signals showed evidence of dense eigenmodes suggesting excitation of the fast wave.

I. Introduction

There is considerable interest in the investigation of alternate ICRF couplers that present potential advantages over conventional antennas. Waveguide launchers are attractive because of their simplicity, possible lack of need for a Faraday shield, and potential compatibility with reactor scenarios. The major difficulty in investigating this class of couplers in current experiments is the impractically large size necessary for a straightforward waveguide implementation. A number of methods have been considered for reducing the transverse dimensions of a transmission structure necessary to carry a waveguide mode of a given frequency [1,2]. This paper describes a coupler employing a dielectric-loaded ridged waveguide that was recently installed on the PLT tokamak in the fast-wave orientation. The experimental results reported here represent the first measurements of ICRF plasma coupling in a tokamak with a ridged waveguide.

II. Ridged Waveguide Coupler

The ridged waveguide coupler is shown in Fig. 1. In order to adequately lower the cutoff frequency, it was necessary to load the waveguide with dielectric. TiO_2 was chosen for its high dielectric constant ($\epsilon \approx$ 85-95). Single pieces of TiO_2 were available only in limited sizes, so the guide was partially filled with a slab of dielectric in the trough region where the electric field was concentrated. The width of the waveguide ridge was tapered in order to provide an impedance transformation within the guide itself. Coupling to the waveguide was accomplished with a probe that passed all the way through the dielectric and made electrical contact with the top of the ridge.

Measurements of the vacuum fields produced by the open-ended waveguide were made using a miniature, 3-axis, electrostatically shielded magnetic probe. Figure 2 shows the magnitude of the three components of the magnetic field along an axial line inside the trough of the guide. The fields are shown for two frequencies (72.6 and 95.0 MHz) and a node is apparent in the guide for the higher frequency case. The fields along two lines transverse to the guide and 2 cm in front of its face are plotted in Fig. 3 for the 72.6 MHz case. Apart from an overall scale factor, no discernible difference was found in the field structure measured outside the guide at the two frequencies. This results demonstrates that the device was, in fact, propagating a single waveguide mode. Note that B_y becomes significant in front of the dielectric slab; this represents the effect of the ridge. The fringing fields near the edges of the guide (indicated by peaks in B_x) are also appreciable and result from surface currents that flow on the outside of the structure. It is evident that these fields are of significant magnitude and should be considered in theoretical models.

III. Experimental Arrangement

The ridged waveguide was installed in PLT in a port of square cross section at the outer midplane of the vessel. The port was equipped with a bellows permitting radial movement of the coupler. A current-voltage (I-V) coupler was installed in the transmission line which fed the

---

[†] Present address: McDonnell Douglas Corp./LANL, Los Alamos, N.M.

[‡] Present address: University of California at Davis

ridged waveguide. This device provides RF signals proportional to the voltage and current on the transmission line at a given point. Phase information is preserved, permitting analysis of the complex impedance. The RF signals were conveyed to the PLT control room via wide-band analog fiber optic links and were processed there with amplitude and phase detectors and then digitized. From the complex impedance thus obtained, the series resonant loading resistance R_s was calculated. We define R_s as the real impedance which would result if the measured complex impedance were transformed along a uniform line in the direction of the antenna by a distance equal to the electrical length from the I-V coupler to the position of the resonance that occurs near the waveguide mouth.

Two magnetic probes which sampled the wave B_z were located on the bottom of the vessel: one was directly beneath the ridged waveguide and the other was located 90 degrees torodially away. In addition, a single Langmuir probe was used to investigate plasma parameter profiles in the edge region near the coupler.

IV. Experimental Results

Experiments were performed in deuterium discharges with $\overline{n}_e = 1 - 3 \times 10^{13}$ cm^{-3}, $I_p =$ 400 – 500 kA, $B_t = 26 - 32$ kG, $T_e \approx 2000$ eV, $T_i \approx 800$ eV, and $P_{rf} \approx 0.01 - 5$ W. Calculated values of the loading were averaged over a 100 msec window during which the plasma parameters were approximately constant. In general, the loading measurements were quite reproducible.

The variation of R_s with the position of the waveguide face is shown in Fig. 4 for the two frequencies investigated. The loading resistance increases monotonically as the coupler approaches the plasma, while the background (vacuum) loading remains nearly constant. The functional form of the loading for the two cases is similar, but the magnitude differs by a large factor (~ 30). For the low frequency case, the loading is large and passes through Z_0 at a particular coupler position ($r \approx 43.5 - 44.0$ cm). For this case, the complex impedance measured at the I-V coupler location is plotted on the complex ρ-plane for the various coupler positions in Fig. 5. Except for the point at $r = 47$ cm, the data fall on a relatively straight line which passes through the origin. The points can be modeled as resulting from a real load impedance, terminating a uniform line, which increases monotonically as the coupler approaches the plasma.

To emphasize the practical consequence of the large loading observed at 72.6 MHz, the magnitude of the reflection coefficient is plotted as a function of coupler position in Fig. 6. The reflection coefficient exhibits a minimum value of less than 8%.

Varying the toroidal field to scan the $2\Omega_H$ resonance layer across the mouth of the waveguide did not affect the observed loading, indicating that damping due to residual hydrogen was not significant. There was also no change in loading observed as the incident power level was varied.

A Langmuir probe scan revealed a nearly linear density profile in front of the coupler face. Moving the coupler into the vessel therefore resulted in increasing plasma density at the waveguide face. In another experiment, the waveguide loading at 72.6 MHz and at fixed position ($r = 43$ cm) increased by a factor of nearly two as the line-averaged density was raised from 1.1 to 3.2×10^{13} cm^{-3} (Fig. 7). The two observations are independent, however, because previous work in PLT has shown little correlation between line-averaged and edge electron density.

Magnetic probe signals showed eigenmode-like structure which did not vary significantly in amplitude around the vessel (Fig. 8). Dense fast wave eigenmodes would be expected in PLT at these frequencies and fields, particularly since there was no resonance layer providing strong damping in the machine.

V. Conclusion

An important conclusion of this work is the demonstration that a simple ridged waveguide structure can provide nearly reflection-free coupling in the ICRF to a tokamak plasma when the waveguide-plasma separation is adjusted appropriately. The frequency at which this situation occurs can be varied by adjusting the length of the guide. The lack of necessity for an external impedance-matching system is a significant advantage over conventional ICRF launchers.

References

[1] Perkins, F.W., in Proc. 4th Int. Symp. on Heating in Toroidal Plasmas, Rome (1984).
[2] Owens, T.L., in Proc. 6th Top. Conf. on RF Heating in Plasmas, Pine Mountain (1985).

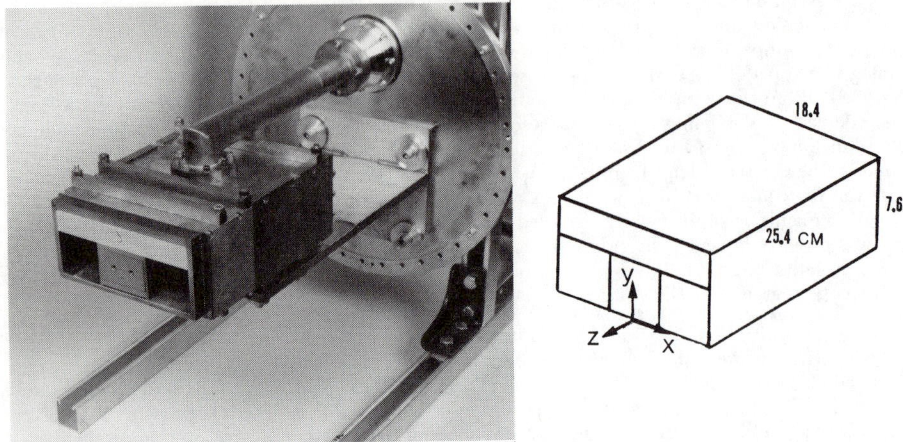

FIG. 1. Ridged waveguide coupler mounted on test stand.

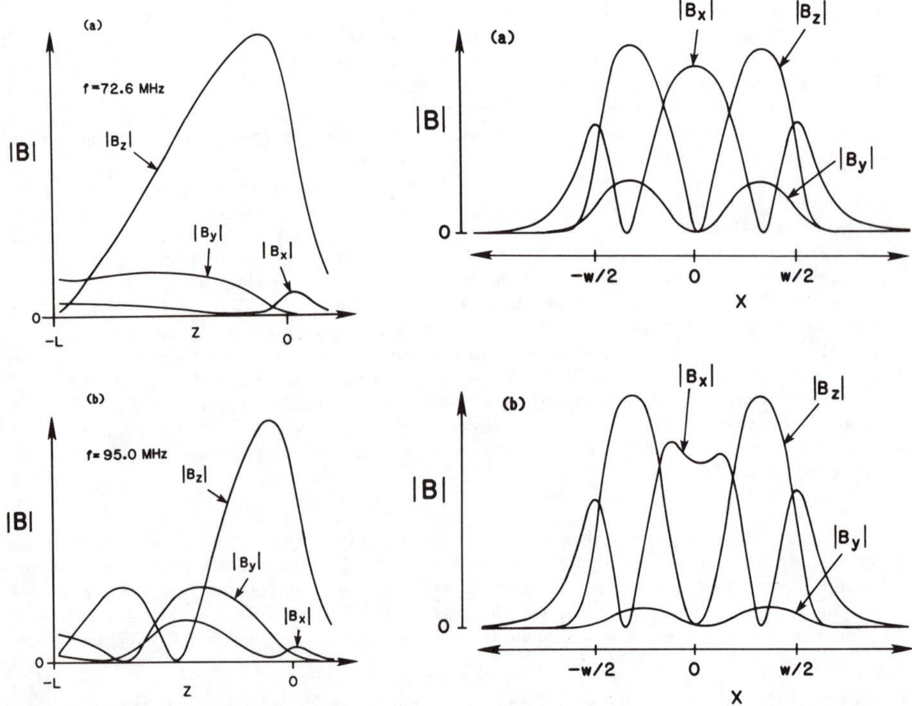

FIG. 2. Magnetic field inside waveguide for two frequencies. Probe scanned along line $x = 6.5$ cm, $y = 2.6$ cm.

FIG. 3. Magnetic field in front of waveguide. (a) Probe scanned along line $y = 6.4$ cm, $z = 2.0$ cm. (b) Probe scanned along line $y = 2.6$ cm, $z = 2.0$ cm.

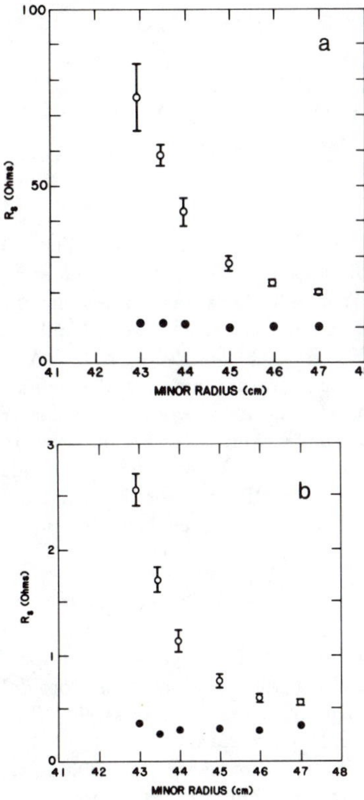

FIG. 4. Loading as a function of waveguide position. (a) f = 72.6 MHz. (b) f = 95.0 MHz.

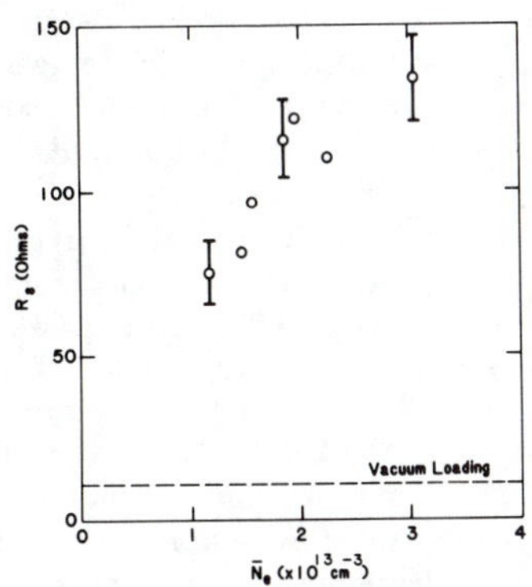

FIG. 7. Loading as a function of plasma density (f = 72.6 MHz and r = 43 cm).

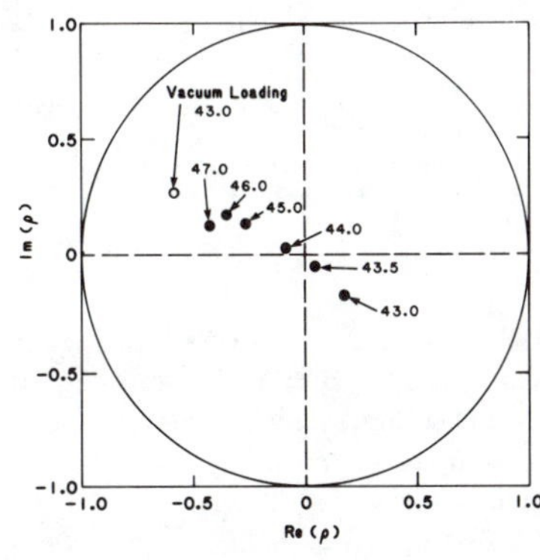

FIG. 5. Complex impedance at $I-V$ coupler location for various waveguide positions.

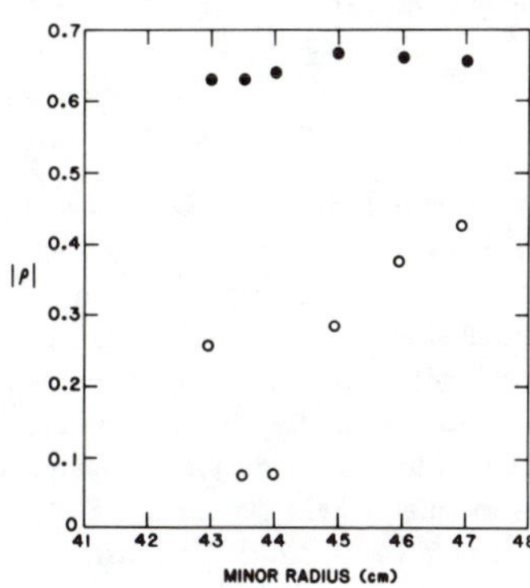

FIG. 6. Reflection coefficient as a function of waveguide position for f = 72.6 MHz.

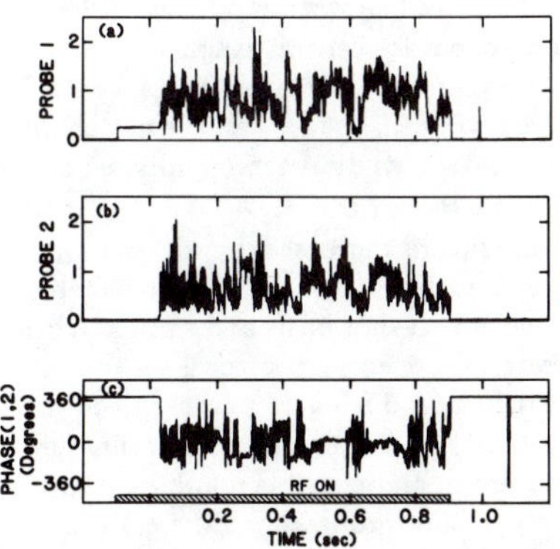

FIG. 8. Magnetic probe signals (f = 72.6 MHz).

ION BERNSTEIN WAVE HEATING EXPERIMENTS ON ALCATOR C

J. D. Moody, M. Porkolab, C. L. Fiore, F. S. McDermott,
Y. Takase, J. Terry, and S. M. Wolfe
Plasma Fusion Center, M.I.T., Cambridge, MA 02139

ABSTRACT

ICRF ion Bernstein wave heating has been studied on the Alcator C tokamak in both the hydrogen 3/2 (B = 7.6 T) and deuterium 5/2 (B = 9.3 T, $\omega = 3\omega_{cD}$ at the plasma edge, $x/a \simeq 0.87$) subharmonic regimes. Rf power of up to 180 kW at 183.6 MHz was coupled into the plasma for up to 150 ms using a toroidal current loop with a loading resistance between 1 and 2Ω. Ion heating efficiencies, $\bar{n}_e \Delta T_i / P_{\rm rf}$, of up to 5.5 eV m^{-3}/kW at $B = 9.3$ T and 2.5 eV m^{-3}/kW at $B = 7.6$ T at $\bar{n}_{e({\rm Oh})} = 0.8 \times 10^{20}$ m^{-3} were observed. A two temperature minority (deuterium) ($T_D < 2000$ eV for $n_D/n_{H+D} < 20\%$) ion distribution as well as improvements by a factor of 2–3 in global particle confinement and central impurity confinement times are observed in both cases. The energy confinement time, τ_E, increased from 5 ms to 7.7 ms with an average incremental confinement time, $\Delta W_{\rm total}/\Delta P_{\rm total}$, of 11 ms for the 9.3 T case. For the $B = 7.6$ T case, τ_E changed from 5 ms (Ohmic) to between 3 ms and 8 ms. At higher densities, $\bar{n}_e > 2 \times 10^{20}$ m^{-3}, the heating becomes inefficient.

INTRODUCTION

Ion Bernstein waves (IBW) were launched from a stainless steel, center fed, T–shaped, movable loop antenna with the center conductor aligned along the direction of the toroidal magnetic field and surrounded by a double layer, molybdenum coated Faraday shield. The outer dimensions of the antenna structure are: width 0.04 m, length 0.25 m, and height 0.04 m. The experiments were conducted under the following conditions: rf frequency $f = 183.6$ MHz; plasma minor radius (set by molybdenum limiters) $a = 0.115$ m, 0.12 m, 0.125 m; major radius $R_0 = 0.64$ m; hydrogen majority plasma with a deuterium minority $0.1\% \lesssim n_D/n_{H+D} \lesssim 20\%$; toroidal magnetic field strength $4.8 \le B_0 \le 11$ T; line-averaged electron density $0.5 \le \bar{n}_e \le 4 \times 10^{20}$ m^{-3}; $P_{\rm rf} \le 180$ kW; plasma current 160 kA $\le I_p \le$ 260 kA; and $Z_{\rm eff} \simeq 2 - 4$. Ion Bernstein wave power absorption can occur either linearly at integral ion-cyclotron harmonic layers or nonlinearly at odd half-integral harmonic layers.[1,2]

HEATING RESULTS

Figure 1 shows a summary of the central ion temperature increase of the hydrogen majority component for central magnetic fields in the range $5\,{\rm T} \le B_0 \le 11$ T. Although the ion temperature increase is greatest at $B_0 = 9.3$ T, heating occurs over a broad range of magnetic fields ($2.4 \le \omega/\omega_{cH(0)} \le 1.1$) and is not confined to a particular resonance location. The majority of the heating studies were conducted at either $B_0 \simeq 7.6$ T or $B_0 \simeq 9.3$ T.

At $B_0 = 7.6$ T, $\bar{n}_e \sim (0.8 - 1) \times 10^{20}$ m^{-3}, $n_D/n_{H+D} \sim 10\%$, $P_{\rm rf} = 100$ kW, the hydrogen temperature increased by an amount $\Delta T_H \simeq 200$ eV and the line-averaged electron density increased by $\sim 20\%$. A mass resolving charge exchange neutral analyzer was scanned radially on a shot-to-shot basis and showed a centrally peaked radial hydrogen temperature profile. The deuterium energy spectrum, on the other hand, exhibited a centrally peaked two temperature distribution during rf power injection;[3] one temperature equal to the background hydrogen (0.9 keV) and the other somewhat hotter (1.3 keV). This result indicates linear heating of deuterium and possible nonlinear heating of hydrogen. It is possible to estimate from the radial temperature profiles that the collisional power transfer from the deuterium to the hydrogen is $\lesssim 30$ kW (for $n_D/n_{H+D} \sim 10\%$); however this estimate is strongly dependent on the fraction of deuterium at the high temperature and also on the radial temperature profile and is somewhat uncertain. At $B_0 = 7.6\,\mathrm{T}$ we expect nonlinear rf power absorption at the $\omega/\omega_{cH} = 3/2$ resonance layer located approximately 3.4 cm to the high field side of the plasma axis.

The most efficient heating is observed at magnetic fields $B_0 \geq 9\,\mathrm{T}$. For example, at a field of $B_0 = 9.3\,\mathrm{T}$ the $\omega/\omega_{cD} = 5/2$ minority subharmonic resonance layer is located on the high magnetic field side of the plasma, approximately 2 cm from the plasma axis. The $\omega/\omega_{cD} = 3$ layer, however, is located approximately 2 cm in front of the antenna Faraday shield. Linear power absorption calculations show that ion Bernstein wave absorption should be complete at the plasma edge ($\omega/\omega_{cD} = 3$). In addition, CO_2 laser scattering measurements[4,5] indicate a strong attenuation in wave power across the $\omega/\omega_{cD} = 3$ layer. However, an ion heating rate of 4.1 eV/kW at $\bar{n}_e = 1 \times 10^{20}\,\mathrm{m}^{-3}$ is observed with a very low power threshold ($0 \lesssim P_{th} \lesssim 10\,\mathrm{kW}$). During rf injection, the hydrogen component exhibits a centrally peaked thermal energy spectrum and the deuterium component shows a two temperature energy spectrum. Figure 2 shows the central ion temperature increases as a function of time for both the thermal hydrogen and the superthermal deuterium at a field of $B_0 = 9.3\,\mathrm{T}$. The temporal temperature evolution indicates that the deuterium achieves its maximum temperature before the hydrogen, and both exhibit a slow decay time of about 5 ms after rf power shut-off possibly indicating central nonlinear heating of the deuterium at the $\omega/\omega_{cD} = 5/2$ layer.

A systematic study of ion heating versus density was carried out and showed a strong decrease in the plasma heating rate for plasma densities $\bar{n}_e \gtrsim 1.5\times10^{20}\,\mathrm{m}^{-3}$. As shown in Fig. 3, significant increases in the ion temperature were not seen for target densities $\bar{n}_e > 2.5\times10^{20}\,\mathrm{m}^{-3}$ and rf powers $P_{rf} \lesssim 100\,\mathrm{kW}$ ($4.8\,\mathrm{T} < B_0 < 11\,\mathrm{T}$). The increases in T_i for $\bar{n}_e > 2.5\times10^{20}\,\mathrm{m}^{-3}$ are within the experimental error of $\Delta T_i = 0$. Scattering measurements of edge density fluctuations suggest that the decrease in ion heating may partially be attributed to inaccessibility of the IBW power to the plasma center due to scattering from edge density fluctuations as the target density is increased. In addition, at high densities it was difficult to inject more than $\sim 100\,\mathrm{kW}$ of power into the plasma (the antenna power density was $P/A \sim 1\,\mathrm{kW/cm^2}$) because of the large increase in radiated power during rf heating.

PARTICLE CONFINEMENT

Improvements in global particle confinement time τ_p, of up to 3 times its value in the Ohmically heated plasma are often observed for $\bar{n}_e < 2.5 \times 10^{20}\,\mathrm{m}^{-3}$. Figure 4 shows a strong dependence of $\tau_{p(rf)}/\tau_{p(Ohmic)}$ on target plasma density with a maximum of 3.4 at $\bar{n}_e = 0.6 \times 10^{20}\,\mathrm{m}^{-3}$ and decreasing to unity for $\bar{n}_e > 2.5\times10^{20}\,\mathrm{m}^{-3}$. Improvements in τ_p are observed over a wide range of toroidal fields $4.8\,\mathrm{T} \leq B \leq 10.4\,\mathrm{T}$; however, the most significant improvement in τ_p occurs in the 9.3 T regime where the $\omega = 3\omega_{cD}$ resonance layer is located at the plasma edge. Here, the line-averaged electron density typically increased by up to 100%. Several non-rf injected discharges in which the density was increased by gas puffing alone were compared with rf discharges. These all showed decreasing particle confinement in the absence of rf wave injection.

Central impurity confinement times were also observed to increase by factors of 2–3 during IBW injection. This was measured by injecting trace amounts of Silicon using the laser blow-off technique. The brightness of He-like Si (an ionization state which exists in the center of these discharges) is observed to decay at a significantly slower rate in rf heated plasmas. Typical values of τ_{Si} corresponding to before (Ohmic), during, and after IBW injection in the 9.3 T regime are 7, 16–20, and 6 ms. The Z_{eff} is typically constant or decreasing during rf injection in these discharges.

ENERGY CONFINEMENT

Figure 5 shows the change in total stored energy, ΔW_t, as a function of the change in total input power, ΔP_t, during IBW heating at $B = 9.3\,\mathrm{T}$. The temperature and density profiles are assumed constant in both the rf and Ohmic heated portions of the discharge when calculating the total stored energy and input power. All discharges shown in the figure have initial Ohmic energy confinement times, $4.8 \leq \tau_{E(\mathrm{Oh})} \leq 6.5\,\mathrm{ms}$, and rf confinement times, $5.2 \leq \tau_{E(\mathrm{rf})} \leq 7.7\,\mathrm{ms}$. The improvement in τ_E during rf heating results from the large increase in total energy caused by both strong heating of ions and nearly doubling the plasma density. When the energy confinement time of Ohmic plus rf heated discharges, $\tau_{E(\mathrm{rf})}$ (at densities $\bar{n}_{e(\mathrm{rf})}$), is compared with the energy confinement time of Ohmic heated discharges, $\tau_{E(\mathrm{Oh})}$ (at densities $\bar{n}_{e(\mathrm{Oh})} = \bar{n}_{e(\mathrm{rf})}$) we find that $7.5\mathrm{ms} \leq \tau_{E(\mathrm{Oh})} \leq 10\,\mathrm{ms}$. A comparison of radiated power during rf and Ohmic heated discharges (at similar densities) indicates that the radiated power increase during rf injection may explain the difference in energy confinement times between the two cases. The data at $B = 7.6\,\mathrm{T}$ also shows improvements in τ_E during rf injection.

SUMMARY AND CONCLUSIONS

We have demonstrated very efficient Ion heating ($\bar{n}_e \Delta T_i(0)/P_{\mathrm{rf}} \lesssim 55\,\mathrm{ev/kW}\,10^{19}\,\mathrm{m}^{-3}$) via ion Bernstein wave injection for $P_{\mathrm{rf}} < P_{\mathrm{Oh}}$. Improvements in global particle, and central impurity confinement times (by a factor of up to three) accompany this type of rf heating. As a consequence of these results the global energy confinement time increases relative to the initial Ohmic confinement time. At higher densities the heating becomes inefficient and may result from the inaccessibility of wave power to the plasma center as well as a limit to the power density at the antenna Faraday shield.

ACKNOWLEDGEMENT

This work was supported by the U. S. Department of Energy under U. S. DOE Contract No. DE-AC02-78ET51013.

REFERENCES

1. M. Porkolab, *Phys. Rev. Lett.* **54**, 434 (1985).
2. M. Porkolab et al., *Plasma Phys. and Contr. Nucl. Fusion Res.* (IAEA 11th Int. Conf., Kyoto, 1986) IAEA-CN-47/F-II-2.
3. C. L. Fiore, *et al.*, this conference.
4. Y. Takase, *et al.*, MIT Plasma Fusion Center Report PFC/JA-86-60 (1986); to be published.
5. Y. Takase, *et al.*, this conference.

FIGURE CAPTIONS

Fig. 1. Hydrogen ion temperature increase as a function of central magnetic field. $P_{\mathrm{rf}} \leq 180\,\mathrm{kW}$.

Fig. 2. Time history of bulk hydrogen and superthermal deuterium temperatures during rf heating.

Fig. 3. Hydrogen ion heating rate, $\Delta T_i(0)/P_{\mathrm{rf}}$, as a function of line-averaged density of the rf heated discharge.

Fig. 4. Comparison of particle confinement time for Ohmic and Ohmic plus IBW heated discharges as a function of the line-averaged density of the pre-rf, Ohmically heated plasma.

Fig. 5. Increase in plasma thermal energy, ΔW_t, as a function of ΔP_{tot}, the change in total (rf plus Ohmic) input power. The dashed lines show the incremental energy confinement time $\tau_{\mathrm{inc}} \equiv \Delta W_t/\Delta P_t$.

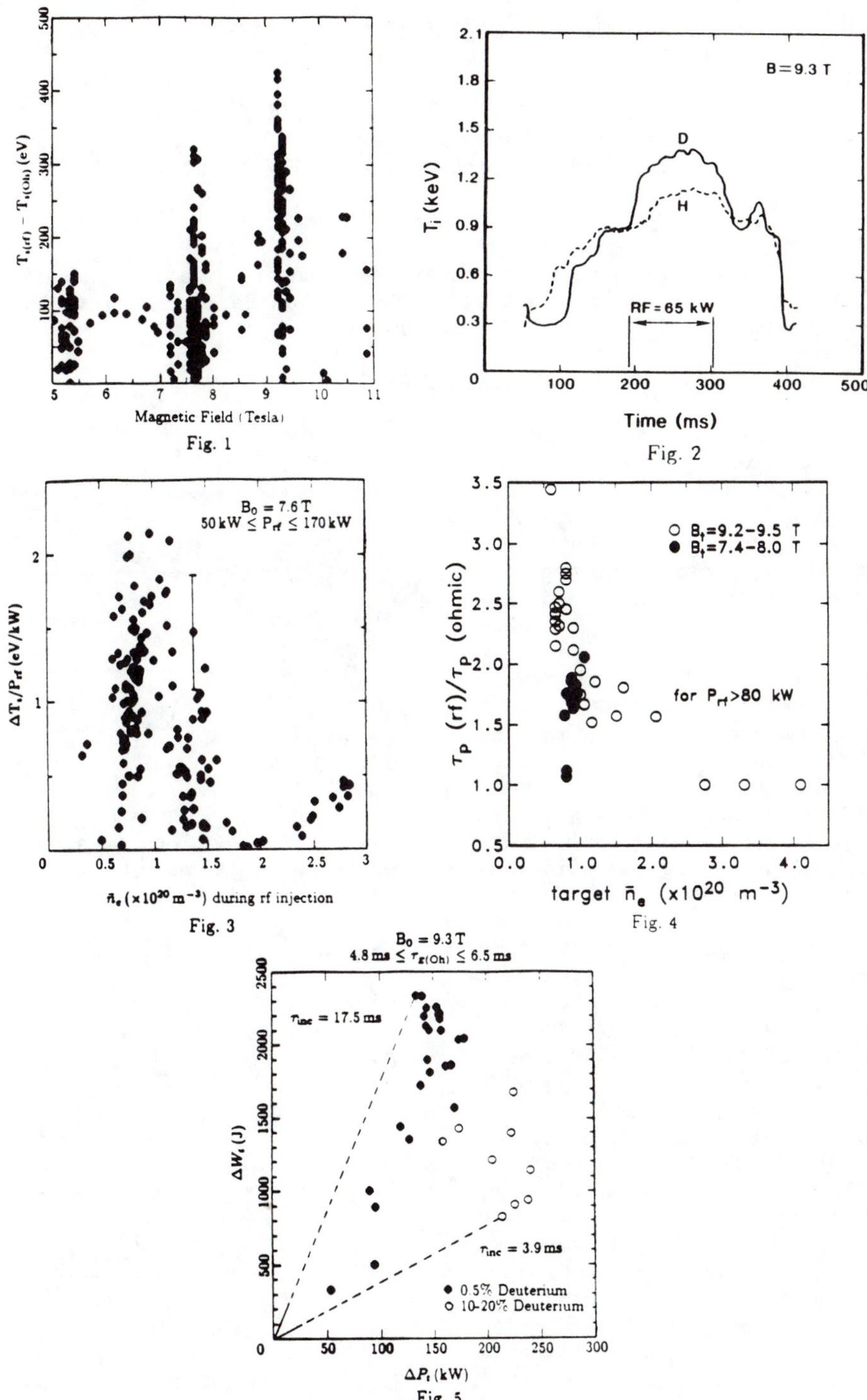

Fig. 1

Fig. 2

Fig. 3

Fig. 4

Fig. 5

EXCITATION, PROPAGATION AND HEATING FOR THE ION BERNSTEIN WAVE IN THE TNT-A TOKAMAK

S. Shinohara, O. Naito, H. Toyama and K. Miyamoto,
Department of Physics, Faculty of Science,
University of Tokyo, Tokyo 113, Japan

ABSTRACT

The parameter dependence of antenna loading, wave propagation and subharmonic resonance heating for an ion Bernstein wave have been investigated in the TNT-A tokamak.

INTRODUCTION

Recently, an ion Bernstein wave (IBW)[1] has shown attractive means of additional heating in tokamaks.[2-5] However, basic phenomena such as antenna-plasma coupling[6] and wave propagation have not yet been studied fully in tokamaks. Moreover, we have a few examples of subharmonic resonance heating both experimentally[2-5] and theoretically.[7,8] Here, studies on antenna loading, direct measurements of wave propagation and subharmonic resonance heating including new proposed heating scheme are presented.

Experimental parameters are as follows; major radius R = 40 cm, minor radius a = 8.8 cm, plasma current I_p = 4-8 kA, loop voltage V_l = 3-7 V, mean plasma density $\overline{n}_e$ = (3-10) x 10^{12} cm^{-3}, central electron and ion temperatures, $T_e(0)$ and $T_i(0)$, are in the ranges of 30-120, 15-80 eV, respectively, discharge duration time = 20-30 ms, ratio of radiated power P_r to ohmic power P_{oh} = 0.5-0.7, frequency of the generator f = 5-10 MHz, input RF power P_{inp} < 50 kW, typical RF pulse width = 4 ms (IBW antenna is shown in Fig.1).

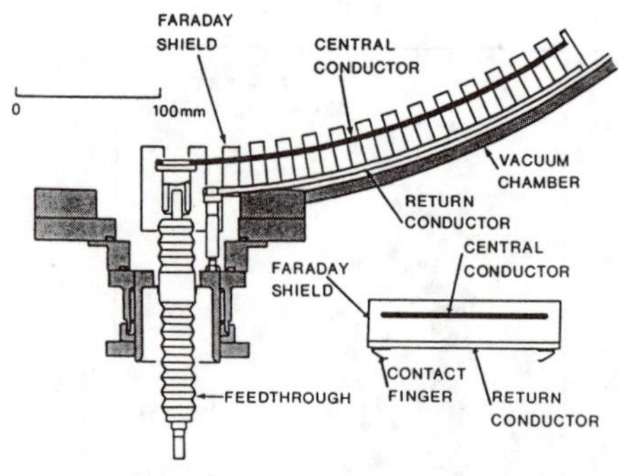

Fig.1. Cross-sectional view of ion Bernstein wave antenna.

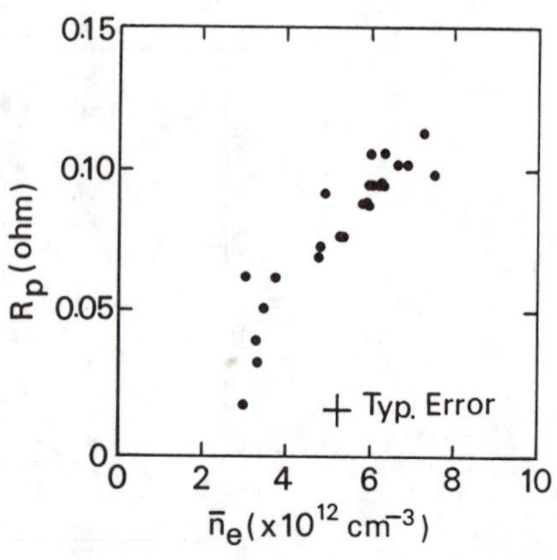

Fig.2. Relation between R_p and $\overline{n}_e$.

ANTENNA LOADING

The dependence of antenna loading resistance R_p, which reflects power absorption by the plasma, on various parameters was studied. This resistance increased with mean plasma density strongly (Fig.2) in comparison to fast magnetosonic wave case, but it was weakly dependent on the toroidal field (wave frequency was fixed) and concentration ratio (ratio of deuterium to hydrogen densities). These results are consistent with those of the small rotatable antenna,[6] but absolute value of loading was increased by a

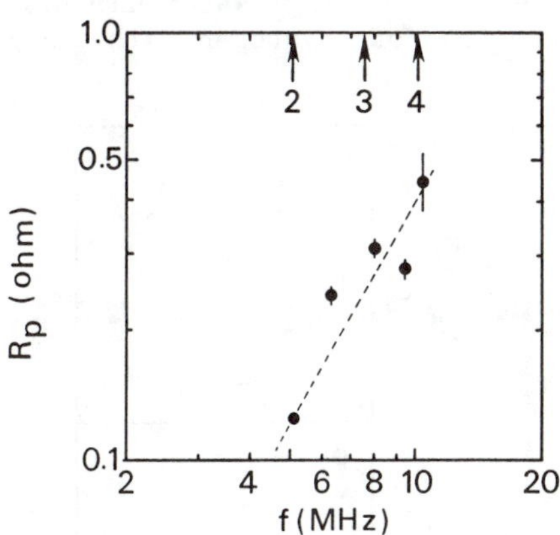

Fig.3. Dependence of R_p on frequency f.

factor of more than two. Inverse dependent on the input power was found, which was also similar with the previous case[6] and fast magnetosonic wave case[9] (plasma parameters near the plasma surface may play an important role). As shown in Fig.3, antenna resistance R_p increased with the wave frequency f, i.e., R_p was roughly proportional to $f^{1.7}$, whereas R_p was $f^{3.3}$ for the fast magnetosonic wave in in TFR.[10] Preliminary calculation shows the same dependence for IBW, but precise calculation needs to be done.

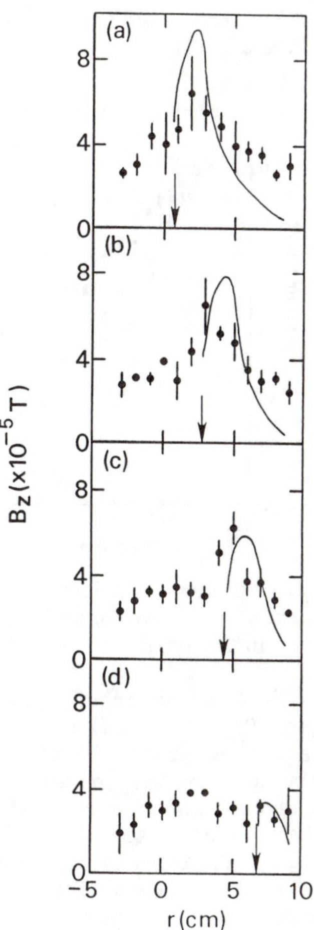

Fig.4. Radial profiles of B_z (B_t = (a)3.50, (b)3.67 (c)3.83 and (d)4.00 kG). Arrows show $\omega = 4\omega_D$ layer and solid lines are from ray tracing calculations.

WAVE PROPAGATION

Wave propagation was studied by inserting magnetic probes into the plasma. Radially movable probe was separated toroidally by 90 degree from the antenna and a reference probe was set near this antenna. The amplitude of wave magnetic field B_z (toroidal component), as shown in Fig.4, increased as the wave advanced and became maximum slightly before the $\omega = 4\omega_D$ layer. Note that B_z is nonvanishing in the limit that n_z (parallel refractive index) is zero. This peak of amplitude moves with the resonance layer as the static toroidal field B_t was varied. These results agree well with ray tracing[11] calculations.

ION HEATING

The ion heating as well as electron heating[3] was found at $\omega = 2.5\,\omega_D$. Figure 5 shows energy spectra of hydrogen with (P_{inp} = 17 kW and absorbed power by the plasma P_{net} = 6 kW) and without RF pulse. The proton heating can be explained by the power flow from deuterons, which were heated at the subharmonic resonance frequency. Heating efficiency is about twice larger than that for the fast magnetosonic wave.[12] Little increase in mean plasma density $\overline{n}_e$ and ratio of total radiated power P_r to total input power (P_{oh} + P_{net}) were found under the condition of P_{net} < 13 kW = 0.5 P_{oh}.

As for the nonlinear heating,

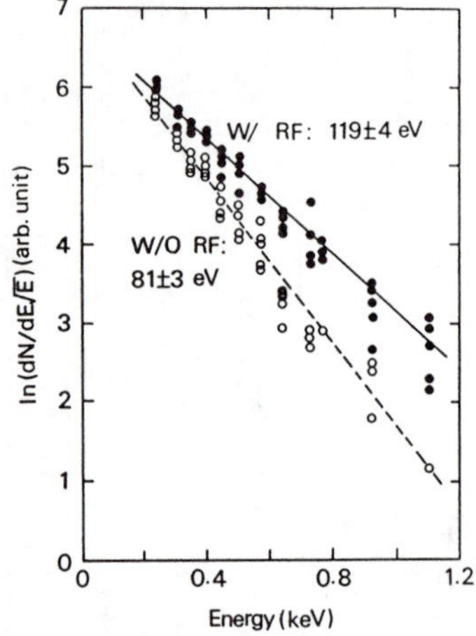

Fig.5. Charge exchange energy spectra of hydrogen with and without RF pulse at $\omega = 2.5\,\omega_D$.

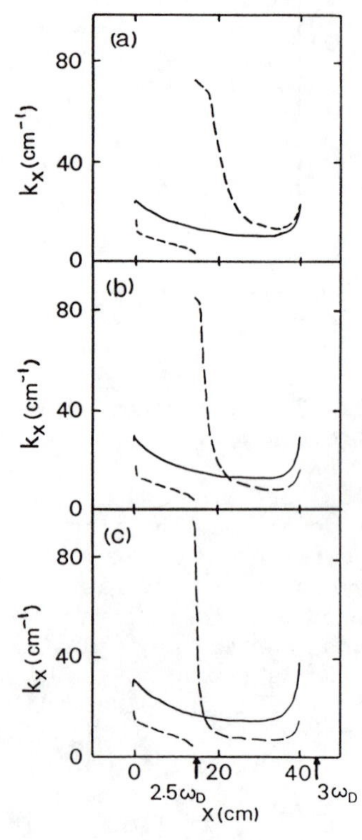

Fig.6. Radial profiles of k_x for different concentration ratio $n_H/(n_D+n_H)$ = (a)0.1, (b)0.5 and (c)0.8 in PLT. Solid and dotted lines show the wave number of launched fundamental and that of second harmonic in half, respectively.

there are a few theories to account for experiments.[7,8] Here, we propose another heating scheme; excitation of the second harmonic (SH) of the launched wave (LW) as a real wave due to the nonlinearity (self interaction), and ion damping at $2\omega = n\omega_{ci}$ (n: integer, ω_{ci}: ion cyclotron frequency) and/or electron Landau damping near the subharmonic resonance layer.

The conditions of exciting the second harmonic due to the self interaction are (1) the frequency and wave number of SH are twice as large as those of LW (matched condition), and (2) the coupling between two waves are strong. Figure 6 shows the perpendicular wave number of SH and LW for PLT parameters,[13] by the use of the ray tracing. Here, twice of parallel wave number and also twice of frequency f of LW is used for SH. Crossed points between two curves show the presence of matched condition, the place of which depends on the concentration ratio. From this crossed point, the second harmonic wave, generated as a real wave, can propagate into the plasma and damps at the harmonic resonance layer for SH. A rough estimate of the threshold power governing nonlinearity has been done, using equations in Ref.14. This power is about 1 kW in Fig.6 (a) and 10 kW in Fig.6 (b) and (c), which are smaller by more than one order of magnitude than the input power experimentally executed.

CONCLUSIONS

The basic phenomena for the ion Bernstein wave were investigated in the TNT Tokamak; the antenna loading increased with the mean plasma density and wave frequency, but weakly dependent on the toroidal field and inverse dependent on the input power were found. The wave propagation was studied by measuring wave magnetic field and we have good agreements with the ray tracing calculations. Subharmonic ion heating at $\omega = 2.5\omega_D$ was found and new heating scheme, i.e., excitation of the second harmonic as a real wave due to the self interaction and harmonic damping, was proposed.

REFERENCES

1. I. B. Bernstein, Phys. Rev. 109, 10 (1958).
2. M. Ono et al., Phys. Rev. Lett. 54, 2339 (1985).
3. S. Shinohara et al., J. Phys. Soc. Jpn. 55, 2648 (1986).
4. M. Ono et al., Plasma Physics and Controlled Nuclear Fusion Research, IAEA-CN-47/F-I-3.
5. M. Porkolab et al., ibid. IAEA-CN-47/F-II-2.
6. S. Shinohara et al., Nucl. Fusion 26, 1097 (1986).
7. H. Abe et al., Phys. Rev. Lett. 53, 1153 (1984).
8. M. Porkolab, ibid. 54, 434 (1985).
9. S. Shinohara et al., J. Phys. Soc. Jpn. 53, 1746 (1984).
10. K. Theilhaber and J. Jacquinot, Nucl. Fusion 24, 541 (1984).
11. S. Shinohara et al., Jpn. J. Appl. Phys. 26, 94 (1987).
12. Y. Ueda et al., J. Phys. Soc. Jpn. 55, 806 (1986).
13. J. Hosea et al., Proc. 12th European Conf. on Controlled Fusion and Plasma Physics 2, 120 (1985).
14. C. S. Liu and V. K. Tripathi, Phys. Rep. 130, 143 (1986).

ROLE OF ICRF-INDUCED PONDEROMOTIVE FORCE ON THE TOKAMAK EDGE PLASMA

J. R. Myra, D. A. D'Ippolito, and G. L. Francis
Science Applications International Corporation,
Plasma Research Institute, 1515 Walnut St., Boulder, CO 80302

ABSTRACT

The ponderomotive force due to applied ICRF waves can influence MHD equilibrium and stability. Conditions on $|E|^2$ for stabilization of kink and ballooning modes are summarized. The ponderomotive force is found to be particularly important in the edge plasma when strongly evanescent surface waves with a significant $E_{\parallel}$ component are produced (e.g. by ion Bernstein wave couplers). The solution of a simple model for the ICRF field penetration indicates that a stabilizing ponderomotive force layer about 1 cm in thickness is achievable for an edge density of 2×10^{12} cm^{-3} and a parallel electric field of about 250 V/cm. Order-of-magnitude estimates of the corresponding ICRF power are discussed.

INTRODUCTION

When large amplitude ICRF fields are applied to a plasma the resulting ponderomotive force $\vec{F}$ can be substantial enough to modify the MHD properties of the plasma. This has been observed experimentally in mirrors[1] and theory indicates the possibility of interesting applications in tokamaks.[2-4] The ponderomotive force merits examination both to assess its role in ICRF heating experiments and to deliberately exploit it to enhance tokamak performance.

The portion of the ponderomotive force density on a fluid element which contributes to MHD stability may be written in the form

$$\vec{F} = \frac{1}{16\pi} \sum_{\mu=L,R,\parallel} \varepsilon_\mu \vec{\nabla}|E_\mu|^2 , \tag{1}$$

where μ is the polarization index (left circular, right circular, or parallel) and ε_μ is the dielectric function

$$\varepsilon_\mu = - \sum_{j=i,e} \frac{\omega_{pj}^2}{\omega(\omega + \sigma_\mu \Omega_j)} . \tag{2}$$

Here $\omega_{pj}^2 = 4\pi N q_j^2/m_j$, $\Omega_j = q_j B/m_j c$, $\sigma_\mu = 0,\pm1$ for $\mu = \parallel,R,L$ respectively and ω is the ICRF wave frequency. Referring to Eq. (1) and noting that $\varepsilon_\parallel \gg \varepsilon_\perp \gg 1$ for ICRF, it can be seen that a highly evanescent parallel electric field at the plasma surface can generate a large local ponderomotive force. This type of wave structure is characteristic of an ion Bernstein wave coupler.

The influence of the ponderomotive force on MHD stability is characterized by a parameter[2-4]

$$\alpha_{rf} \equiv \frac{R^2q^2}{4B^2} \sum_{\mu} \left\{ \varepsilon_{\mu} \frac{d\ell nN}{dr} \frac{d|E_{\mu}|^2}{dr} + 2\left(\frac{d\ell nN}{dr}\right)^2 \tau_s \right\} \tag{3}$$

where τ_s is a "sideband term" which may be neglected when $|L_{rf}| \ll |L_N|$. Here $L_{rf} \equiv (d\ell n|E_{\mu}|^2/dr)^{-1}$, $L_N = (d\ell nN/dr)^{-1}$, R is the major radius, q the safety factor and N the density. High-n (n is the toroidal MHD mode number) ballooning modes are stable when[2]

$$\alpha_{rf} \gtrsim \alpha \tag{4a}$$

where $\alpha \equiv -Rq^2 d\beta/dr$ is the usual ballooning pressure gradient drive. For lower mode number ballooning instabilities satisfying an intermediate-n ordering[2], a stable quiescent layer at the plasma edge is achieved when

$$\alpha_{rf} \gtrsim (a/nqL_{rf})^2 \tag{4b}$$

where a is the minor radius. Finally, external kink modes are ponderomotively stabilized[3] when

$$\int dr\, \alpha_{rf} \gtrsim a/nq \,. \tag{4c}$$

Of course α_{rf} in Eq. (3) is not necessarily positive. It is the aim of the present paper to show that in a simple self-consistent model for the ICRF fields at the plasma edge, positive α_{rf}'s of the required magnitude can be achieved.

ICRF MODEL AND NUMERICAL RESULTS

As a simple model problem we consider the self-consistent solution for the ICRF-generated wave fields in the edge plasma, obtained by solving Maxwell's equations with the cold fluid Stix dielectric tensor of Eq. (2). In cylindrical plasma and antenna geometry with waves proportional to $\exp(im\theta + ik_{\parallel}z - i\omega t)$, this procedure yields a 4th order system of radial ODE's. These are solved subject to the boundary conditions of regularity at $r = 0$ and vanishing $E_{\parallel}$ and E_{θ} on a conducting wall at $r = r_w$. An ion Bernstein wave antenna (i.e. with parallel current flow) is modelled by demanding continuity of E_{θ} but a jump in $E_{\parallel}$ at the antenna radius $r_a < r_w$. In this paper we restrict discussion to cases where the specified density profile N(r) drops smoothly from its on-axis value $N(0) = 5{\times}10^{13}\ cm^{-3}$ to a wall value $N(r_w) = 5{\times}10^{10}\ cm^{-3}$, so that there is no strict vacuum region. Other parameters are $m = 5$, $k_{\parallel} = 0.1\ cm^{-1}$, $B = 10$ kG, $r_a = 20$ cm, $r_w = 30$ cm, $N(r_a) = 1.9{\times}10^{12}$ and we normalize the ICRF fields so that $E_{\parallel}(r_a) = 500$ V/cm. To calculate the MHD parameters α and α_{rf} we also specify $q(0) = 1$, $q(r_a) = 4$, $T(0) = 2$ keV, $T(r_a) = 30$ eV and $R = 160$ cm.

The numerical results verify that for $\omega/\Omega_i \sim 1$ all three components of the electric field decay exponentially into the plasma with a scale length given by $L_{rf} = (c/2\omega_{pe})(1 - k_{\parallel}^2c^2/\varepsilon_{\perp}\omega^2)^{-1/2}$ where $\varepsilon_{\perp} \equiv (\varepsilon_L + \varepsilon_R)/2$. Because the driving antenna current is parallel to the magnetic field, the waves would have primarily $E_{\parallel}$ and B_{θ} polarizations for $k_\parallel = 0$, but when $k_\parallel \neq 0$ a strong radial component of $\vec{E}$ (and hence a strong E_L and E_R) is generated by Ampere's law. When $k_\parallel L_{rf} \sim (m_e/m_i)^{1/2}$, $\varepsilon_\perp |E_\perp|^2$ is of order $\varepsilon_\parallel |E_\parallel|^2$ so all three components of $\vec{E}$ must be retained when calculating α_{rf}.

Our results indicate that the rf field is stabilizing over a broad parameter range in m, $k_\parallel$ and ω. Typically we find that $\alpha_{rf} \gg 1$ at the antenna and that it decays into the plasma, until at some radius $r = r_*$ we have $\alpha_{rf} = \alpha$. We can define a penetration length $L_p \equiv r_a - r_*$, which is the approximate width of the MHD stabilized edge layer. Figure 1 shows that $L_p \sim 1$ cm over a range of frequencies. Near $\omega/\Omega_i = 0.93$ a surface wave with substantial $E_\parallel$ propagates and drives $\alpha_{rf} < 0$. (Note that an $|E_L|^2$ which decays into the plasma is destabilizing for $\omega < \Omega_i$.) For $\omega/\Omega_i \lesssim 0.92$ the presence of the Alfven resonance ($k_\perp^2 c^2/\varepsilon_\perp \omega^2 = 1$) in the plasma is responsible for large destabilizing $E_\perp$ fields. For other frequency ranges, we find that α_{rf} is positive and dominated by the $|E_\parallel|^2$ component decaying into the plasma. Of course a warm plasma model would be expected to modify Fig. 1 in regions where additional modes appear. Note that $L_p \simeq L_{rf} \ln(\alpha_{rfa}/\alpha)$ where $\alpha_{rfa} \equiv \alpha_{rf}(r_a)$, so that halving $E_\parallel(r_a)$ in Fig. 1 to 250 V/cm only decreases L_p by 0.2 cm. For $E_\parallel(r_a) = 500$ V/cm a value of $\int dr\, \alpha_{rf}/a \sim 5$ is achieved.

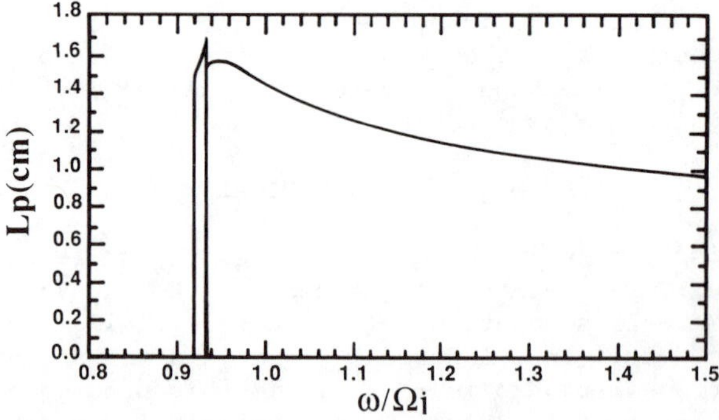

Fig. 1. Width of the MHD stabilized layer L_p, for $m = 5$, $k_\parallel = 0.1$ cm^{-1} as a function of ICRF frequency. Stabilization occurs over a broad parameter range.

DISCUSSION AND CONCLUSIONS

From these results it can be seen that the ponderomotive force generated by parallel ICRF fields of a few hundred V/cm can be effective in stabilizing the edge plasma. When the rf is nonuniform toroidally it can be shown[4] that under some restrictions (including $q(r_a) \gg 1$) the required rf fields must be increased by $(2\pi R/L_A)^{1/2}$ where L_A is the toroidal length of the near field of the antenna. Poloidally, the rf at $\theta = 0$ is what matters most for ballooning mode control.

The ICRF power corresponding to a given antenna current I_A may be estimated from $P = I_A^2 R_L$ where R_L is the loading resistance. Employing Maxwell's equations to relate I_A to $E_\parallel$ at the antenna, we obtain

$$P(MW) \sim 2.8\ 10^{-4} g^2 R_L(Ohms) E_\parallel (V/cm)^2 / B(kG)^2 \qquad (5)$$

where g is an antenna-plasma coupling factor which we estimate to be about 3 for present IBW experiments. Thus for $E_\parallel$ = 200 V/cm, R_L = 2 Ohms, and B = 10 kG, Eq. (5) gives P ~ 2 MW.

Presently, rf systems are designed to maximize the coupling of power into the plasma. For purposes of stabilization it would instead be desirable to maximize $\vec{E}$ for fixed P, i.e. design antennae with smaller g and R_L. This raises the key question of what physics determines R_L when waves are chosen that do not interact resonantly with the plasma (i.e. when deliberate heating by Landau damping or cyclotron resonance is avoided). Also, in addition to the ponderomotive effects discussed here, other nonlinear effects and equilibrium plasma modifications[4] driven by the rf merit further study.

ACKNOWLEDGEMENT

This work was supported by the U. S. Department of Energy under contract DE-AC03-76-ET53057.

REFERENCES

1. J. R. Ferron, N. Hershkowitz, R. A. Breun, S. N. Golovato, and R. Goulding, Phys. Rev. Lett. 51, 1955 (1983); Y. Yasaka and R. Itatani, Phys. Rev. Lett. 56, 2811 (1986); and refs. therein.
2. D. A. D'Ippolito, J. R. Myra, and G. L. Francis, Science Applications International Corporation Report #SAIC-86-1981/PRI-108 (1986).
3. D. A. D'Ippolito, presented at the 1987 Sherwood Controlled Fusion Theory Conference, San Diego, California, paper 3B31.
4. J. R. Myra, D. A. D'Ippolito, and G. L. Francis, presented at the 1987 Sherwood Controlled Fusion Theory Conference, San Diego, California, paper 2C29.

Boundary Effects on Minority Ion Resonance Heating

J. A. TATARONIS AND J. WANG
University of Wisconsin, Madison, WI 53706

ABSTRACT

Minority ion resonance heating in a bounded plasma column with weak spatial inhomogeneities along the plasma axis is explored. This geometrical configuration simulates plasma confined in a mirrror device. In the presence of boundaries, plasmas may support waves that have no counterparts in unbounded geometry. It is this class of rf modes that we examine for possible coupling to minority ions through the cyclotron resonance. As our basic plasma configuration, we examine a cold, uniform, magnetized, plasma cylinder, separated radially from a vacuum region by a free interface. The plasma is composed of electrons, a dominant ion species, and one or more species of heavier, low density (minority) ions. The rf wave dispersion characteristics and coupling to the heavy minority ions in this sharp boundary model are explored. To simulate a magnetic beach, which is a characteristic of a mirror magnetic field, weak axial inhomogeneities are introduced, and the implications are examined.

INTRODUCTION

An effective method to heat magnetically confined plasma is based on resonant coupling of electromagnetic waves to ions in the ion cyclotron range of frequencies (ICRF). Coupling can occur either at the fundamental cyclotron frequency of the ions via interaction with the slow magnetosonic wave, or at harmonics of the ion cyclotron frequency if ion thermal effects are sufficiently strong. In current ICRF heating schemes at the fundamental ion cyclotron frequency, the fast magnetosonic wave, which is readily excited by an external antenna, couples rf energy to the slow wave in the presence of a minority ion species. The energy in the slow wave is then dissipated at a local ion resonance. Effective plasma heating, attributed to this sequence of wave coupling and energy absorption by ions, has been observed in a varity of experiments, principally of the Tokamak type.

In this paper, we explore elements of minority ion resonance heating in mirror plasmas. An eventual goal of this study is to determine if this form of ICRF heating might be effective in the Phaedrus mirror experiment at the University of Wisconsin-Madison. In our theoretical study, we rely on the long-thin approximation of a mirror equilibrium, which means that spatial variations along the mirror axis are weak in comparison to spatial variations in the radial direction. It is assumed that the ICRF waves are excited by a localized antenna situated in a vacuum region exterior to the plasma. Because the antenna is localized, we cannot specify an axial wave number. With the frequency of the rf source fixed, it is necessary to solve a boundary value problem to determine the dispersion characteristics and resonances of ICRF waves that propagate along the mirror axis. This approach is different from that taken in modeling ICRF wave phenomena in Tokamaks, where, because of the spatial periodicity of the torus, a toroidal wave number can be identified, assuming that wave damping in the toroidal direction is sufficiently weak.

Previous studies[1,2] of minority resonance heating in mirror geometry are limited, and the modeling of the mirror in those studies may not be realistic since the treated plasma equilibrium is typically uniform and infinite in extent. The effects of plasma boundaries are simulated simply by fixing the component of the wave vector perpendicular to the magnetic field. The present treatment is more realistic since we consider a radially bounded plasma.

MAGNETIZED PLASMA CYLINDER

In this section, we explore the dispersion properties of ICRF waves along a circular cylinder of cold magnetized plasma, which is separated from a vacuum region by a free interface. The magnetic field, **B**, is directed parallel to the cylinder axis, **B**=Be_z, where e_z is a unit vector. For purposes of simplifying the mathematical analysis, we assume that the vacuum region extends radially to infinity. The plasma is composed of electrons with density N_0 and two ion species: hydrogen, with density N_h, and deuterium, with density N_d. Our interest is principally in the case where hydrogen is the dominant ion, although we also show results with hydrogen as the minority ion. It should be noted that the case of a heavy minority ion (deuterium in a hydrogen plasma) \is opposite to the typical condition in Tokamaks. Studies of plasmas with heavy minority ions, as in the present analysis, have recently appeared in the literature[3,4].

We use cylindrical coordinates, (r,θ,z), and a Fourier decomposition of the form, exp[i(ωt-mθ-kz)], for all wave amplitudes. The linearized plasma dynamics is modeled by the cold plasma dielectric tensor, which has perpendicular components ε_1, in the two diagonal positions, and $\pm i\varepsilon_2$ in the off-diagonal positions. Neglecting finite electron mass effects, and combining the components of Maxwell's equations, we find the following differential equation for E_θ, the azimuthal componet of the wave electric field,

$$r\frac{\partial}{\partial r}\left[\frac{rA^2}{A^2r^2+m^2}\frac{\partial\varphi}{\partial r}\right] + Q(r)\varphi = 0 \qquad (1)$$

$$Q(r) = -A^2 + A^2\frac{(rA)^2\gamma^2}{(rA)^2+m^2} - \frac{m}{r}\frac{\partial}{\partial r}\left[\frac{(rA)^2\gamma}{(rA)^2+m^2}\right] + 2m\frac{A^2\gamma}{(rA)^2+m^2}$$

where $\varphi(r)$ has been written in plasce of, $rE_\theta(r)$, and the following variables have been introduced: $k_0 = \omega/c$, $A^2 = k^2 - k_0^2\varepsilon_1$, and $\gamma = k_0^2\varepsilon_2/A^2$. Equation (1), which is consistent with results obtained elsewhere[5], allows radial variations in the equilibrium magnetic field and particle densities. For a uniform plasma, Eq.(1) can be solved in terms of order m Bessel Functions, $J_m(z)$. To find a dispersion relation, we invoke two conditions at the plasma-vacuum interface at r=a: continuity of E_θ and continutity of the z-component of the wave magnetic field. For a uniform equilibrium, the following relastion between ω and k results:

$$\frac{J_m'(R)}{RJ_m(R)} + \frac{m\gamma}{R^2} + \frac{K_m'(ka)}{(ka)K_m(ka)} = 0 \qquad (2)$$

$$R = a\sqrt{A^2(\gamma^2 - 1)}$$

where $K_m(ka)$ is a modified Bessel function of order m, and a prime designates differentiation with respect to the argument of the function. The third term containing the modified Bessel function represents effects due to the free plasma-vacuum interface.

Figures (1) and (2) show numerical solutions of Eq.(2) for ω in terms of k, where, as indicated previously, we assume a plasma with hydrogen and deuterium ions. In both figures, the frequency is normalized to Ω_h, the hydrogen cyclotron frequency, the wave number is normalized to the length, V_{Ah}/Ω_h, where V_{Ah} designates the Alfven speed computed with the hydrogen mass, and m=1 symmetry is assumed. Two ion concentrations are considered: $(N_d/N_h) = 0.2$, in Fig. (1), which corresponds to a heavy minority ion (deuterium), and $(N_d/N_h) = 5.0$, in Fig. (2), which corresponds to a light minority ion (hydrogen).

There are two types of modes present in the dispersion diagrams. The modes labeled **S** and $\mathbf{C_1}$-$\mathbf{C_4}$ are slow magnetosonic waves that would be present in a plasma cylinder with a fixed boundary. In an unbounded plasma, the **S mode** would also exist , with a dispersion relation given by $A^2 = 0$, where A^2 is the coefficient that appears in Eq.(1). The modes labeled **f** are fast waves that are related to the free interface separating the plasma and vacuum regions. The **f** modes, which can be classified as surface modes, are not present in a plasma cylinder with a ridged boundary. The ICRF surface waves have been examined theorectically elsewhere[6], and are readily excited in the Phaedrus mirror experiment. In addition to the surface wave and the slow wave, there is present another class of waves identified as $\mathbf{C_1}$, $\mathbf{C_2}$, $\mathbf{C_3}$ and $\mathbf{C_4}$. The curves so labeled actually represent a dense set of dispersion curves that accumulate about the labeled curves. These densely packed modes are connected with the factor A^2 that multiplies the second derivative of $\varphi(r)$ in Eq.(1). When A^2 is small, the effective wave number in the radial direction is large, implying that many radial wavelengths can fit in the cross section of the cylinder. Thus, the modes $\mathbf{C_1}$-$\mathbf{C_4}$ correspond to cylinder modes that have radial wavelengths much smaller than the plasma radius.

The important point to note in both dispersion diagrams is the coupling that occurs between the surface wave and a slow ICRF wave. Comparing the two cases presented in the figures, it is seen that the coupling occurs between the fundamental cyclotron frequencies of the ion species. If one species is a minority species, the coupling occurs close to the the cyclotron frequency of the minority ion. For the case of a deuterium plasma with a hydrogen minority , as shown in Fig.(2), there are additional fast modes, but these modes have cut-off frequencies and hence do not propagate for sufficiently low frequencies. For purposes of comparison, note that the surface waves has no low frequency cut-off. In Fig.(1), the smallest cut-off frequency of the higher order fast modes is larger than the hydrogen cyclotron frequency and hence are beyond the frequency scale of the dispersion diagram.

As a possible scenario for mode coupling of this type in a mirror plasma, let us consider the behavior of an m=1 fast surface excited in the central region of the mirror with the wave frequency slightly greater than the local cyclotron frequency of hydrogen.. We assume a hydrogren plasma with a deuterium minority so that Fig. 1 applies. As the wave propagates toward a plug of the mirror, an increasing magnetic is experienced by the wave, corresponding to displacement toward smaller values of the normalized frequency, (ω/Ω_h), along the fast branch in Fig. 1. According to Fig. (1), when a point in the mirror is attained where $(\omega/\Omega_h) \approx 0.6$, mode conversion to the slow branch occurs. The excited slow wave then propagates in the opposite direction toward a local cyclotron resonance of hydrogen where absorption of energy takes place.

In conclusion, we have illustrated mode coupling been fast surface waves and slow magnetosonic waves in a bounded plasma cylinder with multiple ion species. Although we have emphasized the mirror plasma, there are possible applications of surface wave mode coupling in Tokamaks with multiple ion species. Further analysis is clearly needed to obtain a reliable estimate of the efficiency of the mode coupling and the level of plasma heating. Kinetic effects must be included in a more detailed analysis. This work is currently underway, and the results will be reported elsewhere.

This work is supported by U.S. DOE contract No. DE-AC02-78ET51015.

REFERENCES

1. R. B. White, S. Yoshikawa and C. Oberman, Phys. Fluids 25, 384 (1982).

2. C. N. Lashmore-Davies, V. Fuchs and R. A. Cairns, Phys. Fluids 28, 1791 (1985).

3. A. V. Longinov, S. S. Pavlov and K. N. Stepanov, Proc. 13th European Conf. on Contr. Fusion and Plasma Heating (Schliersee, 1986) Part 2, p. 185.

4. A. V. Longinov and K. N. Stepanov, Proc. IV Intern. Symp. on Plasma Heating in Toroidal Systems, (Rome, 1986) Part 1, p. 488.

5. D. L. Gregov, K. N. Stepanov and J. A. Tataronis, Sov. J. Plasma Phys. 7, 411 (1981).

6. F. J. Paoloni, Phys. Fluids 18, 640 (1975).

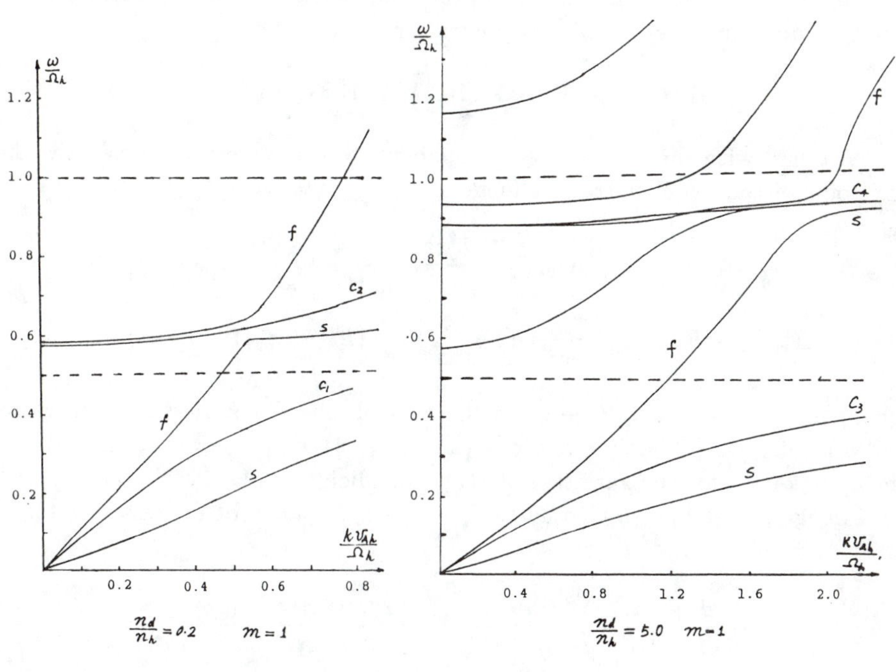

$\frac{n_d}{n_h} = 0.2 \quad m = 1$

Figure 1

$\frac{n_d}{n_h} = 5.0 \quad m = 1$

Figure 2

EXACT ORDER REDUCTION IN MODE CONVERSION USING SINGULAR INTEGRAL EQUATIONS

D. G. Swanson
Auburn University, AL 36849

P. L. Colestock and C. K. Phillips
Plasma Physics Laboratory, Princeton University, Princeton, NJ 08544

ABSTRACT

A new numerical approach to solving the tunneling equation has been found which includes all of the complexity of collisionless absorption and mode conversion but which eliminates the difficulties of solving fourth order equations. A procedure is outlined which involves solving two second order wave equations numerically, one of which is precisely the equation one would solve for the fast wave in the WKB approximation with absorption but no coupling to the slow wave. The coupling is established by constructing a Green function from the second order solutions which converts the coupled fourth order system into an integral equation which is solved by iteration.

DIAGONALIZING THE WAVE EQUATION

We begin with the two coupled components of the wave equations[1] near the harmonic of the ion cyclotron frequency

$$\begin{aligned} K_{xx}E_x + K_{xy}iE_y - \frac{k_z^2c^2}{\omega^2}E_x + \frac{d}{dx}\left[g\frac{d}{dx}(E_x + iE_y)\right] &= 0 \\ K_{xy}E_x + K_{xx}iE_y - \frac{k_z^2c^2}{\omega^2}iE_y + \frac{d}{dx}\left[g\frac{d}{dx}(E_x + iE_y)\right] + \frac{c^2}{\omega^2}iE_y'' &= 0 \end{aligned} \tag{1}$$

where $K_{xx} = S$ and $K_{xy} = -D$ are the cold plasma dielectric tensor elements and $g = -\omega_{pi}^2\rho_L^2 Z(\varsigma)/4\omega k_z v_i$ and $\varsigma = (\omega - 2\omega_{ci}(x))/k_z v_i = -\omega x/k_z v_i L$ and L is the scale length for the variation of B_0 perpendicular to $\mathbf{B}_0$.

Changing variables so that $\omega\mu x/V_A \to z - z_0$, this can be expressed in terms of $E_+ = E_x + iE_y$ and iE_y as

$$\begin{aligned} \frac{1}{w}\frac{d}{dz}\left(w\frac{dE_+}{dz}\right) + \frac{\lambda^2 z + \gamma}{F}E_+ &= \frac{2(1+\gamma)(\lambda^2 z + \gamma)}{\gamma F}iE_y \\ \frac{d^2 iE_y}{dz^2} - \gamma iE_y &= -\frac{\gamma}{2}E_+ \end{aligned} \tag{2}$$

where $w = F/(\lambda^2 z + \gamma)$ and $F = -\varsigma Z(\varsigma)$. The other normalization variable changes are found in ref. 1. This set of equations is diagonalized by choosing the matrix of coefficients so that $a_{11} = 1/w$, $a_{12} = -2(1+\gamma)/\gamma w$, $a_{21} = \gamma/2$ and $a_{22} = -\gamma$. Then the eigenvalues for the diagonalization are

$$\Lambda_{1,2} = \frac{a_{22}+a_{11}}{2} \pm \sqrt{\left(\frac{a_{22}-a_{11}}{2}\right)^2 + a_{12}a_{21}} = \frac{1-\gamma w}{2w} \pm \sqrt{\left(\frac{1+\gamma w}{2w}\right)^2 - \frac{1+\gamma}{w}}. \tag{3}$$

The roots are identical to those from the WKB analysis of this set, namely

$$k^4 - \left(\frac{\lambda^2 z+\gamma}{F} - \gamma\right)k^2 + \frac{\lambda^2 z+\gamma}{F} = 0\,.$$

The matrix which diagonalizes this system is

$$\mathsf{S} = \begin{pmatrix} 1 & 1 \\ -\frac{a_{21}}{\eta} & \frac{\eta}{a_{12}} \end{pmatrix}, \qquad \mathsf{S}^{-1} = -\frac{a_{12}}{\Delta}\begin{pmatrix} \frac{\eta}{a_{12}} & -1 \\ \frac{a_{21}}{\eta} & 1 \end{pmatrix} \tag{4}$$

where $\eta \equiv a_{22} - \Lambda_1 = \Lambda_2 - a_{11}$ and $\Delta = \Lambda_1 - \Lambda_2$. A useful identity is $\Delta\eta + \eta^2 + a_{12}a_{21} = 0$. The transformation is then defined as

$$\mathbf{E} = \begin{pmatrix} E_+ \\ iE_y \end{pmatrix} = \mathsf{S}\begin{pmatrix} \eta\Phi \\ \phi \end{pmatrix} = \mathsf{S}\Psi\,. \tag{5}$$

Representing the original equation as

$$\mathsf{D}\mathbf{E} + \mathsf{A}\mathbf{E} = 0$$

then the transformed equation is

$$\mathsf{S}^{-1}\mathsf{D}\mathsf{S}\Psi + \mathsf{S}^{-1}\mathsf{A}\mathsf{S}\Psi = \mathsf{D}'\Psi + \Lambda\Psi = 0 \tag{6}$$

where $\Lambda = \mathsf{S}^{-1}\mathsf{A}\mathsf{S}$ is diagonal and

$$\mathsf{D}' = \mathsf{S}^{-1}\begin{pmatrix} \frac{1}{w}\frac{d}{dz}w\frac{d}{dz} & 0 \\ 0 & \frac{d^2}{dz^2} \end{pmatrix}\mathsf{S}$$

with the resulting pair of equations

$$\frac{1}{\eta}\frac{d}{dz}\left(\eta\frac{d\Phi}{dz}\right) + \Lambda_1\Phi = U(z) \equiv \frac{\eta'}{\eta}\Phi' + \frac{1}{\Delta}(P\Phi - Q\phi) \tag{7}$$

$$\frac{d^2\phi}{dz^2} + \Lambda_2\phi = u(z) \equiv -\frac{w'}{w}\phi' - \left(1+\frac{\eta}{\Delta}\right)P\Phi + \frac{\eta}{\Delta}Q\phi \tag{8}$$

where P and Q are operators given by

$$P = \left(2\eta' + \frac{w'}{w}\eta\right)\frac{d}{dz} + \eta'' + \frac{w'}{w}\eta'$$

$$Q = \left(\frac{2\eta'}{\eta} + \frac{w'}{w}\right)\frac{d}{dz} + \frac{(\eta w)''}{\eta w},$$

The separation between terms on the left and right may appear arbitrary, but they are separated so that the terms on the left determine the asymptotic forms and the terms on the right are asymptotically vanishing, ensuring a bounded kernel as we construct the Green functions.

Since Δ is the difference between the roots, $\Delta \to 0$ at the turning points, and the kernel will have singular terms. With any finite damping, however, these are not on the path of integration, so they will not prevent numerical integration but may make small k_z cases numerically difficult.

USING THE GREEN FUNCTION TO FORM THE INTEGRAL EQUATIONS

Defining the homogeneous solutions of Eq. (7) to be F_1 and F_2 and the combination $F_- = F_1 - iF_2$, and the solutions of Eq. (8) to be f_1 and f_2, where the asymptotic forms of each are given by

$$\begin{gathered} Te^{-iz} \leftarrow f_1 \rightarrow e^{-iz} \\ e^{iz} + Re^{-iz} \leftarrow f_2 \rightarrow Te^{iz} \\ (-z)^{-3/4}\exp[\tfrac{2}{3}\lambda(-z)^{3/2}] \leftarrow F_1 \rightarrow z^{-3/4}\cos(\tfrac{2}{3}\lambda z^{3/2} + \tfrac{\pi}{4}) \\ \tfrac{1}{2}(-z)^{-3/4}\exp[-\tfrac{2}{3}\lambda(-z)^{3/2}] \leftarrow F_2 \rightarrow z^{-3/4}\sin(\tfrac{2}{3}\lambda z^{3/2} + \tfrac{\pi}{4}) \\ (-z)^{-3/4}\exp[\tfrac{2}{3}\lambda(-z)^{3/2}] \leftarrow F_- \rightarrow z^{-3/4}\exp(-\tfrac{2i}{3}\lambda z^{3/2}), \end{gathered} \tag{9}$$

then the Green function may be represented by

$$\begin{aligned} G(y,z) &= \frac{1}{\eta(F_2F_1' - F_1F_2')} \begin{cases} F_2(y)F_-(z) & z > y \\ F_-(y)F_2(z) & z < y \end{cases} \\ g(y,z) &= Af_k(y)f_{k'}(z) + \frac{1}{(f_1f_2' - f_2f_1')} \begin{cases} f_1(y)f_2(z) & z > y \\ f_2(y)f_1(z) & z < y \end{cases} \end{aligned} \tag{10}$$

where A is a normalizing constant and $k = 1, 2$ and $k' = 2, 1$, depending upon the boundary conditions (incident from left or right).

The integral equations are therefore given by

$$\Phi_-(y) = \frac{1}{\eta(F_2F_1' - F_1F_2')}\left[F_-(y)\int_{-\infty}^{y} F_2(z)U(z)\,dz + F_2(y)\int_{y}^{\infty} F_-(z)U(z)\,dz\right] \tag{11}$$

and

$$\phi_k(y) = f_k(y) + \frac{1}{(f_1f_2' - f_2f_1')}\left[f_2(y)\int_{-\infty}^{y} f_1(z)u(z)\,dz + f_1(y)\int_{y}^{\infty} f_2(z)u(z)\,dz\right]. \tag{12}$$

ASYMPTOTIC FORMS AND COUPLING COEFFICIENTS

Using the asymptotic forms from Eq. (9), we may establish the asymptotic forms of the solutions of the integral equations. The results are

$$\begin{gathered} f_1(1+I_{21}) \leftarrow \phi_1 \rightarrow f_1 + f_2 I_{11} \\ f_2 + f_1 I_{22} \leftarrow \phi_2 \rightarrow f_2(1+I_{12}) \\ C_- F_2 \leftarrow \Phi_j \rightarrow C_j F_- \end{gathered} \tag{13}$$

where numerically obtained integrals are given by

$$\begin{aligned} I_{ij} &= \frac{1}{2iT} \int_{-\infty}^{\infty} f_i u(z, \phi_j)\, dz \\ C_j &= \frac{1}{\lambda^3} \int_{-\infty}^{\infty} F_2 U(z, \phi_j)\, dz \end{aligned} \tag{14}$$

since $f_1 f_2' - f_2 f_1' \rightarrow 2iT$ and $\eta(F_2 F_1' - F_1 F_2') \rightarrow \lambda^3$. C_- is of no concern since it is just a constant and F_2 is exponentially small as $z \rightarrow -\infty$. From these we obtain the coupled transmission and reflection coefficients,

$$\begin{aligned} T_1 &= T(1+I_{21})\,, & R_1 &= T I_{11}\,, \\ T_2 &= T(1+I_{12})\,, & R_2 &= R + T I_{22}\,, \end{aligned} \tag{15}$$

and the conversion coefficient comes from the definition of iE_y so that

$$-\frac{2i}{\gamma} E_y = \Phi + \frac{w\eta}{1+\gamma}\phi \rightarrow \Phi - \frac{\phi}{1+\gamma}\,. \tag{16}$$

CONCLUSIONS

A scheme has been described which requires the numerical solution of two ordinary differential equations which are each individually easily handled by ordinary techniques. We have then constructed a pair of coupled integral equations which solve the original fourth order system exactly, using only second order techniques. The coupling coefficients are given in terms of integrals which have singular terms in the coefficients. It may be that these can be calculated using contour methods, since there are poles at the turning points, potentially eliminating the need for solving the integral equations numerically. From earlier numerical work[2], it believed that $I_{11} = I_{12} = I_{21} = I_{22} = 0$ and this type of analysis may lead to a formal proof of this conjecture. The way the conversion coefficient is obtained by this method is novel, and may provide more insight into the coupling problem.

REFERENCES

1. D. G. Swanson, Phys. Fluids **24**, 2035(1981).
2. D. G. Swanson, Phys. Fluids **28**, 2645(1985).

TWO-DIMENSIONAL EFFECTS ON THE ABSORPTION AND MODE CONVERSION OF FAST MAGNETOSONIC WAVES IN TOKAMAKS

C. K. Phillips, P. L. Colestock
Plasma Physics Laboratory, Princeton University
Princeton, NJ 08544

and

D. G. Swanson
Auburn University, Auburn, AL

ABSTRACT

Two-dimensional refractive effects on the absorption and mode conversion of fast waves in tokamaks have been included in the wave equation using a simplified two-dimensional model for the kinetic absorption and mode conversion layer in tokamaks. The equilibrium magnetic field is assumed to be oriented in the z-direction and to vary linearly in the x-direction, while the equilibrium density is allowed to vary in the y-direction. Using the parabolic approximation, the mode conversion-tunneling equation has been derived from the linearized Vlasov-Maxwell system of equations. In addition, the corresponding two-dimensional form of the power conservation equation will be discussed.

INTRODUCTION

Two-dimensional effects on ICRF wave propagation, absorption, and mode conversion within the kinetic layer in tokamaks are not well understood. Most of the existing models are limited to one-dimensional inhomogeneous equilibria,[1] though some effort has been made to extend the models to include two-dimensional gradients.[2-4] Recently, the parabolic approximation method was utilized to gain some insight into the diffraction effects associated with a finite-size wavefront propagating in the kinetic layer, assuming one-dimensional equilibrium gradients.[5] In this paper, the model will be extended to include two-dimensional refractive effects on the wave fields by allowing equilibrium density variations in the vertical (or $\hat{y}$) direction within the kinetic layer. The basic model and the derivation of the equivalent dielectric tensor are described in Sec. 1. A two-dimensional form of the power conservation equation is presented in Sec. 2. The parabolic approximation is used in Sec. 3 to extract the generalized form of the mode conversion-tunneling equation.

SECTION 1

During RF heating of tokamak plasmas, mode conversion and absorption processes are generally important only in a thin layer located about the cyclotron resonance layer for the waves. In the simplest two-dimensional model of the layer, the equilibrium magnetic field is assumed to be of the form $\bar{B} = B_o(1 + x/L)\hat{z}$, while the equilibrium density is allowed to vary weakly in the $\hat{y}$-direction. For notational ease, only second harmonic heating of a single ion species plasma will be considered. A Fourier decomposition of the wavefront in time and in the $\hat{z}$-direction is taken, with only a single value of $k \equiv k_z$ treated explicitly in the remaining sections. By neglecting effects related to finite electron inertia, the parallel component of the wave electric field, E_z, vanishes.

The wave equation is derived from the linearized Vlasov-Maxwell equations, given by

$$[k^2 - \omega^2 c^2]\, E_x - \frac{\partial^2 E_x}{\partial y^2} + \frac{\partial^2 E_y}{\partial x \partial y} = \frac{4\pi i\omega}{c^2} J_x \quad , \tag{1}$$

$$[k^2 - \omega^2/c^2]\, E_y - \frac{\partial^2 E_y}{\partial x^2} + \frac{\partial^2 E_x}{\partial x \partial y} = \frac{4\pi i\omega}{c^2} J_y \,, \tag{2}$$

where the perturbed plasma currents, J_i, $i = x,y$, are given in terms of the perturbed particle distribution functions and ω is the wave frequency. Following procedures used by Swanson,[6] the perturbed particle distribution function can be written in the form

$$f_1 = \exp(-i\Omega\phi) \int^{\phi} d\phi' \exp(i\Omega\phi')\Big\{\frac{v_\perp \cos\phi'}{\omega_c} \frac{\partial f_1}{\partial x} + \frac{v_\perp \sin\phi'}{\omega_c} \frac{\partial f_1}{\partial y} - \frac{2 f_o v_\perp c}{v_o^2} \Big[\frac{E_x \cos\phi' + E_y \sin\phi'}{B}\Big]\Big\}, \tag{3}$$

where $\Omega = \omega - kv_z/\omega_c$, ϕ is the gyrophase angle, $v_\perp$ is the particle's perpendicular velocity, v_o is the particle's thermal velocity, and the equilibrium distribution function, f_o, is given by

$$f_o = \frac{n(y)}{\pi^{3/2} v_o^3} \exp\{-(v_x^2 + v_y^2 + v_z^2)/v_o^2\} \,. \tag{4}$$

Effects related to equilibrium particle drifts have been neglected by assuming Eq. (4). The integral equation for f_1 may be solved to second order in $(v_\perp/\omega_c)$ using a perturbation expansion.

The results of the expansion procedure may be expressed in terms of an equivalent dielectric tensor, $\bar{K}$, defined by:

$$\bar{\bar{K}}\cdot\bar{E} = \bar{E} + \frac{4\pi i}{\omega}\bar{J} = \bar{\bar{K}}_o\cdot\bar{E} + \bar{\bar{K}}_1\cdot\frac{\partial\bar{E}}{\partial x} + \bar{\bar{K}}_2\cdot\frac{\partial^2\bar{E}}{\partial x^2} + \bar{\bar{M}}_1\cdot\frac{\partial\bar{E}}{\partial y} + \bar{\bar{M}}_2\cdot\frac{\partial^2\bar{E}}{\partial y^2} \,, \tag{5}$$

$$\bar{\bar{K}}_o = \bar{\bar{K}}_{oc} + \frac{\omega_{pi}^2 \rho_L^2 F'}{4\omega^2 X^2} \left(1 + i\,\frac{L}{n}\frac{\partial n}{\partial y}\right) \bar{\bar{A}}, \qquad \bar{\bar{A}} = \begin{bmatrix} 1 & i \\ -i & 1 \end{bmatrix} , \tag{6}$$

$$\bar{\bar{K}}_1 = \frac{\partial\bar{\bar{K}}_2}{\partial x} \,, \quad \bar{\bar{K}}_2 \; \frac{-\omega_{pi}^2 \rho_L^2 L\, F}{4\,\omega^2\, X} \bar{\bar{A}} \,, \quad \bar{\bar{M}}_1 = i\,\bar{\bar{K}}_1 \,, \quad \bar{\bar{M}}_2 = \bar{\bar{K}}_2 \,, \tag{7}$$

$F = -\zeta_{-2}\, Z(\zeta-2)$, $\zeta_{-2} = (\omega - 2\omega_c/kv_o)$, $Z(\zeta_{-2})$ is the plasma dispersion function, and $\bar{K}_{oc}$ is the cold plasma dielectric tensor. The main complication introduced by the equilibrium density gradient is that the terms are now functions of y as well as x. Using Eqs. (5-7), Eqs. (1) and (2) may be rewritten in the form:

$$\left[\gamma_1 - L_1 - L_2 - \frac{\partial^2}{\partial y^2}\right] E_x + \left[\frac{\partial^2}{\partial x \partial y} - i\left(\gamma_2 + L_1 + L_2\right)\right] E_y = 0 \;, \quad (8)$$

$$\left[\frac{\partial^2}{\partial x \partial y} + i\left(\gamma_2 + L_1 + L_2\right)\right] E_x + \left[\gamma_1 - L_1 - L_2 - \frac{\partial^2}{\partial x^2}\right] E_y = 0 \;, \quad (9)$$

where $L_1 = \omega^2/c^2\, \partial/\partial x\, K_{xx2}\, \partial/\partial x$, $L_2 = \omega^2/c^2\, [iK_{xx1}\, \partial/\partial y + \partial/\partial y\, K_{xx2}\, \partial/\partial y]$, $\gamma_1 = k^2 - \omega^2/c^2\, K_{xx0}$, $\gamma_2 = -i\omega^2/c^2\, K_{xyo}$. Two-dimensional refraction and diffraction effects have been retained to lowest order in (x/L).

SECTION 2. POWER CONSERVATION RELATION

The local form of the power conservation equation can be constructed directly from the linearized Vlasov-Maxwell equations. Real power flow within the layer is given by a generalized Poynting's theorem in the form:

$$\text{Real } \nabla\cdot\bar{S} = -\frac{1}{2}\,\text{Real}\left(\bar{E}\cdot\bar{J}^{\,*}\right) \equiv Q = -\text{Real } P(x,y) - \text{Real } \nabla\cdot\bar{T} \;, \quad (10)$$

where $\bar{S} = c/8\pi(\bar{E}\times\bar{B}^{\,*})$ is Poynting's vector. The separation of Q into the local power deposition P(x,y) and a kinetic flux $\bar{T}$ is motivated by earlier studies.[7,8] In terms of the effective dielectric tensor, $\bar{\bar{K}}$, Q is given explicitly in the form:

$$\frac{i\,16\pi}{\omega} Q = \left[\bar{E}\cdot\bar{\bar{K}}^{\,*}\cdot\bar{E}^{\,*} - \bar{E}^{\,*}\cdot\bar{\bar{K}}\cdot E\right]$$

$$= \frac{\partial}{\partial x}\left\{\left[\frac{\partial\bar{E}^*}{\partial x}\cdot\bar{\bar{K}}_2^{\,+}\cdot\bar{E} - \bar{E}^{\,*}\cdot\bar{\bar{K}}_2\cdot\frac{\partial\bar{E}}{\partial x}\right] - i\left[\frac{\partial\bar{E}^*}{\partial y}\cdot\bar{\bar{K}}_2^{\,+}\cdot\bar{E} + \bar{E}^{\,*}\cdot\bar{\bar{K}}_2\cdot\frac{\partial\bar{E}}{\partial y}\right]\right\}$$

$$+ i\frac{\partial}{\partial y}\left\{\left[\frac{\partial\bar{E}^*}{\partial x}\cdot\bar{\bar{K}}_2^{\,+}\cdot\bar{E} + \bar{E}^{\,*}\cdot\bar{\bar{K}}_2\cdot\frac{\partial\bar{E}}{\partial x}\right] - i\left[\frac{\partial\bar{E}^*}{\partial y}\cdot\bar{\bar{K}}_2^{\,+}\cdot\bar{E} - \bar{E}^{\,*}\cdot\bar{\bar{K}}_2\cdot\frac{\partial\bar{E}}{\partial y}\right]\right\}$$

$$+ \bar{E}^{\,*}\cdot\left(\bar{\bar{K}}_o^{\,+} - \bar{\bar{K}}_o\right)\cdot\bar{E} + \frac{\partial\bar{E}^*}{\partial x}\cdot\left(\bar{\bar{K}}_2^{\,+} - \bar{\bar{K}}_2\right)\cdot\frac{\partial\bar{E}}{\partial x} + \frac{\partial\bar{E}^*}{\partial y}\cdot\left[\bar{\bar{K}}_2 - \bar{\bar{K}}_2^{\,+}\right]\cdot\frac{\partial\bar{E}}{\partial y}$$

$$- i\frac{\partial\bar{E}^*}{\partial x}\cdot\left[\bar{\bar{K}}_2^{\,+} - \bar{\bar{K}}_2\right]\cdot\frac{\partial\bar{E}}{\partial y} + i\frac{\partial\bar{E}^*}{\partial y}\cdot\left[\bar{\bar{K}}_2^{\,+} - \bar{\bar{K}}_2\right]\cdot\frac{\partial\bar{E}}{\partial x} \;. \quad (11)$$

where * denotes complex conjugate and + denotes adjoint. The first two lines contain the generalized form of the kinetic flux, $\bar{T}$, while the remaining terms yield the two-dimensional form of the localized power deposition. In the limit that the $\hat{y}$-dependence of the electric field depends on only a single value of $k_y \sim 1/E\ \partial E/\partial y$, Eq. (11) reduces in form to that derived earlier by Colestock and Kashuba.[7]

SECTION 3. 2-D MODE CONVERSION-TUNNELING EQUATION

Within the kinetic layer, the vertical ($\hat{y}$) dimension of the incident wavefront is limited due to focussing induced by the launcher geometry and the refractive properties of the equilibrium. For wavefronts which vary slowly in the $\hat{y}$-direction relative to the $\hat{x}$-direction, the paraxial propagation constraint, in which $k_y \sim 1/E\ \partial E/\partial y << k_x \sim 1/E\ \partial E/\partial x$, may be used to solve Eqs. (8) and (9) iteratively for E_x and E_y. This procedure effectively decouples the two equations, leading to an algebraic equation for E_x and the 2-D mode conversion-tunneling equation for E_y, given below:

$$\frac{1}{\gamma_1} L_1 \frac{\partial^2 E_y}{\partial x^2} - 2\left(\frac{\gamma_1+\gamma_2}{\gamma_1}\right) L_1 E_y - \left(\frac{\gamma_1^2-\gamma_2^2}{\gamma_1}\right) E_y - \frac{\partial^2 E_y}{\partial x^2} + i\left(\frac{\partial}{\partial x}\frac{\gamma_2}{\gamma_1}\right)\frac{\partial E_y}{\partial y} - \frac{\partial^2 E_y}{\partial y^2}$$
$$+ i\, a_1(\gamma_1 + \gamma_2)\frac{\partial^3 E_y}{\partial x^3} - i\, a_2\,(\gamma_1 + \gamma_2)\frac{\partial E_y}{\partial x} + a_3 \frac{\partial E_y}{\partial y} = 0, \quad (12)$$

where a_1, a_2, and a_3 depend explicitly on the $\hat{y}$-gradients in the equilibrium. When equilibrium $\hat{y}$-gradients are neglected, Eq. (12) reduces to the tunneling equation given in Ref. [5], while if all $\hat{y}$-derivative terms are neglected, the results of Ref. [6] are recovered. As can be seen from Eq. (12), the primary complication introduced by vertical density gradients is coefficients which depend on both x and y; the order of the tunneling equation is not increased. Solution methods for Eq. (12) will be discussed in future reports.

ACKNOWLEDGMENT

This work was supported by the U.S. Department of Energy Contract No. DE-AC-02-76-CHO-3073.

REFERENCES

1. D.G. Swanson, Phys. Fluids 28, 2645 (1985).
2. D.J.Gambier and A. Samain, Nucl. Fusion 25, 283 (1985).
3. A. Fukuyama, K. Itoh, and S.-I. Itoh, Comput. Phys. Rep. 4, 137 (1986).
4. D.N. Smithe, P.L. Colestock, R.J. Kashuba, and T. Kammash, Princeton Plasma Physics Laboratory Report, PPPL-2400 (April 1987), to be published in Nucl. Fusion.
5. D.G. Swanson, S. Cho, C.K. Phillips, D.Q. Hwang, W. Houlberg, L. Hively, IAEA-CN-47/F-IV-4, Kyoto, Japan, November 1986.
6. D.G. Swanson, Phys. Fluids 24, 2035 (1981).
7. P.L. Colestock and R.J. Kashuba, Nucl. Fusion 23, 763 (1983).
8. B.D. McVey, R.S. Sund, and J.E. Scharer, Phys. Rev. Letts. 55,507 (1985).

KINETIC TREATMENT OF PARALLEL GRADIENT NONUNIFORMITY FOR ICRF HEATING - A FOURIER INTEGRAL APPROACH

D. Smithe, T. Kammash
University of Michigan, Ann Arbor, MI 48109

P. Colestock
Princeton Plasma Physics Laboratory, Princeton, NJ 08544

ABSTRACT

We study the generally nonlocal Vlasov-Maxwell wave propagation and absorption problem for an arbitrarily nonuniform plasma. The Fourier transform of the nonlocal dielectric response kernel, **K**(r,k), is constructed by integration along particle orbits in the nonuniform field. Although a finite Larmor radius expansion of the transverse particle motion still applies, the phase integrals which comprise the usual plasma dispersion function are altered, containing an additional parameter characterizing the parallel field gradient. The use of realistic phase decorrelation estimates over a single bounce orbit leads to a reduction of the phase integrals to a tractible form. We numerically solve a 1-D sheared field version of the resultant integral equation describing the mode conversion physics. Significant changes are found for small $k_{\|}$ values. In addition, local absorption in the resonance zone appears to be stratified in conjunction with the rf-particle phase correlation which occurs for particles passing through the localized resonance.

INTRODUCTION

This paper addresses two difficulties arising in the treatment of wave propagation in a nonuniform plasma possessing both parallel and perpendicular field gradients. The first, exact treatment of particle orbits in the lineraized Vlasov theory, is needed for an accurate description of the particle-wave phase integral near the ion cyclotron resonance and its harmonics. Our work furthers an endpoint expansion method previously employed by Itoh. *et al.*[1]; we analytically reduce the velocity-integrated phase to a single simple integral. In doing so it is found that collisional phase-diffusion plays an important role in preserving the analyticity of velocity-integrated phase in the uniform limit. We have adopted the methods of Kerbel and McCoy[2] and Cohen *et al.*[3] to estimate phase damping for uniform or weakly nonuniform conditions.

Second, we treat questions concerning formulation and solution of the wave equation, especially in what concerns the correct implemtation of the nonlocal plasma response. For a fixed frequcney oscillation, the linear response has a general form $\underline{J}_p(\underline{r})=\int d^3r\ e^{i\underline{k}\cdot\underline{r}}\ \boldsymbol{\sigma}(\underline{r},\underline{k})\cdot\underline{E}(\underline{k})$, where $\underline{J}_p$ is the plasma currnt, and $\underline{E}$ is the electric field in the Fourier domain. When there is shear the parallel wavenumber, $k_{\parallel}$, which appears in the phase integral, is no longer ignorable with respect to the nonlocal integration. To produce a differential formulation, one is forced to approximate the phase integral with a Taylor expansion about some fixed value of $k_{\parallel}$.[4] However, even for small values of shear, this expansion is inappropriate when the spectrum of $\underline{E}(\underline{k})$ is fairly wide, such as for rapid variation of the Bernstein wave's wavelength, and when the phase integral has significant structure. Both conditions occur when there is a parallel gradient.

We avoid this second difficulty entirely by solving the nonlocal integral equation directly. The wave equation is formulated as

$$\int d^3r\ e^{i\underline{k}\cdot\underline{r}}\ \mathbf{D}(\underline{r},\underline{k})\cdot\underline{E}(\underline{k}) = \underline{J}_{ant}(\underline{r}) \tag{1}$$

where $\mathbf{D}(\underline{r},\underline{k})$ is the dispersion kernel, which in form resembles the dispersion tensor of uniform plasma theory. A nonsparse matrix equation is produced which is the discrete version of integral equation (1), and we invert directly for the quanities $\underline{E}(\underline{k})$.

PHASE INTEGRAL

The most important effect of the parallel gradient on the conductivity kernel, $\boldsymbol{\sigma}(\underline{r},\underline{k})$, is the alteration of the phase integral which arises from integration along the particle orbits in the nonuniform field. This integral is:

$$\int_{-\infty}^{\infty} dv_{\parallel} \int_0^{\infty} d\tau\ \exp\left(i \int_0^{\tau} d\tau'\,(\omega - n\Omega' - k_{\parallel}v_{\parallel}')\right)\ \exp(-v_{\parallel}^2/v_{Th}^2)$$

where primed quantities follow the unperturbed orbits. The values of $v_{\perp}$ and $v_{\parallel}$ change slowly, and are expanded around the endpoint, $\tau=0$, in a Taylor series, to give "quasilocal" particle orbits which depend on the local parallel gradient through the single parameter $L_{\parallel}^{-1} = \nabla_{\parallel}B/B$. The $v_{\parallel}$ integral can be done analytically, leaving just the integral over τ, which replaces the standard plasma dispersion function, $Z((\omega-n\Omega)/|k_{\parallel}|v_{Th})$. After inluding collisional phase-damping, we find its replacement to be:

$$Z(\zeta,\alpha;\gamma) \equiv i\int_0^{\infty} dx\ \exp(-x^2(1-\alpha x/2)^2/4 + i\zeta x - \gamma x^3/8),$$

where $\zeta=(\omega-n\Omega)/|k_{\parallel}|v_{Th}$, and $\alpha=n\Omega/k_{\parallel}|k_{\parallel}|L_{\parallel}v_{Th}$; and $\gamma=\nu/|k_{\parallel}|v_{Th}$ is the phase diffusion parameter, where ν is the ion-ion deflection frequency.

For comparison we note that the new parameter, α, measuring the parallel gradient, strongly resembles the quantity $(k_{\parallel}\lambda_c)^{-2}$, which Gambier and Samain[5] use to gauge parallel gradient effects in their nonlocal variational treatment. We also note analytical agreement in all limits with Itoh *et al.*'s phase integral.[1]

Plots of $Z(\zeta,\alpha;\gamma)$ versus ζ for a large negative and a small positive value of α are shown in Figure 1. In the $k_{\parallel}\to 0$ limit (large α) there is damping, with $Z \propto i/\sqrt{|\alpha|}$, in contrast to the uniform plasma result. For small positive α, there appears a modulation of integrated phase due to systematic particle-rf phase accumulation along the field gradient. The average value of this modulation is the uniform plasma value. As $\alpha \to 0^+$, the modulation becomes finer, and is reduced in magnitude by collisional phase-diffusion.

a)

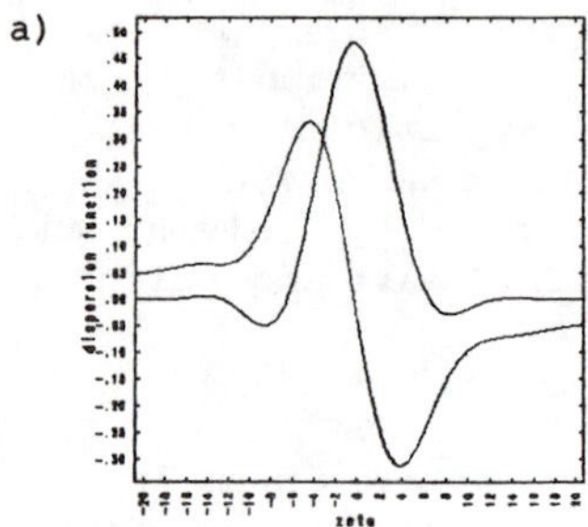

b)

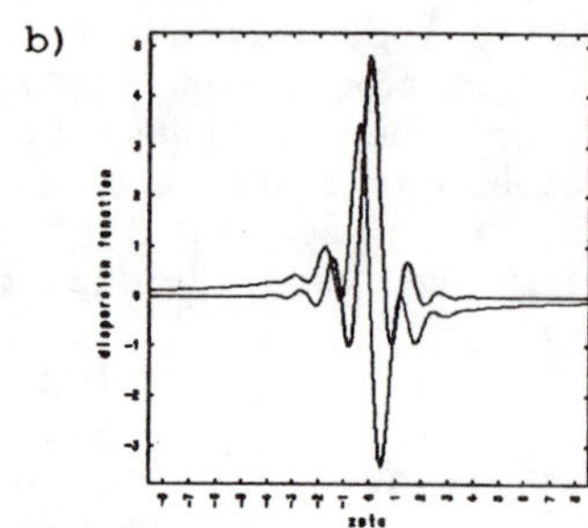

Figure 1. $Z(\zeta,\alpha;\gamma)$ vs. ζ for: a) α=-10. and b) α=.5; real part is solid line, imaginary part is dotted line.

SOLUTIONS OF THE INTEGRAL WAVE EQUATION

The phase integral is employed in wave equation (1), which in its discretized form requires $O(N^2)$ evalutions of the replacement Z-function, where N = # grid points = # Fourier components. In our 1-D calculations we use $N \approx 100$. For sheared field runs we used $B_p/B=1/10$. Several series of runs were made showing the wavenumber spectrum, the values $E_r(x)$, and local energy fluxes for TFTR-like conditions (f=60MHz, B_o=40kG, $n_o=4\times10^{13}cm^{-3}$, T=5keV, $k_y=0m^{-1}$, R_{maj}=3m, a_{wall}=1m, 95%D-5%H minority heating, or 100%D 2nd harmonic heating). Individual modes were easily identifiable on the wavenumber spectrum. When there was strong cyclotron damping at positive k_z, a highly damped local oscillation was barely visible in the E_r plot around the cyclotron resonance, with wavelength corresponding to the replacement Z-function's modulation for Hydrogen at the prominent $k_\parallel$'s of the fast wave. The local energy balances showed stratification of the energy deposition, indicative of a local standing wave at resonance. Our energy-like quantities derive from extrapolation of self-adjoint methods applied to the shear-free geometry. Here we point out that the correct distinction between reactive and dissipative power in nonlocal systems is still an active area of inquiry.

In any case, the form of local energy quantities does not affect the validity of the scattering coefficients which are calculated in the asymptotic regions. Figure 2 compares the scattering coeffients versus k_z of a minority heating outside launch scheme, with and without shear, for the above parameters. Transmission appears to be systematically shifted in k_z, caused simply by the relation $k_\parallel=b_T k_z+b_p k_x$. Reflection and Absorption are altered in more complicated fashion, with significant absorption at $k_\parallel \approx 0$ being clearly evident. Comparisons were also made for minority heating inside launches, and 2nd harmonic heating, both inside and outside launches. The implication from this study, for experiments, is that the ion cyclotron resonance is more absorptive than predicted by gradient-free theory, thus relaxing the constraints on $k_{||}$ for adequate wave damping.

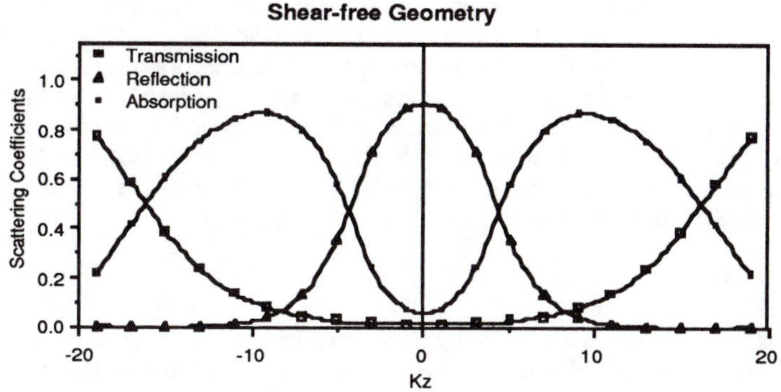

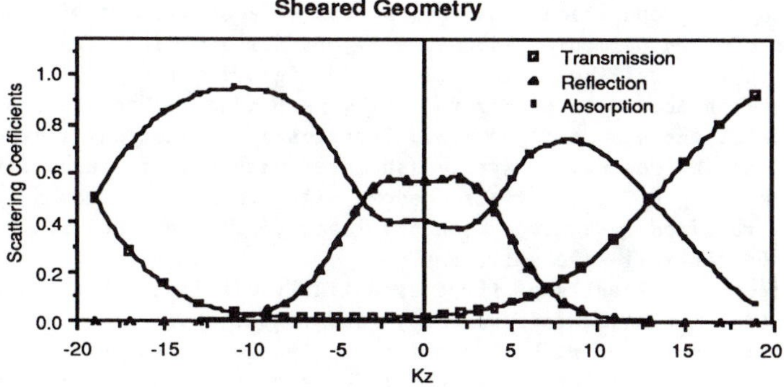

Figure 2. Scattering coefficients vs. k_z for a minority heating inside launch scheme, in shear-free and sheared geometries. Plasma parameters are given in text.

REFERENCES

1. S.-I.Itoh, A.Fukuyama, K.Itoh, and K.Nishikawa, Journal of the Physical Society of Japan 54, 1800 (1984).
2. G.D.Kerbel and M.G.McCoy, Comp. Phys. Comm. 40, 105 (1986).
3. B.I.Cohen, R.H.Cohen, and T.D.Rognlien, Phys. Fluids 26, 808 (1982).
4. A.Fukuyama, K.Itoh, S.-I.Itoh, Comp. Phys. Reports 4, 137 (1986).
5. D.J.Gambier and A.Samain, Nuclear Fusion 25, 283 (1985).

Plasma Heating near Ion Cyclotron Resonance Harmonic Layers*

Kaya Imre** and Harold Weitzner
Courant Institute of Mathematical Sciences, NYU, New York, N.Y. 10012

ABSTRACT

A survey of results is presented for plasma heating and wave propagations for incident waves with frequencies matching ion cyclotron resonance harmonics. It is assumed that outside the resonance layer geometrical optics is applicable, and the plasma is stratified perpendicularly to the equilibrium magnetic field. A boundary layer expansion is used near the resonance layer. Numerical results for the case of pure third and fundamental resonances are given. Ion-ion hybrid resonance case with the minority species at fundamental resonance is also presented.

I. INTRODUCTION

We consider plasma heating by incident waves at frequencies matching ion cyclotron resonance frequency or one of its harmonics. The success of this method depends on the energy absorption efficiency, as well as the accessibility of the incident waves to the resonance region. This paper presents a study of propagation, absorption, and mode conversion of waves crossing ion cyclotron harmonic resonances layers for a medium stratified perpendicular to the equilibrium magnetic field, ignoring the problem of accessibility. When the waves propagate nearly perpendicular to the equilibrium magnetic field, the standard geometrical optics approximation is expected to breakdown near the resonance layer. Associated with this failure, reflection and mode coupling effects may also become significant, so that a full wave analysis is required. We use a boundary layer expansion about the resonance surface to obtain full wave solutions.

We present the results of three essentially distinct, but closely related problems: Nth harmonic resonance, fundamental cyclotron resonance, and ion-ion hybrid resonance with a majority ion species at second harmonic resonance and minority species at fundamental resonance. We obtain wave equations appropriate to each case. In all these problems we have a fast mode which propagates on both sides of resonance, and a slow mode which propagates on only one side of resonance. We study the parameter ranges corresponding to PLT, JET, and CIT devices for the case of ion-ion hybrid resonance.

The boundary layer expansion parameter we use is the ratio of the ion thermal speed to the phase speed of the nonresonant cold plasma waves. By assuming that the equilibrium parameters vary slowly within a wavelength of cold plasma waves, we solve the linearized Vlasov equation in terms of the field variables within the boundary layer containing the resonance surface. This solution allows us to evaluate the current density in terms of the field variables. When we substitute the current into Maxwell's equations, we obtain field equations valid within the boundary layer. The field equations are solved analytically, when possible, or numerically within the layer. The asymptotic solutions are then connected to the geometrical optics solution, which is assumed to be valid outside the resonance layer. We thus

*This work is supported by the U. S. Department of Energy
**Permanent Address: College of Staten Island, CUNY

obtain the full wave solution for our problem.

II. FIELD EQUATIONS

We assume that $\delta=c/(n_c L\omega)\ll 1$, where L is the equilibrium characteristic length, and we use $\varepsilon=v_{th}n_c/c$ as boundary layer expansion parameter. These two parameter are related by the consistency of the boundary layer expansion. We use (+,-) representation where for any vector $\mathbf{V}$, $V_{\mp} = V_x \mp iV_y$, where the equilibrium magnetic field is $\mathbf{B}_0 = B_0(x)\hat{\mathbf{z}}$. We find that to the lowest significant order in the expansion parameter only ++ component of the dielectric tensor is affected by the contribution of the boundary layer. We substitute the current into Maxwell's equations, and find that $E_z = O(m_e/m_i)$, thus it may be ignored. We also eliminate E_- between the remaining field equations, and thus obtain the field equation for E_+ which is valid within the resonance layer. We give only the final results of these calculations here.

$$[1 + d^2 + (1 + d^2 + \mu)d^{N-1} A_N d^{N-1}]E_+ = 0,$$

where $d = d/ds$, $s = -\xi n_c \delta N L\omega/c$, and ξ is the stretched boundary layer variable: $\xi = (1/N - \Omega/\omega)/\delta$. Here $n_c^2 = (\varepsilon_{--}-n_{||}^2)(\varepsilon_{++}-n_{||}^2)/(\varepsilon_{xx}-n_{||}^2)$, $\varepsilon_{xx} = (\varepsilon_{++}+\varepsilon_{--})/2$, $\varepsilon_{++} = 1-N^2X_i/(N-1)$, $\varepsilon_{--} = 1+N^2X_i/(N+1)$, $\mu = (\varepsilon_{--}-n_{||}^2)/(\varepsilon_{++}-n_{||}^2)$, $X_i=(\omega_{pi}/\omega)^2$, and

$$A_N = \frac{X_i}{(\varepsilon_{xx} - n_{||}^2)} \left[\frac{n_c^2}{X_i}\right]^{N-1} \frac{(-1)^N (N/2)^{2N-1}}{(N-1)!} \quad \frac{n_c\delta}{\varepsilon n_{||}} Z\left(\frac{N\xi n_c\delta}{\varepsilon n_{||}}\right).$$

The consistency of the expansion is established by choosing $\delta=\varepsilon^{2N-2}$.

For the case of majority fundamental ion heating the scaling must be selected so that $\delta=\varepsilon^4$, $E_+=O(E_-\varepsilon^2)$ which leads to

$$[1 + d^2 + d^2 A_F d^2]E_- = 0,\ A_F = [n_F^2\sigma/(16X_F)]Z(\xi\sigma),\ \sigma=\varepsilon^3 n_F/n_{||},$$

where n_F is the cold plasma refractive index for the magnetosonic waves evaluated at the resonance, namely $n_F^2 = 2+X_F-n_{||}^2$. We find that this equation couples the slow and fast branches of the magnetosonic mode in the high field side.

When a majority ion species, S, is at second harmonic resonance and a minority ion species, F, is at the fundamental resonance in the same resonance surface, two-ion hybrid resonance occurs. We also investigate this problem by our boundary layer method, and find that

$$[1 + d^2 + (1 + d^2 + \mu)(d\, A_2\, d - B_1)]E_+ = 0,$$

where A_2 and ξ are evaluated for the species S, $\delta=\varepsilon_S^2$, and

$$B_1 =\{-X_F/[2\delta(\varepsilon_{xx}-n_{||}^2)]\}\sigma_F Z(2\xi\sigma_F),\ \sigma_F = (\delta\, n_c)/(\varepsilon_F\, n_{||}).$$

For each of the above problems we prove reciprocity theorems and show that the fast mode transmission coefficient is the same as that given by geometrical optics, and there is no reflected fast mode when that mode is incident from the high field side. We now give the numerical results.

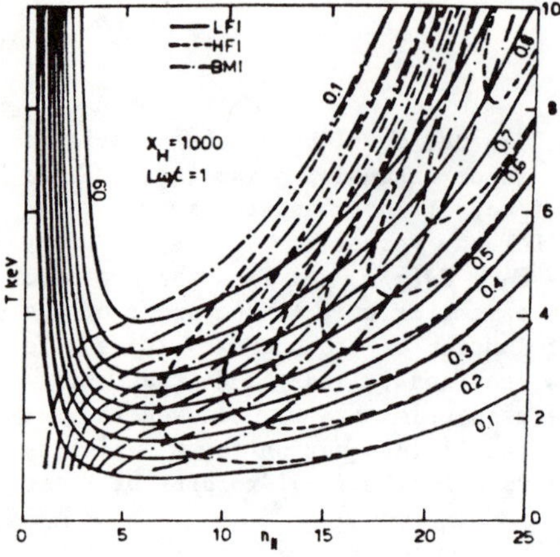

Figure 1. Absorption contours for the third harmonic resonance for X_H=1000, $L\omega/c$=1 in the $(T,n_{||})$ plane. LFI: low field side incidence of fast mode, HFI: high field incidence of fast mode, BMI: Bernstein mode incidence. Geometrical optics apply when LFI absorption rate is the same as HFI absorption rate.

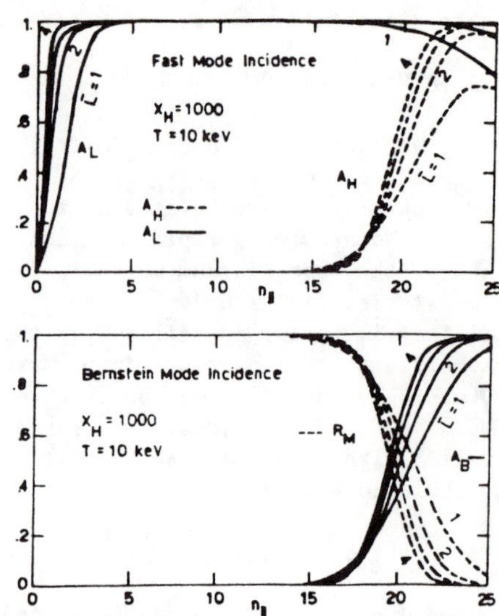

Figure 2. The effect of the machine size on the third harmonic heating. $\bar{L}=L\omega/c$ = 1,2,3,4, X_H= 1000, T=10keV. $A_{H,L}$: energy absorption for high and low field incident fast mode. R_M: reflected mode conversion coefficient. A_B: energy absorption for Bernstein mode incidence.

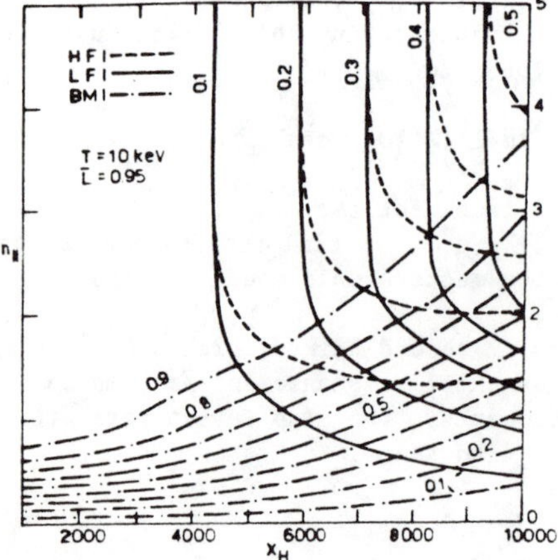

Figure 3. Absorption contours for magnetosonic waves across the fundamental ion resonance layer, in the $(X_H, n_{||})$ plane, for T=10keV, $\bar{L}$=0.95. Geometrical optics apply when A_H=A_L.

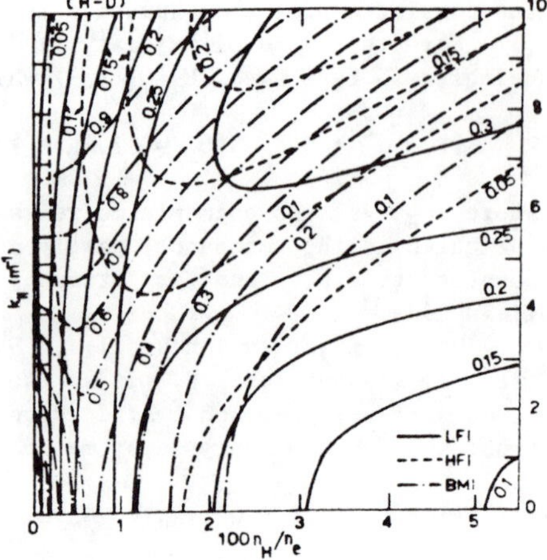

Figure 4. Absorption contours for a PLT example: B=3T, L=130cm, T=2keV, $n_{elec.}$= $3\times10^{13}cm^{-3}$, in the plane of $k_{||}$(1/m) and minority density percentage.

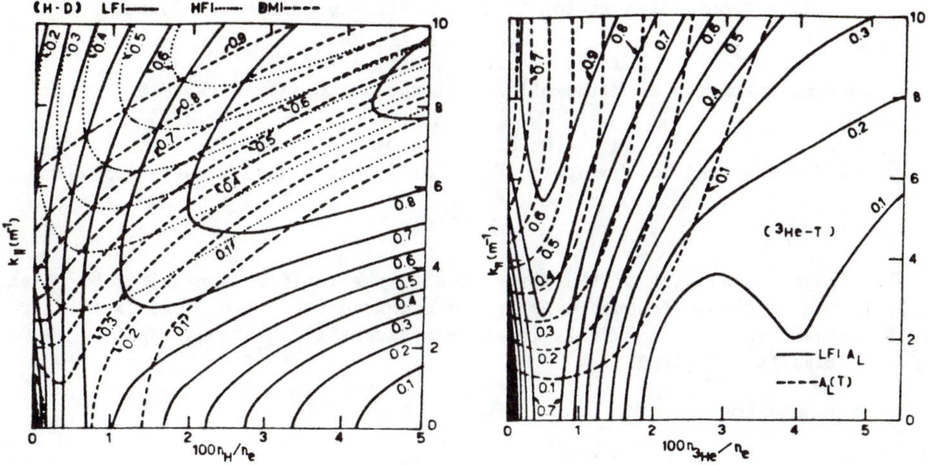

Figure 5. Absorption contours for ion-ion hybrid resonance for a JET example: B=3.45T, L=3m, T=5keV, $n_{elec.}=3.5\times10^{13}cm^{-3}$, in the plane of $k_{||}(1/m)$ and minority density percentage. Geometrical optics is established for $n_H< 0.01\ n_{elec.}$ and $k_{||}>10(1/m)$. For small fixed values of $k_{||}$, the absorption rate is peaked at $n_H\approx0.01n_{elec.}$.

Figure 6. Absorption contours for ion-ion hybrid resonance for a CIT example: B=10T, L=122cm, T=10keV, $n_{elec.}=5\times10^{14}cm^{-3}$. Dashed lines denote energy absorption rate by the majority species T for the low field incident fast mode. The total absorption rate is peaked when the minority ion species ^3He density is about 0.6% of the electron density.

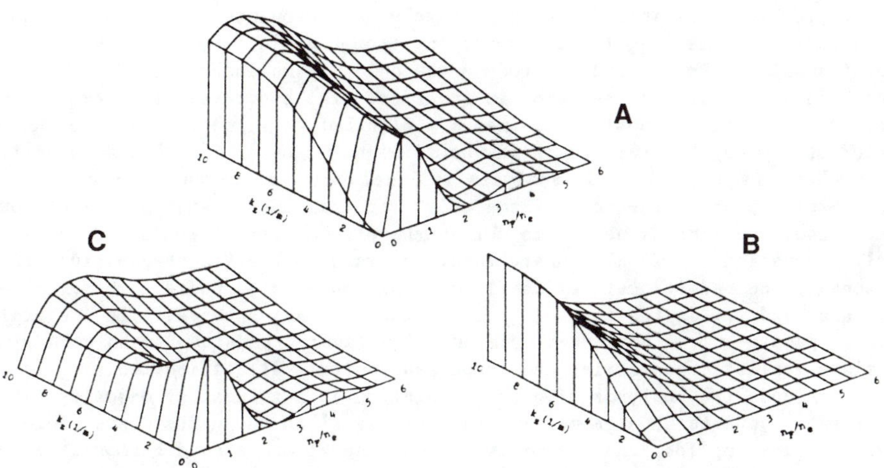

Figure 7. Absorption surfaces for the CIT example given in Figure 6. A: total ion absorption, B: majority ion (tritium) absorption, C: minority ion (^3He) absorption. In this example the high field side incident fast mode, and also the high field side incident Bernstein mode are almost totally mode converted.

Wave Propagation at Ion Cyclotron Resonance in Large Toroids*

Harold Weitzner and Kaya Imre**
Courant Institute of Mathematical Sciences, NYU, New York, N.Y. 10012

D. B. Batchelor
Oak Ridge National Laboratory, P.O. Box Y, Oak Ridge, Tennessee 37831

ABSTRACT

The range of validity of the perpendicularly stratified approximation near cyclotron harmonic resonance layers is examined for large toroids. Where the usual approximation fails integro-differential equations apply. Some exact results are given.

I. INTRODUCTION

For wave propagation in large toroids at frequencies on the order of ion cyclotron frequencies the geometrical optics approximation generally holds away from cyclotron harmonic resonance. Near these resonances the system is usually represented by waves propagating in a perpendicularly stratified medium. Here we examine the validity of the perpendicular stratification approximation for two problems. We consider a large toroidal device whose equilibrium toroidal field B_T is much larger than the equilibrium poloidal field B_P. We define α such that $B_P/B_T = \tan\alpha$, where α is usually small. In the first problem we assume that there is a minority density species at fundamental resonance on some surface and that no other ion species are resonant near that surface. In the second problem we assume that there is a majority species with second harmonic resonance on a surface and that no other ion species is resonant nearby. By combining these problems we could examine ion-ion hybrid resonance. In each case we start to characterize the waves by introducing an entirely non-resonant cold plasma dielectric tensor by omitting the terms arising from the minority fundamental resonant species. We may then introduce the cold plasma index of refraction n_c for a fast wave at the resonance in question. The hypothesis that geometrical optics generally applies becomes $\delta \equiv c/(n_c L\omega) \ll 1$, where L is the characteristic length over which the equilibrium magnetic field and density vary. Here $L = B/|\nabla B|$. Our analysis is local near the resonance surface.

Near resonance the cold plasma model may fail and an analysis based on the Vlasov equation is needed to obtain the plasma current to insert in Maxwell's equations. In a boundary layer analysis of the Vlasov equation near resonance an additional critical small parameter ε appears, where $\varepsilon = n_c v_{th}/c$ and v_{th} is the thermal speed of the relevant species. In several other papers[1-4] we have developed boundary layer expansion techniques in various realistic geometries and in perpendicularly stratified media. This note extends the analysis of wave propagation in a tokamak at majority fundamental resonance when geometrical optics fails[2] to the cases where geometrical optics applies, and for minority fundamental and second harmonic resonance.[4] We intend to publish the details of the relatively straightforward derivations together with numerical work later; here we summarize the re-

*This work is supported by the Department of Energy
**Permanent Address: College of Staten Island, CUNY

sults.

We start the analysis of the Vlasov-Maxwell system by writing it in cylindrical coordinates r, θ, z, where **θ** is the toroidal direction, **b** ($\mathbf{b}_p$) is the unit vector in the direction of the equilibrium (poloidal) magnetic field. We change coordinates from (r,θ,z) to (ξ,η,θ) where $\delta\xi(r,\theta,z) = e[B_0 - B(r,\theta,z)]/(m_i c\omega)$, η is the coordinate of the orthogonal trajectories to the surface ξ(r,θ,z) = const., and B_0 is the resonant magnetic field value. We find that there is consistent scaling and expansion of the Vlasov-Maxwell system in the neighborhood of the resonance layer such that geometrical optics applies in the coordinates η and θ, but not in the variable ξ. Thus we may consider an incoming or outgoing wave with wavenumber **k** or index **n**, and we may define a parallel wavenumber $k_{||} = \mathbf{b}\cdot\mathbf{k}$, (or a parallel index $n_{||}$) both of which are constant in the resonant boundary layer region. Equally we may introduce a wavenumber and index in the η direction k_η and $n_\eta = k_\eta c/\omega$. We may also introduce the geometric quantities that determine the direction of $\nabla\xi$, $K_\mp = (\mathbf{b}_p \times \boldsymbol{\theta} \mp i\mathbf{b}_p)\cdot\nabla\xi$, $K_- K_+ = K^2$ and Maxwell's equations in the resonance layer reduce to

$$[K^2 d^2 + 1 - n_\eta^2 + (K^2 d^2 + 1 - n_\eta^2 + \varepsilon^c_{--}/\varepsilon^c_{++})/(2\varepsilon^c_{xx})\ \varepsilon^H_{++}]E_+ = 0,$$

where $d = d/d\xi$, $n_c^2 = \varepsilon^c_{++}\varepsilon^c_{--}/\varepsilon^c_{xx}$, $\varepsilon^c_{xx} = (\varepsilon^c_{++} + \varepsilon^c_{--})/2$, and ε^H_{++} is an as yet unspecified operator. In order to obtain the above system we have assumed $E_{||} = \mathbf{E}\cdot\mathbf{b}$ is approximately zero. The determination of ε^H_{++} depends on the truncation and solution of the Vlasov equation after it has been expanded in a Fourier series in the gyrophase angle, cf. Refs.1-4. The boundary layer expansion and stretching is carried out in such a way that the boundary layer thickness is on the order of $c/(n_c\omega)$. We introduce two new constants, $\bar{n}_{||} = n_{||}\,\varepsilon\delta/a$ and $a = \varepsilon\sin\alpha\ \mathbf{b}_p\cdot\nabla\xi/\delta$.

MINORITY FUNDAMENTAL RESONANCE

We describe two cases of interest for minority fundamental resonance. In both cases we assume that $n_{||} \lesssim O(n_c\delta/\varepsilon)$. If $n_{||}$ is larger than this estimate we can show that geometrical optics applies across the resonance layer and heating is weak. In the first case we require that $\varepsilon \sin\alpha \ll \delta$. Provided that $X_F \lesssim O(\delta X_m)$, where $X_i = (\omega_{pi}/\omega)^2$ and the subscripts F and m denote, respectively, the minority fundamental and majority ion species, we obtain

$$\varepsilon^H_{++}E_+ = -X_F n_c/(\varepsilon n_{||})Z[-n_c\delta\xi/(\varepsilon n_{||})]E_+,$$

where Z(z) is the standard plasma dispersion function. This system is an ordinary differential equation of the type studied in Ref.4. If $X_F > O(\delta X_m)$ we expect that the resonant wave E_+ is very small near resonance and little heating occurs.[4] Alternatively if $\varepsilon \sin\alpha \gtrsim \delta$ but $X_F = O(X_m \varepsilon\sin\alpha)$ then

$$\varepsilon^H_{++}E_+ = iX_F/(\delta a)\int_{-\infty}^{\infty} d\xi'\ E_+(\xi')F\left(\frac{|\xi^2-\xi'^2|}{2a}\right)\exp[i\bar{n}_{||}\ (\xi-\xi')],$$

where F(λ) was introduced in Refs.2-3: $2F(\lambda) = \int d^3u[G_0(\mathbf{u})/|u_{||}|]\exp(i\lambda/|u_{||}|)$, and G_0 is the normalized equilibrium distribution function. Thus, we see explicitly that if sinα is not small enough then the conventional local representation fails. The current operator is essentially the same as in a parallel stratified medium, see Ref.3, in which wave absorption is strong. There we introduced an integral operator M such that after appropriate definition of constants $[1/(2\varepsilon^c_{xx})]\ \varepsilon^H_{++}\ E_+ = \exp(i\bar{n}_{||}\xi)M\circ[E_+ \exp(-i\bar{n}_{||}\xi)] \equiv (X_F/\delta)N\circ E_+$. The integro-differential equation is equivalent to the system

$$(K^2d^2 + 1 - n^2_\eta + \varepsilon^c_{--}/\varepsilon^c_{++})V = (\varepsilon^c_{--}/\varepsilon^c_{++})E_+$$

$$V = E_+ + (X_F/\delta)N\circ E_+,$$

which is a minor modification to the low density case studied in Ref.3. With slightly varied estimate for $F(\lambda)$ to guarantee that $F(\lambda)\exp[i2n_{||}(\xi-\xi')]$ is uniformly bounded in the appropriate half-plane it is easy to show that for high field side incident waves no wave is reflected and that the transmission coefficient is the same as in geometrical optics or in the case $\varepsilon \sin\alpha \ll \delta$, or $T = \exp[-\pi X_F/(2K\delta\varepsilon^c_{xx})(\varepsilon^c_{--}/\varepsilon^c_{++})(1 - n^2_\eta)^{-1/2}]$.

The reciprocity theorems require different proofs than in Ref.3. The operator N is a function of $\bar{n}_{||}$, which we denote as $N(\bar{n}_{||})$, and provided that appropriate integrals converge and one can exchange orders of integration then $\int d\xi\ \tilde{E}_+ N(\bar{n}_{||})\circ E_+ = \int d\xi\ E_+ N(-\bar{n}_{||})\circ \tilde{E}_+$. Thus, the problem adjoint to a given problem has $\bar{n}_{||}$ (or $\mathbf{b}_p$) replaced by $-\bar{n}_{||}$ (or $-\mathbf{b}_p$). The argument in Ref.3 then states that a wave incident from the high field side with parallel index $\bar{n}_{||}$ has the same transmission coefficient as a wave incident from the low field side with parallel index $-\bar{n}_{||}$. Since the former transmission coefficient is independent of $\bar{n}_{||}$, so is the latter, and a true reciprocity result occurs: the transmission coefficient is independent of the direction of incidence and given by the above formula. These results are equally valid, but much easier to obtain for $\varepsilon \sin\alpha \ll \delta$.

SECOND HARMONIC RESONANCE

In order to truncate consistently the system derived from the Fourier expansion of the Vlasov equation in gyrophase angle we take $\delta = \varepsilon^2$. In order that the system not reduce in the boundary layer to geometrical optics we must assume $n_{||} \lesssim O(2\varepsilon n_c)$. We distinguish three cases $\varepsilon \gg \sin\alpha$, $\varepsilon \sim \sin\alpha$, and $\varepsilon \ll \sin\alpha$. In the last case we find that geometrical optics based on the cold plasma model applies in leading order, and the small effect of plasma heating may be calculated as a perturbation induced by warm plasma effects.

When $\varepsilon \gg \sin\alpha$, we obtain a reasonably well understood system of ordinary differential equations as is treated in Ref.4, and

$$\varepsilon^H_{++}E_+ = (2X_S\varepsilon n_c/n_{||})(Kd - n_\eta)Z(-2n_c\varepsilon\xi/n_{||})(Kd + n_\eta)E_+,$$

This system allows fast mode to propagate on both sides of resonance, and Bernstein modes on the low field side. It is unlike Ref.4 in the presence of n_η, and it gives precise means of interpreting the system in terms of a perpendicularly stratified medium. We have the same type of exact results as in Ref.4. For incident fast waves the transmission coefficient is independent of the side of incidence. For mode conversion of reflected or of transmitted waves the coefficients are the same whether the incident wave is a fast wave converted to a Bernstein wave or a Bernstein wave converted to a fast wave. Further, no fast wave is reflected with an incident high field side fast wave, and the transmission coefficient for the incident fast wave is $T = \exp[-\pi X_S/(2K\delta\varepsilon^c_{xx})(\varepsilon^c_{--}/\varepsilon^c_{++})(1 - n^2_\eta)^{-1/2}]$. For the special case of $n_{||} = 0$, it is possible to represent the solutions as integral transforms just as in Ref.4, and one concludes that $R_M = 1-T$, $T_M = T(1-T)$, $R_F = (1-T)^2$, $R_B = T^2$, where R_M and T_M are reflected and transmitted mode conversion coefficients, R_F is reflection coefficient for low field side incident fast waves, and R_B is reflection coefficient for the Bernstein mode incidence.

For the case likely to be of some physical interest, $\varepsilon \sim \sin\alpha$, we have again an integro-differential equation, and

$$[1/(2\varepsilon^c_{xx})]\varepsilon^H_{++}E_+ = -2X_S(Kd - n_\eta)N\circ[(Kd + n_\eta)E_+].$$

Thus, in this parameter range in which $B_p/B = \sin\alpha \sim \varepsilon$ the current is not a local function and the problem of wave propagation becomes an integro-differential equation. The system is not too different from that studied in Refs.2-3, and we expect to be able to carry out numerical solutions and further analysis. The presence of the fourth order derivative indicate the existence of the Bernstein mode, in addition to the fast mode. We may formally integrate by parts in ξ' and recover the local operator given for the case when $\sin\alpha \ll \varepsilon$. This contribution is expected to be the dominant part for $|\xi|$ large. The real adjoint problem is again obtained by reversing the direction of the poloidal field. We obtain reciprocity theorems just as above, which connect the properties of the two problems. For technical reasons the proofs stated for the minority fundamental resonance case,[2,3] which determines the transmission coefficient and the high field side incident fast wave reflection coefficient fail here. We suspect the results still hold.

CONCLUSIONS

We have examined the validity of the perpendicularly stratified medium approximation as it is commonly used near ion cyclotron resonance layers. We find that only for B_p/B extremely small is the usual type of local approximation valid. For other values we obtain an integro-differential equation with many properties of a parallel stratified medium. We have given the equations for the cases of isolated minority fundamental resonance, and isolated second harmonic majority resonance. In the former case we have many exact results, while in the latter we can give many fewer exact results. In the future we hope to provide further analysis for the latter case, and we expect to carry out numerical calculations for these cases, as well as the ion-ion hybrid resonance case, which is a combination of the cases studied here.

REFERENCES

[1]H. Weitzner, Phys. Fluids **26**, 998 (1983).

[2]A. Fruchtman and H. Weitzner, Phys. Fluids **29**, 1620 (1986).

[3]A. Fruchtman, K. Riedel, H. Weitzner, and D.B. Batchelor, Phys. Fluids **30**, 115 (1987).

[4]K. Imre and H. Weitzner, submitted to Phys. Fluids.

THE EFFECT OF HOT α-PARTICLES ON ION CYCLOTRON ABSORPTION

R A Cairns and A Kay
University of St Andrews, St Andrews KY16 9SS, UK

C N Lashmore-Davies
Culham Laboratory, Abingdon, OX14 3DB, UK
(UKAEA/Euratom Fusion Association)

ABSTRACT

Ion cyclotron heating involves mode conversion of the incident fast wave to a Bernstein mode, which is generally strongly damped and only propagates in a localised region of the plasma. The usual theoretical approach to this problem involves solution of fourth or higher order equations, but by treating the Bernstein mode as a driven response to the fast wave we obtain a simple second order equation. Comparison of its solutions with those of higher order equations shows that this approach gives good results. The same method can be used to consider the effect of a small population of hot α-particles on ion cyclotron heating, since it does away with the need for a small larmor radius expansion. The results indicate that a small concentration of α-particles can absorb a substantial fraction of the incident energy.

INTRODUCTION

Equations which describe the propagation of the fast and Bernstein modes in both directions must necessarily be of fourth or higher order, and solution of such equations is complicated by the need to avoid spurious growing solutions in regions where there are evanescent waves[1-5]. In many cases of interest the Bernstein mode is strongly damped and only propagates in a localised region around where it is generated by mode conversion of the fast wave. The essence of our method is to treat this damped Bernstein mode as a driven response to the fast wave, a procedure closely analogous to the theory of parametric decay involving a highly damped quasi-mode[6]. In this way we obtain a simple second order equation, solution of which for the transmission and reflection coefficients of the fast wave, together with calculation of the absorption profiles, is a simple matter, easily handled by any standard integrations routine. We have already demonstrated that it gives good results for absorption at the second and third ion cyclotron harmonics[7,8], while others have recently adopted the same procedure to look at minority heating, again obtaining good results[9].

Having established that the technique works, our main concern here is to use it to investigate the effect on ion cyclotron heating of a small population of very hot fusion α-particles. By taking the response of the α's to be driven by the fast wave we do not need to make a small larmor radius expansion, but can evaluate exactly the Bessel functions which appear in the α-particle conductivity tensor.

The essential features of our method are described in the next section, followed by a summary of the results obtained on α-particle absorption.

DERIVATION OF THE SECOND ORDER EQUATION

We shall illustrate the procedure by looking at propagation near the second harmonic of the ion cyclotron frequency. If a small larmor radius expansion is made then we may eliminate all but the y-component of electric field to obtain

$$\left(p^2 + \frac{1}{3} - \frac{\beta^{\frac{1}{2}} v_A^2}{4\omega^2 p} Z(\zeta) k_\perp^2\right)\left(k_\perp^2 - \frac{\omega^2 \mu^2}{v_A^2}\right) E_y = \frac{\beta^{\frac{1}{2}}}{4p} \frac{\left(\frac{1}{3} - p^2\right)^2}{\left(\frac{1}{3} + p^2\right)} Z(\zeta) k_\perp^2 E_y \qquad (1)$$

where the magnetic field is in the z-direction and $k_\perp$ in the x-direction. The various parameters which appear are

v_A = Alfvén speed

$$p = k_{\parallel} \frac{v_A}{\omega}$$

β = thermal/magnetic energy density

$$\mu^2 = \frac{\left(\frac{1}{3} - p^2\right)(1 + p^2)}{\frac{1}{3} + p^2}$$

$$\zeta = \frac{\omega - 2\Omega}{k_{\parallel} v_i}$$

Ω = ion cyclotron frequency

v_i = ion thermal velocity.

Usually, to describe a plasma with the magnetic field varying in the x-direction, a fourth order system is derived, reducing to (1) if if $\frac{d}{dx} \to ik_\perp$. What we do is to keep the exact spatial dependence in the $\underline{\nabla} \times (\underline{\nabla} \times \underline{E})$ term, but in the hot plasma contributions to the plasma current, which involve $k_\perp$ through the argument of Bessel functions, we simply put the $k_\perp$ appropriate to the fast wave. Thus we regard these currents as being forced by the fast wave. This leads to the equation

$$\frac{d^2E_y}{dx^2} + \left(\frac{\mu\omega}{v_A}\right)^2 E_y = -\left(\frac{\mu\omega}{v_A}\right)^2 \frac{\dfrac{\beta^{\frac{1}{2}}}{4p}\dfrac{\left(\frac{1}{3}-p^2\right)^2}{\frac{1}{3}+p^2} Z(\zeta)}{p^2+\dfrac{1}{3}-\dfrac{\beta^{\frac{1}{2}}\mu^2}{4p}Z(\zeta)} E_y. \qquad (2)$$

If this procedure is adopted then it is clearly unnecessary to make a small larmor radius expansion in the hot plasma terms since the various terms involving Bessel functions can be evaluated exactly. This is what we do for the hot α-particle case. This description has the desirable property that the absorbed energy density, as calculated from the divergence of the Poynting flux associated with the fast wave, is the same as that calculated from the average of $\underline{J}.\underline{E}$. Also integration of (2) to investigate the transmission, reflection and absorption of the fast wave is straightforward.

RESULTS FOR α-PARTICLE ABSORPTION

We present here some representative results, simply to indicate the substantial effects produced by a hot α population. More detailed results will appear elsewhere[8]. In Fig 1 we show the reflection coefficient at the 2nd harmonic for parameters used by Swanson[2], showing very close agreement with his results, together with effect of adding 1% of α-particles at 1MeV. Fig 2 shows the absorption profile for $k_{\parallel}R_0 = 12$ for the same parameters. The broad absorption width for the α's means that the plasma density profile has to be taken into account. The results of Fig 2 assume a Gaussian density profile whose width is consistent with a torus with aspect ratio around 3. These results indicate that a small proportion of high temperature α's can absorb a disproportionately large fraction of the incident energy.

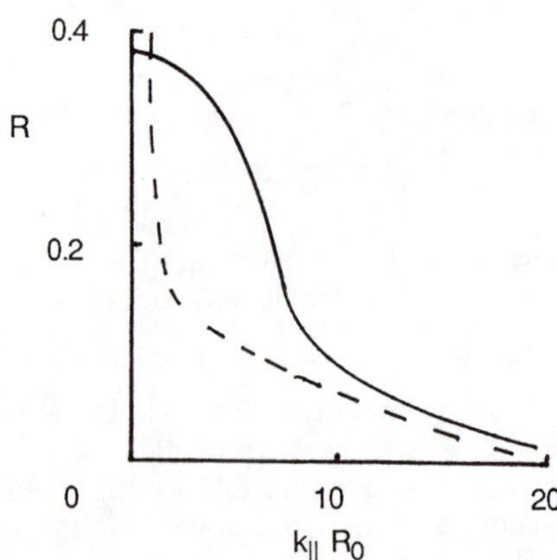

Fig. 1 Reflection coefficient with (dotted) and without (full line) α-particles

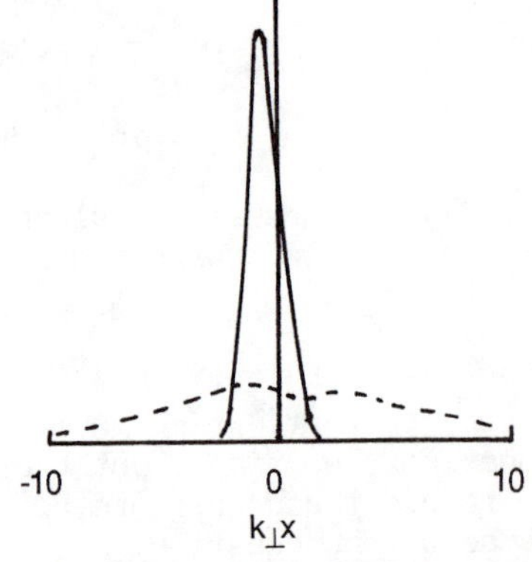

Fig. 2 Absorption profile for deuterium (full line) and α-particles (dotted)

CONCLUSIONS

We have shown how to derive a simple second order equation to describe the transmission, reflection and absorption of an incident fast wave in the ion cyclotron frequency range. This approach avoids the numerical problems and analytic complications associated with other approaches, while yielding essentially the same results. In addition, since the theory does not rely on a small larmor radius expansion it has been used to look at the effect of a small population of α-particles in the MeV range of energies. The results indicate that a small population of α's may absorb a substantial fraction of the incident energy and that this absorption takes place over a much broader region than in the absence of α-particles. Although we have used a Maxwellian distribution of α's, for simplicity, there is no reason why a more realistic distribution should not be used, and we are currently extending the work in this way. Also the basic method should be applicable to other mode conversion regimes, for example the absorption of the second harmonic electron cyclotron wave.

REFERENCES

1. D G Swanson, Nuclear Fusion 20, 949 (1980).
2. D G Swanson, Phys. Fluids 28, 1800 (1985).
3. D G Swanson, Phys. Fluids 28, 2645 (1985).
4. P L Colestock and R J Kashuba, Nuclear Fusion 23, 763 (1983).
5. T Hellsten, K Appert, J Vaclavik and L Villard, Nuclear Fusion 25, 99 (1985).
6. K Nishikawa, in Advances in Plasma Physics, Vol.6, ed. A Simon and W B Thompson (John Wiley and Sons
7. A Kay, R A Cairns and C N Lashmore-Davies, Proc. 13th European Conference on Controlled Fusion and Plasma Heating (European Physical Society, 1986) p.93.
8. A Kay, R A Cairns and C N Lashmore-Davies, submitted to Plasma Phys. and Contr. Fusion (1987).
9. C N Lashmore-Davies, V Fuchs, L Gauthier, G Francis, A K Ram and A Bers, Proc. International Conference on Plasma Physics, Kiev, 1987, to be published.

FAST-WAVE TRANSMISSION AND REFLECTION IN ION-CYCLOTRON HEATING*

V. Fuchs,[a] C.N. Lashmore-Davies,[b] G. Francis, A.K. Ram,
and A. Bers, MIT, Plasma Fusion Center, Cambridge, MA02139

ABSTRACT

A second-order differential equation for the ion-cyclotron fast wave propagating in a hot, two ion species plasma is obtained. The equation includes the effects of both minority fundamental and majority second harmonic cyclotron damping. The approximation is based upon replacing the coupling to the ion Bernstein wave by a modified propagation description of the fast wave. For the case of perpendicular propagation, the second order equation reduces to the Budden equation giving the well known transmission coefficient for both two ion hybrid and second harmonic resonance. The solutions of the second order equation as a function of $n_\parallel$ give transmission and reflection coefficients which agree very well with the results based on models using higher order differential equations.

INTRODUCTION

Absorption of the fast wave in the ICRF involves mode conversion by coupling to ion Bernstein waves, cyclotron damping due to all the ion species present in the plasma, electron Landau and transit time damping. In general, these effects are not readily separable. It is therefore desirable to look for ways of simplifying the description of the fast wave while still retaining the essential physics. On either side of the central core of the plasma the fast wave has a simple description given, approximately, by a cold plasma model. On the other hand, inside the central core of the plasma, where mode conversion and dissipation take place, the dynamics must be described by the much more complex Vlasov, kinetic plasma model. A simplification is obtained by seeking a description of the fast wave transmission and reflection in the presence of the combined effects of mode conversion and dissipation. As far as the fast wave is concerned, at the boundaries of the central core where its incident, reflected and transmitted power flows are well defined, the combined effects of mode conversion and dissipation appear as power "absorbed" within these boundaries. A description of the fast wave in this manner requires only a second order ordinary differential equation to be established which matches asyptotically to the fast wave propagation outside the central core. Such a simplification, in addition to reducing the order of the wave propagation equations would also benefit considerations of scaling.

*Work supported in part by Hydro-Quebec Project No. 01584-575-10143 and in part by DOE Contract No. DE-AC02-78ET-51013.
(a) Permanent address: Project Tokamak, Institut de Recherche d'Hydro-Quebec, Varennes, Quebec, Canada, J0L 2P0.
(b) Permanent address: UKAEA Culham Laboratory, Abingdon, Oxfordshire, OX14 3DB, UK (Euratom/UKAEA Fusion Association).

In this paper we shall attempt to construct such a global description for the fast wave absorption in the heating of a two ion species plasma. The physical effects we must include in this analysis are of course coupling to ion Bernstein waves and the collisionless damping at the fundamental resonance of the minority species and at the second harmonic resonance of the majority species. We shall ignore the rotational transform (poloidal magnetic field) and electron Landau damping.

THE FAST WAVE APPROXIMATION OF THE DISPERSION RELATION

For the central region of the plasma we consider a slab model, inhomogeneous in only the x-direction and assume the fast wave is launched with a wave vector $\underset{\sim}{k} = (k_\perp, 0, k_\parallel)$ where $\perp$ and $\parallel$ denote perpendicular and parallel to the equilibrium magnetic field. Assuming a harmonic variation $\exp i\,(\underset{\sim}{k}\cdot\underset{\sim}{x} - \omega t)$ the Vlasov-Maxwell system of equations give

$$[n^2 \underset{\sim}{I} - \underset{\sim}{n}\,\underset{\sim}{n} - \underset{\approx}{\varepsilon}(\omega, \underset{\sim}{k})]\ \underset{\sim}{E}(\omega, \underset{\sim}{k}) = 0 \qquad (1)$$

where $\underset{\sim}{n} = c\,\underset{\sim}{k}/\omega$, $\underset{\approx}{\varepsilon}$ is the dielectric tensor of the plasma and $\underset{\sim}{E}$ is the electric field vector. Neglecting electron inertia gives

$$(n_\parallel^2 - \varepsilon_{xx})\, E_x - \varepsilon_{xy}\, E_y = 0\ , \quad \varepsilon_{xy}\, E_x + (n^2 - \varepsilon_{yy})\, E_y = 0 \qquad (2)$$

To first order in $k_\perp^2\, v_{Ti}^2/\omega_{ci}^2$ we have

$$\varepsilon_{xx} = \varepsilon_{yy} \simeq \varepsilon_\perp - \alpha n_\perp^2\ , \quad \varepsilon_{xy} \simeq ig - i\alpha n_\perp^2 \qquad (3)$$

$$\varepsilon_\perp \simeq - c^2/3c_A^2 + \omega_{p2}^2\ Z\,(z_{12})/2\sqrt{2}\ \omega k_\parallel v_{T2} \qquad (4)$$

$$\alpha \simeq - \omega_{p1}^2\ \omega\ v_{T1}\ Z\,(z_{21})/2\sqrt{2}\ k_\parallel\, \omega_{c1}^2\, c^2 \qquad (5)$$

$$g \simeq \varepsilon_\perp - c^2/3c_A^2 \qquad (6)$$

where $z_{nj} \equiv (\omega - n\omega_{cj})/\sqrt{2}\,k_\parallel v_{Tj}$ and $n = 1, 2$. $j = 1$ corresponds to the majority species and $j = 2$ to the minority species; $\omega_{cj} = eZ_jB_0/m_j$ with e positive, and $c_A^2 = B_0^2/n_1 m_1 \mu_0$. Only the finite Larmor radius terms coming from the majority second harmonic have been retained.

The dispersion relation resulting from Eqs. (2) is

$$n_\perp^2 = [(\varepsilon_{xx} - n_\parallel^2)\,(\varepsilon_{yy} - n_\parallel^2) + \varepsilon_{xy}^2]/(\varepsilon_{xx} - n_\parallel^2) \qquad (7)$$

We observe that in the limit of a cold plasma Eq. (7) is the well known approximate fast wave solution. In the case of a hot plasma we approximate by substituting the cold plasma solution

$$n_\perp^2 = [(\varepsilon_\perp - n_\parallel^2)^2 - g^2]/(\varepsilon_\perp - n_\parallel^2) \equiv (n_\perp^2)_A \qquad (8)$$

into the thermal correction terms in the dielectric tensor elements appearing in Eq. (7). Thus, substituting (8) into the finite Larmor radius corrections in Eq. (7) we obtain the "fast wave approximation" appropriate to a hot, degenerate two-ion species plasma (as in D(H))

$$n_\perp^2 \simeq \frac{(\varepsilon_\perp - g - n_\parallel^2)(\varepsilon_\perp + g - n_\parallel^2 - 2\,\alpha(n_\perp^2)_A)}{(\varepsilon_\perp - n_\parallel^2 - \alpha\,(n_\perp^2)_A)} \qquad (9)$$

A similar approximation has been used by Kay et al[1] to describe heating at the second and third ion-cyclotron harmonics. The interpretation of Eq. (9) is that the coupling of the fast wave to a propagating ion Bernstein wave has been represented as a localised perturbation of the fast wave. By ensuring that the approximate dispersion relation retained the correct structure we have preserved the most important characteristics of this coupling, i.e. the coupling region occurs on the high field side of the degenerate second harmonic majority resonance and not at the resonance itself.

We emphasize that the approximate fast wave solution (9) contains the full effects of fundamental (minority) and second harmonic (majority) cyclotron damping.

RESULTS

The dispersion relation (9) can be transformed into a differential equation by identifying $k_\perp$ with $-\,id/dx$. We get

$$\frac{d^2\phi}{d\xi^2} + \frac{(1 - 3\,N_\parallel^2)\left\{1 + N_\parallel^2 - \dfrac{\eta}{2\sqrt{2}}\dfrac{c_A}{N_\parallel v_{T2}}\,Z(a_2\xi) - \dfrac{f v_{T1}}{\sqrt{2}\,N_\parallel c_A}\,Z(a_1\xi)\right\}}{\left\{1 + 3\,N_\parallel^2 - \dfrac{3\eta}{4\sqrt{2}}\dfrac{c_A}{N_\parallel v_{T2}}\,Z(a_2\xi) - \dfrac{3f}{2\sqrt{2}}\dfrac{v_{T1}}{N_\parallel c_A}\,Z(a_1\xi)\right\}}\,\phi = 0 \qquad (10)$$

where, given $B = B_0\,(1-x/R_0)$, with R_0 the tokamak major radius, $a_{1,2} = c_A/\sqrt{2}N_\parallel v_{T1,2}\,R_A$, $R_A = R_0\omega/c_A$, $\eta = n_2/n_1$, $\xi = x\omega/c_A$, $N_\parallel = c_A\,k_\parallel/\omega$, and $f = (1 + N_\parallel^2)\,(1 - 3N_\parallel^2)/(1 + 3N_\parallel^2)$.

In the limit $N_\parallel \to 0$ equation (10) reduces to

$$\frac{d^2\phi}{d\xi^2} + \frac{(\xi + \eta\,R_A/2 + v_{T1}^2\,R_A/c_A^2)}{(\xi + 3\,\eta\,R_A/4 + 3\,v_{T1}^2\,R_A/2c_A^2)}\,\phi = 0 \qquad (11)$$

which will be recognised as Budden's equation. We may therefore immediately write down the power transmission coefficient[2]

$$T = \exp\left\{-\,\pi\,(\eta/4 + v_{T1}^2/2c_A^2)\,R_A\right\} \qquad (12)$$

which agrees with the well known results for the ion-ion hybrid resonance or the second harmonic in a pure plasma.[3]

We have solved Eq. (10) numerically and compared it with results obtained from much more complicated systems of equations.[4,5] Table I gives a sample comparison of our results with those of Ref. 5.

TABLE I
PLT minority heating "benchmark" case [D(H), η=0.05]. Comparison of our results (first two rows) with those of Ref. 5 (last two rows).

$k_{\parallel}$ [m^{-1}]	1	2	3	4	5	6	7	8	9	10
T	16.3	18.3	18.3	20.2	22.3	26.0	29.8	34.5	40.0	46.0
R	69.1	65.2	59.3	52.1	44.2	36.3	28.8	22.1	16.5	11.9
T	15.4	16.2	17.5	19.5	22.1	25.4	29.5	34.4	40.0	46.4
R	61.2	57.9	52.9	46.7	39.8	32.9	26.4	20.4	15.3	11.1

In conclusion, we have verified that the second order fast wave approximation presented in this paper produces transmission and reflection coefficients compatible with results previously obtained from numerical integration of fourth or higher order equations.[4,5] The coupling of the fast wave to the ion Bernstein wave was treated as resulting in a modified propagation description of the fast wave, so that the power not transmitted or reflected on the fast wave appears as "absorbed". This "absorption" is clearly a combination of mode conversion and dissipation.

REFERENCES

(1) A. Kay, R.A. Cairns, and C.N. Lashmore-Davies, in Proceedings of the 13th European Conf. on Controlled Fusion and Plasma Heating, Schliersee, 1986, Vol. 10C, part 2, p. 93.
(2) K.G. Budden, the Propagation of Radio waves (Cambridge U.P., Cambridge, 1985) Chap. 19.
(3) D.G. Swanson, Nucl. Fusion 20, 949 (1980).
(4) P.L. Colestock and R.J. Kashuba, Nucl. Fusion 23, 763 (1983).
(5) K. Imre and H. Weitzner, submitted to The Physics of Fluids.

NON-RESONANT MODE-COUPLING MODEL OF ICRF HEATING

G. Francis,† A. Bers, and A. K. Ram
M.I.T. Plasma Fusion Center, Cambridge, MA 02139

ABSTRACT

The theory of non-resonant coupling of a well-defined mode to a quasimode is generalized to include the effects of weak inhomogeneity. The resulting formalism is used to determine an analytic form for the transmission coefficient of the fast wave in ICRF heating; this entails the solution of only a first order differential equation. The analytical results, valid for arbitrary $k_{\parallel}$, are compared with numerical results of usual models involving higher-order differential equations, and found to be in excellent agreement.

INTRODUCTION

Mode-coupling between an incident fast wave and an ion-Bernstein mode is usually treated by numerically solving a fourth-order differential equation which represents the two modes in a plasma system[1,2]. The theory of non-resonant coupling of a well defined mode to a quasimode[3] has recently been generalized to include inhomogeneous plasmas[4]. In this paper we make use of this formalism to treat analytically the fast wave transmission through the mode-conversion region in ICRF heating. Cairns and Lashmore-Davies, through a somewhat different approach, have given a similar treatment for wave transmission at the second harmonic of the electron cyclotron frequency[5].

NON-RESONANT COUPLING OF MODES

We assume a slab geometry with a weak spatial inhomogeneity in the x direction. The Vlasov-Maxwell system of equations may be written in the form[6]:

$$\int_0^{\infty} dt' \int_{-\infty}^{+\infty} d^3\bar{r}' \,\overline{\overline{D}}(\bar{r}-\bar{r}', t-t'; \underbrace{\frac{x+x'}{2}}_{\text{inhomogeneity}}) \cdot \overline{E}_M(\bar{r}', t') = \overline{E}_p(\bar{r}, t), \tag{1}$$

where $\overline{E}_M$ is the mode electric field of the well defined wave, and $\overline{E}_p$ is a localized perturbation due to the presence of a quasimode. We seek a solution of Eq.(1) of the form:

$$\overline{E}_M(\bar{r}, t) = a(\bar{r}, t) E_0(x) \bar{\varepsilon}_0(x) e^{i\int k_x(x)dx + ik_z z - i\omega t} \equiv a(\bar{r}, t)\overline{E}_0(x) \tag{2}$$

where $E_0(x)$, $\bar{\varepsilon}_0(x)$, $k_x(x)$, and ω are the unperturbed amplitude, polarization vector, wavenumber, and frequency, respectively, of the mode in the absence of $\overline{E}_p(\bar{r}, t)$. The perturbing effects of the coupling are accounted for in the slowly varying amplitude function $a(\bar{r}, t)$. We expand the left hand side of Eq.(1) about a local solution to first order in all the quantities associated with the weak inhomogeneity. The perturbation $\overline{E}_p(\bar{r}, t)$ is treated locally. We assume that dissipation in the unperturbed system is weak, such that the anti-hermitian part of the local dispersion tensor, $\overline{\overline{D}}^a$, is ordered with the inhomogeneity. We first solve for the unperturbed quantities by setting $\overline{E}_p = 0$. The local wavenumber and the orientation of the polarization vector are the determined by the local dispersion relation:

$$\overline{\overline{D}}^h(k_x(x), \omega; x) \cdot \bar{\varepsilon}_0(x) = 0 \tag{3}$$

† **Permanent Address: Lawrence Livermore National Laboratory.**

where $\overline{\overline{D}}^h$ is the hermitian part of the local dispersion tensor. If we choose the (power flux) normalization of $\overline{\varepsilon}_0(x)$ such that:

$$\overline{\varepsilon}_0^* \cdot \frac{\partial \overline{\overline{D}}^h}{\partial k_x} \cdot \overline{\varepsilon}_0 = -\frac{\zeta}{k_x}; \quad s_x = \frac{\omega \epsilon_0}{4} \frac{\zeta}{k_x} |E_0(x)|^2 \tag{4}$$

where $\zeta = +1$ for a forward wave and -1 for a backward wave, then the equation for the unperturbed amplitude becomes:

$$\frac{dE_0}{dx} = \left[\frac{1}{2k_x} \frac{dk_x}{dx} + i\zeta k_x \overline{\varepsilon}_0^* \cdot \overline{\overline{D}}^a \cdot \overline{\varepsilon}_0 \right] E_0(x). \tag{5}$$

We then make use of Eq.(5) to solve for the perturbed, slowly varying amplitude function $a(\overline{r}, t)$:

$$\left[\frac{\partial}{\partial t} + \overline{v}_{g0}(x) \cdot \nabla \right] a(\overline{r}, t) = \frac{-\frac{1}{4} \overline{E}_0^*(x) \cdot \overline{J}_p(\overline{r}, t)}{\mathrm{w}_0(x)}, \tag{6}$$

where $\overline{v}_{g0}$ and w_0 are the local group velocity and energy density, respectively, of the unperturbed wave, and $\overline{J}_p(\overline{r}, t)$ is the quasimode current density associated with $\overline{E}_p$.

SECOND HARMONIC HEATING

We treat first the mode conversion of a fast magnetosonic wave in a plasma of a single ion species near $\omega = 2\omega_{ci}$. We assume a slab geometry where the inhomogeneity is in x, perpendicular to the toroidal magnetic field $\overline{B}(x) = B(x)\,\hat{z}$. We also assume a linear variation in the magnetic field, and position the second harmonic cyclotron resonance at $x = 0$. We allow for arbitrary oblique incidence of the fast wave, $\overline{k}(x) = k_\perp(x)\hat{x} + k_\parallel \hat{z}$, and seek the fast wave power transmission coefficient T_{2H}. The plasma dispersion tensor describing both the fast wave and the Bernstein mode is adequately represented by neglecting the electron inertia ($E_z = 0$) and expanding to first order in ion finite Larmor radius (FLR). This approximate dispersion tensor can be split up into a zero-order FLR portion, $\overline{\overline{D}}_0$, which describes the fast wave, and a first-order FLR portion, $\overline{\overline{\chi}}_1$, which we identify with the quasimode perturbation of the fast wave:

$$\overline{\overline{D}}_0(\overline{k}, x) = \begin{pmatrix} -n_\parallel^2 - \frac{1}{3}\frac{c^2}{v_A^2} & -i\frac{2}{3}\frac{c^2}{v_A^2} \\ i\frac{2}{3}\frac{c^2}{v_A^2} & -n_\perp^2 - n_\parallel^2 - \frac{1}{3}\frac{c^2}{v_A^2} \end{pmatrix} \tag{7}$$

$$\overline{\overline{\chi}}_1 \approx \frac{n_\perp^2}{4\alpha_D} \frac{v_{tD}^2}{v_A^2} Z\left(\frac{x}{R_0 \alpha_D} \right) \begin{pmatrix} 1 & i \\ -i & 1 \end{pmatrix}, \tag{8}$$

where $Z(\xi)$ is the plasma dispersion function, v_A is the Alfven velocity, R_0 is the major radius, $v_{tD}^2 = 2T_D/m_D$, and $\alpha_D = n_\parallel v_{tD}/c$. The perturbing current density in Eq.(6) is thus given by:

$$\overline{J}_p = -i\omega\epsilon_0 \overline{\overline{\chi}}_1 \cdot \overline{E}_M \tag{9}$$

The transmission coefficient is found by solving Eqs.(5) and (6) using Eqs.(7)-(9). The result is:

$$\mathrm{T}_{2H} = \frac{\left|\overline{E}_M(x \to -\infty)\right|^2}{\left|\overline{E}_M(x \to +\infty)\right|^2} = e^{-2\pi\mu_{2H}} \tag{10}$$

where

$$\mu_{2H} = \frac{1}{\pi} \int_{-\infty}^{+\infty} |k_\perp| \mathrm{Im}\left\{ \overline{\varepsilon}_0^* \cdot \overline{\overline{\chi}}_1 \cdot \overline{\varepsilon}_0 \right\} dx = \frac{R_0 \omega_{pD}}{4c} \frac{N_{\perp 0}^5}{(1 + N_\parallel^2)^2} \frac{v_{tD}^2}{v_A^2} \tag{11}$$

$$N_{\perp 0}^2 \equiv n_{\perp 0}^2 \frac{v_A^2}{c^2} = \frac{(1-3N_\parallel^2)(1+N_\parallel^2)}{(1+3N_\parallel^2)}; \quad N_\parallel \equiv n_\parallel \frac{v_A}{c}. \tag{12}$$

This simple analytic result reduces to the Budden coefficient[7] in the limit $k_\parallel \to 0$, as it should. In Fig. 1 we compare the $k_\parallel$ dependence of our analytical result to the numerical results of Colestock and Kashuba[1] for the typical PLT parameters listed in Fig. 13 of their paper.

MINORITY HEATING

We now treat the more general case when a second ion species is present in the plasma. In particular, we look at the degenerate case of a D(H) plasma, such that the second harmonic of the deuterium and the fundamental of the minority hydrogen are coincidental in physical space. The presence of the minority species modifies the zero-order FLR portion of the plasma dispersion tensor:

$$\overline{\overline{D}}_0 \to \overline{\overline{D}}_0 + \frac{\eta}{4\alpha_H}\frac{c^2}{v_A^2} Z\left(\frac{x}{R_0\alpha_H}\right)\begin{pmatrix} 1 & i \\ -i & 1\end{pmatrix}, \tag{13}$$

where $\eta = n_h/n_d$ is the minority concentration ratio, and $\alpha_H \equiv n_\parallel v_{tH}/c$. If the hermitian part of this local dispersion tensor is used to calculate the local wavenumber, we find the cold ion-ion hybrid resonance at $x \approx 3\eta R_0/4$. This artificial resonance does not exist in the actual warm plasma we are modeling, but is simply a consequence of the way we have separated out the first-order FLR effects from the local dispersion tensor. We remove this singularity by expanding $\overline{\overline{D}}_0^h$ about the minority resonance:

$$\overline{\overline{D}}_0^h(x) \approx \overline{\overline{D}}_0^h(x=0) + \left.\frac{\partial \overline{\overline{D}}_0^h}{\partial x}\right|_0 x \tag{14}$$

and renormalizing the expansion of Eq.(1) such that the second term on the right hand side of (14) is ordered with other terms of first order in the weak inhomogeneity. With this choice of renormalization and $\overline{\overline{\chi}}_1$ as in (8), the resulting transmission coefficient is now found to have the simple form:

$$\mathrm{T}_{ii} = \frac{|\overline{E}_M(x\to -\infty)|^2}{|\overline{E}_M(x\to +\infty)|^2} = e^{-2\pi\mu_{ii}}, \tag{15}$$

where

$$\mu_{ii} = \frac{1}{\pi}\int_{-\infty}^{+\infty} |k_\perp| \mathrm{Im}\left\{\overline{\varepsilon}_0^* \cdot \left(\overline{\overline{D}}_0^a + \overline{\overline{\chi}}_1\right)\cdot \overline{\varepsilon}_0\right\} dx = \frac{R_0\omega_{pD}}{4c}\frac{N_{\perp 0}^5}{(1+N_\parallel^2)^2}\left[\frac{v_{tD}^2}{v_A^2} + \frac{\eta}{N_{\perp 0}^2}\right]. \tag{16}$$

In the limit $\eta \to 0$, this reduces to the second harmonic transmission coefficient (10). For minority concentrations much larger than the plasma beta $\beta = v_{tD}^2/v_A^2$, the transmission coefficient is independent of temperature. In Figures 2-4, we show the dependence of T_{ii} on the various plasma parameters $k_\parallel$, η, and n_e, respectively, and compare them with the numerical results of Colestock and Kashuba[1] for typical PLT parameters.

REFERENCES

1. P. L. Colestock and R. J. Kashuba, Nuclear Fusion, **23**, 763 (1983).
2. H. Romero and J. Scharer, Nuclear Fusion, **27**, 363 (1987).
3. A. Bers, in Plasma Physics - Les Houches 1972, (C. DeWitt and J. Peyraud, eds.), Gordon and Breach Science Publishers, NY, 1975.
4. G. Francis, Ph.D. Thesis, Physics Department, M.I.T., Cambridge, MA, 1987.

5. R. A. Cairns and C. N. Lashmore-Davies, Phys. Fluids, **29**, 3639, (1986).

6. H. L. Berk and D. L. Book, Phys. Fluids, **12**, 649 (1969).

7. K. G. Budden, Radio Waves in the Ionosphere, Cambridge University Press, Cambridge, 1961.

Work supported by DOE Contract No. DE-AC02-78ET-51013.

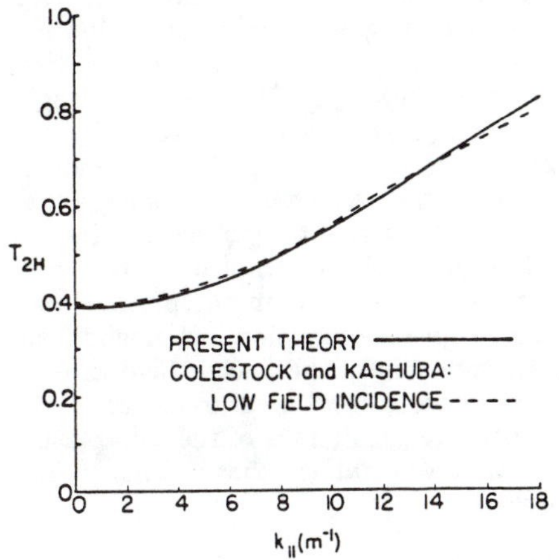

Fig. 1: Comparison of the $k_{\|}$ dependence of T_{2H} with the results of Colestock and Kashuba[1].

PRESENT THEORY
COLESTOCK and KASHUBA
Tii
k||(m⁻¹)

Fig. 2: Comparison of the $k_{\|}$ dependence of T_{ii} with the results of Colestock and Kashuba[1].

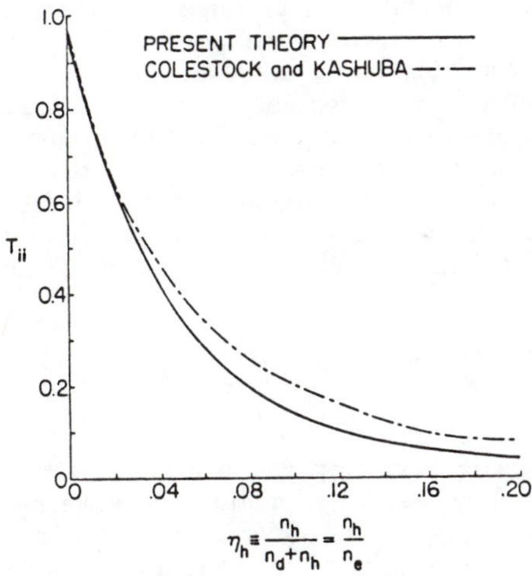

Fig. 3: Comparison of the minority concentration dependence of T_{ii} with the results of Colestock and Kashuba[1].

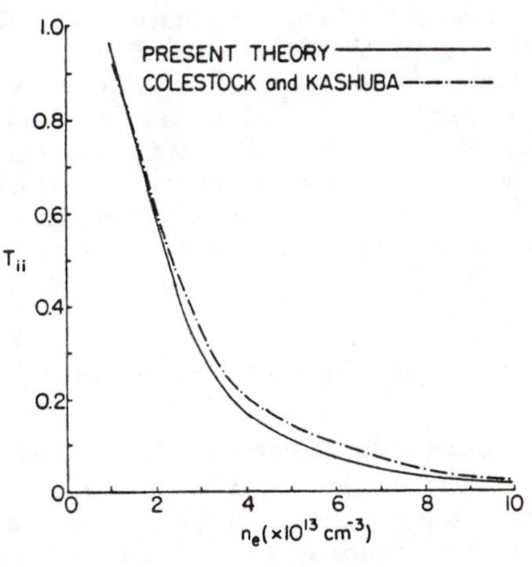

Fig. 4: Comparison of the plasma density dependence of T_{ii} with the results of Colestock and Kashuba[1].

Modeling of ICRH Experiments in the Tara Tandem Mirror*

R. C. Myer and S. N. Golovato
Plasma Fusion Center, M.I.T., Cambridge, MA 02139

I. ABSTRACT

The production and heating of the central cell plasma in Tara is provided by a slot antenna located on the midplane bump of the axial magnetic field profile. Slow ion cyclotron waves excited by the slot propagate down a magnetic beach to ion cyclotron resonance layers located on either side of the bump where the RF power is strongly damped by the ions. Two different theoretical models are being used to study the efficiency of coupling to slow waves in this configuration. Wave propagation models which are based on the infinite plasma dispersion relation for a cold plasma indicate that radially propagating left hand polarized slow waves are converted to right hand polarized fast waves at the Alfvén resonance layer due to the radial density gradient. If this were to occur we would expect a lower coupling efficiency to the ions in the plasma core. On the other hand, a nonlocal kinetic model of RF wave propagation in a nonuniform plasma slab indicates that a significant left hand component of the electric field extends beyond the Alfvén resonance layer. Preliminary experimental measurements of the radial inductive field profile agree qualitatively with the predictions of the cold plasma model, however, there is insufficient data at this time to establish that a density limit for slow wave accessibility to the plasma core exists.

II. INTRODUCTION

Plasma heating in the ion cyclotron range of frequencies (ICRF) has been used throughout the Tara tandem mirror experiment for producing, heating, and stabilizing (RF ponderomotive stabilization) the central cell plasma, potential modification in the plug cell to reduce the particle endloss (RF plugging), and stabilizing the plasma by increasing the beta in the anchor cells. Of the different ICRF antennas that are used, the central cell slot antenna plays the dominant role in the overall power balance and stability of the plasma. The slot antenna is located just off the midplane of the machine in a local maximum of the magnetic field and tuned to a frequency below the local ion cyclotron frequency. This configuration is ideal for coupling to left hand polarized (ie. rotating in the same sense as the ions) slow ion cyclotron waves which propagate down the "magnetic beach" to be strongly damped by the ions at the resonance layers located on either side of the antenna. Theoretical models of the wave coupling, propagation, and damping are being used in order to better understand the operation of the slot antenna.

III. DESCRIPTION OF THEORETICAL MODELS AND RESULTS

A. Local Cold Plasma Dielectric Models

The simplest description of electromagnetic wave propagation in a magnetized plasma is obtained in the cold plasma limit where the characteristic field wavelengths are large compared to the ion Larmor radius ($k_\perp \rho_i \ll 1$). McVey's code ANTENA[1] solves the field equations in this limit for the electromagnetic field in a radially stratified plasma cylinder whose dielectric properties are given locally by the infinite cold plasma dielectric function. The wave damping mechanisms included in this model are cyclotron

* Supported by U.S. DOE Contract DE-AC02-78ET51013

damping at the fundamental ion cyclotron frequency and electron Landau damping. The model also retains the effect of finite electron inertia ($E_z \neq 0$) which is crucial in modeling the slot antenna since the large axial electric field that it generates is responsible for heating the electrons which, in turn, provide the energy for ionizing the central cell plasma. This code is used to predict radial profiles of the electromagnetic field and power absorption, as well as the antenna impedance, for various antenna geometries. The McVey code provides a good estimate of the antenna impedance despite the fact that it assumes a uniform axial magnetic field whereas the slot antenna is located on a local maximum of the magnetic field. The good agreement between the code and the observed loading is presumably due to the fact that the uniform magnetic field model is adequate for calculating the wave coupling in the vicinity of the antenna. However, one would expect the code to underestimate the ion absorption due to slow wave heating since the effect of the ion resonance layer is not included in the model.

The physics of the coupling to a cold plasma column is contained in the infinite plasma dispersion relation,

$$Sk_\perp^4 + k_\perp^2[k_z^2(S+P) - k_0^2(SP+RL)] + P(k_z^2 - k_0^2R)(k_z^2 - k_0^2L) = 0 \qquad (1)$$

where S, P, R, and L are the usual Stix dielectric tensor elements and $k_0 = \omega/c$. The dependence of $k_\perp$ on k_z for conditions of slow wave heating ($\omega/\Omega_i < 1$) is shown in Fig. 1. The two perpendicular cutoffs at the points labeled 1 and 2 in Fig. 1, occur at $k_z^2 = k_0^2R$ and $k_z^2 = k_0^2L$, respectively. The point $k_z^2 = k_0^2S$, which lies between these two cutoffs, corresponds to the Alfvén wave resonance and separates the fast and slow branches. If we consider the radial propagation of a wave for a given $k_z = k_s$, which is originally on the slow branch, then there is a density at which the perpendicular wavenumbers switch from the slow branch to the fast branch with a concomitant change in polarization from left hand polarized to right hand polarized. The density at which this occurs is given by $\omega_{pi}^2 \approx k_s^2c^2[1-(\omega/\Omega_i)^2]$. For the slot parameters in Tara this corresponds to a density of $n_e = 3.0 \times 10^{12}\mathrm{cm}^{-3}$. Results from the antenna code show a decrease in the power coupled to the core by a factor of 10 when the density is varied (keeping all other plasma parameters fixed) from $3.0 \times 10^{12}\mathrm{cm}^{-3}$ to $5.0 \times 10^{12}\mathrm{cm}^{-3}$. Experimental evidence of a density limit of about $4.0 \times 10^{12}\mathrm{cm}^{-3}$ has been observed for some shots when the axial confinement time is enhanced by double ended plugging. However, it has not been conclusively demonstrated that radial accessibility of the slow wave is the mechanism responsible for the density limit in these shots.

B. Nonlocal Electromagnetic Model

A nonlocal kinetic model of electromagnetic wave propagation in a nonuniform plasma slab developed previously[2] is being used to study the propagation of slow waves through the plasma edge where the conversion from slow waves to fast waves should occur. This model assumes a plasma slab which is infinite in the y and z directions and has an arbitrary density variation in the x-direction. The nonlocal model includes finite Larmor radius corrections to all orders and the effect of diamagnetic drifts without making an assumption regarding the ratio of the density gradient scalelength to the field wavelength. In this model, we numerically solve an integral equation for the Fourier transformed electric field and then reconstruct the spatial variation of the fields by performing the inverse Fourier transform.

The Alfvén wave resonance appears most clearly in the radial profile of the y-component of the RF magnetic field as shown in Figure 2. In this case a current sheet placed at $x = +15\,\mathrm{cm}$ is used to excite slow waves in a plasma slab whose density profile is shown by the dashed curve. The real and imaginary parts of the complex B_y field are labeled. For the parameters chosen in this case ($n_e = 1.0 \times 10^{12}\mathrm{cm}^{-3}, k_z = .05\,\mathrm{cm}^{-1}$, and $\omega/\Omega_i = .85$), the Alfvén resonance occurs in the edge plasma and appears as a bump in the amplitude of B_y. The bump in the RF magnetic field is due to a dipole distribution of the axial plasma current which is driven by a rapid change in the sign of the axial electric field in the plasma edge. As a result, the electron power absorption profile develops the double peaked structure as shown in Figure 3. The surprising result is that the left hand polarized electric field, E^+, shown in Fig. 4, does not show a dramatic reduction in magnitude as one might have expected based on the cold plasma model.

IV. CONCLUDING REMARKS

The theoretical predictions of the local cold plasma model and a nonlocal kinetic model of radially propagating slow waves are in strong disagreement. The reduction in E^+ across the Alfvén wave resonance layer that one would expect from the cold plasma model is not observed in the nonlocal model. Furthermore, the electron power absorption profile is radically different from that predicted by the cold plasma model. The resolution of the discrepancy of these models will require a comparison of the theoretical predictions to fairly detailed measurements of the inductive field profile.

REFERENCES

1. B. D. McVey, MIT Report PFC/RR-84-12 (1984).
2. R. C. Myer and B. D. Fried, MIT Report PFC/JA-87-20 (1987).

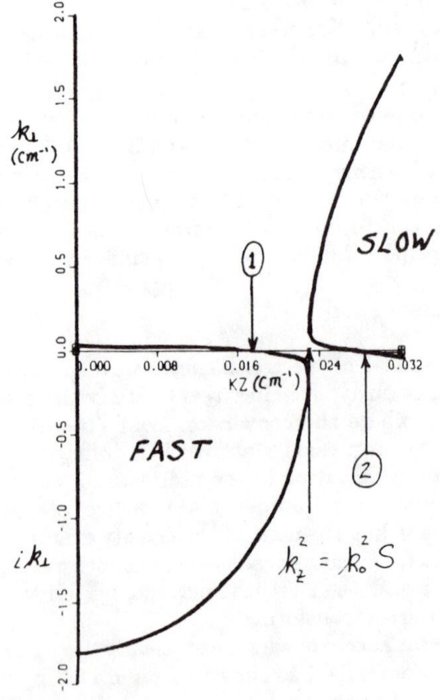

Figure 1. Cold Plasma Dispersion Curve ($\omega/\Omega_i < 1$)

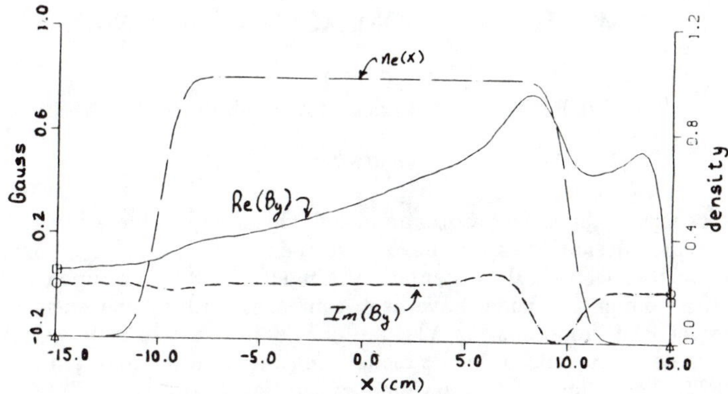

Figure 2. B_y and density vs. x

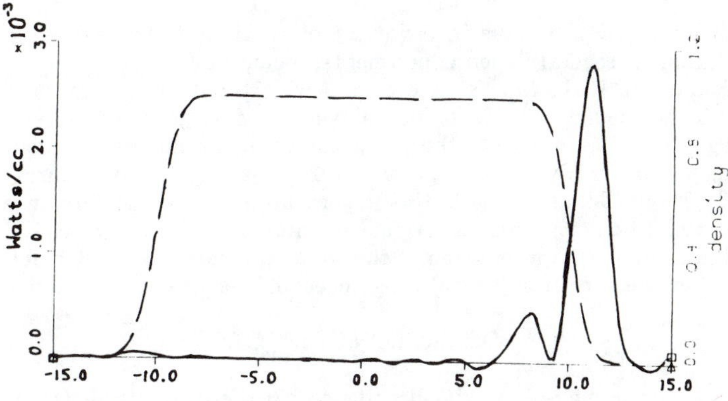

Figure 3. Electron power absorption profile

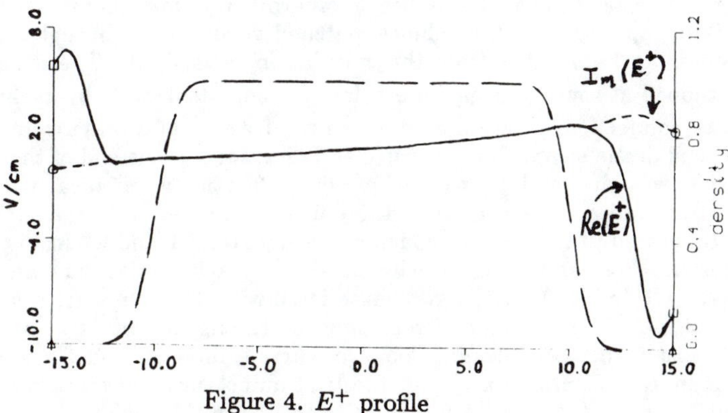

Figure 4. E^+ profile

MODELING OF COMPACT LOOP ANTENNAS*

F. W. Baity
Oak Ridge National Laboratory, Oak Ridge, TN 37831

ABSTRACT

A general compact loop antenna model which treats all elements of the antenna as lossy transmission lines has been developed. In addition to capacitively tuned resonant double loop (RDL) antennas, the model treats stub-tuned RDL antennas. Calculations using the model have been compared with measurements on full-scale mock-ups of RDL antennas for ATF and TFTR in order to refine the transmission line parameters. Results from the model are presented for RDL antenna designs for ATF, TFTR, Tore Supra, and the Compact Ignition Tokamak (CIT).

INTRODUCTION

Compact loop antennas in a variety of configurations are being planned for ICRF heating on several fusion experiments, including CIT. Compact loop antennas, in particular the RDL, can operate over a wide frequency range at high rf power levels, can be moved relative to the plasma, and require no external impedance matching. In order to predict the performance of the antennas under a variety of operating conditions, over as large as a 3:1 frequency range, accurate models are required. The model described in this paper can predict the performance of antennas over the full frequency and load range of interest. The model is thus usable for determining the optimum position of the input coaxial feed and for determining the actual capacitance (or stub length) range needed for a specific application.

DESCRIPTION OF THE CIRCUIT

A schematic diagram of the circuit used in the model is shown in Fig. 1. The configuration shown is that of an RDL with capacitor tuning,[1] the configuration for the ATF, TFTR, and Tore Supra designs. RF power is fed to the antenna through an arrangement consisting of a gas break of arbitrary impedance, a fixed stub for cooling fluid input, and an impedance-matched vacuum feedthrough. Any of these components can be omitted from the calculation, if desired. The antenna current strap is tapped at some intermediate point α along its length by means of a low-inductance connection, where α represents the fraction of total current strap length from one end of the strap. A dielectric disk is included at the end of the vacuum feed coax and serves the dual purpose of providing partial mechanical support for the current strap and canceling most of the effect of the feed inductance. The current strap is connected to the tuning capacitors, designated C1 and C2 in Fig. 1, by short transmission line segments. Each variable capacitor also has an inductance, designated LC1 and LC2 in the figure, associated with it which is not negligible for the designs considered. The option of replacing the tuning capacitors with shorted stubs has been provided. Stub tuning looks attractive for CIT and for high-frequency operation on Tore Supra. Except for the lead inductance and the tuning capacitors, each element is treated as a lossy transmission line in the model.

*Research sponsored by the Office of Fusion Energy, U. S. Department of Energy, under contract DE-AC05-84OR21400 with Martin Marietta Energy Systems, Inc.

Circuit Model

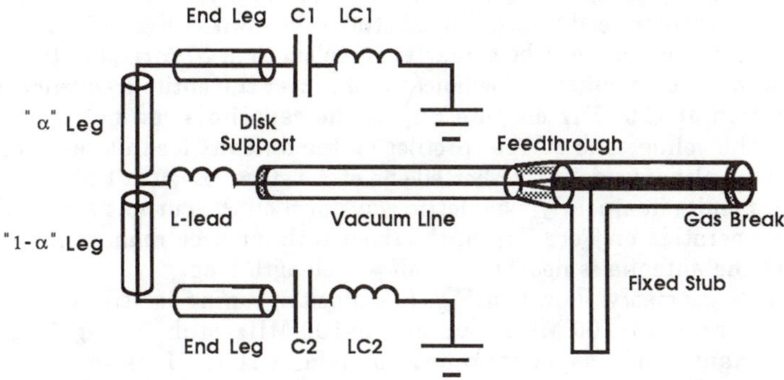

Fig. 1. Schematic diagram of the circuit used in the model.

The two capacitors tune the antenna to the desired frequency and adjust the input impedance to match the transmitter impedance. No additional matching components are required. The effect of the fixed stub can be eliminated over a 3:1 bandwidth by slight retuning of the capacitors.

The model requires the characteristic impedance and phase velocity for each transmission line segment. For the current strap these data can be obtained from 2-D calculations[2] which include the effect of the Faraday shield or from experimental measurements on a mock-up.

The model can be used to calculate the voltage and current distribution along the current strap and tuning elements and the tuning positions required as a function of antenna load resistance. This information is required during the design of an antenna in order to determine (1) the optimum tap position along the current strap for the frequency and load range of operation anticipated and (2) the power handling capability of the antenna resulting from voltage and current limitations on the various components.

RESULTS

RDL antennas for fast wave heating have been designed for TFTR[3] at PPPL, for Tore Supra[3] at CEN-Cadarache, and for ATF at ORNL; conceptual designs have been studied for CIT. The antennas for TFTR and for Tore Supra have several similarities. Both have two current straps side by side with a common Faraday shield, are designed to couple 2 MW per loop into the plasma, and work in about the same frequency range (35 to 80 MHz for Tore Supra). The antenna for the ATF stellarator has a single loop and operates over a lower frequency range (10 to 30 MHz) than TFTR and Tore Supra. The CIT antenna has two current straps stacked end to end in a single port. The design frequency will depend on the final CIT parameters, but should be in the range of 70 to 120 MHz.

Current and voltage profiles for the Tore Supra antenna at 35 MHz and at 80 MHz are shown in Figs. 2 and 3, respectively. The plot extends from one capacitor terminal to the other. The dotted vertical line represents the position of the feed line, and the two solid vertical lines are drawn at the ends of the current strap. Both the cases plotted are near the minimum loading required to couple 2 MW into the plasma for the frequency under consideration, using capacitor limits of 1-kA rms current and 50-kV peak voltage. Note that the tap position, α, of 0.65 is nearly optimal for loads in this frequency range, resulting in fairly symmetric profiles. The voltage and current in the vacuum feed line and fixed stub are shown in Fig. 4 for the 80-MHz case. The fixed stub was chosen to be a quarter wavelength at 57 MHz, but the antenna can be matched to the transmitter impedance of 30 Ω over the entire frequency band.

For operation at 120 MHz on Tore Supra, the capacitors can be replaced with shorted stubs. The voltage and current profiles on the antenna for this case are shown in Fig. 5, where the plot extends to the shorted end of the stubs. Even at 120 MHz the tap position of 0.65 results in a nearly symmetric waveform on the current strap. For this high-frequency operation on Tore Supra, the fixed stub must be reduced in length by half. Note that the antenna is nearly one-half wavelength long.

For the CIT geometry shown in Fig. 6, capacitor tuning is still feasible up to frequencies greater than 100 MHz, but above 100 MHz stub tuning looks more attractive. An example of a capacitor-tuned antenna at 100 MHz is shown in Fig. 7. For a loading of 10 Ω/m and a maximum capacitor voltage of 50 kV, 1.3 MW per loop can be achieved. Nevertheless, with higher loading or voltage limits, it would appear that 1.6 MW per loop is a realistic goal for CIT antennas operating near 100 MHz.

REFERENCES

1. T. L. Owens, F. W. Baity, and D. J. Hoffman, p. 95 in Radiofrequency Plasma Heating (AIP Conf. Proc. 129, New York, 1985).
2. G. L. Chen, P. M. Ryan, D. J. Hoffman, F. W. Baity, D. W. Swain, and J. H. Whealton, "Resonant Loop Antenna Design with a 2-D Steady State Analysis," this conference.
3. D. J. Hoffman, F. W. Baity, W. E. Bryan, G. L. Chen, K. H. Luk, T. L. Owens, J. M. Ray, P. M. Ryan, D. W. Swain, and J. C. Walls, "The Design of High Power ICRF Antennas for TFTR and Tore Supra," this conference.

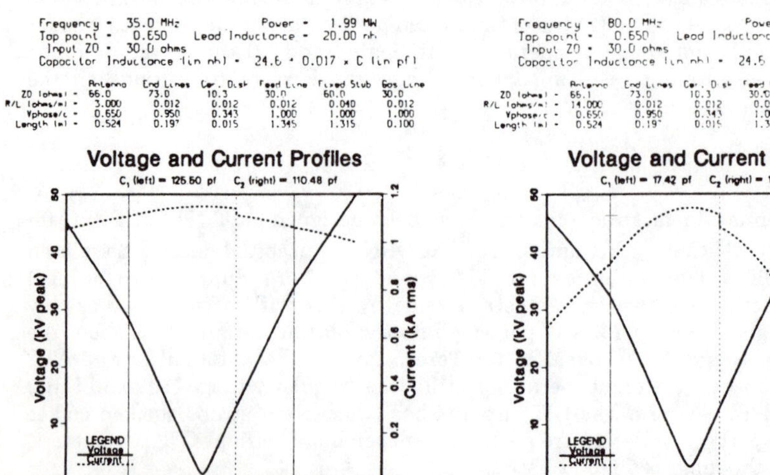

Fig. 2. Current and voltage profiles for the Tore Supra antenna at 35 MHz.

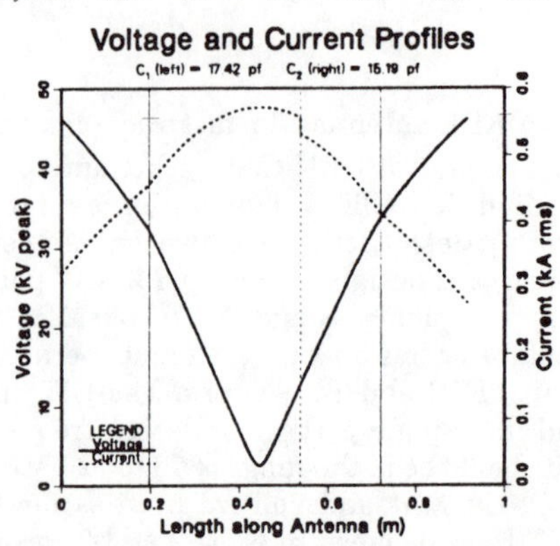

Fig. 3. Current and voltage profiles for the Tore Supra antenna at 80 MHz.

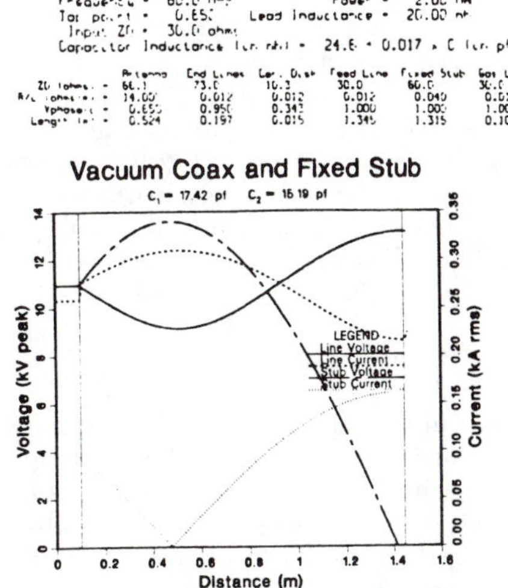

Fig. 4. Current and voltage profiles in the vacuum coax and fixed stub for the Tore Supra antenna at 80 MHz.

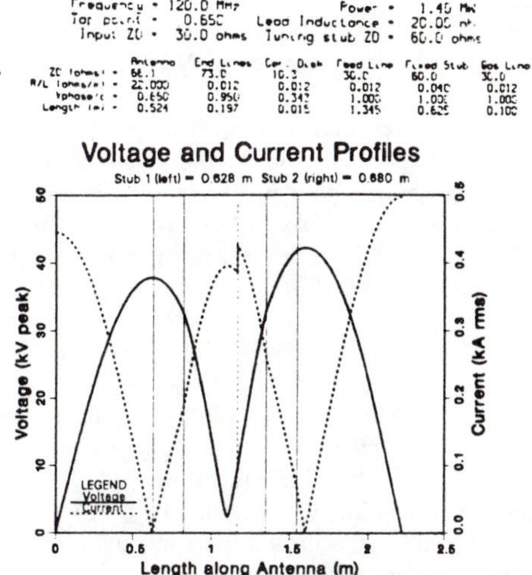

Fig. 5. Current and voltage profiles for the Tore Supra antenna at 120 MHz with stub tuning.

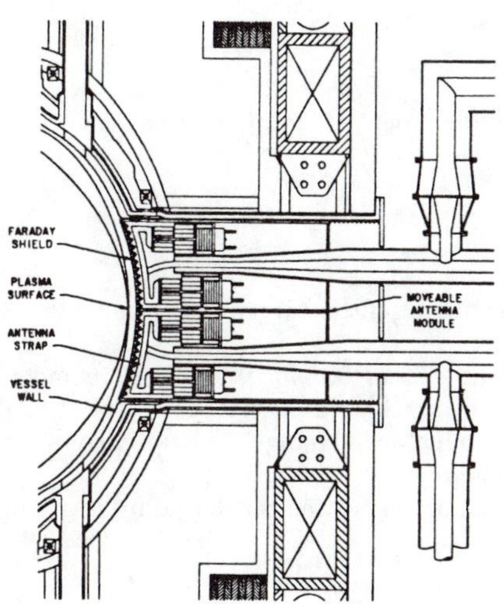

Fig. 6. Conceptual CIT ICRF antenna with capacitor tuning.

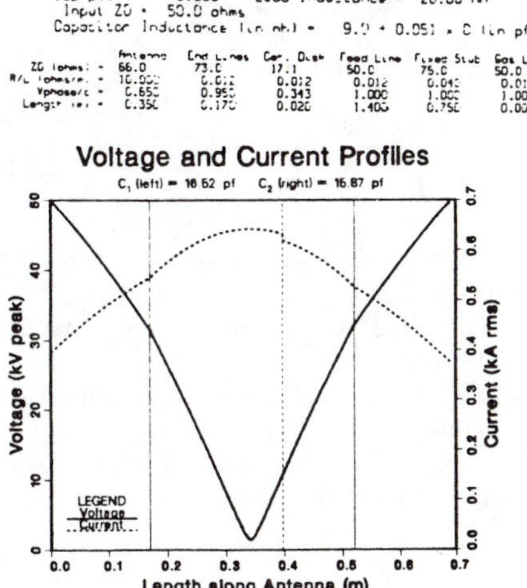

Fig.7. Current and voltage profiles for the CIT antenna at 100 MHz.

RESONANT LOOP ANTENNA DESIGN WITH A 2-D STEADY-STATE ANALYSIS*

G. L. Chen, P. M. Ryan, D. J. Hoffman, F. W. Baity, D. W. Swain, J. H. Whealton
Oak Ridge National Laboratory, Oak Ridge, TN 37831

Evaluation of resonant loop antenna designs for ICRF heating of plasmas requires information concerning the electrical characteristics of the structure. The 2-D steady-state model described here provides current strap inductance and capacitance, surface current distributions, and flux linkage to the plasma. These are used to determine the current and voltage requirements, ohmic dissipation, frequency limits and matching requirements, maximum electric fields, and plasma loading in order to compare antenna designs.

THE MAGNETOSTATIC MODEL

We consider a resonant loop structure consisting of a current strap, the surrounding conducting cavity, and the plasma surface; it is similar to that treated by Mau et al.[1] and is shown in Fig. 1. The antenna and plasma are considered uniform in the ignorable y-direction (poloidal direction in tokamak geometry). The plasma and cavity walls are considered perfectly conducting surfaces, enabling the system to be described by the Laplace equation for A_y:

$$\nabla^2 A_y = \frac{\partial^2 A_y}{\partial x^2} + \frac{\partial^2 A_y}{\partial z^2} = 0 \ , \tag{1}$$

with the boundary conditions $A_y = A_{y0}$ on the current strap surface and $A_y = 0$ on the cavity walls and the plasma surface.

The magnetic field can be expressed as

$$\mathbf{B} = \nabla \times \mathbf{A} \ \rightarrow \ B_x = \frac{-\partial A_y}{\partial z} \ , \ \ B_z = \frac{\partial A_y}{\partial x} \ . \tag{2}$$

Knowledge of A_y and $\mathbf{B}$ enables us to solve for the inductance per unit length of the current strap,

$$L' = \frac{A_{y0}}{I_0} = \frac{A_{y0}}{\mu_0} \left| \oint \mathbf{B} \cdot d\mathbf{l} \right|^{-1} \ , \tag{3}$$

where the line integral of $\mathbf{B}$ is taken around the perimeter of the current strap to yield the antenna current I_0.

The calculated inductance per unit length was 2.23 nH/cm for the DIII-D geometry and 3.08 nH/cm for the TFTR geometry; the measured values were 2.0 nH/cm and 3.1 nH/cm, respectively. We also verified that the inductance changes only slightly when the antenna is moved closer to the plasma in both machines.

The surface current distribution $\mathbf{J}_s$ on the conducting surfaces is found by invoking the boundary conditions on $\mathbf{B}$ at a perfect conductor,.

$$\mathbf{J}_s = \frac{\mathbf{n} \times \mathbf{B}}{\mu_0} = \frac{\mathbf{n} \times \nabla \times A_y}{\mu_0} \ . \tag{4}$$

*Research sponsored by the Office of Fusion Energy, U.S. Department of Energy, under contract DE-AC05-84OR21400 with Martin Marietta Energy Systems, Inc.

Typical plots of J_s on the strap and the walls for the TFTR geometry are shown in Figs. 2 and 3. It is apparent that the current is concentrated on the ends of the strap and is returned in the wall sections that are closest to them.

This current distribution can then be used to determine the average resistance per unit length in the current strap and the overall ohmic power dissipation in the structure,

$$R' = \frac{\eta\delta}{2} \frac{\oint J_s^2 dl}{(\oint J_s dl)^2} , \tag{5}$$

$$P'_{diss} = R' I^2 , \tag{6}$$

where R' is the effective resistance per unit length, η is the resistivity and δ the skin depth of copper, and I is the rms current in the strap.

The cooling requirements of an antenna become critical as rf pulses extend to multisecond duration and as current densities increase to provide the high power densities required by tokamaks. The ohmic resistance per unit length of the current strap was calculated to be 0.028 Ω/m at 47 MHz and 150°C; the power dissipation using this value was found to be in good agreement with the cw calorimetric measurements made on the TFTR prototype.

For given plasma conditions, the power loading of an ICRH antenna is related to the amount of toroidally directed magnetic flux that is linked to the plasma. If one neglects the k_z dependence and takes the plasma surface impedance $Z_p = E_y/H_z$ to be a constant, the power delivered to the plasma per unit length of antenna is approximated by

$$P'_{plasma} = \frac{Z_p}{\mu_0^2} \int_{-\infty}^{\infty} \left| B_z \right|^2 dz . \tag{7}$$

This may be rewritten in terms of the rms strap current by use of (3), normalized to $A_{y0} = 1$ A-H/m:

$$P'_{plasma} = \frac{Z_p}{\mu_0^2} I^2 \left(L'^2 \int_{-\infty}^{\infty} \left| B_z \right|^2 dz \right) . \tag{8}$$

For a given antenna current, maximizing the term in parentheses maximizes the coupling efficiency of the antenna. Figure 4 shows this coupling factor versus distance to the plasma for the TFTR single loop and the DIII-D antenna geometries.

THE ELECTROSTATIC MODEL

The current distribution along the strap and the frequency range of operation are determined by the wavelength of the antenna, $\lambda = v_p/f$, where the phase velocity is determined by the inductance and capacitance per unit length, $v_p = (L'C')^{-1}$. In the absence of the Faraday shield the phase velocity is the speed of light and $C' = (c^2 L')^{-1}$, the capacitance being primarily determined by the current strap ends and the proximity of the side walls in an analogous fashion to the inductance L'. However, the addition of a perfect Faraday shield ($\sigma_z = \infty$, $\sigma_y = 0$) will influence C' but will not affect L'.

We may estimate the total capacitance in two ways. The first, method I, models a section of a two-tier periodic Faraday shield (Fig. 5) to obtain the strap-to-shield capacitance. The equation to be solved is $\partial^2\Phi/\partial x^2 + \partial^2\Phi/\partial y^2 = 0$, subject to the boundary conditions in Fig. 5. The capacitance per unit length is then

$$C_s' = \frac{Q'}{\Phi_1 - \Phi_2} = \frac{W}{a}\int_0^a E_x dy = \frac{W}{a}\int_0^a \frac{\partial\Phi}{\partial x} dy \ , \tag{9}$$

where W is the effective width of the current strap. The phase velocity is then $v_p = [(C_s' + C_w')L']^{-1/2}$ where $C_w' = (c^2L')^{-1}$, the capacitance due to the side walls.

An alternative method, method II, replaces the Faraday shield with a conducting plane and solves the 2-D Laplace problem in (x, z) space, $C'' = \oint \nabla\Phi \cdot d\mathbf{l}$, with the integration taken around the current strap. The phase velocity is $v_p = (L'C'')^{-1/2}$ with L' calculated without a Faraday shield. Method II is more accurate for large gaps. The phase velocities from both methods are plotted in Fig. 6 as a function of the distance between strap and shield for TFTR geometry.

1. T. K. Mau, S. C. Chiu, and D. R. Baker, "Coupling Analysis for the ICRF Resonant Cavity Launcher on DIII-D," to be published in IEEE Trans. Plasma Sci. (1987).

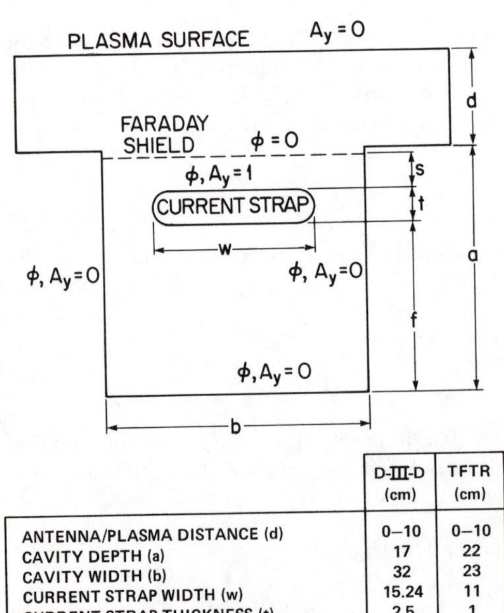

	D-III-D (cm)	TFTR (cm)
ANTENNA/PLASMA DISTANCE (d)	0–10	0–10
CAVITY DEPTH (a)	17	22
CAVITY WIDTH (b)	32	23
CURRENT STRAP WIDTH (w)	15.24	11
CURRENT STRAP THICKNESS (t)	2.5	1
CURRENT STRAP REAR WALL (f)	7.5	16
CURRENT STRAP TO FARADAY SHIELD (s)		1.5

Fig. 1. The geometry and boundary conditions for the steady-state analyses.

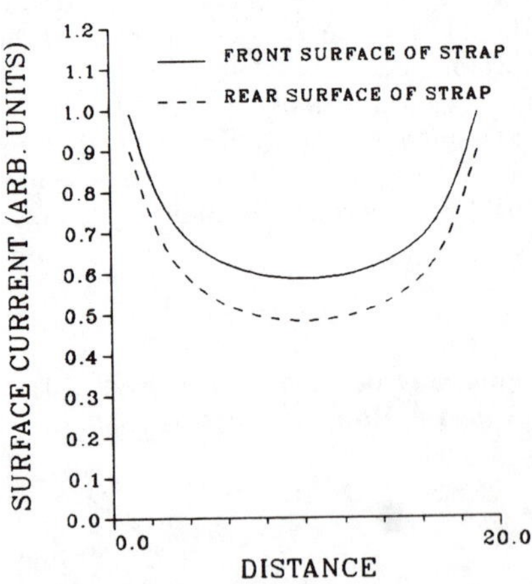

Fig. 2. The surface current distribution on the antenna strap for the TFTR geometry.

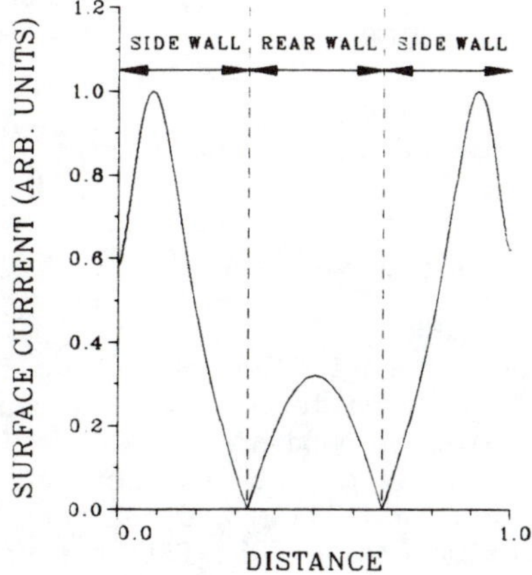

Fig. 3. The surface current distribution on the cavity walls for the TFTR geometry.

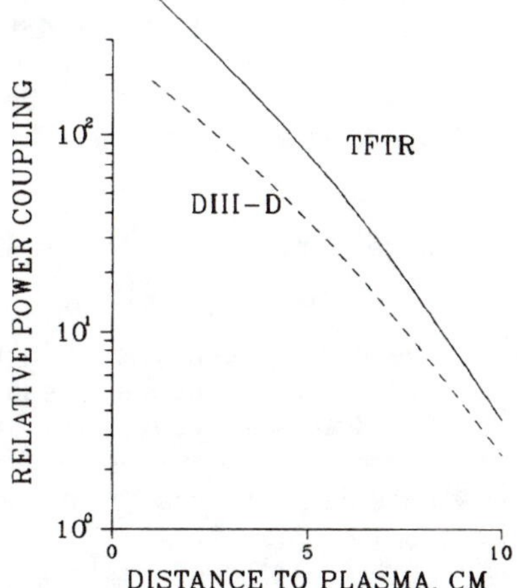

Fig. 4. The power coupling factor as defined in the text versus the strap to plasma distance.

ORNL-DWG 87-2292 FED

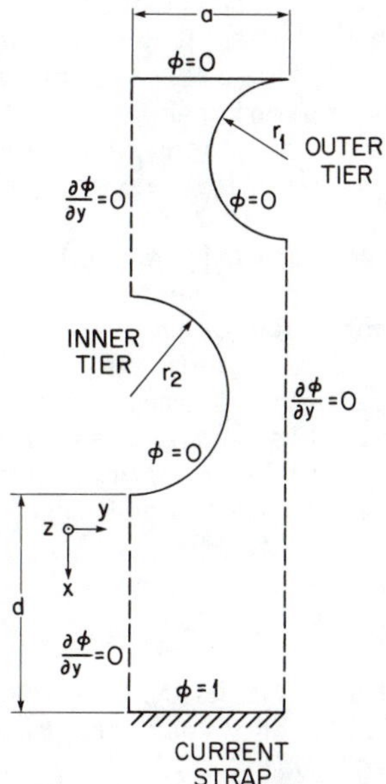

Fig. 5. The geometry for calculating the capacitance per unit periodic area (method I in text).

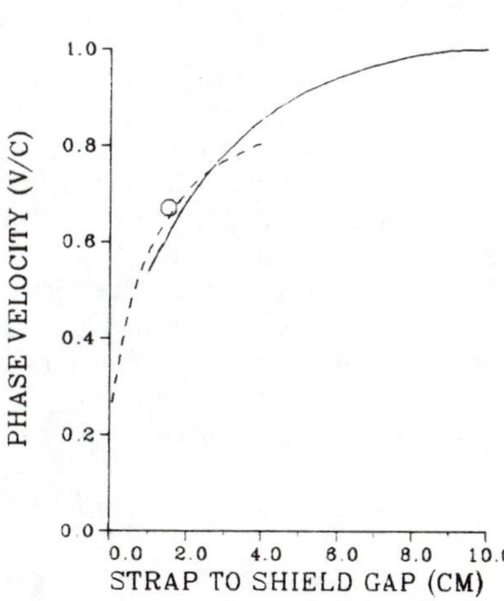

Fig. 6. Phase velocity versus the strap to Faraday shield gap. The dotted line and solid line are methods I and II as mentioned in the text. The data point was measured on the TFTR development antenna.

ICRF CURRENT DRIVE FOR INTOR/ETR

David A. Ehst
Argonne National Laboratory, Argonne, IL 60439

T. K. Mau
University of California-Los Angeles, Los Angeles, CA 90024

MOTIVATION FOR ICRF

There has been encouraging success in small tokamaks using the ion cyclotron resonance frequency (ICRF) range to achieve ion heating with the fast wave, and the associated rf hardware is attractive for reactor applications as bends and windows may be incorporated in the transmission line to effectively contain tritium and neutrons in the reactor vessel. It is likely therefore that ICRF will be a primary candidate for auxiliary heating in an experimental tokamak reactor (ETR).

It is worthwhile, thus, to consider whether wave generated toroidal currents can be driven in an ETR using ICRF technology. For practical purposes, at frequencies less than a few hundred MHz the slow wave only exists near the very low density surface of the plasma, but any fast wave launched with a parallel index of refraction $n_\parallel \gtrsim 1.1$ will propagate unimpeded towards the magnetic axis. The spatial profile of generated current density will depend on the power spectrum and on the variation of $n_\parallel$ and the plasma parameters along the fast wave's ray trajectories. We present here our initial studies of ICRF current drive in which electron Landau and transit time magnetic damping are used to create centrally peaked current profiles in two dimensional MHD equilibrium.

We note that additional current drive opportunities exist with the ICRF system. These alternatives are minority heating (the rf equivalent to neutral beam current drive) and mode conversion of the fast wave to a strongly electron-damped ion Bernstein wave. Furthermore, we suggest that the ETR be provided with a small amount of lower hybrid (LH) slow wave power at ~3 GHz; this system can provide current density near the surface to provide control over the current profile, and it has already shown promise for current ramp or transformer recharge during low density transients in tokamaks.

ETR SIZE SELECTION

Our premise for optimizing an ICRF-driven ETR is that a minimum cost device should be designed which, however, provides the maximum flexibility to achieve a neutron wall load $W_n \geq 1.0$ MW/m^2. In studying different size reactors we used the current drive figure of merit $\gamma \equiv \bar{n}_e R_o I_o / P_{CD}$ (MA/MW/m^2) found in a survey of first stability regime equilibrium generation[1] with the high frequency (GHz) fast wave:

$$\gamma = \bar{\bar{T}}_e^{\,0.77}(0.034 + 0.196\,\beta),$$

where $\bar{T}_e$ is the average electron temperature (keV) and β is the toroidal beta; $\bar{n}_e$ is the average electron density, R_o is the major radius, I_o is the toroidal current, and P_{CD} is rf power absorbed in the plasma. The formula is valid for a broad range of pressure profiles, aspect ratios, and cross sectional shapes and assumes Z_{eff} = 1.5. We expect this formulation to hold as well for lower (ICRF) frequencies.

The comparison of ETR designs involved additional constraints. Beta was limited to the first regime value; the inboard blanket and shield thickness is $\Delta_{B/S}$ = 0.80 m for W_n = 1.0 MW/m^2; the average current density in the toroidal field coil (TFC) is $\bar{J}_{TF} \leq 1.4$ kA/cm^2; and, if present, an ohmic transformer (OHC) is limited to a maximum field $B_{OH} \leq 8.9$ T. In each reactor design I_o is maximized with the safety factor constrained to q = 2.3. The power balance calculation takes credit for the external heating provided by P_{CD}, which eases the electron energy confinement time requirement. We monitored the TFC parameters and the current drive power, which represent capital costs and find that there is a trade-off between P_{CD} and the cost of the TFC. The table summarizes the parameters of two ETR candidates optimized for different burn cycles.

Referring to the table we see that the tokamak designed for steady state (CW) operation has the smaller major radius with minimum stored energy in the coils. It has no OHC; power supplies for the equilibrium field coils (EFC) are minimized assuming leisurely current ramp up is possible. The number of full power burn cycles and fatigue problems are minimized. Disruptions are possibly reduced in frequency, and the duty factor and availability are maximized. Of course, the steady state ETR has a large power requirement, $P_{CD} \simeq 55$ MW. The alternative design, optimized for the hybrid burn cycle, has a fusion burn of a few minutes and has a transformer which is recharged while $\lesssim 20$ MW of rf power sustains the full current, at a low density. Relative to an inductively driven ETR of the INTOR (R_o=5m) size, this option retains an acceptable duty factor (70%) but offers substantial cost savings with reduced size, less fusion power, smaller poloidal coil power supplies, and reduced mechanical fatigue in the magnet support structure.

Evidently it is possible to design an ETR with $R_o \simeq 4.0$ m which can be operated in a number of modes. Continuous operation may be tested if ICRF and LH combined heating power exceed ~55 MW. The hybrid operation is likely using only ~20 MW of LH power. If necessary, this ETR could also have a long purely inductive burn, especially if the superconducting cable current densities can exceed the 1.4 kA/cm^2 which we postulated. Moreover, low β current ramp via LH current drive would reserve most of the OHC flux to sustain the burn. In addition, ICRF/LH current drive may be combined with inductive drive in a synergistic fashion to achieve burn periods of ~5-10 minutes.

ICRF IMPLEMENTATION

The figure displays a steady state current density profile in MHD equilibrium for an ETR-class tokamak. This reactor has a centrally peaked current generated by the ICRF/LH combination, and β is typical of values within the first stability regime. The ICRF frequency is selected such that the second deuterium harmonic is just outside the low-field plasma region while the second tritium harmonic lies inboard of the magnetic axis. By designing the appropriate power spectrum most of the wave power is transferred to electrons without significant ion heating. The calculation uses the cold plasma dispersion relation to predict the ray trajectories and provides the heating profiles assuming weak damping. Ray tracing is done on the actual flux surfaces generated by noninductive current drive[1], so the ray tracing, current drive, and MHD equilibrium are self-consistent.

There are several attractive features of this ICRF system. It seems probable, for example, that this same system can provide ample power to heat the fuel ions to ignition by slightly tuning the source frequency to bring a harmonic resonance to the magnetic axis. Moreover, the substantial external power (P_{CD} ~ 55 MW) nearly equals the alpha heating power ($P_\alpha \gtrsim 70$ MW), and this may provide an effective burn control mechanism, as the reactor might be operated subignited. Such a driven fusion reaction provides better assurances that the nuclear mission of the ETR will be achieved, even if poor energy confinement does not allow ignition. Of equal importance is the potential to control or modify MHD behavior if noninductive current drive is provided. Already experiments have shown that LH waves can eliminate sawtoothing, for example, and raise the axis safety factor above one. With the ICRF/LH system envisioned here, the ETR will provide an unparalleled opportunity to test stability theory by attempting operation in the second stability regime. Preliminary calculations with the fast wave[2] have found current profiles which are qualitatively similar to those tested stable to ballooning modes in the second regime.

CONCLUSIONS

Incorporating rf power, both ICRF (~ 40 MW) and LH (~ 20 MW), into the ETR design will provide an attractive experimental program with a high degree of flexibility. Several types of current drive mechanisms may be tested: electron damping of fast and LH waves; mode conversion electron heating; and minority ion heating. By selecting $R_o \simeq 4$ m a variety of operating modes are plausible: CW operation at high density with ICRF; hybrid burn cycle with LH-transformer recharge; and inductively driven cycle with LH/ICRF ramp-up and assist. Additional flexibility will be available to test high beta operation and otherwise modify the plasma behavior. Finally, provision for high power ICRF heating will better assure achievement of the requisite neutron production to carry out the ETR's nuclear testing.

Work supported by the U.S. Department of Energy, Office of Fusion Energy.

REFERENCES

1. D. A. Ehst, K. Evans, Jr., Nucl. Fusion (to be published).
2. D. A. Ehst, K. Evans, Jr., Tokamak Concept Innovations, IAEA-TECDOC-373, IAEA (Vienna) 1986, p191.

Table I: Minimum Size ETR Designs for Two Burn Cycle Options

Quantity	Variable	Units	CW	Hybrid
Major radius	R_o	m	3.8	4.2
Aspect ratio	A		3.0	3.6
Toroidal beta	β		0.071	0.057
Toroidal current	I_o	MA	10.2	8.3
Maximum toroidal field	B_M	T	11.5	10.6
Average electron temperature	$\bar{T}_e$	keV	20	10
Average electron density	$\bar{n}_e$	$10^{20} m^{-3}$	0.69	1.58
Neutron wall load	W_n	MW/m^2	1.0	1.3
Alpha power	P_α	MW	71	95
Current drive power	P_{CD}	MW	55	$\lesssim 20$
OHC power supply	P_{OH}	MVA	0	27
EFC power supply	P_{EF}	MVA	<10	210
Burn volt-seconds	ϕ_{OH}	Vs	-	11.0
Fusion burn period	t_f	s	CW	169
Cycle lifetime to $3 MW\text{-}y\text{-}m^2$	N_f		$\lesssim 10^3$	4.3×10^5

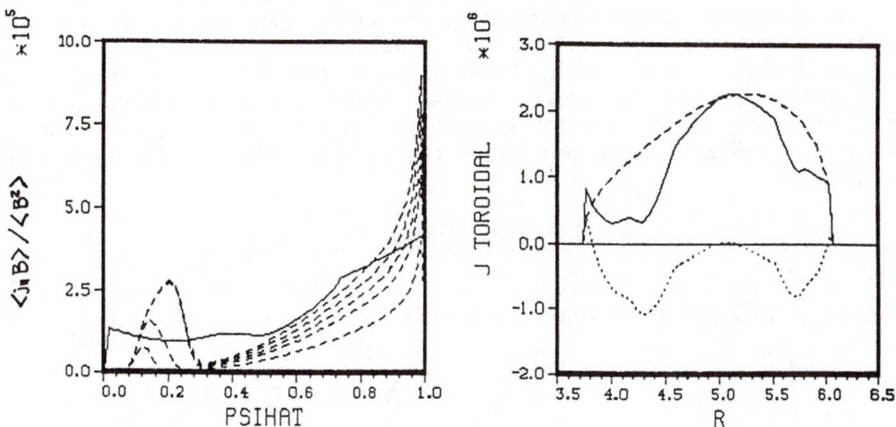

Figure 1 INTOR with equilibrium $\beta = 0.031$, generated with ICRF at the magnetic axis (PSIHAT=1) and LH on the surface; f = 64 MHz, $3.0 \leq n_\parallel \leq 4.5$, $P_{CD} = 52$ MW, $I_o = 5.9$ MA, $R_o = 5$ m, $n_e = 0.6 \times 10^{20}$ m^{-3}, $T_e = 17$ keV.

ION CYCLOTRON EMISSION CALCULATIONS USING A 2D FULL WAVE NUMERICAL CODE*

D. B. Batchelor, E. F. Jaeger
Oak Ridge National Laboratory, Oak Ridge, TN 37831

P. L. Colestock
Princeton Plasma Physics Laboratory, Princeton, NJ 08544

ABSTRACT

Measurement of radiation in the HF band due to cyclotron emission by energetic ions produced by fusion reactions or neutral beam injection promises to be a useful diagnostic on large devices which are entering the reactor regime of operation.[1] A number of complications make the modelling and interpretation of such measurements difficult using conventional geometrical optics methods. In particular the long wavelength and lack of high directivity of antennas in this frequency regime make observation of a single path across the plasma into a viewing dump impractical. Pickup antennas effectively see the whole plasma and wall reflection effects are important. We have modified our 2D full wave ICRH code[2] to calculate wave fields due to a distribution of energetic ions in tokamak geometry. The radiation is modeled as due to an ensemble of localized source currents distributed in space. The spatial structure of the coherent wave field is then calculated including cyclotron harmonic damping as compared to the usual procedure of incoherently summing powers of individual radiators. This method has the advantage that phase information from localized radiating currents is globally retained so the directivity of the pickup antennas is correctly represented. Also standing waves and wall reflections are automatically included.

INTRODUCTION

A 2-D finite difference code has been developed to solve the full wave ICRF problem in tokamak and stellarators geometry.[2,3] the basic equation is

$$\vec{\nabla} \times \vec{\nabla} \times \vec{E} - \frac{\omega^2}{c^2}\vec{E} - \frac{4\pi i\omega}{c^2}\sum_s \vec{J}_s = \frac{4\pi i\omega}{c^2}\vec{J}_{\text{ext}} \tag{1}$$

Where the plasma current due to species s, $\vec{J}_s$ is obtained from the warm plasma conductivity expanded to lowest significant order in $(k_\perp \rho_i)^2$, $\vec{J}_s = \sum_s \overleftrightarrow{\sigma}^s$

* Research sponsored by the Office of Fusion Energy, U. S. Department of Energy, under contract No. DE-AC05-840R21400 with Martin marietta Energy Systems, Inc.

$(\ell, k_\perp, k_\|) \cdot \vec{E}$ where $\ell =$ harmonic number $k_\perp$ is obtained from the local dispersion relation and $k_\|$ is approximately k_z. The source current $\vec{J}_{\text{ext}}$, normally taken to be the antenna current, is now the random, fluctuating current due to the radiating particles distributed throughout the plasma. We assume the equilibrium to be uniform in z and Fourier analyze in that direction. Since $\omega \ll \omega_{pe}$ we assume $E_\| = 0$ which reduces (1) to two-coupled second order equation for components of $\vec{E}$ in the $\vec{\nabla}\psi$ and $\vec{B} \times \vec{\nabla}\psi$ directions which are solved in a poloidal cross section, $\vec{x}_\perp$. The primary task now is to determine $\vec{J}_{\text{ext}}$ in terms of the assumed distribution of radiating α particles.

CALCULATIONS OF THE SOURCE CURRENT, $\vec{J}_{\text{ext}}$

At each grid point, p, we must determine the real and imaginary parts of $\vec{J}$ spatially averaged in a small area A_p about p. These are zero mean normal random variables whose statistics we must compute. For purposes of calculating these statistics we neglect the poloidal field and work in terms of $\vec{J}_\perp$ the x, y components of $\vec{J}$. We introduce the four-dimensional vector $\vec{I}(\vec{x}_\perp, k_\|, \omega)$ $= (Re\{\vec{J}_\perp\},\ Im\{\vec{J}_\perp\}) = (Re\{J_x\},\ Re\{J_y\},\ Im\{J_x\},\ Im\{J_y\})$. The average current, I_p associated with grid point p is then

$$\vec{I}_p(k_\|, \omega) = \frac{1}{A_p} \int_{A_p} d^2x_\perp\, \vec{I}(\vec{x}_\perp, k_z, \omega) \tag{2}$$

The statistics of $\vec{I}_p$ are thus determined by the covariance matrix $\overleftrightarrow{Q}_{p,p'}\,(k_z, \omega)$ where

$$\overleftrightarrow{Q}_{p,p'}\,(k_z, \omega) = \left\langle \frac{1}{A_p A_p'} \int_{A_p} d^2x_\perp \int_{A_{p'}} d^2x'_\perp \vec{I}^{\top}(x_\perp, k_z, \omega)\vec{I}(x'_\perp, k_z, \omega) \right\rangle \tag{3}$$

$\langle\rangle$ denotes ensemble average and $\vec{I}^{\top}$ denotes matrix transpose. Expressing $\vec{I}(\vec{x}_\perp, k_z, \omega)$ in terms of its k space transform $I(\vec{k}, \omega)$ we have

$$\begin{aligned} \overleftrightarrow{Q}_{p,p'}\,(k_z, \omega) = \frac{1}{A_p A_p'} \int_{A_p} d^2x_\perp \int_{A_{p'}} d^2x' \\ \int d^2k_\perp d^2k'_\perp e^{i(\vec{k}_\perp\cdot\vec{x}_\perp + \vec{k}'_\perp\cdot\vec{x}'_\perp)}\, \overleftrightarrow{R}\,(\vec{k}, \vec{k}', \omega, \omega') \end{aligned} \tag{4}$$

where $\overleftrightarrow{R}\,(\vec{k}, \vec{k}', \omega, \omega') = \left\langle \vec{I}^{\top}(\vec{k}, \omega)\vec{I}(\vec{k}', \omega') \right\rangle$ is the cross-spectral density matrix.

In order to proceed we make a number of simplifying assumptions.

1. We neglect correlations between different grid points ($\overleftrightarrow{Q}_{p,p'} = 0$ if $p \neq p'$).
2. We neglect particle-particle correlations (the probability density functions for the N particles of the ensemble is of the form $f^N(\vec{x}, \cdots, \vec{x}_N, \vec{v}, \cdots, \vec{v}_N)$ $= \prod_{n=1}^{N} f^1(\vec{x}_n, \vec{v}_n)$)

3. Gyroangles are uniformly distributed
4. Over the region A_p the plasma can be considered uniform ($f'(\vec{x},\vec{v}) = 1/V\, f(\vec{v})$ where V is the plasma volume)
5. The domain of averaging A_p is a cylinder of radius r_A about the point p.

After a somewhat tedious calculation, one can express Eq (4) as

$$\overleftrightarrow{Q}_{p,p}\,(k_z,\omega) = \frac{\pi}{2}(2\pi)^3\frac{N}{V}(2\pi q)^2\frac{1}{\Delta\omega}\frac{Lz}{2\pi}\sum_{\ell}\int_0^{\infty} dk_\perp k_\perp$$

$$\int d^3v\, f\,(\vec{v})\,\frac{1}{\pi^2 k_\perp^2 r_A^2} J_1^2\,(k_\perp r_A) \tag{5}$$

$$\times \overleftrightarrow{[}\; AS\left(\ell, k_{\|}, k_\perp, v_{\|}, v_\perp\right)\delta\left(\omega - \ell\Omega - k_{\|}v_{\|}\right)$$

Where $\overleftrightarrow{S}$ is the symmetric matrix

$$\overleftrightarrow{S} = \begin{bmatrix} M_1 & 0 & 0 & M_2 \\ 0 & M_1 & -M_2 & 0 \\ 0 & -M_2 & M_1 & 0 \\ M_2 & 0 & 0 & M_1 \end{bmatrix}$$

$M_1 = \pi v_\perp^2\left[\frac{\ell^2}{k_\perp^2\rho^2}J_\ell^2(k_\perp\rho) + J_\ell'^2(k_\perp\rho)\right]$ and $M_2 = 2\pi v_\perp^2\frac{\ell}{k_\perp\rho}J_\ell(k_\perp\rho)J_\ell'(k_\perp\rho)$.

COMPUTATIONAL APPROACH

The discretization process converts Eq (1) into a linear algebraic system of the form

$$\overleftrightarrow{T}\;\vec{E} = \vec{J}_{\text{ext}} \tag{6}$$

where now $\vec{E}$ and $\vec{J}_{\text{ext}}$ are 4N dimensional vectors (2 field components $\times$ 2 complex components $\times$ N grid points and $\overleftrightarrow{T}$ is a $4N \times 4N$ matrix. After numerically inverting $\overleftrightarrow{T}$ the solution is obtained $\vec{E} = \overleftrightarrow{T}^{-1}\vec{J}_{\text{ext}}$. We can now take either of two computational approaches. (1) A Monte-Carlo approach in which random samples for $\vec{J}_{\text{ext}}$ are generated for each grid point from a multivariate normal distribution using the covariance matrix of Eq (5). The field $\vec{E}$ is computed and the power p coupled to the pickup antenna is computed from $\vec{E}$. Since $\vec{E}$ is a random variable many sample E's should be generated and average $\langle P\rangle$ determined. Of course, the matrix $\overleftrightarrow{T}^{-1}$ need only be computed once to generate the multiple $\vec{E}$ samples.

Alternatively, since the quantities of interest to us, such as the power P induced in the pickup antenna, can be written as a quadratic form in $\vec{E}$

$$(P = E_i P_{ij} E_j) \quad . \tag{7}$$

The ensemble average of P, can be written in terms of the covariance matrix of $\vec{E}$ ($\langle P\rangle = P_{ij}\langle E_i E_j\rangle$). We now take advantage of the fact that a linear transformation of a multivariate normal distribution is again a normal distribution. The covariance matrix of $\vec{E}$, $\left\langle \vec{E}\vec{E} \right\rangle$, can be written

$$\left\langle \vec{E}\vec{E} \right\rangle = \overleftrightarrow{T}^{-1} \overleftrightarrow{Q} \left[\overleftrightarrow{T}^{-1} \right]^{T} \quad , \tag{8}$$

where $\overleftrightarrow{Q} = \left\langle \vec{J}^{T}\vec{J} \right\rangle$ is the direct sum of matrices $\overleftrightarrow{Q}_{p,p}$ for each point.

CONCLUSIONS

We can model ion cyclotron emission in plasmas with our existing 2D full-wave code provided the statistics of the local fluctuating current spectrum $\vec{J}(\vec{x}_{\perp}, k_z, \omega)$ can feasibly be computed. The advantage of this method over the use of geometrical optics is that global properties of standing waves and plasma reabsorption (i.e., cavity Q) are included and antenna directivity is naturally included. Many model source current distributions can be computed without need to recompute the inverse system matrix $\overleftrightarrow{T}^{-1}$. Such an approach is computationally feasible provided not too many values of ω and $k_{\|}$ need be considered. The principle limitation of the method is that the finite grid spacing places an upper limit on the frequencies (or lower limit on plasma wavelengths) which can be considered. The same method can of course be adapted to other radiation sources such as that generated by instabilities.

ACKNOWLEDGEMNT

The authors would like to thank Professor Allan Kaufman for pointing out an error in an earlier version of this paper.

REFERENCES

[1] G. A. Cottrell, P. P. Lallia, G. Sadler, P van Belle, 13th European conference on "Controlled Fusion and Plasma Heating," Schliersee, 1986 Part II p.37.

[2] E. F. Jaeger, D. B. Batchelor, H. Weitzner and J. H. Whealton, Computer Physics Communications 40,33 (1986).

[3] E. F. Jaeger, D. B. Batchelor, H. Weitzner and P. L. Colestock, "Global ICRF Modeling in Large Non-Circular Tokamak Plasmas with Finite Temperature," Proceedings of this conference.

GLOBAL ICRH MODELING IN LARGE NONCIRCULAR TOKAMAK PLASMAS WITH FINITE TEMPERATURE*

E. F. Jaeger and D. B. Batchelor
Oak Ridge National Laboratory, Oak Ridge, TN 37831

H. Weitzner
Courant Institute of Mathematical Sciences
New York University, New York, NY 10012

P. L. Colestock
Princeton Plasma Physics Laboratory, Princeton, NJ 08544

ABSTRACT

Full-wave ion-cyclotron resonance heating (ICRH) coupling calculations[1] in two and three dimensions have been extended to treat tokamaks with noncircular flux surfaces and conducting boundaries. The magnetic field configuration is derived from a Solov'ev equilibrium[2] with finite poloidal magnetic fields. The conducting boundary may be of arbitrary shape. The mode conversion model is that of Colestock et al.[3] in which the fourth-order finite-temperature wave equation is reduced to a second-order equation that describes the effects of mode conversion on the fast wave but neglects the detailed structure of the ion Bernstein wave. Results show the effect of a noncircular cross section on excitation, wave propagation, and absorption in Doublet III-D and the Joint European Torus (JET). Also, in the limit of a circular cross section, toroidal phasing of the resonant double-loop antenna design for the Tokamak Fusion Test Reactor (TFTR) is studied.

INTRODUCTION

High-power ICRH experiments are already under way at JET and are planned in the near future for TFTR and Doublet III-D. To provide reliable and accurate estimates of wave fields and power deposition in these experiments, we extend the full-wave solution of Maxwell's equations in Ref. 1 to include approximate warm-plasma effects and noncircular magnetic cross section with conducting boundaries of arbitrary shape.

MODEL EQUATIONS

An approximate magnetic equilibrium of the Solov'ev type is assumed, where

$$B^0(r,\theta) = \frac{\partial\psi}{\partial\theta}\frac{\hat{r}}{r} - \frac{\partial\psi}{\partial r}\hat{\theta} + \chi\,\hat{z} \tag{1}$$

and the poloidal and axial flux functions are

* Research sponsored by the Office of Fusion Energy, U.S. Department of Energy, under contract DE-AC05-84OR21400 with Martin Marietta Energy Systems, Inc.

$$\psi(r,\theta) = \frac{-\iota_0 B_0}{2R_T}\left\{\frac{[1+(x/R_T)]^2}{\kappa^2}y^2 + \left(1+\frac{x}{2R_T}\right)^2 x^2\right\}$$
$$\chi(r,\theta) = \frac{B_0}{1+(x/R_T)} \tag{2}$$

Since this equilibrium is independent of z, the wave electric field can be Fourier decomposed as

$$\vec{E}(r,\theta,z) = \sum_{k_z} \vec{E}_{k_z}(r,\theta)e^{ik_z z} \tag{3}$$

to allow a two-dimensional solution for $\vec{E}_{k_z}(r,\theta)$ from the vector wave equation

$$-\nabla\times\nabla\times\vec{E}_{k_z} + \frac{\omega^2}{c^2}\vec{E}_{k_z} + i\omega\mu_0\sum_s \vec{J}_s = -i\omega\mu_0\,\vec{J}_{\text{ext}} \tag{4}$$

In the mode conversion model of Colestock et al.,[3] the plasma current $\sum_s \vec{J}_s$ is represented with the warm-plasma dielectric tensor

$$\overleftrightarrow{K} = \overleftrightarrow{\epsilon}^{(0)} + ik_{\perp,\text{fast}}\,\overleftrightarrow{\epsilon}^{(1)} - k^2_{\perp,\text{fast}}\,\overleftrightarrow{\epsilon}^{(2)} \tag{5}$$

where $k_{\perp,\text{fast}}$ is the fast-wave root of the local dispersion relation for an infinite uniform plasma with $\vec{B}=$ const and $k_{\|}\to k_z$:

$$\text{Det}\left[\vec{n}\vec{n} - |\vec{n}|^2\,\overleftrightarrow{I} + \overleftrightarrow{\epsilon}^{(0)} + ik_\perp\,\overleftrightarrow{\epsilon}^{(1)} - k_\perp^2\,\overleftrightarrow{\epsilon}^{(2)}\right] = 0 \tag{6}$$

Assuming $E_z = 0$ for ω on the order of Ω_{ci} (zero electron mass limit), this is fourth-order in $k_\perp$ and quadratic in $k_\perp^2$. Equation (6) is solved locally at each point on the (r,θ) finite difference mesh. Then, discarding the Bernstein wave root and keeping $k_{\perp,\text{fast}}$ only, we formally reduce the wave equation to a second-order partial differential equation that describes the fast wave only:

$$-\frac{1}{k_0^2}(\nabla\times\nabla\times\vec{E}) + \underbrace{\left[\overleftrightarrow{\epsilon}^{(0)} + ik_\perp\,\overleftrightarrow{\epsilon}^{(1)} - k_\perp^2\,\overleftrightarrow{\epsilon}^{(2)}\right]}_{\overleftrightarrow{K}}\cdot\vec{E} = -\frac{i}{\omega\epsilon_0}\vec{J}_{\text{ext}} \tag{7}$$

Dotting this vector equation with the unit vectors $\hat{e}_1 = \nabla\psi/|\nabla\psi|$ and $\hat{e}_2 = \hat{b}\times\hat{e}_1$, we obtain two equations for the two components of $\vec{E}$ perpendicular to $\vec{B}$. Local energy deposition is computed from

$$\dot{W}_{k_z}(r,\theta) = \frac{\omega\epsilon_0}{2}\text{Im}\left[\vec{E}^*\cdot(\overleftrightarrow{K} - \overleftrightarrow{I})\cdot\vec{E}\right] \tag{8}$$

Because of the approximate model used, both mode converted and dissipated energy appear in Eq. (8) as power dissipation.

NUMERICAL RESULTS

Figure 1(a) shows density contours and the conducting wall for the noncircular tokamak geometry of Doublet III-D. Elongation is 1.8; the antenna, 40 cm high, is located on the low-field side of the plasma. The dashed line shows the second harmonic resonance in a pure hydrogen plasma with $f = 55\,\text{MHz}$. Keeping 80 toroidal modes ($k_z R_T = n_{\text{toroidal}} = -40 \rightarrow 40$) to represent an antenna half-width of 17.5 cm in the toroidal (z) direction, we find contours of $\text{Re}(E_x)$ and power absorbed at the antenna midplane as shown in Figs. 1(b) and 1(c), respectively. The temperature assumed was $T_e = T_i = 3\,\text{keV}$. Figure 2 shows the toroidal spectrum for this result. The solid curve shows the unweighted two-dimensional results for power absorbed versus toroidal wave number k_z. The dashed curve shows the weighted contribution of each toroidal mode in the Fourier sum which represents the antenna assumed in Fig. 1. Similar calculations have been done for JET with an elongation of 1.66 and a 6.8% ^{3}He minority in deuterium. The JET results show Alfvén resonance[4] ($\omega^2 = k_{\|}^2 v_A^2$) near the high-field edge and the associated rapid spatial increase in $k_\perp$.

Resonant double-loop antennas have been studied for the circular geometry of TFTR with a 5% ^{3}He-D mixture, $n_e = 10^{14}\,\text{cm}^{-3}$, $T_e =$ and $T_i = 10\,\text{keV}$. Figure 3 compares contours of $\text{Re}(E_x)$ in the equatorial (r, z) plane for a double-loop antenna where the loops are in phase [Fig 3(a)] and out of phase [Fig 3(b)]. Note the more extensive absorption and hence lower transmission in the out-of-phase case. Higher transmission, as shown in Fig 3(a), can lead to greater susceptibility to cavity modes at low density because of reflections from the opposite wall.

REFERENCES

1. E. F. Jaeger et al., Comput. Phys. Commun. **40**, 33 (1986).
2. L. S. Solov'ev, Sov. Phys. JETP **26**, 400 (1968).
3. P. L. Colestock et al., Princeton Plasma Physics Laboratory, private communication (1986).
4. K. Appert et al., Comput. Phys. Commun. **43**, 125 (1986).

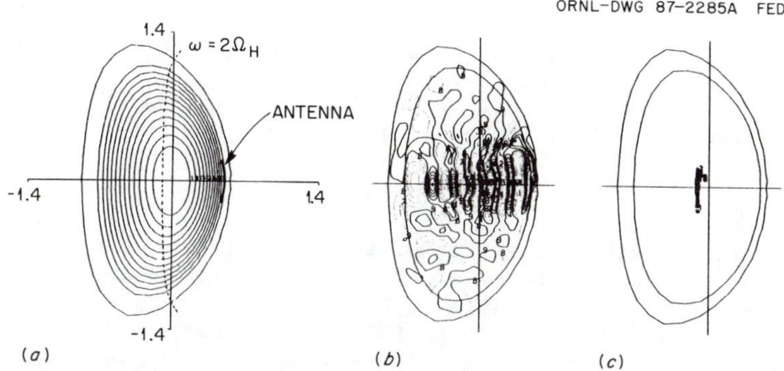

Fig. 1: a. Noncircular tokamak geometry: Density contours (solid) and second harmonic resonance surface (dashed) for hydrogen in Doublet III-D. **b.** Contours of constant RE(E_x) at antenna midplane. **c.** Contours of constant absorbed power at antenna midplane.

Fig. 2: Toroidal spectrum for antenna half-width of 17.5 cm in z: Solid lines show unweighted two-dimensional power absorption versus k_z, and dashed curves show weighted spectrum for the assumed antenna.

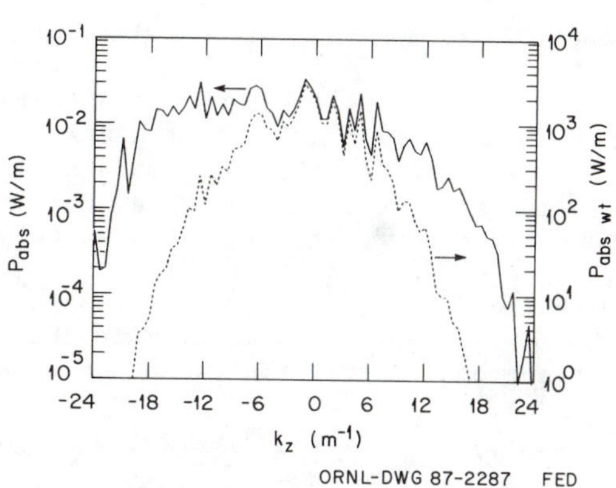

Fig. 3: Contours of constant Re(E_x) in equatorial plane for TFTR resonant double-loop antenna with loops **a.** in phase and **b.** out of phase.

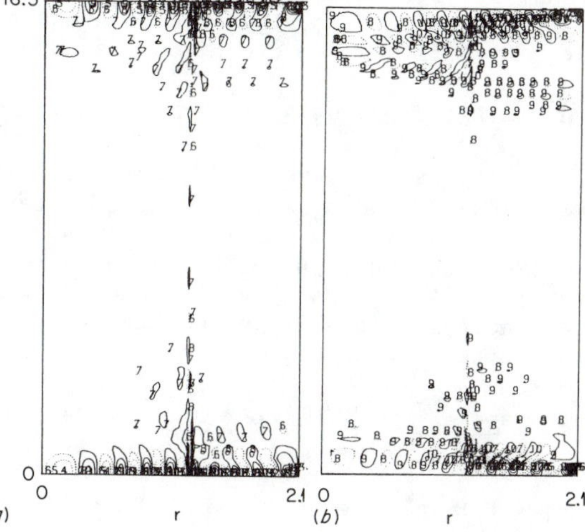

EFFECT OF POLOIDAL FIELD ON ION BERNSTEIN WAVES

J. Milovich, B. D. Fried and G. J. Morales
University of California, Los Angeles, CA 90024-1547

ABSTRACT

We have analyzed the effect of a small poloidal field on the propagation and absorption of electrostatic ion Bernstein waves in cylindrical geometry, assuming $r_{ci}/L \ll 1$, L being the spatial scale length of the electrostatic potential Φ. For uniform axial current density, the helical field lines cause a shift in the effective $k_{\parallel}$, resulting in an asymmetry between waves propagating in opposite axial directions and a modification in absorption at the cyclotron resonance layer. Also, the radial dependence of $k_{\perp}$ limits the radial extent of propagation. Solutions of $\Phi(r)$ for excitation by a localized antenna are given.

ANALYSIS

In view of the interest in ion Bernstein waves (IBW) for plasma heating and current drive in tokamaks[1], it is important to understand the effects of the poloidal component of $\underline{B}$ on the wave properties. Since IBW are not very sensitive to plasma density variations, we assume a uniform cylindrical plasma column, of radius a, in a shearless magnetic field $\underline{B} = B_o\, \underline{e}_z + J_1\, r\underline{e}_\theta$, where B_o and J_1 are constant and $aJ_1/B_o \ll 1$. The ion distribution function is assumed to be an isotropic Maxwellian with thermal velocity $v_i = (2T_i/m_i)^{1/2}$. The electrons distribution function is a shifted Maxwelllian,

$$f_{oe}(\underline{v}) = (a_e\, \pi^{1/2})^{-1} \exp[-(v_{\parallel} - v_d)^2/v_e{}^2]\ \delta(\underline{v}_{\perp}).$$

Analysis of the single ion motion in the assumed magnetic field shows that for an ion moving with constant speed $v_{\parallel}$ along a helical magnetic field line, the cyclotron resonance frequency is

$$\Omega = \omega_c(R) + 2J_1\, v_{\parallel}/B_o \tag{1}$$

plus corrections of higher order in $\varepsilon = aJ_1/B_o$ or r_{ci}/a, where $\omega_c(R) = eB(R)/Mc$ and R is the guiding center location. (In Fig. 1, which shows the orbit of an ion driven at the lab frame resonant frequency $\omega_c(R) + J_1 v_{\parallel}/B_o$, the monotonic increase of energy is apparent; excitation at other frequencies results in oscillation of the ion energy.) In solving the linearized Vlasov equation we use a guiding center expansion, setting

$$\underline{r} + z\underline{e}_z = \underline{R} + (V/\Omega)(\cos\phi\ \underline{e}_2 - \sin\phi\underline{e}_1) + Z\underline{e}_z$$

$$\underline{e}_1 = \underline{R}/R \qquad \underline{e}_2 = (B_o\ \underline{e}_\theta - B_\theta\ \underline{e}_z)/B$$

and drop terms quadratic in ε and r_{ci}/a. Integrating the Vlasov equation over orbits (or characteristics), expanding Φ in the integrand in powers of r_{ci}/r and substituting the density into Poisson's equation leads to a self-adjoint (hence energy conserving) second order differential equation for $\Phi(r)$,

$$r^{-1}(\partial/\partial r)(rD_2\partial\Phi/\partial r) + D_o\Phi = S(r) \qquad (2)$$

where S(r) is the external source,

$$D_2 = -(r_{ci}{}^2/2)\partial D_o/\partial b$$

$$D_o = \ell^2/r^2 - k_z{}^2 - k_{Di}^2 + (k_{De}{}^2/2)\ Z[(\omega - k_\parallel v_d)/k_\parallel a_e]$$

$$-k_{Di}^2\ \omega \sum_n \{(\Lambda_n/k_{eff}\ a_i)\ Z_n + (\varepsilon r_{ci}/r)\ [b\Lambda_n' Z_n' -$$

$$-\Lambda_n Z_n' - nk_z r_{ci}\Lambda_n' Z_n]\} \quad ,$$

and $b = (k_\perp r_{ci})^2/2$, $\Lambda_n = I_n(b)e^{-b}$, $\varepsilon = J_1 a/B_o$, $x_n = (\omega - n\omega_{ci})/\ k_{eff}a_i$, $Z_n = Z(x_n)$. Here, we have assumed $\exp[i(\ell\theta + k_z z)]$ for the θ and z dependence of Φ. Since the Doppler shifted resonance condition is

$$\omega - n\Omega - k_\parallel v_\parallel = \omega - n[\omega_c(R) + 2J_1 v_\parallel/B] - k_\parallel v_\parallel = 0$$

the effective parallel wave number for the nth harmonic is $k_{eff} = k_\parallel + 2nJ_1/B$.

Solutions of (1) for a radially localized source $S(r) = \delta(r-r_o)/r$ (thin cylindrical shell with oscillating charge) are shown in Figs. 2 and 3. For $\varepsilon = 0$ and other parameters the same as in Fig. 2, there is negligible wave damping. Fig. 3 shows the striking difference between waves propagating in opposite directions, resulting from the variation of electron Landau damping with k_{eff}.

REFERENCES

1. M. Ono, Ion Bernstein Wave Heating, 7th APS Topical Conference on Applications of Radio-Frequency Power to Plasmas, May, 1987.

ACKNOWLEDGEMENT

This research was partially supported by the Office of Fusion Energy, U.S. Department of Energy.

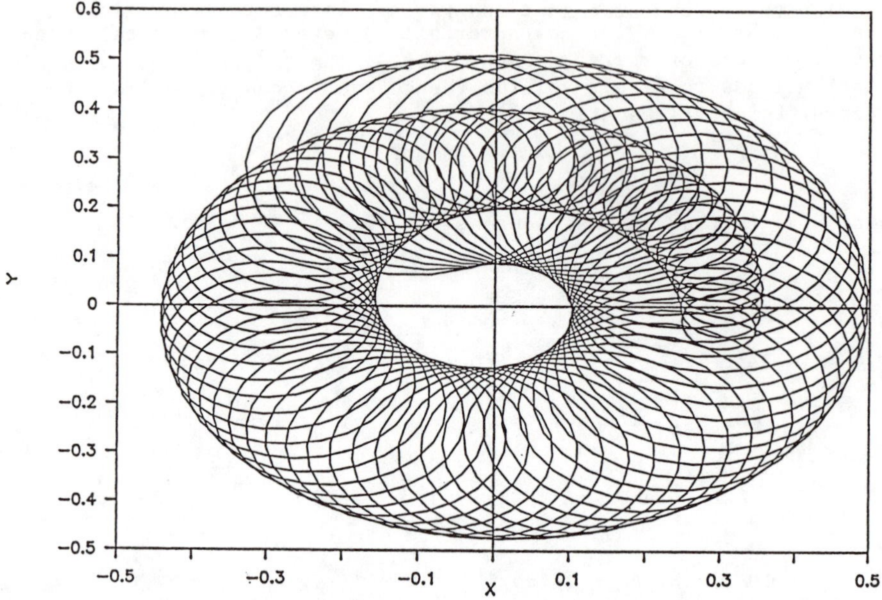

Fig. 1: Single particle motion in the presence of a poloidal magnetic field driven by an electrostatic electric field at the shifted frequency $\omega = \omega_c + J_1 v_\parallel / B$. The motion in the x–y plane shows the continuous absorption of energy due to perfect resonance. $|E_{ext}|^2/m\omega_c^2 a^2 = 10^{-3}$; $B_\theta(a)/B = \cdot 3$; $v_{zo}/\omega_c a = .05$; $\omega/\omega_{ci} = 1.015$

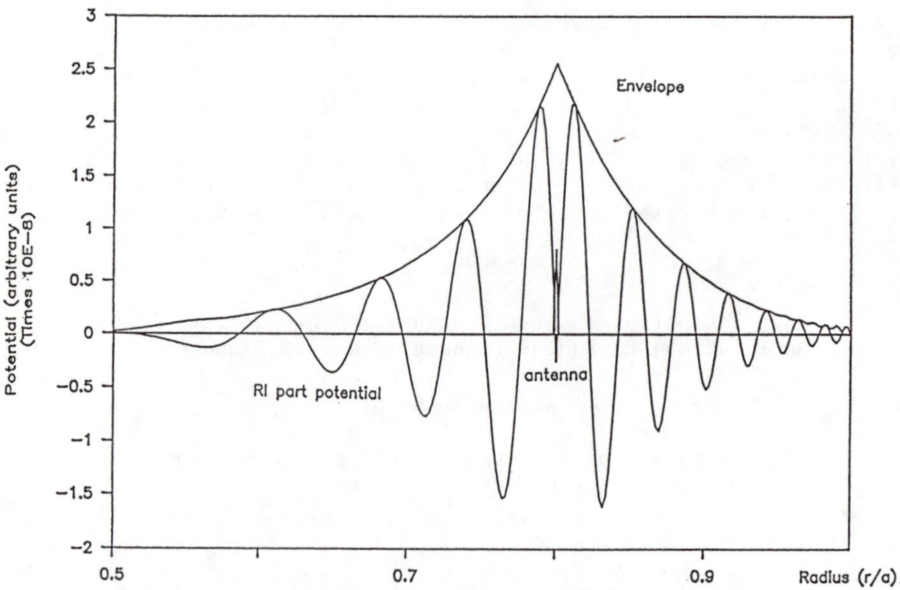

Fig. 2: Electrostatic potential as a function of radius showing the increased damping due to the presence of the poloidal magnetic field. $\omega/k_\parallel a e = 4$, $\omega/\omega_{ci} = 1 \cdot 7$, $x_{antenna}/a = \cdot 8$, $B_\theta(a)/|B| = \cdot 1, T_e/T_{i2} = 1$

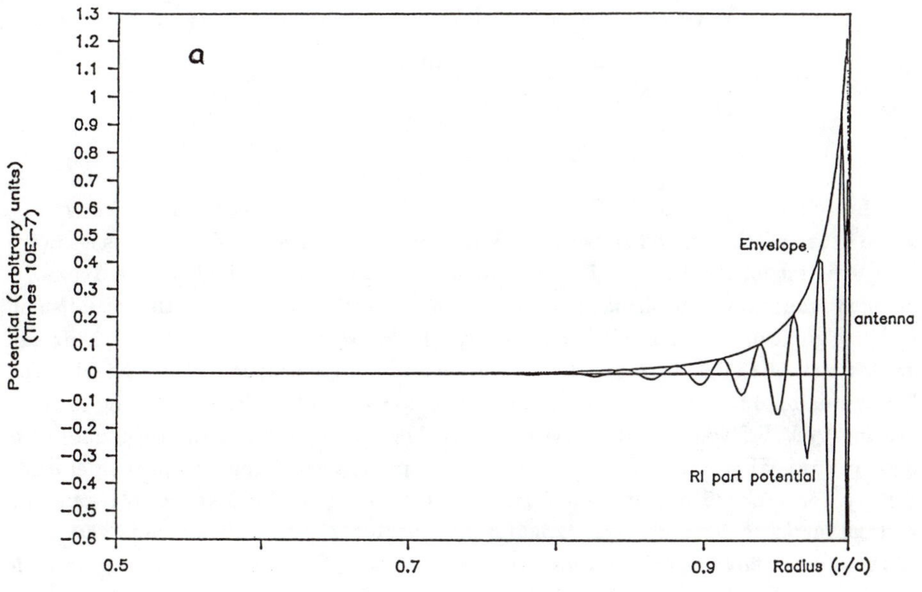

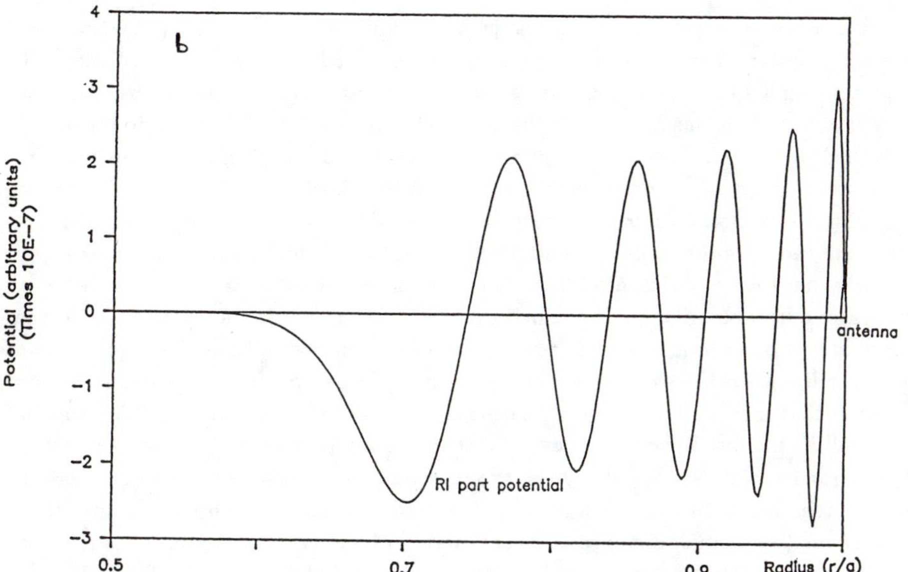

Fig. 3: Real part of the electrostatic ion Bernstein wave potential as a function of radius for different axial directions of propagation. (a) The wave propagates in the same direction as the pitch of the helical field line with $\omega/k_{\parallel}a_e = 4$. (b) The wave propagates opposite to the magnetic field pitch, $\omega/k_{\parallel}a_e = -4$. $B_\theta(a)/B = \cdot 1$, $x_{antenna}/a = 1$, $T_e/T_i = 1$, $\omega/\omega_{ci} = 1\cdot 7$

Electron Heating by Mode-Converted Ion-Bernstein Waves in ICRF Heating of Tokamak Plasmas

A. K. Ram and A. Bers
Plasma Fusion Center, M.I.T., Cambridge, MA 02139

ABSTRACT

In a tokamak plasma, ion-Bernstein waves (IBW) can be excited by mode-conversion of the externally launched fast wave for ICRF heating. This conversion process is known to be efficient for low $k_{\|}$'s which carry substantial power from a single loop antenna. A detailed numerical analysis of the propagation of the IBW shows that the initial small $k_{\|}$ are significantly enhanced along the rays due to toroidal effects. The upshift can occur for short radial distances of propagation and is large enough so that the IBW can Landau damp onto the electrons. This could help explain the observed strong electron heating by ICRF waves in tokamak plasmas. The numerical ray trajectory analysis is done in toroidal geometry for a hot Maxwellian plasma with gradients in temperature, density, toroidal and poloidal magnetic fields included in a WKB sense. A simple analytical model is derived which explains the upshift in $k_{\|}$ and gives results very close to the numerically obtained values. Approximate analytical conditions for appreciable electron Landau damping of the IBW are also given.

INTRODUCTION

It has been observed in various tokamak experiments (PLT [1], JET [2,3]) that there is strong electron heating when the externally launched fast component of the ICRF waves is coupled to a two ion-species plasma. There are several scenarios by which one may try to explain this heating. If the excited spectrum has substantial power at high $k_{\|}$'s the fast wave could heat the electrons by transit time damping or Landau damping. However, in most of the experiments mentioned the power spectrum is peaked around $k_{\|} = 0$ (for instance, for loop or monopole antennas) and substantial electron heating is still observed. Another possible mechanism could be the slowing down of the energetic minority ion tails onto the electrons. It is experimentally observed that the fast wave generates a hot minority species tail. If these ions are energetic enough they could heat the electrons by slowing down on them. This would require that their slowing down time be much less than the electron confinement time. In this paper we present a different scenario that may explain electron heating in these experiments. It is well known that for small $k_{\|}$ the fast wave can mode-convert to the ion-Bernstein wave near the ion-ion hybrid resonance layer. We study the propagation of this IBW and show that toroidal effects, together with the poloidal magnetic field, can lead to substantial upshifts in $k_{\|}$ over relatively small distances, so that electron Landau damping becomes effective. Results of a detailed numerical and analytical ray trajectory analysis of the IBW are presented.

NUMERICAL RAY TRAJECTORY ANALYSIS

A numerical code has been developed which solves for the propagation of rays in three-dimensional toroidal geometry. The local dispersion function, D, used for the rays is that for a hot Maxwellian plasma [4] and includes all the nine elements of the dielectric tensor. The spatial profiles of the magnetic field, density and temperature are explicitly included in D. The toroidal ray equations [5] are set up for an axisymmetric tokamak (but, can be easily generalized) so that D is independent of ϕ (the toroidal angle). In terms of the toroidal coordinates $k_{\parallel}$ and $k_{\perp}$ are given by:

$$k_{\parallel} = \frac{1}{|\mathbf{B}|} \left[k_r B_r + \left(\frac{m}{r}\right) B_\theta + \left(\frac{n}{R + r\,cos\theta}\right) B_\phi \right] \quad (1)$$

$$k_{\perp}^2 = k_r^2 + \left(\frac{m}{r}\right)^2 + \left(\frac{n}{R + r\,cos\theta}\right)^2 - k_{\parallel}^2 \quad (2)$$

where r is the radius measured from the magnetic axis of the torus, θ is the poloidal angle, R is the major radius, k_r, $k_\theta = m/r$ and $k_\phi = n/(R + rcos\theta)$ are the radial, poloidal and toroidal wave numbers, respectively; B_r, B_θ and B_ϕ are the radial, poloidal and toroidal components of the magnetic field, respectively. We choose a special equilibrium where $B_r = 0$, B_θ is given by the toroidal current profile given in [6] and $B_\phi = B_{\phi o}/(R + rcos\theta)$. The electron temperature profile is as given in [6] while the ion temperature and the electron and ion densities are assumed to vary like $[1 - (r/a)^2]^{1.5}$ (a being the minor radius) with their peak values being on the magnetic axis. In the numerical results shown in figs.(1-4) we used JET-type parameters [2] where we have assumed a deuterium plasma with a 4% hydrogen minority, peak electron density of 2.8×10^{13} cm.$^{-3}$, peak electron temperature of 1.8 keV, peak ion temperature of 1.7 keV, toroidal current of 2 MA, $a = 125$ cm., $R = 300$ cm., $B_{\phi o} = 2$ Tesla, $q = 1$ surface at $r = 50$ cm. and the wave frequency $\omega_o = 1.58 \times 10^8$ sec^{-1}. Fig.(1) shows the poloidal projection of an IB ray trajectory which started off at $r = 60$ cm., $\theta = 0.47$ radians and with $k_{\parallel} = 0$ (and $m = 0$). The ray initially propagates radially in the direction shown. By the time it turns in the poloidal direction the ray has Landau damped substantially. Fig.(2) shows the temporal damping along the ray. (The horizontal axis in figs.(2-4) is $\omega_o t$ where t is the time of propagation obtained by dividing the path length along the ray by its group velocity). The ray bends in the poloidal direction around $\omega_o t \approx 310$. The major changes in m and $k_{\parallel}$ occur before this time. The rate of increase of m along the ray is shown in fig.(3). The concomitant increase in $|k_{\parallel}|$ is shown in fig.(4) in terms of the parameter $\omega_o/(\sqrt{2}|k_{\parallel}|v_{te})$ which determines the effectiveness of electron Landau damping ($v_{te} = \kappa T_e/m_e$). Similar results have been obtained for different starting points of the IB rays and for different values of ω_o. Since we have assumed an axisymmetric tokamak, n does not change, along the ray. The enhancement in $k_{\parallel}$ is clearly due to the rapid increase in m. If B_θ had been ignored there would have been no enhancement in $k_{\parallel}$. The fraction of incident energy that is damped on the electrons by this IBW is $(1 - exp(-2\gamma))$ where γ is the area under the curve in fig.(2). For the case considered $\gamma \approx 2.5$ so that most of the wave energy has damped within a radial propagation distance of about 12 cm.

ANALYTICAL MODEL

In the vicinity of the mode-conversion region where $k_\perp \rho_i < 1$ (ρ_i is the ion Larmor radius) an approximate form for the local dispersion function, D^{app}, is obtained by expanding the full D to fourth order in $k_\perp \rho_i$. Since $k_\perp \rho_e \ll 1$, D^{app} contains only the cold electron contributions. For a deuterium-hydrogen plasma, we obtain:

$$D^{app} = \alpha_0 \, k_\perp^4 \, + \, \alpha_1 \, k_\perp^2 \, + \, \alpha_2 \tag{3a}$$

where:

$$\alpha_0 = -2\epsilon y_0 \frac{c^4}{\omega_{pd}^2 \omega_o^2} \left[Z(y_2) + \frac{1}{y_1} + \frac{1}{y_{-1}} \right] \tag{3b}$$

$$\alpha_1 = 2\epsilon y_0 \frac{c^2}{\omega_o^2} \left[2 \frac{\omega_o}{\omega_{cd}} Z(y_2) + \frac{1}{\epsilon} \frac{\omega_o^2}{\omega_{pd}^2} \left\{ \frac{1}{y_1} + \frac{1}{y_{-1}} - \sqrt{2}\eta Z\left(\frac{y_2}{\sqrt{2}}\right) \right\} \right] \tag{3c}$$

$$\alpha_2 = 4y_0^2 \left[\frac{1}{y_1 y_{-1}} - \frac{\omega_o}{\omega_{cd}} \frac{1}{y_0} \left(\frac{1}{y_1} - \frac{1}{y_{-1}} \right) - \frac{\omega_o^2}{\omega_{cd}^2} \frac{1}{y_0^2} + \sqrt{2}\eta Z\left(\frac{y_2}{\sqrt{2}}\right) \left(\frac{\omega_o}{\omega_{cd}} \frac{1}{y_0} - \frac{1}{y_{-1}} \right) \right] \tag{3d}$$

and, $y_l = (\omega_o - l\omega_{cd})/(\sqrt{2}|k_\parallel| v_{td})$, $l = 0, \pm 1$; ω_{cd} (ω_{pd}) is the local deuterium cyclotron (plasma) frequency, Z is plasma dispersion function, c is the speed of light, $\epsilon = v_{td}^2 c^2/(\omega_o^2 \omega_{cd}^2)$ and η is the ratio of hydrogen to deuterium ion densities. Eqn.(3a) can be written in a coupled mode form and the propagating IBW is then given by $(k_\perp)_{IB}^2 = -(\alpha_1/\alpha_0)$ away from the mode-conversion region. We use this form of $(k_\perp)_{IB}$ with the added assumption that $k_\parallel$ is small (so that the asymptotic form of Z-functions can be used). Since the numerical results show that the IB ray propagates for short radial distances we ignore the variation in temperatures and density along the ray and include only the poloidal and radial variations of the magnetic field. The rate of change of m is related to the poloidal variations of D so that the toroidicity included in the magnetic fields is important. The resulting equation relating the change in m to change in r is given by:

$$\Delta m \approx \frac{2}{3} \frac{\omega_{cd}^2}{k_r v_{td}^2} \left(2 + 11 \, \frac{v_{td}^2}{c^2} \frac{\omega_{pd}^2}{\omega_{cd}^2} \right) \frac{r sin\theta}{(R + r cos\theta)} \, \Delta r \tag{4}$$

where we assume that $k_r = (k_\perp)_{IB}$. If the initial value of m is zero the change in m required for electron Landau damping is:

$$\frac{(\Delta m)_L}{r} > \frac{1}{2} \frac{\omega_{cd}}{v_{td}} \frac{|\mathbf{B}|}{B_\theta} \sqrt{\frac{m_e}{m_d}} \tag{5}$$

Eqns.(4,5) can be combined to give the radial distance required for IBW propagation before it damps on the electrons:

$$\frac{\Delta r}{a} > \frac{3}{8} \frac{k_r v_{td}}{\omega_{cd}} \frac{|\mathbf{B}|}{B_\theta} \sqrt{\frac{m_e}{m_d}} \left(\frac{R + r cos\theta}{a sin\theta} \right) \tag{6}$$

For the parameters shown in figs.(1-4) we find that $\Delta m \approx 1.7\Delta r$. With $\Delta r \approx -12$cm. this gives a $\Delta m \approx -20.4$ which is in good agreement with the numerical results. The condition for Landau damping in eqn.(6) gives $\Delta r/a > .05$. Thus, the analytical model gives good insights into the numerical results.

This work is supported by DOE contract number DE-AC02-78ET-51013.

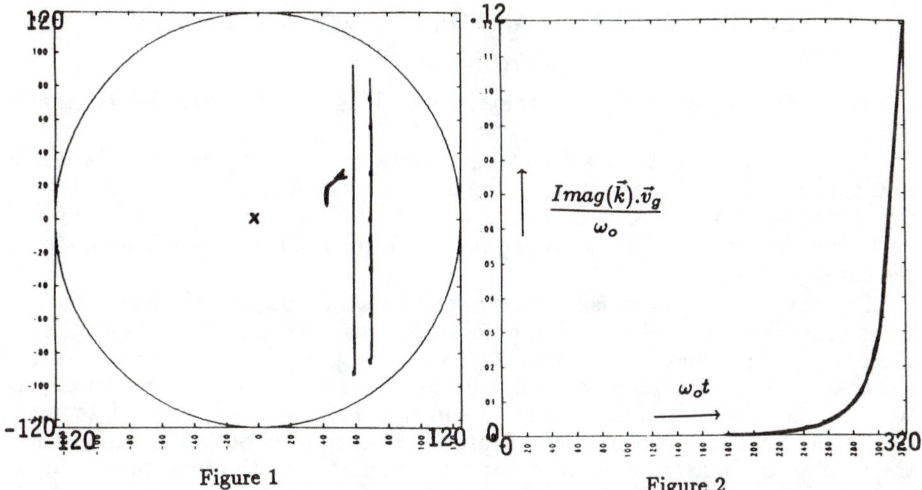

Figure 1

IB ray trajectory in the poloidal plane. $\omega_o = 2\omega_{cd}$ at C; I is the ion-ion hybrid resonance layer.

Figure 2

Temporal damping along the IB ray. $\vec{v}_g$ is the group velocity of the ray.

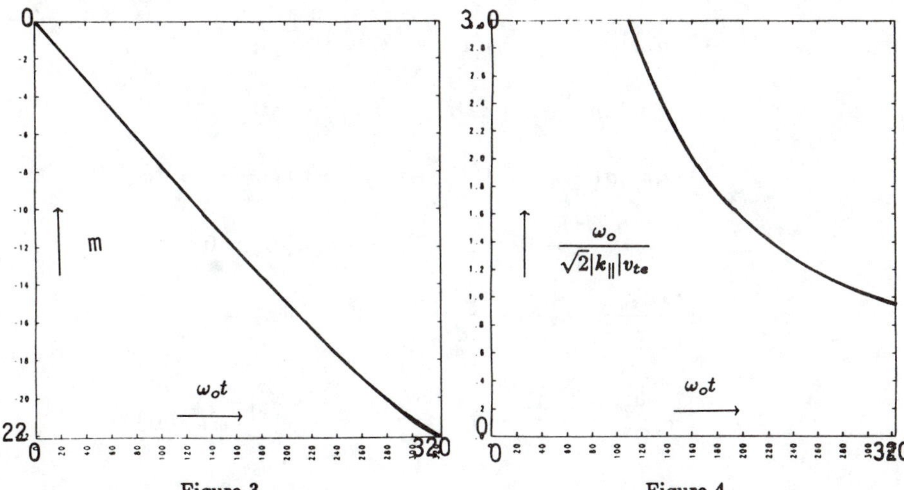

Figure 3

Poloidal mode numbers along the IB ray

Figure 4

Electron Landau resonance parameter along the IB ray.

REFERENCES

1. J. Hosea et al., Heating in Toroidal Plasmas, Grenoble, Vol. I, 213 (1982)
2. J. Jacquinot et al., Plasma Phys. and Controlled Fus. 28, 1 (1986)
3. P. Lallia et al., Plasma Phys. and Controlled Fus. 28, 1211 (1986)
4. T. H. Stix, "The Theory of Plasma Waves," McGraw-Hill, New York (1962)
5. D. W. Ignat, Phys. Fluids 24, 1110 (1981)
6. W. M. Tang, Nucl. Fus. 26, 1619 (1986)

THEORY OF ANTENNAS FOR ION BERNSTEIN WAVES

Marco Brambilla

Max-Planck Institute für Plasmaphysik, Garching bei München, W. Germany

Plasma heating with ion Bernstein Waves has been proposed as a viable alternative to Fast Wave heating in the ion cyclotron frequency domain [1,2]. A two-dimensional theory of antennas designed for Bernstein wave launching was developed by Sy et al. [3]. Here we present a fully three dimensional code for the same purpose.

The real configuration is approximated by a slab model: Cartesian coordinates (x, y, z) simulate the radial, poloidal and toroidal directions, respectively, of a tokamak; the plasma parameters vary only in the x direction, curvature and shear are neglected. The wave field is decomposed in a double Fourier sum over toroidal and poloidal wavenumbers n_z and n_y , discretised as appropriate in the equivalent toroidal problem. The solution in vacuum is obtained analytically; the field at the plasma boundary and the antenna radiation resistance are expressed in terms of the antenna current, and of the surface plasma impedance tensor $Z_{ij}(n_y, n_z)$. To evaluate the latter, we solve the FLR equations [4,5] $(c/\omega = 1)$:

$$-(\frac{d}{dx} + n_y)\left[\sigma(\frac{d}{dx} - n_y)(E_x + iE_y)\right] + (n_y^2 + n_z^2 - S)E_x$$
$$+in_y\frac{dE_y}{dx} + iDE_y - in_z\frac{dE_z}{dx} = 0$$

$$i(\frac{d}{dx} + n_y)\left[\sigma(\frac{d}{dx} - n_y)(E_x + iE_y)\right] + in_y\frac{dE_x}{dx} - iDE_x$$
$$-\frac{d^2E_y}{d^2x} + (n_z^2 - S)E_y - n_yn_zE_z = 0$$

$$in_z\frac{dE_x}{dx} - n_yn_zE_y - \frac{d^2E_z}{d^2x} + (n_y^2 - P)E_z = 0$$

$$L = 1 + \frac{\omega_{pe}^2}{\Omega_{ce}^2}\left(1 - \frac{\Omega_{ce}}{\omega}\left(1 - i\frac{\nu_e}{\omega}\right)\right) - \sum_i \frac{\omega_{pi}^2}{\Omega_{ci}^2}x_{oi}Z(x_{1i})$$

$$R = 1 + \frac{\omega_{pe}^2}{\Omega_{ce}^2}\left(1 + \frac{\Omega_{ce}}{\omega}\left(1 - i\frac{\nu_e}{\omega}\right)\right) - \sum_i \frac{\omega_{pi}^2}{\Omega_{ci}^2}\frac{\omega}{\omega + \Omega_{ci}}$$

$$P = 1 - \frac{\omega_{pe}^2}{\omega^2}\left(1 - 2i\frac{\nu_e}{\omega}\right)x_{oe}^2 Z'(x_{oe})$$

$$S = \frac{R + L}{2} \qquad D = \frac{R - L}{2}$$

$$\sigma = \frac{1}{4}\sum_i \frac{\omega_{pi}^2}{\Omega_{ci}^2}\frac{v_{thi}^2}{c^2}\left(-x_{oi}Z(x_{2i})\right)$$

Here $Z(x)$ is the Plasma Dispersion Function, with $x_{n,i} = \frac{\omega - n\Omega_{ci}}{k_z v_{thi}}$, and ν_e the electron collision frequency. A finite element discretisation with cubic Hermite interpolation functions has been used. The solution is made unique by imposing the outward radiation conditions at sufficiently high density in the plasma. At the plasma-vacuum boundary, in addition to the continuity of E_y, E_z, B_y, B_z, the condition

$$\sigma(0)\Big\{\big(\frac{dE_x(0)}{dx} + i\frac{dE_y(0)}{dx}\big) - n_y\big(E_x(0) + iE_y(0)\big)\Big\} = 0$$

must be imposed: it ensures that the kinetic part of the power flux vanish and the total flux is continuous at the plasma edge.

As an application, we have simulated BW coupling in the Alcator C tokamak [5]. These experiments were performed at a fixed frequency (183.6 Mhz), varying the intensity of the static magnetic field. The results of a similar code scan are shown in Fig. 1. The central plasma parameters were $n_e(0) = 10^{14}$ cm^{-3}, $T_e(0) = 1.5$ keV, $T_i(0) = 1$ keV; at the limiter $n_e(a) = 0.6\ 10^{13}$ cm^{-3}, $T_e(a) = T_i(a) = 50$ eV. The scrape-off plasma (0.5 cm thickness) extends out to the Faraday shield with a density decay length of 0.5 cm. The radiation resistance R_a is close to the measured one when the first harmonic resonance is just behind the antenna; this is also the range in which efficient heating was observed. Outside this range, however, the calculated value is almost an order of magnitude smaller than the observed one. Multiplying the Coulomb collision frequency by a factor 10 increases the computed R_a by 10 to 20% only. Optimum coupling to BW occurs just above the resonance peak in the ion absorption; at still higher magnetic field matching to BW is increasingly difficult as their wavelength becomes shorter. The power launched in the fast mode (at a constant antenna current) is roughly constant over the whole range explored, except for a dip corresponding to the domain where ion damping is strong and partial waves with small n_z are evanescent in the immediate vicinity of the antenna. The power deposited in the electrons is roughly proportional to the power in the Bernstein waves; it also includes about 5% collisional damping localised very close to the plasma boundary. Harmonic IC damping near the antenna is much stronger when BW are excited, as its efficiency is proportional to $k_\perp^2 \rho_i^2$.

Fig. 2 shows the spectral distribution of power among partial waves with different n_z (summed over n_y), for $\omega/\Omega_{ci} = 1.96$ at $r = a$. The power deposition profile and the field components E_x and E_z along a radius from the antenna are shown for the same case in Fig. 3 and 4, respectively.The width of the spectrum corresponds roughly to that of the Fourier spectrum of the antenna current, because BW are propagative from the very edge of the plasma; in this respect they differ from the Fast Wave, which is evanescent up to the R cut-off, a feature which strongly suppresses waves with large $|n_z|$ from the radiated spectrum. By contrast, BW with $n_z = 0$ cannot be directly coupled from outside, because E_z and B_y vanish identically. This has the same consequences for the antenna design as accessibility at higher frequencies. Thus the radiation resistance of the Alcator antenna (central feeder and shorts at the ends, quadrupole current distribution with antisymmetric n_z spectrum) is found to be more than three time larger than that of an equivalent dipole antenna (feeders at the extremes and central short in push-pull, symmetric n_z spectrum).

/1/ S. Puri, Proc 3d Top. Conf. on RF Plasma Heating, Pasadena (Calif.) 1978, paper E1.

/2/ M. Ono, Proc. Course and Workshop on Applications of RF Waves to Tokamak Plasmas, Varenna 1985, p. 197.
/3/ W. N.-C. Sy, T. Amano, R. Ando, A. Fukuyama, T. Watari, Nucl. Fusion, **25** (1985) 795.
/4/ D.G. Swanson, Proc. 3d Joint Varenna-Grenoble Int. Symp. on Heating in Toroidal Plasmas, 1982, Vol 1, p. 285.
/5/ P.L. Colestock, R.J. Kashuba, Nucl. Fusion, **23** (1983) 763.
/6/ Y. Takase et Al. Report PFC/JA-86-60, MIT 1986.

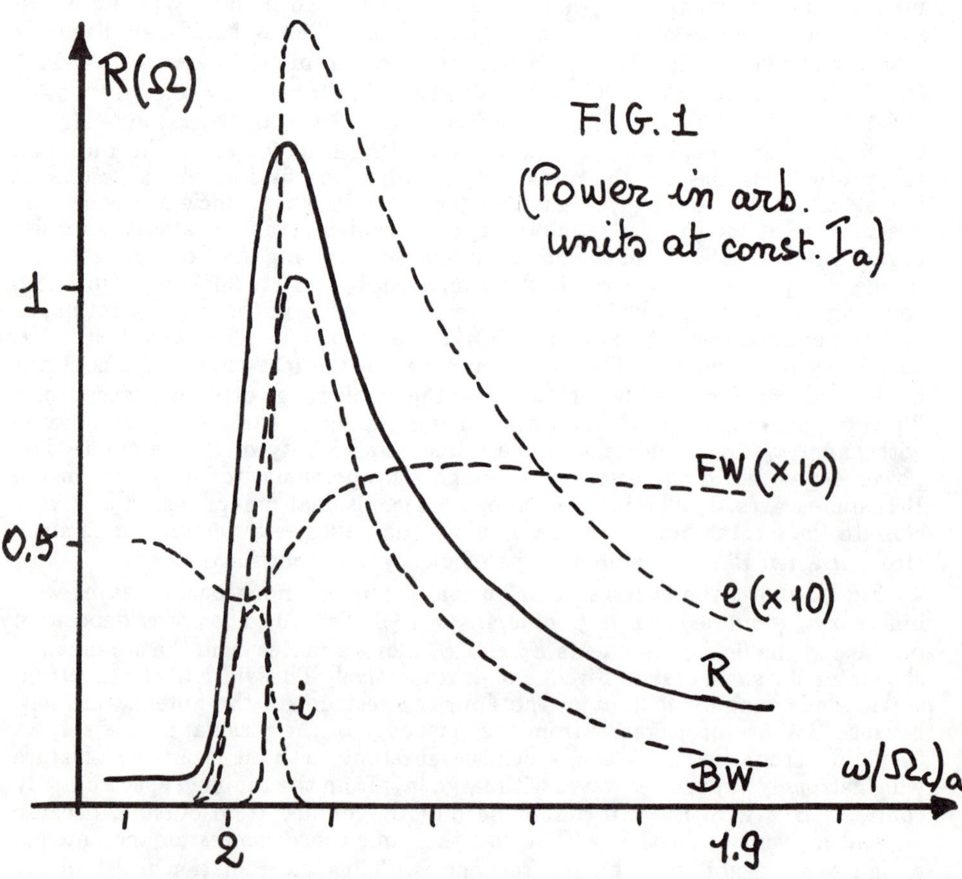

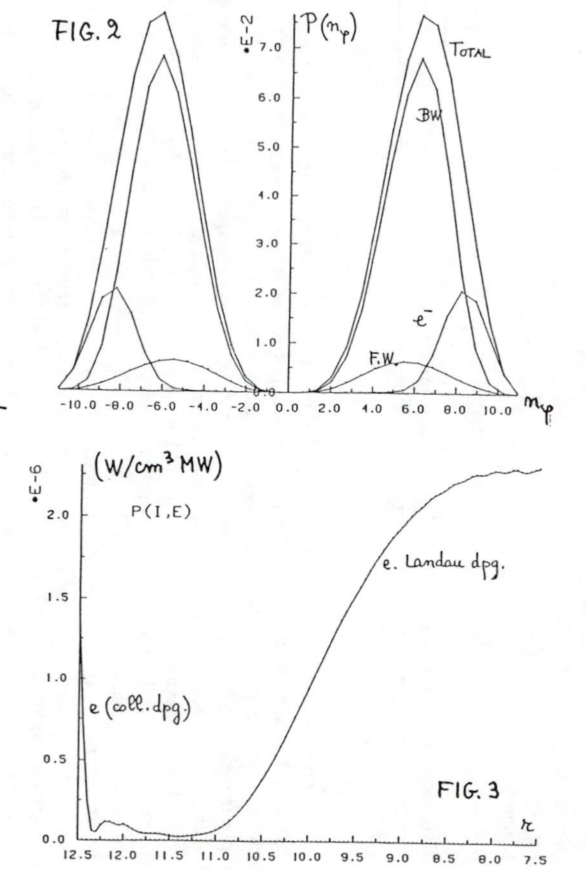

FIG. 2

FIG. 3

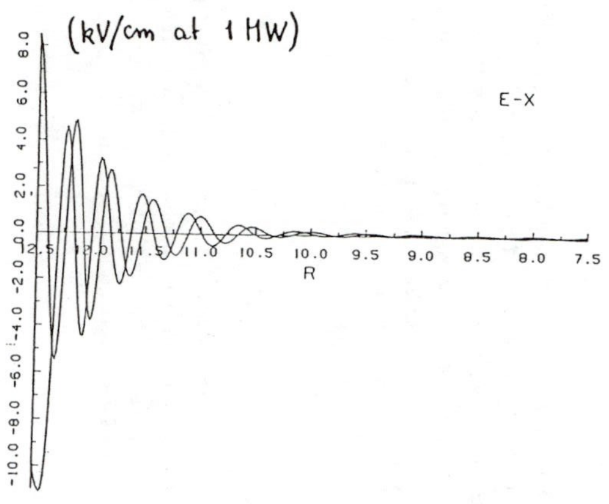

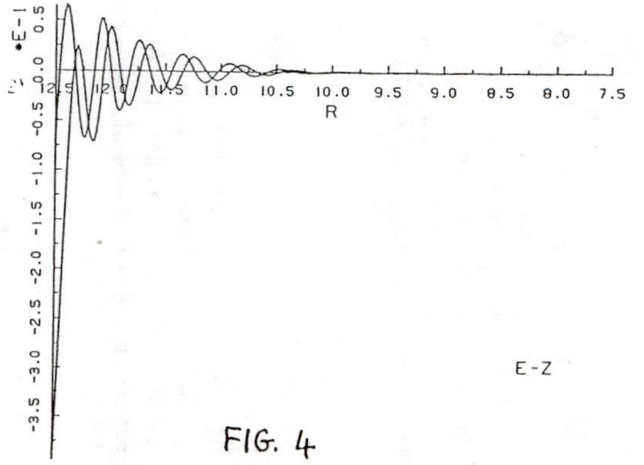

FIG. 4

ALFVÉN WAVE PROBE STUDIES ON PRETEXT.

M. E. Oakes and W. D. Booth
Department of Physics and Fusion Research Center
The University of Texas at Austin, Austin, Texas 78712

ABSTRACT

Alfvén waves launched with two toroidal antennas have been studied in the PRETEXT tokamak. Impedance measurements and laser interferometry have been used to investigate the structure of global modes. The radial profiles measured with the laser were limited to radial positions with adequate density. In order to complement these results we have inserted magnetic probes to measure the wave fields in the plasma. The probes can safely scan from the limiter to 5 cm inside the plasma. The results are compared with a cylindrical kinetic code.[1]

INTRODUCTION

Here we present an investigation of the structure of the global Alfvén eigenmodes (GAE) in the PRETEXT tokamak. The waves are excited by two phased toroidal antennas driven by a 1 kW broadband amplifier at a frequency of 2.1 MHz. The PRETEXT parameters are summarized in Table I. A typical shot is shown in Fig. 1.

TABLE I Pretext Parameters

Minor Radius (r_a) = 17.4 cm	Plasma Current = 20 $\sim$ 35 kA
Major Radius (R) = 53 cm	Central Density = 12 $\sim$ 20 × 10^{12} cm^{-3}
Limiter Radius = 15 cm	B_{tor} = 8 kG
Antenna Radius = 16 cm	$T_e \simeq 200$ eV
Driver Frequency = 2.1 MHz	$q_{edge} \simeq 6$
Antenna Currents $\simeq 50 \sim 100$ Amps per antenna	

Two insulated quarter-turn toroidal antennas connected in parallel with no Faraday shields

In a previous study[3] the structure of a global mode was measured using CO_2 laser interferometry on the density fluctuations. Only the central portion of the plasma was accessible to the interferometer. To complement these results we have measured the wave magnetic fields from the plasma edge to 5 cm past the limiter. To measure the wave magnetic fields, we can use a set of two probes which measure all three components of the fields at the top of the vessel near the end of one antenna and at the midplane between the two antennas. The midplane probe is inserted into the plasma and the top probe is left at the limiter and is used to monitor the extent of the plasma perturbation caused by the first probe. We have successfully scanned to four centimeters into the plasma with this probe. In an effort to insert the probe farther without perturbing the plasma we have constructed a smaller diameter probe which has six coils spaced one centimeter apart and can be inserted into the plasma at the midplane position. This probe provides the ability to obtain a limited radial profile of a given field component (either $\mathbf{B}_\theta$ or $\mathbf{B}_z$) during a single shot. The signals from the probes are bandpass-filtered around the driver frequency and the electrostatic signal is eliminated with a hybrid combiner.[4]

RESULTS

The results obtained from the first set of probes is shown in Figures 1c and 1d for the side probe and in Figures 1f, 1g, and 1h for the top probe. In Figure 1d the upper curve is $\mathbf{B}_z$ and the lower is $\mathbf{B}_r$. Figure 1b shows GAE which is the ratio of the predicted global eigenfrequency to the driver frequency.[2] The two curves shown are calculated for $l = 1$, $m = -1$ and $l = 2$, $m = -1$ with $M_{eff} = 1.5$. As shown in Figure 1c resonances in both the antenna resistance and probe signals occur when GAE=1 (dotted line in figure).

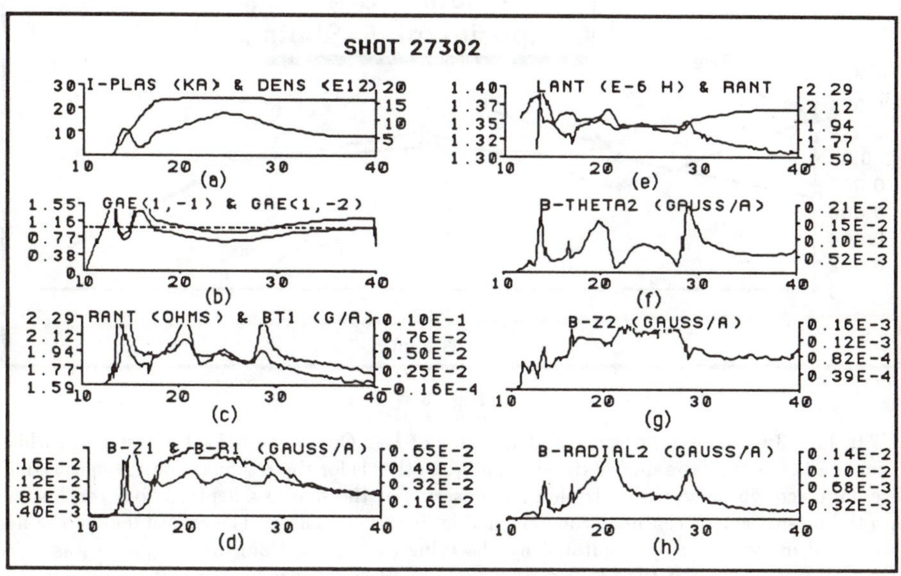

Fig. 1

In Figure 2 we show calculated B-field profiles for the $l = 1, m = -1$ global eigenmode in PRETEXT. These were calculated using the kinetic cylindrical code of Ross, Chen, and

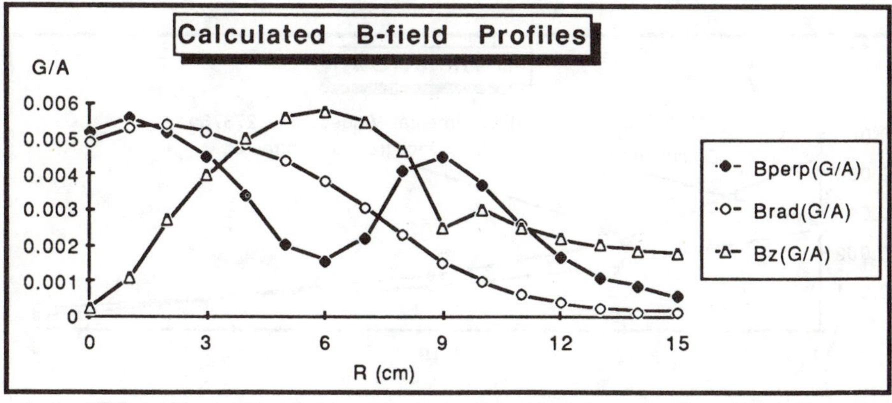

Fig. 2

Mahajan[1] which includes density, temperature, q profiles and the effect of the antenna feeders. The values of the paramaters used in the computation were selected to model PRETEXT at conditions corresponding to an observed value of GAE = 1. It has been suggested in the literature that for toroidal antennas of the type used in this experiment the fields would be strongly localized at the surface of the plasma.[5] It is evident from the figure that the calculated fields are global in nature and are not confined simply to the plasma surface.

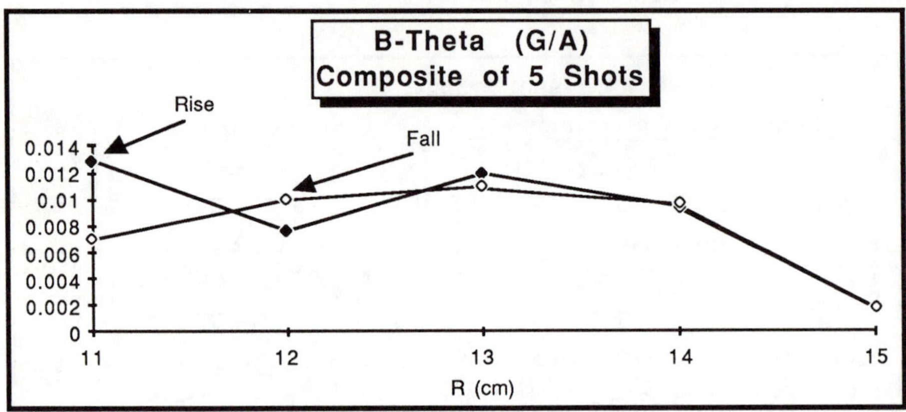

Fig. 3

In Fig. 3 we show a measured radial profile of $\mathbf{B}_\theta$. One curve is for the resonance which occurs on the density rise during the shot and the other is for the resonance on the density fall. The data were obtained with a triple probe inserted at the side to a depth of four centimeters into the plasma at one centimeter increments on five similar shots. The consistancy of the five shots used in the figure as monitored by observing the signals from the top probe was quite high. Attempts to insert the probe farther into the plasma caused significant perturbation of the plasma.

Figures 4 and 5 show radial profiles of $\mathbf{B}_\theta$ obtained on single shots using the six-coil probe inserted at the side. The plotted field values have had the nonresonant background field (typically 10% to 25% of the peak value) subtracted from them. Also shown on the plots is the calculated value for $\mathbf{B}_\theta$ over the same range of radii. Note that the experimental fields in both

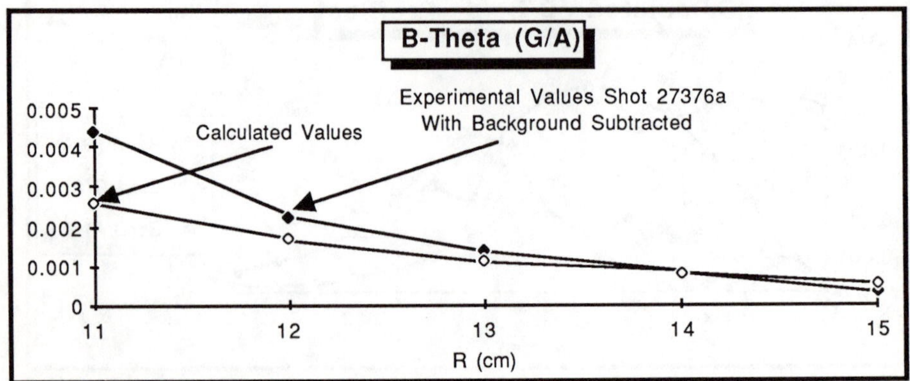

Fig. 4

figures follow the general trend of increasing in value with decreasing radius as predicted by the kinetic code and suggested by extrapolation of the CO_2 laser measurement.[3] Also note that the measured fields agree fairly well in magnitude with the code predictions near the edge.

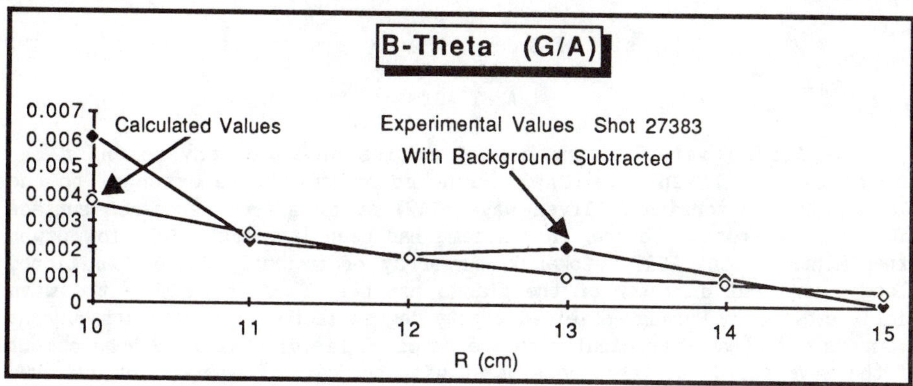

Fig. 5

CONCLUSION

The field values measured by the probes agree well with those predicted by the code. This agreement, along with the previously reported laser measurements, gives resonable confidence in the code's wave field radial profiles. These profiles show small surface attenuation and thus we conclude that if strong surface coupling is present it is due primarily to electrostatic effects.

ACKNOWLEDGMENT

We would like to express our thanks to Keith Carter and Jimmy Vaughan for technical assistance. This work is supported by The Texas Atomic Energy Research Foundation and The Department of Energy.

[1]D. W. Ross, G. L. Chen and S. M. Mahajan, Phys. Fluids **25**, 652 (1982)
[2]S.M. Mahajan, Phys. Fluids **27**, 2238 (1984).
[3]T.E. Evans, et al., Phys. Rev. Letters **53**, 1743 (1984).
[4]Rodney Cross (private communication).
[5]G. G. Borg, M. H. Brennan, R. C. Cross, L. Giannone and I. J. Donnelly, Plasma Phys. Contr. Fusion **27**, 1125 (1985)

TORSIONAL ALFVEN WAVE EXCITATION BY MODE CONVERSION IN THE TORTUS TOKAMAK

A.B. Murphy*
University of Sydney, NSW 2006, Australia

ABSTRACT

The Alfven wave heating scheme relies on mode conversion of the compressional Alfven wave (CAW), launched by an antenna external to the plasma, to the torsional Alfven wave (TAW) at an Alfven resonance surface (ARS) near the plasma centre. This scheme has been investigated in low power experiments in the TORTUS tokamak. An array of magnetic probes positioned along a vertical diameter of the plasma has been used to measure the wave fields excited by Faraday shielded eighty degree sector antennas with current elements oriented perpendicular to the toroidal field. Pronounced enhancement of the wave fields at radii consistent with the calculated radii of ARSs has been observed. The wave fields excited by a number of distinct phasings of an array of three toroidally spaced antennas have been compared. The results provide unambiguous evidence that the wave field enhancement is due to mode conversion of the CAW to the TAW. Measurements of the antenna resistive loading indicate that up to 50% of the total wave energy may be deposited at a selected ARS near the centre of the plasma. Possible means of increasing this already encouraging percentage are discussed.

INTRODUCTION

The Alfven wave heating scheme is a promising candidate for the provision of the supplementary heating required to bring a plasma to thermonuclear temperatures. The scheme involves the excitation of the compressional Alfven wave (CAW) by antennas external to the plasma. The CAW mode converts to the torsional Alfven wave (TAW) at the Alfven resonance surface (ARS), which for a given frequency ω, and poloidal and toroidal mode numbers m and n, lies at radius r_A, given in $2\pi R$ periodic cylindrical geometry by

$$\omega^2(r_A) = \frac{[n + m/q(r_A)]^2}{R^2} \frac{B_\phi^2}{\mu_0 \langle A\rangle m_p n_e(r_A)} \left[1 - \left(\frac{\langle A\rangle\omega}{\langle Z\rangle\omega_{ci}}\right)^2\right] \tag{1}$$

where B_ϕ is the steady axial magnetic field, q is the quality factor, n_e is the electron density, ω_{ci} the ion cyclotron frequency, m_p the proton mass, and $\langle A\rangle$ and $\langle Z\rangle$ respectively the average mass and charge numbers of the ions.

Previous Alfven wave studies in the TORTUS tokamak have shown undesirable direct excitation of the TAW by the antenna [1]. In the present paper, experimental results demonstrating the excitation of the TAW by mode conversion from the CAW are presented. Evidence for the occurrence of the mode conversion process in Alfven wave heating experiments has also been obtained in experiments in the Tokapole II [2] and TCA [3] tokamaks.

*Present address: Max-Planck-Institut für Plasmaphysik, 8046 Garching bei München, Federal Republic of Germany

EXPERIMENTAL DETAILS

Experiments were performed in the TORTUS tokamak [4] (R = 44 cm), using a hydrogen plasma with B_ϕ = 0.80 T and minor radius a = 10 cm. Three Faraday shielded all metal antennas, extending 80° poloidally and 4 cm toroidally, were used. These were positioned on the bottom of the vessel at toroidal angles ϕ = 0°, 225° and 315°, and excited either singly, or simultaneously as a phased array, by 200 W RF amplifiers at frequency 4.20 MHz (so ω/ω_{ci} = 0.35). The poloidal mode spectrum of the antennas is dominated by m = 0, 1 and 2 components, in decreasing order of amplitude. The toroidal mode spectrum is determined by the relative phasing of the antennas: for a single antenna it encompasses approximately equal amplitudes of all n.

The plasma loading of the antenna impedance was determined from measurements of the amplitudes and relative phase of the voltage across the antenna and the current in the antenna. The b_r, b_θ and b_ϕ components of the wave fields excited by the antenna were measured across a vertical diameter of the plasma by magnetic probes consisting of 6 small differential coils, spaced by 2 cm. These were placed in fixed quartz tubes (6 mm O.D.) inserted to the plasma centre from the top and the bottom of the vessel at ϕ = 180°. The presence of tubes in the plasma necessitated a derating of usual TORTUS plasma parameters. The time dependence of the plasma current I_P and the line averaged electron density $\overline{n_e}$ is shown in Fig. 1.

SINGLE ANTENNA RESULTS

Radial profiles, measured at regular time intervals, of b_r excited by the antenna at ϕ = 0°, are shown in Fig. 2. b_θ was found to have similar radial dependence and amplitude, while b_ϕ was approximately ten times smaller, and had little radial variation. Pronounced peaks in b_r in both halves of the plasma are apparent at all times in Fig. 2. The radii of the peaks decreases at successive times; this is correlated with the decrease in $\overline{n_e}$. Fig. 3 compares the time dependence of the radius of the peak in the top half of the plasma with that of the calculated radii of all $|m| \leq 3$ ARSs in the plasma. The radii are calculated using Eq. 1, assuming a parabolic n_e profile (always observed in standard TORTUS plasmas - no measurements have however been made for the derated plasma used here), $n_e(10\text{ cm})/\overline{n_e}$ = 0.25 (from Langmuir probe measurements in the edge plasma), a parabolic current density profile (as measured by low frequency magnetic probes), and $\langle A \rangle$ = 1.2 and $\langle Z \rangle$ = 1.1 (corresponding to 1% oxygen contamination). Note that (+m,+n) and (-m,-n) ARSs are degenerate. The b_r peaks correspond well to the (m,n) = (+1,+2)/(-1,-2) ARS. The width of the peaks in the top half of the plasma is consistent with those expected for the excitation of the TAW at an ARS in the relatively cold ($\langle T_e \rangle$ ~ 15 eV) plasma pertaining here. The relative broadness of the peaks in the b_r profile in the bottom half of the plasma is due to the constructive interference of the wave fields associated with the various ARSs; these tend to cancel in the top half of the plasma.

PHASED ARRAY RESULTS

Vertical radial profiles of b_r excited by a phased array of three antennas have been measured for the antennas phased to preferentially excite

$n = 0$, $+1$, -1, $+2$, -2, $+3$, -3 and ± 4 respectively. At $t \sim 2$ ms, the largest peaks in the radial profiles are produced, at $r = 4$ cm, by the $n = -2$ phasing. This confirms the identification of these peaks as corresponding to TAWs excited at the (+1,+2)/(-1,-2) ARS. b_r profiles for the $n = +2$ and $n = -2$ phasings are compared in Fig. 4. The larger amplitude of the $r \sim 4$ cm peaks for the latter phasing shows that the (-1,-2) TAW is more strongly excited than the (+1,+2) TAW. Theoretical calculations [5] indicate that the (-1,-2) CAW is more strongly excited than the (+1,+2) CAW for the relevant experimental conditions, as the antenna excitation frequency is closer to the (-1,-2) CAW eigenfrequency. It may thus be concluded that the wave field peaks are due to the TAW excited by mode conversion from the CAW at the (+1,+2)/(-1,-2) ARS.

The resistive plasma loading ΔR_A of the antennas phased to excite $n = -2$ shows a significant temporal peak from $t \sim 1.3$ ms to 2.1 ms (corresponding respectively to $r_A \sim 7$ cm to 4 cm for the (+1,+2)/(-1,-2) ARS). Fig. 5 allows a comparison of the time dependence of ΔR_A (of one of the antennas) for the $n = -2$ and $n = +2$ phasings, and for a single antenna. The difference between ΔR_A for the $n = -2$ phasing and for the single antenna may be attributed to the threefold increase in the strength of excitation of the (-1,-2) mode in the former case, if the relatively small alterations of the excitation strengths of other modes are neglected. ΔR_A for the $n = -2$ phasing at $t = 1.7$ ms is 0.09 Ω, 0.03 Ω greater than for the single antenna. It is thus estimated that approximately 50% of the total wave power is transferred to the (-1,-2) TAW, which is at this time situated at $r_A \sim 5$ cm. This is in addition to the power transferred to the (+1,+2) ARS at the same r_A. It is noted however that the percentage decreases as r_A nears the plasma centre.

DISCUSSION AND CONCLUSIONS

Significant excitation of localised wave fields at radii corresponding to the calculated radii of ARSs has been observed. The use of an appropriately phased array of antennas has been shown to preferentially excite a selected mode. An anisotropy between the wave field amplitudes of the (-1,-2) and (+1,+2) TAWs has been observed, providing unambiguous evidence that these waves are excited by mode conversion from the CAW.

Antenna loading data indicate that approximately 50% of the total wave power has been transferred to a given ARS at $r_A/a \sim 0.5$. This is an encouraging result. It could be improved by (a): increasing the number of antennas used in the phased array to allow more precise selection of the CAW modes excited, and (b): decreasing the antenna excitation frequency relative to the ion cyclotron frequency, thereby decreasing the number of Alfven wave modes with ARSs present inside the plasma minor radius.

REFERENCES

1. G.G. Borg et al., Proc. 13th Eur. Conf. Contr. Fusion and Plasma Heating, Schliersee, 1986 (Eur. Phys. Soc., 1986), Part II, p. 53.
2. F.D. Witherspoon, S.C. Prager and J.C. Sprott, Phys. Rev. Lett., 53, 1559 (1984).
3. R. Behn et al., Plasma Phys. Contr. Fusion, 29, 75 (1987).
4. R.C. Cross et al., Atomic Energy Aust., 24 (Parts 3,4), 2 (1981).
5. I.J. Donnelly and N.F. Cramer, Plasma Phys. Contr. Fusion, 26, 769 (1984).

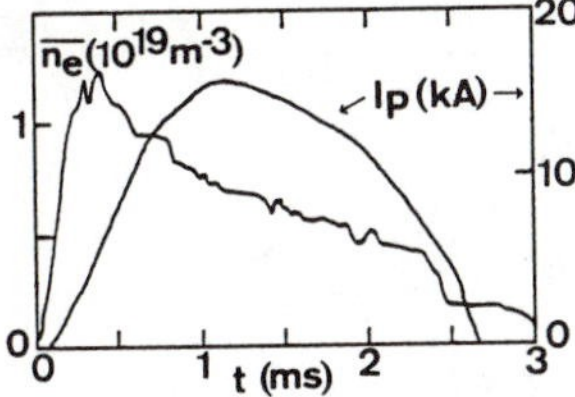

Fig. 1. Time dependence of $\overline{n_e}$ and I_p.

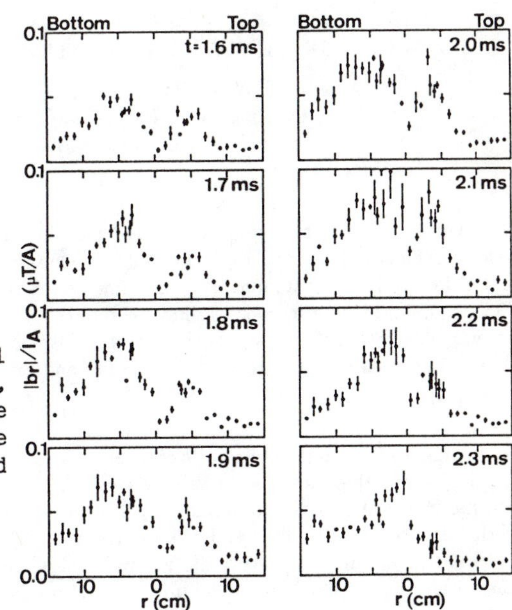

Fig. 2. Vertical radial profiles of $|b_r|$, normalised to the antenna current, for the antenna at 0°, measured at 8 different times.

Fig. 3. Calculated time dependence of ARS radii, compared with observed radii of $|b_r|$ peaks in the top half of the plasma.

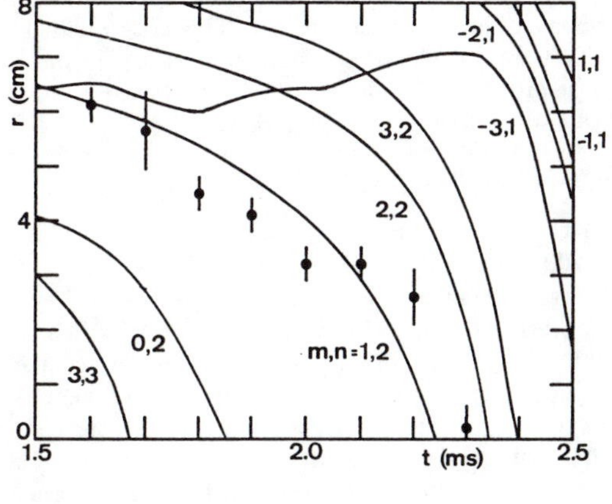

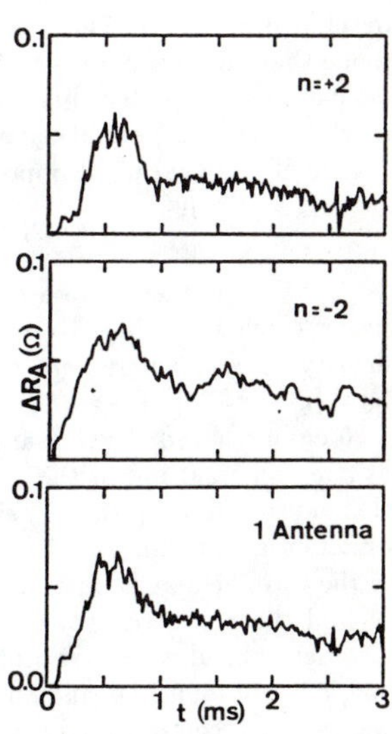

Fig. 5. Time dependence of ΔR_A for two phasings of the three antenna array, and for a single antenna.

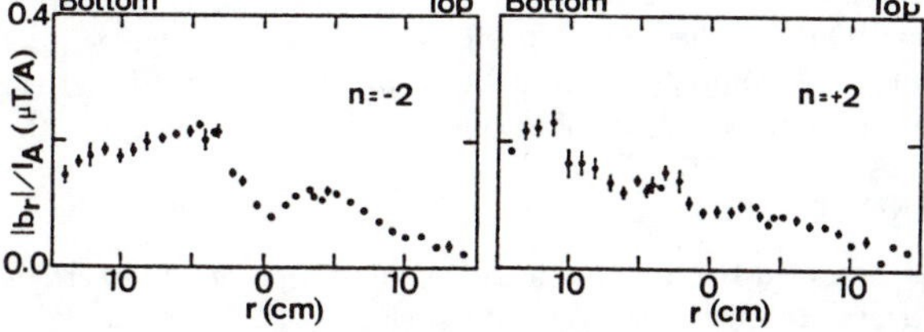

Fig. 4. Vertical radial profiles of $|b_r|$ normalised to the antenna current for two phasings of the three antenna array, at t = 2 ms.

MHD SURFACE WAVE CALCULATIONS FOR A HELICAL ANTENNA

W. D. Booth, M. E. Oakes and L. N. Vu
Department of Physics and Fusion Research Center
The University of Texas at Austin, Austin, Texas 78712

ABSTRACT

Amagishi[1] recently reported magnetic probe radial profile measurements of axial phase velocities and attenuation lengths for 449 and 605 kHz, $m = -1$ MHD surface waves launched by a two element helical antenna. The experiments were carried out in a current free, linear machine (TPH). Using a kinetic code we have computed the plasma response to his antenna Fourier components. From the superposition we calculate an effective phase velocity and compare with Amagishi's results. Also presented are radial and axial profiles of the fields.

INTRODUCTION

Amagishi[1] has reported results of an experimental study of MHD surface waves in a current-free linear machine. Interest in these waves is in part due to their role in Alfvén wave heating schemes.[2] There have been a number of theoretical studies of these waves in cylindrical plasmas with radial density profiles and axial currents.[3,4,5] Stockdale and Cheng[6] have also explored the mode coupling due to toroidal effects. These effects have been seen in TCA.[7] The many theoretical studies have provided quite detailed predictions; however, due to the hostile environment encountered in most local measurements there are few detailed experimental comparisons. Evans et al.[8] have reported radial profiles of global eigenmodes determined with laser interferometry in the PRETEXT tokamak and those results were compared with the kinetic code developed in Ref. 4. The University of Sydney group has operated the TORTUS tokamak in a mode that has permited launching-structures and probes in the plasma. This has provided new insight into the guided character of the Alfvén wave.[9] The low plasma temperature in the TPH machine makes it possible to insert magnetic probes to the plasma center and hence Amagishi's results offer an excellent opportunity to test some details of the kinetic code.

DESCRIPTION OF AMAGISHI'S EXPERIMENT

The experiments were carried out in the TPH device of Shizuoka University. This is a linear machine with a 15 cm diameter, 200 cm length, B=3 kG, $T_e \simeq T_i \sim 7eV$, central density $n_e \sim 5.5 \times 10^{14} cm^{-3}$ and is filled with helium. The density profile is given by $n = n_0 \times [1 - (r/6)^2]/[1 + (r/6)^2]$. The antenna used to launch the waves is a two wire helix with a 20 cm pitch and 12 cm diameter encircling the plasma column. The two wires are connected at one end resulting in the wire currents having opposite phases. Previous experiments with this antenna at frequencies below the ion cyclotron frequency ($f_{ci} = 1.125 \times MHz$) have shown preferential launching of the $m = -1$ mode.[10] Magnetic probes measured the theta component of the wave field at two points downstream from the antenna. Correlation techniques were used to find the phase velocity as a function of radius. Radial scans of attentuation lengths were also determined with this probe arrangement. These experiments were done for frequencies of 449 kHz and 609 kHz and the results were compared with a single mode MHD model. Here we present a comparison with the aforementioned cylindrical kinetic model which includes the effects of feeders and a conducting vessel.

KINETIC CODE RESULTS

In this analysis we have Fourier analysed the antenna in the axial and poloidal directions which are assigned mode numbers (l, m) respectively. We have included $l = 1$ to $l = 32$ in the superposition of axial modes and retained only the $m = -1$ poloidal mode. These modes correspond to the appropriate sense of the helix and do result in preferential fields in the -z direction from the antenna. Below we show the results for the 449 kHz case.

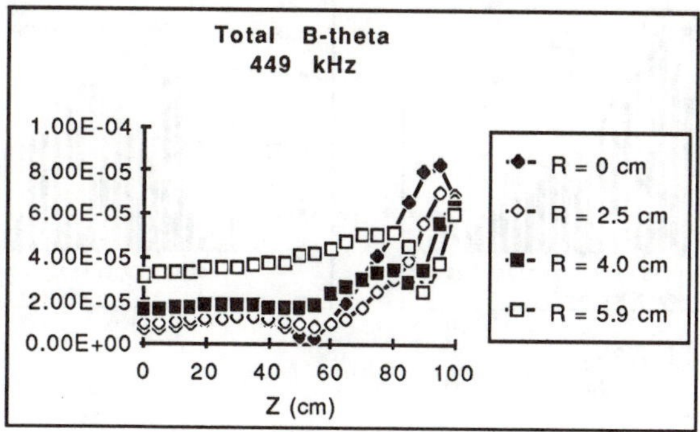

Fig. 1

Figure 1 shows the calculated axial field plots for the mode superposition for radii ranging from the plasma center to the edge. In the coordinate system used here the antenna center is located at z=100 cm and Amagishi's probes are located at z=50 cm and z=70 cm. In Fig. 2 we show a radial profile for the total calculated $\mathbf{B}_\theta$ at a point along the axis between the two probes. Note the nonlocalized character of the fields.

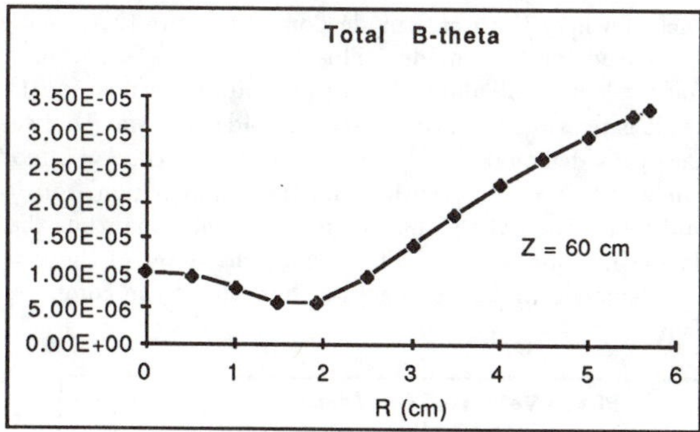

Fig. 2

For the 449 kHz case the amplitude of contributing modes vs axial mode number is plotted in Figures 3 through 6 for R ranging from the center of the plasma to the edge. For the dominant mode at the plasma center ($l = 13$) the parallel phase velocity corresponds to the local Alfvén speed. However the spectrum is quite broad possibly due to damping. This general form continues as we move out to larger radii until at about 2 cm from the plasma center we note the emergence of the $l = 5$ and $l = 6$ modes. These modes continue to grow with increasing radius until finally near the edge they are about $2\frac{1}{2}$ times the next largest mode.

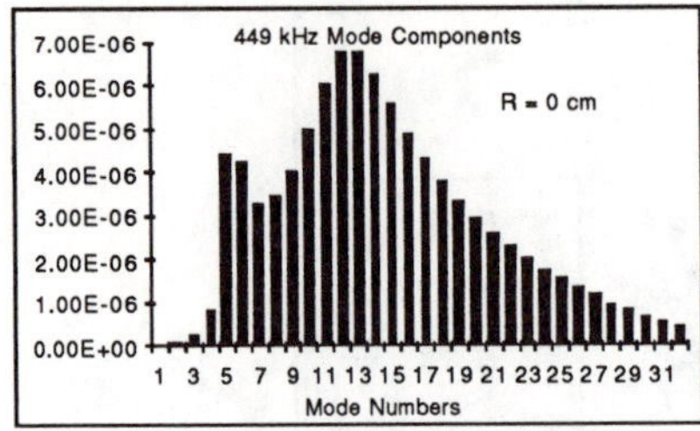

Fig. 3

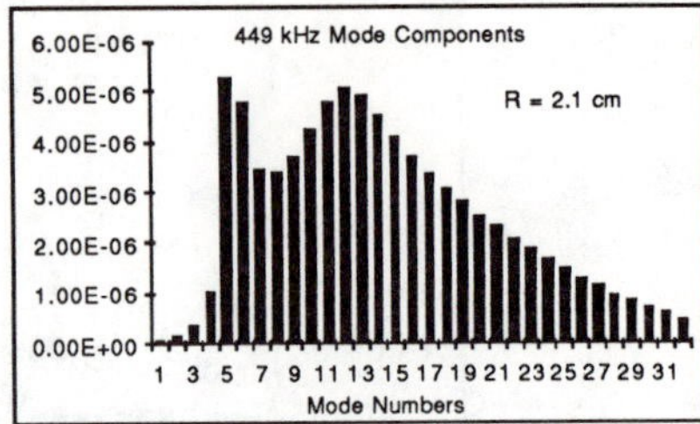

Fig. 4

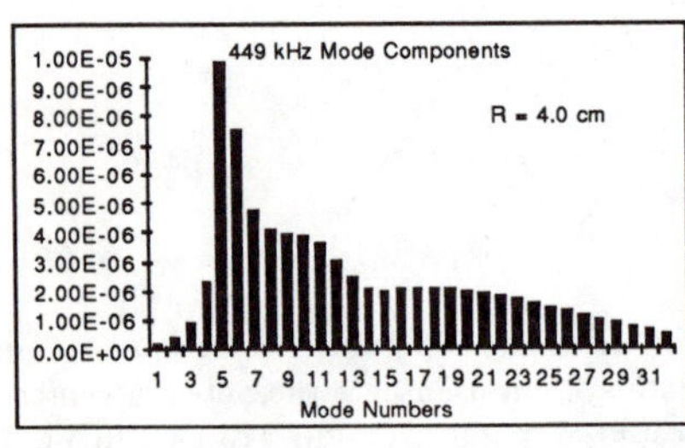

Fig. 5

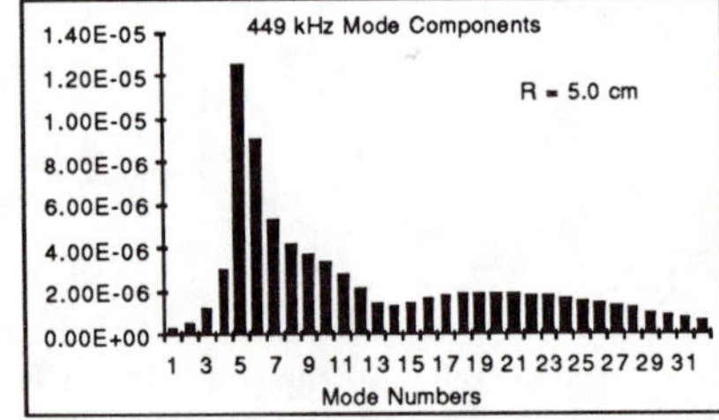

Fig. 6

This trend has the effect of raising the parallel phase velocity over the local Alfvén speed until the point is reached where the $l = 5$ mode dominates. At this point the phase velocity flattens to the Alfvén speed for that mode. This behaviour is apparent in Amagishi's data as well as in the phase velocity calculated from the mode superposition (Fig. 7). Stix[11] has previously shown that it is possible to have a standing cold shear wave between the Alfvén layer and the plasma surface provided that $\beta_e < \frac{m_e}{M_i}$ at the layer. For the $l = 5$ mode the Alfvén layer occures at $r \approx 5.1cm$ which is roughly within the transition region from cold to hot plasma ($r \approx 5.5cm$). It should be noted that this plasma is quite collisional thus the β condition above for a collisionless plasma may be relaxed. The surface character of the $l = 5$ mode is evident in Fig. 8 however the dispersion relation suggests that this is the compressional surface wave discussed by Amagishi.

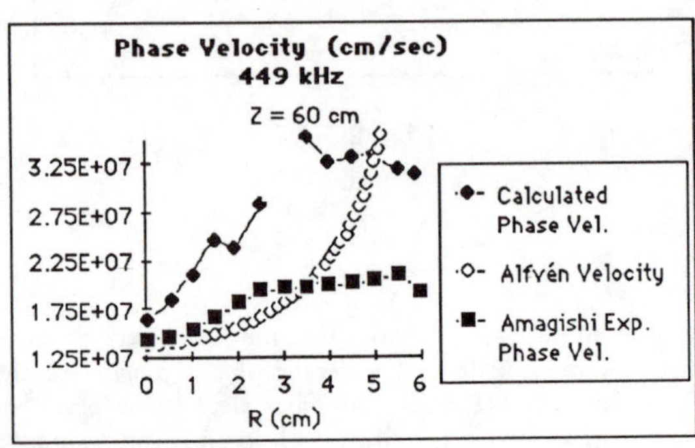

Fig. 7

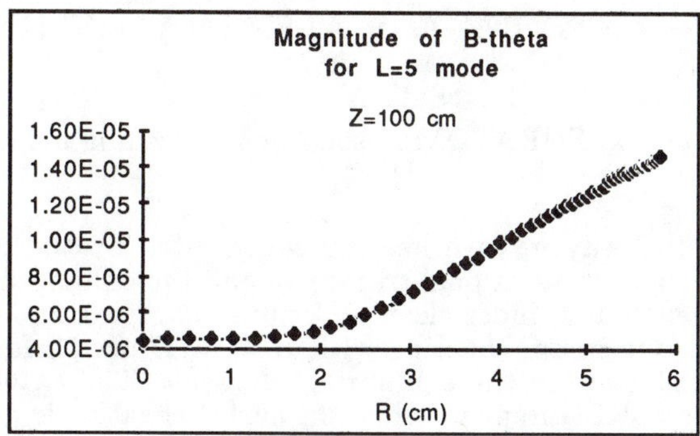

Fig. 8

CONCLUSION

We have used a cylindrical kinetic code to model the conditions in TPH. We have found that at the center of the plasma there is a broad spectrum of modes rather than a single mode and that the dominant mode is the one which has a parallel phase velocity equal to the local Alfvén speed. This changes as the radius increase to about 2.0 cm where modes $l = 5$ and $l = 6$ become large. Since these modes have a phase velocity greater than the local Alfvén speed until $r \simeq 5.1cm$. the effective phase velocity of the superposition of all of the modes is greater than the local Alfvén speed for this region. The $l = 5$ and $l = 6$ modes continue to grow and thus the effective phase velocity is dominated by these two waves and becomes an essentially constant function of radius as the spectrum near the edge narrows to the two dominant modes.

While the general behaviour of the calculations follow those reported by Amagishi it is evident that there is lack of agreement in the magnitude of the experimental and calculated phase velocities. This may be due in part to the fact that the analysis has not included end effects. These corrections are currently under study.

ACKNOWLEDGMENTS

This work is supported by The Texas Atomic Energy Research Foundation and The Department of Energy.

[1]Y. Amagishi, Phys. Rev. Letters **57**, 2807 (1986)
[2]K.Appert, R. Gruber, F. Troyon and J. Vaclavik, in *Heating in Toroidal Plasmas I* , edited by C. Gormezano (Pergamon, New York, 1982), Vol.1, p. 157.
[3]K. Appert and J. Valclavik, Plasma Phys. **25**, 551 (1983).
[4]D.W. Ross, G.L. Chen and S.M. Mahajan, Phys. Fluids **25**, 652 (1982)
[5]I. J. Donnelly and N. F. Cramer, Plasma Phys. and Controlled Fusion **26**,769 (1984)
[6]R. E. Stockdale and C. Z. Cheng, Bull. Am. Phys. Soc. **30**, 1592 (1985)
[7]G. A. Collins, F. Hofmann, B. Joye, R. Keller, A. Lietti, J. B. Lister and A. Pochelon, Phys. Fluids **29**, 2260 1986.
[8]T.E. Evans, et al., Phys. Rev. Letters **53**, 1743 (1984).
[9]G. G. Borg, M. H. Brennan, R. C. Cross, J. A. Lehane and A. B. Murphy, Proc. 13th Eur. Conf. on Contr. Fusion and Plasma Heating, Schliersee, Vol. **10C**, Part II, 53
[10]Y. Amagishi, M. Inutake, T. Akitsu and A. Tsushima, Jpn. J. Appl. Phys. **20**, 2171 (1981)
[11]T. H. Stix, presented at the Second International Symposium on Heating In Toroidal Plasmas, Como, Italy, Sept. 3-12, (1980)

OPTIMUM ANTENNA FOR ALFVÉN WAVE HEATING

Satish Puri
MPI für Plasmaphysik, EURATOM Association, Garching bei München, FRG

Modifications in the Alfvén wave heating contributed by the plasma equilibrium current through the rotational transform and the enhanced Hall effect are studied in a model that includes electron Landau damping and a 3-D Faraday shielded antenna. Optimum coupling with $R \approx 0.7\Omega$ occurs for $n = 8$ and is not significantly affected by the equilibrium current. The antenna $Q \approx 17$ is comparable to the ICRF antennas, while the surface heating is negligibly small. For the large $n \sim 8$ values, mode splitting due to the removal of the poloidal degeneracy combined with the finite electron temperature effects, leads to significant broadening of the energy absorption profile. No evidence of discrete Alfvén wave (DAW) excitation is observed.

1. INTRODUCTION

Efficient coupling to the Alfvén waves requires (i) conditions conducive to high conversion efficiency, and (ii) minimization of the evanescence between the plasma edge and the singular surface, $\gamma_1 = \epsilon_x - n_z^2 = 0$. The first of these conditions demands the choice of a large ω/ω_{ci} ratio, requiring a high toroidal number n, whereas the second condition is best satisfied at low frequencies and therefore for low values of n. Although this predicament has been recognized previously, the precise quantitative implications were brought to a focus in our recent antenna optimization study[1]. It was shown that optimal coupling, possessing high efficiency and low Q necessary for thermonuclear applications, would not be feasible with the low *antenna toroidal wave numbers*, $N \sim 2$ employed in the current experimental practice; acceptable coupling may require using $N \sim 8$ where, $N = N_A/2$, N_A being the number of alternately phased antenna sections along the torus circumference. The roles of the equilibrium current and surface heating via the direct excitation of the torsional Alfvén wave (TOR) are addressed in this paper.

The Alfvén resonance relation in the local coordinates[2]

$$\epsilon_\xi = n_\varsigma^2 = \cos^2\chi \, n_z^2 \left(1 + \frac{m}{nq}\right)^2 , \tag{1}$$

where χ is the angle between the magnetic field and the cylindrical axis, involves both the poloidal (m) and the toroidal (n) wave numbers, causing a multiplicity of radially distributed resonances for a given n. In cylindrical geometry, the Doppler shifted frequency seen by the electrons causes an enhancement of the Hall term by the factor

$$\left|\frac{\tilde{\epsilon}_y}{\epsilon_y}\right| \approx \frac{2}{qk_0 r_T} \frac{\omega_{ci}^2}{\omega\omega_{pi}} \gg 1 . \tag{2}$$

In addition, the rf current accompanying the enhanced Hall effect gives rise to MHD kink modes, also known as the discrete Alfvén waves[2].

Unless stated otherwise, the computational results employ the ASDEX UPGRADE parameters of TABLE I.

TABLE I

r_T	*toroidal radius (m)*	1.65
r_p	*poloidal radius (m)*	0.5
B_0	*toroidal magnetic field (T)*	3.0
$n_e(0)$	*density on axis* $(10^{20}m^{-3})$	3
$T_e(0)$	*temperature on axis (keV)*	7
q_{min}	*safety factor on axis*	1
q_{max}	*safety factor at edge*	3
r_A/r_p	*resonance position*	0.67
N	*antenna toroidal number*	8
2ℓ	*antenna length (m)*	0.8
$2w$	*antenna width (m)*	0.15
$2b$	*antenna wall separation (m)*	0.15
	gas composition	H_2

2. THE COMPUTATIONAL RESULTS

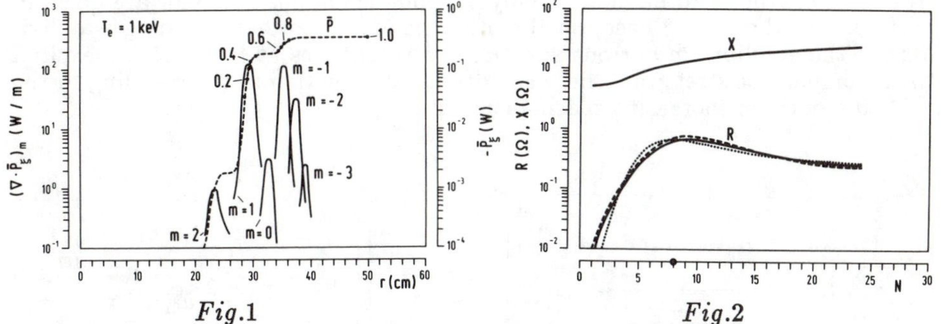

Fig.1 *Fig.2*

Figure 1 shows $-\bar{P}_\xi$ and $(\nabla \cdot \bar{\mathbf{P}})_m$ depicting the Poynting vector and the azimuthal components of the dissipation. Combination of the finite electron temperature and the mode separation due to the removal of the azimuthal degeneracy results in considerable broadening of the absorption profile for the large N values because the dominant pair of the $m = \pm 1$ modes are excited with comparable strength. This advantage does not accompany the low N excitation because of the relatively weak excitation of the $m = +1$ azimuthal mode.

Figure 2 shows the antenna loading in the absence of the plasma equilibrium current (dotted curve), with rotational transform alone (dashed curve) and with the addition of the enhanced Hall effect (solid curve). The equilibrium current produces insignificant changes in the Alfvén wave antenna loading provided a fixed position is maintained for the principal resonance corresponding to $m = \mp 1$, $n/|n| = \pm 1$. The rotational transform does, however, contribute (Fig. 3) to an increased antenna loading for lower values of r_A/r_p, owing to the presence of parasitic resonances occuring closer to the plasma boundary. This effect is further magnified[2] if the position of the secondary resonance corresponding to $m = \pm 1$, $n/|n| = \pm 1$ were to be held fixed, so that the principal resonance with $m = \mp 1$ itself assumes the role of a parasitic mode.

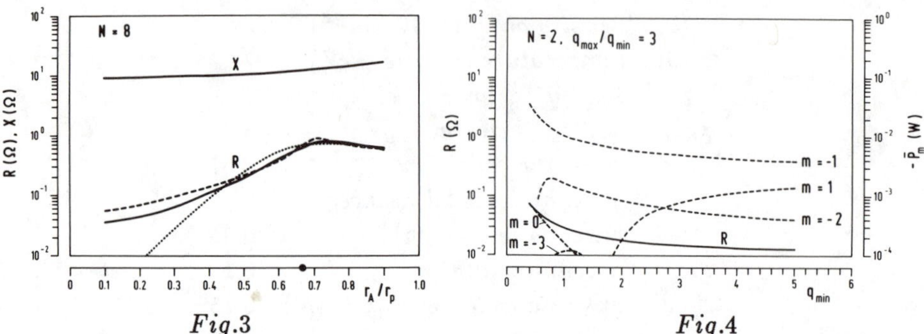

Fig.3 *Fig.4*

The precise effect of the rotational transform is seen in Figs. 4 and 5 for the two antenna configurations $N = 2$ and $N = 8$, respectively. The equilibrium current is simulated by varying the safety factor q_{min} at the axis while keeping $q_{max}/q_{min} = 3$. For the $N = 2$ case, the coupling to the $m = -1$ resonance dominates over the $m = +1$ contribution. Although there is distinct improvement in antenna loading for increasing plasma current, the coupling for the low N values continues to be poor. Analytic grounds for this behavior are given in Ref. 3. For the $N = 8$ case, on the other hand, the antenna loading is oblivious to the equilibrium current and the two resonances at $m = \pm 1$ are excited with comparable strength. There is a dramatic reduction in the coupling to the $m > 0$ modes for increasing plasma current.

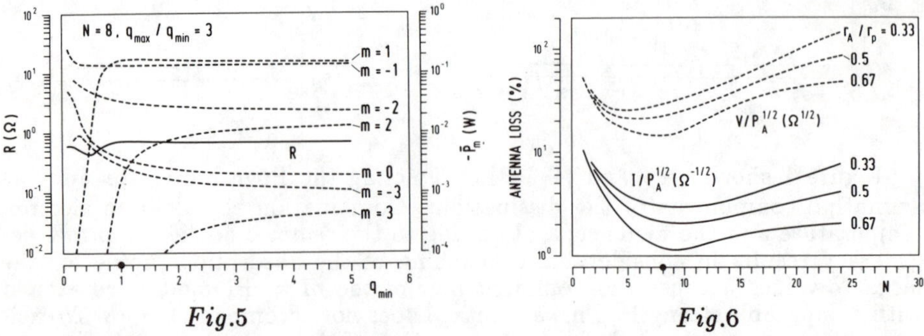

Fig.5 *Fig.6*

Figure 6 shows the antenna voltage and current as a function of N. For 1 MW power input per antenna, the $N = 2$ antenna requires 31 kV and 5.8 kA contrasted with the 14 kV and 1.2 kA needed for the $N = 8$ antenna (assuming $r_A/r_p = .67$). The cooling problems, too, become severe for the low N antennas, 13% of the input power being dissipated in the antenna itself for the $N = 2$ configuration, compared to the more tolerable loss of 1.2% for the $N = 8$ silver plated antenna.

3. DISCUSSION

3.1 Optimal coupling configuration

The primary objective of this work has been to explore the existence of windows in the parameter space conducive for coupling to the Alfvén resonance. The principal conclusion to emerge from the present investigation is that the use of high ($n \sim 8$) toroidal wave numbers is imperative for obtaining acceptable coupling in the context of thermonuclear plasma heating. Assuming an antenna configuration of $N = 8$, a further constraint consists in locating the resonance at approximately two-thirds the plasma radius, which corresponds to depositing the wave energy into the longitudinal motion of the electrons via Landau damping roughly at the center of the plasma volume. Neither of these constraints is significantly altered by the inclusion of the plasma equilibrium current.

3.2 Surface losses due to the direct excitation of TOR

Among the more welcome findings has been the absence of serious surface heating effects which could undermine an otherwise tenable scheme. The low coupling to the surface mode may be indicative of the degree of mismatch between the antenna impedance and the effectual short circuit existing along the magnetic field lines at the relatively low frequencies encountered in Alfvén wave heating. Ironically, the introduction of the Faraday screen enhances coupling to the surface wave. This behavior may possibly arise from the Faraday shield aligned along the magnetic field lines acting as an impedance matching transformer between the azimuthal antenna and the longitudinally conducting plasma. It is apparently preferable to discard the Faraday shield altogether.

4. CONCLUSIONS

The distinctive features of Alfvén wave coupling may be summarized as:

(i) Moderate Q values are comparable to the ICRF antennas. Operation at substantially lower frequencies implies the possibility of coupling more power per antenna for operation at a specified voltage. An input power of 3-4 MW per antenna is conceivable. Coupling 40-60 MW of total power envisaged in a reactor would require the deployment of roughly sixteen antennas which coincides with the optimum $N = 8$ operation.

(ii) The absence of edge heating from the direct coupling of TOR is a most welcome feature in favor of Alfvén wave heating.

(iii) The energy is deposited in the parallel electron motion at two thirds the plasma radius. This may constitute a viable approch for heating the entire plasma volume, particularly if the energy distribution is aided by the enhanced electron heat conductivity.

(iv) The antenna loading is insensitive to variations in density, magnetic field, profile shape, equilibrium current, edge conditions, plasma-antenna separation, and scales favorably to thermonuclear parameters.[1]

REFERENCES

[1] S. Puri, Nucl. Fusion **27**, 229 (1987).

[2] K. Appert, B. Balet and J. Vaclavik, Phys. Lett. **87A**, 233 (1982).

[3] S. Puri *in the Proceedings of the 14th European Conference on Controlled Fusion and Plasma Physics*, Madrid (1987).

EFFECT OF PARALLEL ELECTRIC FIELDS ON THE PONDEROMOTIVE STABILIZATION OF MHD INSTABILITIES

C. Litwin and N. Hershkowitz
University of Wisconsin, Madison, Wi. 53706

ABSTRACT

The contribution of the wave electric field component $E_{\parallel}$, parallel to the magnetic field, to the ponderomotive stabilization of curvature driven instabilities is evaluated and compared to the transverse component contribution. For the experimental density range, in which the stability is primarily determined by the $m = 1$ magnetosonic wave, this contribution is found to be the dominant and stabilizing when the electron temperature is neglected. For sufficiently high electron temperatures the dominant fast wave is found to be axially evanescent. In the same limit, $E_{\parallel}$ becomes radially oscillating. It is concluded that the increased electron temperature near the plasma surface reduces the magnitude of ponderomotive effects.

In recent years there has been a great deal of interest in application of radiofrequency (RF) radiation for stabilization of MHD instabilities. In axisymmetric systems[1,2] it has been demonstrated that that the presence of RF fields can stabilize curvature driven instabilities. It was found that for the RF frequency ω greater than the ion cyclotron frequency ω_{ci}, the plasma is stable. It becomes unstable for $\omega < \omega_{ci}$. This change of stability properties has been initially attributed[1–3] to the singular behavior of the ponderomotive force, associated with the E_+ component of the electric field. However it has been pointed out[4] that this singularity is spurious and in fact, the dominant contribution to the ponderomotive force comes from the E_- component of the electric field. Subsequent analyses[5,6] showed that the ponderomotive force provides a stabilizing contribution for $\omega > \omega_{ci}$. It has been suggested[7] that the lack of stability below the ion cyclotron frequency is caused by the damping of the RF wave by the Alfven wave resonance. In the analyses, referred to above, $E_{\parallel}$ is neglected. While this can be justified, for the fast wave, in the bulk of the plasma, a significant, radially evanescent (in the cold plasma limit) $E_{\parallel}$ exists near the plasma edge, due to the finite electron mass.[8] Following Stix[9] we shall refer to this mode as the ordinary mode. In the present paper we shall investigate its contribution to the ponderomotive stability.

The presence of the oscillating electric field leads to the appearance of the ponderomotive force on the plasma which modifies its stability properties. In the following discussion we shall restrict our attention to the effect the propagating modes which have been observed[10] to be primarily responsible for the observed stability. Eliminating (by an appropriate phasing of antennas) the propagating part of the excitation spectrum causes the plasma to become unstable. This is in an apparent contradiction with the theoretical result,[5,6] that, in the absence of the conducting walls, the far-field contribution vanishes for the lateral displacement, with the azimuthal wavenumber $l = 1$. This mode is expected to be the most unstable mode, since higher l modes are presumably stabilized by finite Larmor radius effects. It is not clear what the resolution of this discrepancy is. It appears possible that it is due to the difference between the idealized theoretical boundary conditions (absence of limiters, infinite length cylinder) and the actual experimental situation.

We calculate the contribution of the ponderomotive force $\mathbf{F}$ to the energy principle by considering the energy change δW_p due to displacement ξ

$$\delta W_p = -\int d^3x \xi \cdot \delta \mathbf{F} \tag{1}$$

The ponderomotive force exerted by an RF wave on a fluid element in a uniform axial magnetic field $\mathbf{B} = B\hat{z}$ has been found by Pitaevskii[11] using the stress tensor formalism of Landau and Lifshitz.[12] It can conveniently be expressed in form

$$\mathbf{F} = \sum_{\sigma=-1}^{1} \frac{\epsilon_\sigma - 1}{8\pi} \nabla |E_\sigma|^2 \tag{2}$$

Here we employ the usual helicity coordinates: $E_{\pm 1} \equiv E_\pm = E_x \pm iE_y$ and $E_0 \equiv E_\parallel = E_z$. In this representation the dielectric tensor is diagonal and ϵ_σ are the diagonal elements.[13] In (2) the contribution of the magnetization force, due to the magnetization induced by the presence of time-varying fields, is omitted. In a homogeneous magnetic field the latter contribution is a total gradient and, as such, it does not influence the stability with respect to incompressible modes. For the $l = 1$ rigid displacement $\xi = \alpha\hat{\mathbf{x}}$, $\delta W_p = \sum_\sigma \delta W_\sigma$ where

$$\delta W_\sigma = \frac{\epsilon_\sigma - 1}{4n} |\alpha|^2 \int dr r \frac{\partial n}{\partial r} \frac{\partial |E_\sigma|^2}{\partial r} \tag{3}$$

In Eq. (3) we have omitted the term corresponding to perturbed RF fields which is responsible for the afore-mentioned cancellation of the far-field contribution.[5,6] Motivated by experimental observations we shall ignore this term.

In order to compute δW_p, the wave electric field must be determined. We shall consider a density profile in which the width d of the nonvanishing density gradient region is much smaller than the plasma radius r_p and can be approximated by a square profile. Then, to lowest order in d/r_p, electric field components can be expressed in terms of Bessel functions:

$$E_\sigma(r, \phi, z, t) = \sum_{i=1}^{2} C_\sigma^{(i)} J_{m+\sigma}(\nu_i r) e^{i[kz - \omega t + (m+\sigma)\phi]} \tag{4}$$

The two radial wavenumbers ν_1 and ν_2 are solutions of a double-quadratic equation which follows from Maxwell's equations. The $i = 1$ term corresponds to the solution when $E_\parallel$ is neglected while the $i = 2$ term is the mode found by Hosea and Sinclair.[8] Since $|\nu_2| \gg \nu_1 \sim k$, we call this mode ordinary. In the cold plasma limit, $\nu_2^2 < 0$ for the fast wave and $\nu_2^2 > 0$ for the slow wave. Thus the ordinary mode is radially evanescent for the fast wave, while it is oscillating for the slow wave. Finite temperature effects make ν_2^2 complex and the ordinary mode acquires an oscillating component. Maxwell's equations provide also a relation between the coefficients $C_\sigma^{(i)}$. By imposing appropriate boundary conditions on the plasma-vacuum interface and at the conducting wall surrounding the vacuum, these coefficients are determined up to an overall constant, dependent on the antenna excitation level. Explicit formulae will be omitted here. In order to obtain a transparent result, exhibiting the relative sizes of various contributions to the ponderomotive force, we shall use asymptotic expansions of Bessel function, appropriate for the experimental parameters. First, we shall restrict our discussion to the vicinity of the ion cyclotron frequency where most of the stabilization experiments have been conducted.[1,10] This assumption simplifies the analysis considerably, since only the lowest radial harmonic of the $m \geq 1$ fast wave can propagate in the vicinity of the ion cyclotron frequency for the plasma density range observed in these experiments. For higher densities, higher radial harmonics can propagate, while below the ion cyclotron frequency the slow wave appears. The latter experiences a strong thermal damping in the vicinity of the ion cyclotron resonance.

For $m \geq 1$ magnetosonic waves in the experimental parameter regime, the long wavelength approximation is applicable[14] and Bessel functions can be expanded to lowest order in $\nu_1 r_p < 1$. Also, $|\nu_2| r_p \gg 1$. Assuming for simplicity that the conducting walls are far from the plasma surface we find, in the cold plasma limit

$$\frac{C_0^{(1)} J_m(\nu_1 r_p)}{C_0^{(2)} J_m(\nu_2 r_p)} = \frac{k r_p}{2m} \frac{k}{|\nu_2|} \tag{5}$$

From Eq. (5) one can show that δW_{-1} is dominated by $i = 1$ term while the main contribution to δW_0 comes from $i = 2$ term. In the vicinity of the cyclotron frequency, the contribution of E_+ can be neglected as small compared to the contribution of E_- for $i = 1$ and to that of $E_\parallel$ for $i = 2$.

Approximating $\partial n / \partial r$ by a constant and using the relation between $C_{-1}^{(1)}$ and $C_0^{(2)}$ we find, for the $m = 1$ fast wave which dominates the antenna spectrum in the discussed experiments,[1,10]

$$\frac{\delta W_{-1}}{\delta W_0} \approx \left(\frac{\omega_{pi} r_p}{c}\right)^2 \frac{d/2r_p}{1 - e^{-2|v_2|d}} \tag{6}$$

By $\omega_{p\alpha}$ we shall denote the plasma frequency of species α. Since for the experimental range of parameters $\omega_{pi} r_p / c \lesssim 1$, δW_0 is the dominant contribution to the energy principle. Thus the contribution of the propagating modes is much larger than previously concluded.[3,4] Note that δW_0 is stabilizing, as can be seen from Eq.(3), since $\partial |E_\parallel|^2 / \partial r > 0$ and $\epsilon_0 = -\omega_{pe}^2/\omega^2 < 0$ in the cold plasma limit. It has to be stressed that in the actual experimental situation the electron thermal velocity v_{te} is of the same order as the wave phase velocity which can have a significant effect of the stability. In the adiabatic regime, for $\omega \ll k v_{te}$, $\epsilon_0 \approx \omega_{pe}^2 / k^2 v_{te}^2 > 0$ and one might expect that the ponderomotive force becomes destabilizing. However, the ordinary mode is no longer purely evanescent, as mentioned earlier. In fact, it easy to show that, in the adiabatic regime, $\mathrm{Re}\nu_2 \gg \mathrm{Im}\nu_2$. Because of this oscillatory behavior, the ordinary mode contribution to the energy integrals largely cancels out. In addition, since $E_\parallel$ is now present in the bulk of the plasma, one might expect that the magnetosonic wave will become damped. The previously derived dispersion relation[14] can be solved, again assuming $\nu_1 r_p < 1$ and $\mathrm{Im}\nu_2 r_p \gg 1$, and the evanescence rate is found

$$\mathrm{Im}kr_p \approx \frac{m}{8\sqrt{2}}\beta_e^{1/2} \tag{7}$$

where $\beta_e \equiv 8\pi n_e T_e/B^2$ is the ratio of thermal and magnetic energy densities. We see that while the damping reduces the fast wave contribution to the far field, it is only significant for long, high β_e devices. It is amusing to note that, for tokamaks, this requires the aspect ratio to be large: $R/r_p \gtrsim 2/m\beta_e^{1/2}$, where R is the major radius. Thus the main effect of the high electron temperature on the ponderomotive stabilization is due to the oscillatory behavior mentioned earlier. It has to be noted that since the ordinary mode is characterized by a short radial wavelength, its properties are determined by the local plasma parameters. Thus the presence of cold electrons near the plasma edge prevents the mode from acquiring an oscillatory behavior and would therefore have a stabilizing influence.

Supported by U.S. DOE contract DE-AC02-78ET51015

REFERENCES

1. J.R. Ferron, N. Hershkowitz, R.A. Breun, S.N. Golovato and R. Goulding, Phys. Rev. Lett. **51**, 1955 (1983)
2. Y. Yasaka, R. Itanani, Nucl. Fusion **24**, 445 (1984)
3. J.R. Myra and D.A. D'Ippolito, Phys. Rev. Lett. **53**, 914 (1984); Phys. Fluids **28**, 1895 (1985)
4. P.L. Similon and A.N. Kaufman, Phys. Rev. Lett. **63**, 1061(1984); Phys. Fluids **29**, 1908 (1986)
5. J.R. Myra, D.A. D'Ippolito and G.L. Francis, Phys. Fluids **29**, 1908 (1986)
6. P.L. Similon, Phys. Rev. Lett. **58**, 495 (1987)
7. C. Litwin, N. Hershkowitz and J.D. Callen, Eur. Conf. Abstracts (Proc. 12th Eur. Conf. Contr. Fusion Plasma Phys., Budapest, Hungary 1985), vol. 9F (part II), p.64, European Physical Society, Budapest (1985)
8. J.C. Hosea and R.M. Sinclair, Phys. Fluids **13**, 701, 1970
9. T.H. Stix, Theory of Plasma Waves, McGraw Hill, New York (1962)
10. R. Majeski, J.J. Browning, S. Maessick, N. Hershkowitz, T. Intrator and J.R. Ferron, submitted to Phys. Rev. Lett.
11. L.P. Pitaevskii, Sov. Phys. JETP, **12**, 1008 (1961)
12. L.D. Landau and E.M. Lifshitz, Electrodynamics of Continuous Media, Pergamon Press, New York (1980)
13. G. Schmidt, Physics of High Temperature Plasmas, Academic Press, New York (1979), p. 251
14. C. Litwin and N. Hershkowitz, Phys. Fluids, May 1987

Coherence Limited Resonant Diffusion *

G. D. Kerbel

National Magnetic Fusion Energy Computer Center
Lawrence Livermore National Laboratory

Abstract

A description of wavepacket driven, collision limited resonant diffusion in tokamaks can be represented mathematically through the use of a phase diffusion kernel in the trajectory integral. A new technique has been implemented which allows the global development of numerical representations of these integrals as the solutions of a certain partial differential equation whose initial and boundary data derive from an asymptotic analysis of the fully coherent system.

Introduction

Conventional Fokker-Planck Quasilinear models of collisional and resonant diffusion in tokamaks rely principally on the mechanism of magnetic decorrelation to mediate the wave-particle interaction dynamics. Though a charged particle must be in resonance to interact strongly with the exciting RF fields, the magnetic correlation time $|\tau_c|$ is *not* the most rational measure of the strength of the interaction in all cases. Finite bandwidth wave spectra and collisional effects can significantly alter the calculation of the effective correlation times - especially in nearly uniform magnetic fields - through the mechanism of interaction phase decorrelation. Better realism, as well as computational convenience and completeness, are among the incentives to generalize the unperturbed collisionless orbits to include these effects on an *a priori* basis.

The mechanism at issue can be most clearly understood by considering the uniform magnetic field case. In collisionless theory, those particles whose gyrocenters lie on resonant orbits remain in resonance for all time; nearby orbits witness no resonant diffusion whatever. In other words, the resonant diffusion is highly localized. This aspect of the theory allows the analytic evaluation of certain quantities of interest once the phase space density is known. In particular, the power absorption can be calculated directly. However, in contradistinction to the analytic theory, this (idealized) structure becomes increasingly difficult to resolve with a numerical scheme in the limit of uniform magnetic field. For such cases, the (collisionless) magnetic correlation time is either longer than the phase decorrelation time or undefined.

---

* Work performed under the auspices of the U.S.D.O.E. by LLNL under contract No. W-7405-ENG-48.

Collisions decorrelate the wave-particle resonance and limit the energy exchange between particles and the wave field. The presence of a large multiplicative factor of the pitchangle scattering rate in the gyro-phase diffusion is responsible for the importance of this effect even deeply within the (collisionless) banana regime. Small angle Coulomb collisions alter the helical pitch of a resonant particle trajectory causing the resonant exchange of energy between the field and the particle to slow, stop, or reverse. The inverse process can also occur: A particle can occupy a resonant trajectory temporarily and experience resonant interaction on the fly.

The mechanism involved in finite bandwidth $k_{||}$ spectra decorrelation can be understood on the basis of a comparison of magnetic or collisional decorrelation times with transit times across a Gaussian wavepacket. The wavepacket serves as a convenient model for a ray bundle, a beam, or a weakly turbulent wave field with coherence on a particular length scale.

A mathematical model of the collisionally broadened resonance was presented in Kerbel and McCoy[1]. In that treatment a simplification was invoked to enable the evaluation of the bounce-averaged trajectory integral: The decorrelation of the induced (micro)current with respect to the driving field was computed as if composed of a product of two independent decorrelations, each computed with respect to a fixed point, the resonant point. The result incorrectly retains coherent interaction structures where phase decorrelation actually destroys them. Decorrelation in fact occurs over a (variable) time interval separating the induced and prompt components. The simplification was thought necessary for fast computation; an efficient algorithm for computing the orbit integrals was unknown. Recently, an algorithm which evaluates these trajectory integrals rapidly has been developed: The remainder of this report will be devoted to a discussion of the technique.

Interaction Distributions

The resonant diffusion tensor $\mathbf{D}$ of Fokker-Planck Quasilinear theory can be represented as

$$\mathbf{D} = \mathcal{I}\mathbf{d}(\vec{\mathbf{v}}_\mathbf{0}; \vec{\mathbf{k}}, \omega; \vec{\mathbf{E}}(\vec{\mathbf{k}}, \omega; \mathbf{t}, \mathbf{t_0}))$$

where $\mathbf{d}$ is a symmetric tensor function in $\vec{v}_0$-space, slowly varying along a gyrocenter orbit parameterized by (t, t_0); $\mathcal{I}$ is a distribution with compact support on (t, t_0) in the vicinity of resonance representing the strength of the wave-particle interaction, or the orbit averaged field autocorrelation along the unperturbed trajectory. $\mathcal{I}$ takes slowly varying functions of orbit time (t, t_0) into real numbers. In the familiar limit of uniform magnetic field and vanishing phase decorrelation, $\mathcal{I}$ becomes a δ distribution at resonance; In the presence of a phase decorrelating mechanism, the distribution support is broadened, viz. the term resonance broadening. That is to say, orbits which in the fully coherent case were non-interacting become interacting; strongly interactive orbits become less so.

$\mathcal{I}$ has the general form

$$\mathcal{I} = \oint d\tau \, e^{-i\Psi(\tau)} \int_{\tau-\tau_B}^{\tau} d\tau' \, e^{i\Psi(\tau')} \mathcal{D}(\tau - \tau') + cc.$$

where $\mathcal{D}(\tau - \tau')$ determines the phase coherence between orbit points at (τ, τ'), and $\mathcal{D} = 1$ is the fully phase coherent case. The eikonal Ψ, given by

$$\Psi(n, \tau) = \int_0^{\tau} d\tau' \, (n\Omega + k_{\|}v_{\|} - \omega) = \int_0^{\tau} d\tau' \, \nu(n, \tau'),$$

represents the advance of the interaction phase, ν, along a given gyro-center trajectory. In large gyration frequency theory, $\Psi \sim \Omega\tau_B \gg 1$, so that Ψ is viewed as rapidly varying, and $\Omega\tau_B$ serves as the large parameter in the asymptotic analysis. Solution of a Langevin equation for the diffusion of the phase due to small angle pitchangle scattering events leads to the collisional phase diffusion kernel

$$\mathcal{D}_c = e^{-\beta_c|\tau-\tau'|^3}.$$

The τ^3 dependence arises due to the secular accumulation of random gyrophase increments along a gyrocenter trajectory. Interaction with a wavepacket can be represented by a phase diffusion kernel of the form

$$\mathcal{D}_k = e^{-\beta_k|\tau-\tau'|^2}.$$

Here the τ^2 dependence arises due to the finite transit time of the gyrocenter across the wavepacket.

Four rather distinct cases for $\mathcal{I}$ arise through combinations of isolated or coalescing resonances, combined with collisional or spectral decorrelation. Employing the technique of differentiation under the integral for the coalescing resonances cases we discover that the spectral decorrelation case solves a diffusion equation with respect to two dimensionless parameters relating to the wavepacket spectral width β, and the degree of detuning from resonance, z,

$$\frac{\partial \mathcal{I}}{\partial \beta} = \frac{\partial^2 \mathcal{I}}{\partial z^2}.$$

For the same decorrelation model, the isolated resonance case can be evaluated analytically in terms of complex error functions.

The limiting case of zero spectral width, $\beta = 0$, no decorrelation, can be evaluated analytically. This provides initial data for the diffusion equation which $\mathcal{I}$ solves. Certain asymptotic limits are also accessible analytically, thus providing a check on the numerical procedure.

The collisional model is much more difficult to evaluate due to the τ^3 dependence in $\mathcal{D}_c$. The approach we have taken with respect to collisional decorrelation is to determine an effective coherence scale which combines both collisional and spectral decorrelation;

$$\beta_e \sim \beta_c^{2/3} + \beta_k \sim \tau_{coh}^{-2}$$

The implication is that it is this finite coherence scale rather than the details of the destruction of coherence which matter for our purposes. Qualitative differences in those limits which are fully accessible to our calculations are indiscernable.

The figure shows the β evolution for increasingly strong decorrelation and the associated broadening and smoothing of the interaction structure. The ordinate measures interaction strength, $\mathcal{I}$, and the abscissa measures the degree of detuning. The computation is incorporated in the RF module of the code family CQL available in public FILEM and requires an additional overhead in computation time of less than 15 seconds for any given run.

References

[1] G.D.Kerbel and M.G.McCoy, Comput. Phys. Commun., **40**, 105, (1986).

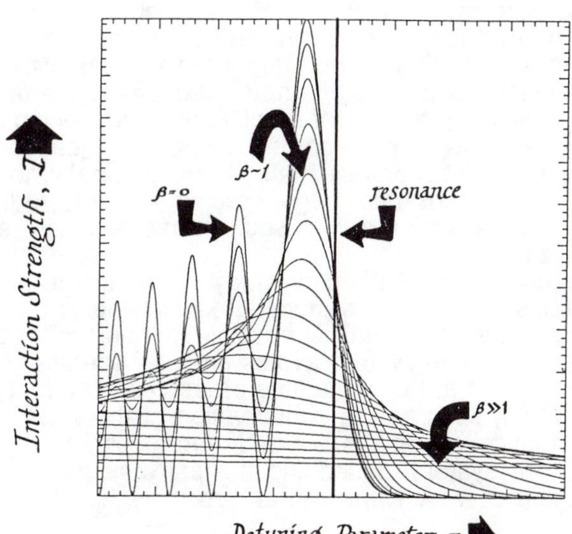

CONGRUENT REDUCTION AND MODE CONVERSION IN 4-DIMENSIONAL PLASMAS

Lazar Friedland and Allan N. Kaufman
LBL and Physics Department, University of California, Berkeley, CA 94720

ABSTRACT

Standard eikonal theory reduces, to N=1, the order of the system of equations underlying wave propagation in inhomogeneous plasmas. The condition for this remarkable reducibility is that only one eigenvalue of the unreduced NxN dispersion matrix **D**(k,x) vanishes at a time. If, however, two or more eigenvalues of **D** become simultaneously small, the geometric optics reduction scheme becomes singular. These regions are associated with linear mode conversion, and are described by higher order systems. We develop a new reduction scheme based on congruent transformations of **D**, and show that, in "degenerate" plasma regions, a partial reduction of order is possible. The method comprises a constructive step-by-step procedure, which, in the most frequent (doubly) degenerate case, yields a second order system, describing the pairwise mode conversion problem, the solution of which in general geometry has been found recently.

INTRODUCTION

Small amplitude waves in inhomogeneous plasmas are described by systems of linear equations for such quantities as electromagnetic fields **E**, **B**, perturbed average velocities $\mathbf{V}_{\alpha}$ of various species in fluid models, etc. The complexity of the problem due to the multicomponent structure of the waves is typically resolved by using some sort of reduction (elimination) of several wave components from the problem. The conventional geometric optics theory,[1] for example, makes use of the reduced dispersion tensor **D**, a 3x3 matrix, describing the components of the electric field **E** alone. The geometric optics theory then uses a perturbation expansion technique to reduce the problem even further. In non-degenerate cases, when only one eigenvalue of **D** vanishes at a time, the theory reduces to the solution of a single first order equation for the scalar amplitude of the electric field. This perturbation scheme, however, becomes singular in cases when an additional eigenvalue of **D** becomes small.[2] More generally, the singular situations are characteristic of near degeneracies of the unreduced (N-th order) system and are associated with such phenomena as resonances and mode converison.

This paper summarizes our recent study of reduction in geometric optics,[3] and describes a constructive procedure which yields the optimally reduced system, describing the smallest possible number of "irreducible" wave components. The final system avoids singular coefficients, has the lowest possible order, and at the same time preserves important properties of the unreduced wave, such as the conservation of the wave-action flux. As an important application, in doubly degenerate plasmas, the theory yields a second order system, describing the pairwise mode conversion problem, the solution of which in general geometry has been obtained recently.[4]

TRANSPORT EQUATION AND CONGRUENCE TRANSFORMATION

Our starting point is the linear integral "evolution" equation

$$\int d^4x_2\ \hat{D}_{ij}(x_1, x_2)\, Z_j(x_2) = 0 \qquad (1)$$

for the (complex) N-component field $\mathbf{Z}(x)$ on space-time. The two point NxN kernel D_{ij} is assumed to be Hermitian and its local Fourier transform

$$\mathbf{D}(k,x) = \int d^4s\, \hat{\mathbf{D}}\,(x+\frac{1}{2}s,\, x-\frac{1}{2}s)\exp(-ik\cdot s) \qquad (2)$$

defines the dispersion matrix of the unreduced problem. The eikonal representation $\mathbf{Z}(x)=\mathbf{A}(x)\exp[i\psi(x)]$, with slowly varying amplitude $\mathbf{A}(x)$ and wave vector $k_\mu(x)=\partial\psi/\partial x^\mu$, leads to (Refs.1,3):

$$\mathbf{D}\bullet\mathbf{A} = i\left[\frac{\partial\mathbf{D}}{\partial k_\mu}\bullet\frac{\partial\mathbf{A}}{\partial x^\mu} + \frac{1}{2}\frac{d(\partial\mathbf{D}/\partial k_\mu)}{dx^\mu}\bullet\mathbf{A}\right] + O(\delta^2) \equiv \mathbf{L}(\mathbf{A}) \qquad (3)$$

where δ is the small parameter describing the slow variation of the background. Eq. (3) is the conventional transport equation, being the starting point of the geometric optics perturbation scheme. This equation also yields the action flux conservation

$$\partial J^\mu(x)/\partial x^\mu = 0, \quad J^\mu(x) \equiv A_i^*(\partial D_{ij}/\partial k_\mu)A_j + O(\delta) \qquad (4)$$

At this stage we make the transformation to a new amplitude $\mathbf{A}$ via

$$\mathbf{A}(x) = \mathbf{Q}(k(x),x)\bullet\overline{\mathbf{A}}(x) \qquad (5)$$

where the matrix $\mathbf{Q}(k,x)$ is assumed to be slowly varying and invertible. Then Eq.(3) can be written for $\mathbf{A}$:

$$\overline{\mathbf{D}}\bullet\overline{\mathbf{A}} = \mathbf{Q}^+\bullet\mathbf{L}(\mathbf{Q}\bullet\overline{\mathbf{A}}) \qquad (6)$$

where the transformed dispersion matrix $\mathbf{D}$ is given by the congruence transformation

$$\overline{\mathbf{D}} = \mathbf{Q}^+\bullet\mathbf{D}\bullet\mathbf{Q} \qquad (7)$$

The purpose of the transformation $\mathbf{Q}$ is to simplify the problem and in particular (see the next Section) to annihilate, if possible, some wave components, thus reducing the order of the system.

REDUCTION THEOREM

The main result of our studies is the following Reduction Theorem: If there exists at least one large (of O(1)) element in the matrix $\mathbf{D}(k,x)$, characterizing the unreduced N-component (N-th order) system, then one of the components of $\mathbf{A}$ can be eliminated from the problem in such a way that the remaining N-1 components of the wave are fully described by a reduced, Hermitian tensor $\mathbf{D}^r$ with the reduced transport equation of form (3), $\mathbf{D}$ being replaced by $\mathbf{D}^r$.

The proof of this theorem can be found in Ref.3. Here we shall briefly describe the reduction algorithm. The reduction proceeds differently in the following two cases :

Case 1. For general N, suppose that the element D_{kk} of **D** is O(1). Then the reduction to N-1 is achieved by $Q_{ij} = \delta_{ij} - \delta_{ik}D_{kj}/D_{kk} + \delta_{ik}\delta_{jk}$. This transformation annihilates the k-th component of **A**, leaving the remaining components of **A** unchanged, so that the reduced wave amplitude is $\mathbf{A}^r = \{A_i, i \neq k\}$. The reduced dispersion tensor, to lowest order, is

$$\mathbf{D}^r = D_{ij} - D_{ik}D_{kj}/D_{kk} , \qquad i,j \neq k \qquad (8)$$

Case 2. When all diagonal elements of **D** are O(δ), but D_{qr} (r≠q) is of O(1), we use the transformation matrix $Q_{ij} = \delta_{ij} + \alpha\delta_{ir}\delta_{jq}$, where α = 1 if Re D_{qr} = O(1) and α=i if Im D_{qr} = O(δ). After this transormation the q-th diagonal element of **D** is $D_{qq} = 2\mathrm{Re}(\alpha D_{qr}) + O(\delta) = O(1)$.Therefore the reduction to N-1 can be achieved as in Case 1. It can be also shown that the reduced matrix then has $D_{rr} \sim O(1)$, so that Case 2 actually allows reduction from N to N-2.

The Reduction Theorem provides a tool for reducing the order, step-by-step, for weakly varying plasmas of arbitrary geometry. The successive application of the algorithm yields a final reduced Hermitian matrix $\mathbf{D}^f$ of rank M≤N, such that all its elements are O(δ). A further attempt to reduce the system would yield singular coefficients, so that the system is "irreducible" within the geometric optics approximation. The simplest and most frequent case is M=1, when all but one of the components of **A** are eliminated from the problem. This is the non-degenerate case, when only one eigenvalue of **D** is small. The final transport equation in this case is a single first order PDE for the remaining wave component A^f.

NORMAL DEGENERACY AND MODE CONVERISON

Less frequent, but nevertheless important in applications, is the situation when M=2, in which case the final system comprises a set of two coupled PDE's. The final reduced matrix $\mathbf{D}^f$ in this case can be written as

$$\mathbf{D}^f = \begin{bmatrix} D_a & \eta \\ \eta^* & D_b \end{bmatrix} \qquad (9)$$

where all elements are of O(δ), and let us denote $\mathbf{A}^f = (A_a, A_b)$.

Now we argue that since the degeneracy implies a simultaneous satisfaction of three smallness conditions ($D_a, D_b, \eta \sim O(\delta)$), the degeneracy is a rare phenonenon. The most probable scenario of the double degeneracy, therefore, is that due to some special physical conditions (existence of a global small parameter, for example) one of the elements of $\mathbf{D}^f$ is O(δ) in an extended region of the phase space. We also argue that since $\mathrm{Det}(\mathbf{D}^f) = D_aD_b - |\eta|^2$, only when η~ O(δ) in an extended region of the phase space does there exist the possibility of having two distinct modes ($D_a \approx 0$, $D_b \approx 0$) away from the degenerate region. We shall refer to this situation as normal degenerancy.

The reduced transport equation (3) in the case of the normal degeneracy becomes[4]

$$V_a \cdot \partial A_a/\partial x - ix' \cdot R^a A_a = i\eta A_b$$

$$V_b \cdot \partial A_b/\partial x - ix' \cdot R^b A_b = i\eta^* A_a \tag{10}$$

where $V^\mu{}_{a,b} = \partial D_{a,b}/\partial k_\mu$ and $R_\mu{}^{a,b} = \partial D_{a,b}/\partial x^\mu$ are the group four-velocity and spatial gradient (or four-refraction). These are constant vectors, evaluated at the crossing point x_o, k_o where $D_a(x_o,k_o) = D_b(x_o,k_o) = 0$. In (10), $x' \equiv x - x_o$. Solution of Eqs. (10) has been found in Ref. 4. It was shown that the action flux J_a associated with the dispersion relation $D_a = 0$ in the non-degenerate region is only partially transmitted through the neighborhood of the crossing point, and that the transmission coefficient is

$$T = \exp(-2\pi|\eta|^2/|B|) \tag{11}$$

where B is the Poisson Bracket

$$B = (\partial D_a/\partial x^\mu)(\partial D_b/\partial k_\mu) - (\partial D_a/\partial k_\mu)(\partial D_b/\partial x^\mu) \tag{12}$$

Summarizing, we have shown that pairwise mode conversion events are typically associated with normal degeneracy of the final 2x2 dispersion matrix (M=2), and that the reduction algorithm automatically provides the characteristic form of the dispersion tensor, describing two easily identifiable coupled modes asssociated with the diagonal elements of the matrix, while its non-diagonal element serves as a small coupling coefficient. These elements can then be used directly in the mode conversion theory[4] for calculating the transmission and mode conversion coefficients in cases of interest.

ACKNOWLEDGEMENT

This research was supported by the U.S. Department of Energy under contract No. DE-AC03-76SF00098.

REFERENCES

1. I. Bernstein and L. Friedland in "Handbook of Plasma Physics", M. Rosenbluth and R. Sagdeev, eds. (North-Holland, Amsterdam,1983) Vol.1, Ch. 2.5.
2. L. Friedland, Phys. Fluids 28, 3260 (1985).
3. L. Friedland, and A. Kaufman, Phys. Fluids (in press).
4. L. Friedland, G. Goldner, and A. Kaufman,Phys.Rev.Lett. 58,1392 (1987); A. Kaufman and L. Friedland, Phys. Lett. A (in press).

COLLISIONAL MAGNETIC PUMPING AS AN EFFICIENT WAY TO HEAT A PLASMA*

M. Laroussi and J. R. Roth
University of Tennessee, Knoxville, Tennessee 37996-2100

ABSTRACT

This paper describes the application of collisional magnetic pumping[1] to plasma heating. Theoretical solutions giving the heating rate for different magnetic field waveforms of the RF perturbing wave are presented. Numerical solutions to the energy transfer problem reveal a plasma parameter regime in which collisional magnetic pumping is most effective, and provide insight into the energy transfer mechanism. It is shown[2] that when the magnetic field waveform is a sawtooth function, the heating rate is proportional to the first power of the field modulation. This is an improvement over the case when a sinusoidal waveform is used, which results in second order heating. To implement experimentally the sawtooth waveform, a switching circuit using solid state power metal-oxide-semiconductor (TMOS) transistors was designed. The circuit is capable of driving a sawtooth shaped unidirectional current in the coil used to generate the RF magnetic field.

INTRODUCTION

Magnetic pumping is achieved by wrapping an exciter coil around a cylindrical plasma and perturbing the confining magnetic field, $B = B_o (1+\delta f(t))$, where B_o is the uniform steady-state background magnetic field, δ is the field modulation, and f(t) is a periodic function. To achieve the collisional heating regime the following inequalities have to be satisfied

$$\tau_{ci} << \tau_{coll} \sim \tau_f << \tau_{tr}, \tag{1}$$

where τ_{ci}, τ_{coll}, τ_f, and τ_{tr} are respectively the cyclotron period, the collision time, the period of the RF driving signal, and the transit time of the particles through the heating region. The equations governing the change in the parallel and perpendicular components of the energy of the particles are[1,3]

$$\frac{dE_\perp}{dt} = \frac{E_\perp}{B}\frac{dB}{dt} - v_c\left(\frac{E_\perp}{2} - E_{11}\right) \tag{2}$$

$$\frac{dE_{11}}{dt} = v_c\left(\frac{E_\perp}{2} - E_{11}\right) \tag{3}$$

*Work supported by contract AFOSR 86-0100 (Roth).

Where v_c is the collision frequency. The above equations have been solved both analytically and numerically[2]. The analytical treatment provides the heating rate dE/dt for specific cases of the perturbing function f(t) and also provides the condition to be met if a heating rate proportional to the first power of the field modulation δ is to be achieved[2,4,5]. The numerical solutions provide additional information on the energy transfer process between the RF perturbed magnetic field and the parallel and perpendicular energy components of the heated species.

ANALYTICAL TREATMENT

When combined together, Equations (2) and (3) yield the following equation

$$\frac{d^2E}{dt^2} - \left[-\frac{3}{2} v_c + \frac{d^2B}{dt^2} \left(\frac{dB}{dt} \right)^{-1} \right] \frac{dE}{dt} - \frac{v_c}{B} \frac{dB}{dt} E = o \tag{4}$$

This is a second order differential equation with periodic coefficients. Such equations appear in astronomical and other applications where the stability and perturbations of periodic systems are at issue. The general solutions of these equations have been given by Floquet,[6] and have the following form

$$E = a_1 e^{\lambda t} p(t) \tag{5}$$

Using a perturbation treatment and solving Equation (4) for the first power of δ, the factor ℓ_1 corresponding to this power in the expansion is found to be[4,5]

$$\ell_1 = \frac{v_c}{T} \int_0^T e^{\frac{3}{2} v_c s} \left\{ \frac{1}{1 - \mathrm{Exp}\left(-\frac{3}{2} v_c T \right)} \int_0^T f'(u) e^{-\frac{3}{2} v_c u} du \right.$$

$$\left. - \int_0^s e^{-\frac{3}{2} v_c u} f'(u) du \right\} ds. \tag{6}$$

If $f(t) = \cos \omega t$, $\ell_1 = 0$ and Equation (4) has to be solved for the coefficient, ℓ_2, of the second power of δ. Carrying out this procedure yields the energy increase rate

$$\frac{dE}{dt} = \frac{\delta^2}{6} \frac{v_c \omega^2}{\frac{9}{4} v_c^2 + \omega^2} E_0 \tag{7}$$

The above result agrees with the one found by Berger et al[1]. If the magnetic field assumes the profile shown in Fig. (1), the heating rate is of first order and is given by

$$\frac{dE}{dt} = \frac{\delta\omega}{3\pi} E_0 \tag{8}$$

Since $\delta << 1$, the above result can represent many orders of magnitude improvement on the preceding case.

NUMERICAL TREATMENT

Equations (2) and (3) have also been solved numerically. The magnetic field profile used for this analysis is shown in Fig. (2). The program has the flexibility of changing the slopes of the rising and decreasing ramps. An important variable is the collisionality parameter $v_c T$, defined as the dimensionless product of the collision frequency, v_c, and the period, T, of the RF signal. The time step chosen is $10^{-4}T$. Fig. (3a) through Fig. (3d) show the plots of the parallel component of the energy and of the total energy of the particles versus time. The units on the energy axis are arbitrary, and on the time-axis are multiples of the period T. The field modulation for all cases is $\delta = 0.1$. As illustrated in Fig. (3b), when the collisionality parameter is less than 1, the plot of energy vs time doesn't show an increase with time in its average value. The explanation of this lies in Fig. (3a) which shows that after an initial transient phase the parallel component of the energy saturates and the magnetic pumping fails. As the collisionality parameter increases the saturation phenomena ceases and an increase in the average value of the total energy is obtained. This is illustrated in Fig. (3c) and Fig. (3d). Carrying out this analysis it is shown that there exists a threshold value of the collisionality parameter below which collisional magnetic pumping ceases to be effective. Also the energy transfer process through collisions is a resonant one where an optimum collisionality parameter exists at which the power absorption is maximized.

DESIGN IMPLICATIONS

The plasma on which the collisional magnetic pumping is to be tested is generated by a classical Penning discharge. The background uniform axial magnetic field can be varied up to 0.44 Tesla. The electron number density is typically $2 \times 10^9/cm^3$ in helium gas, with Te = 5- 10 ev. Fig. (4) is the schematic of the circuit to be used to generate the magnetic field waveform of Fig. (1). The inductor value is 1.6 μH. The circuit should be able to switch on and off a current of 100A at a frequency of several hundred KHz. The RF magnetic field generated is about 20 Gauss. The sudden change of the RF magnetic field from its maximum value to zero is not instanteneous however. Using state of the art TMOS transistors as switches a fall time of few hundreds of nanoseconds is possible. This slight distortion in the magnetic field waveform should not be a problem as long as the fall time is shorter than the collision time.

REFERENCES

1. J. M. Berger, et al., Physics of Fluids, Vol. 1, (1958), pp. 301-307.
2. M. Laroussi and J. R. Roth, APS Bulletin, Vol. 31, No. 9, (1986), p. 1421.
3. M. Laroussi, Proceedings of the 18th Southeastern Symposium on System Theory, ISSN 0094-2898, (1986), pp. 475-479.

4. M. Laroussi and J. R. Roth, APS Bulletin, Vol. 30, No. 9, (1985), p. 1390.
5. M. Laroussi and J. R. Roth, Proc. IEEE International Conf. on Plasma Science, Cat. No. 86CH2317-6, (1986), pp. 91-92.
6. M. G. Floquet, Annales E.N.S., T. 12, (1883), pp. 47-88.

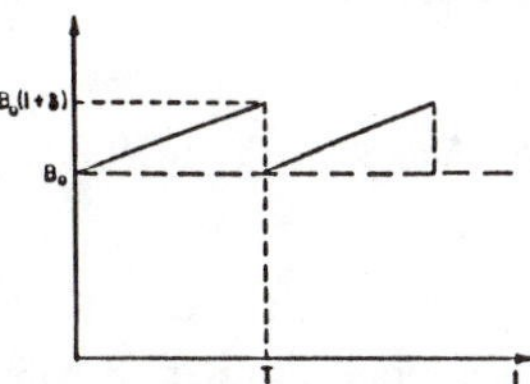

Fig. 1. Sawtooth waveform

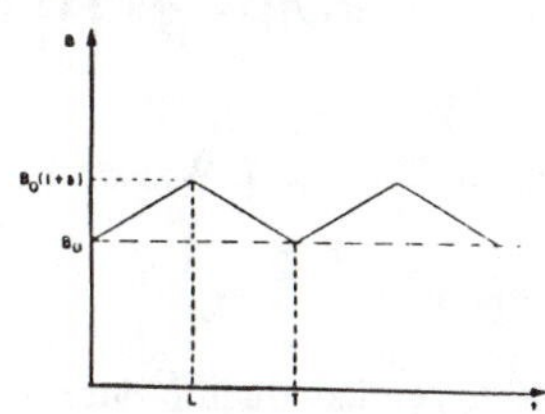

Fig. 2. Triangular waveform model used in the numerical treatment

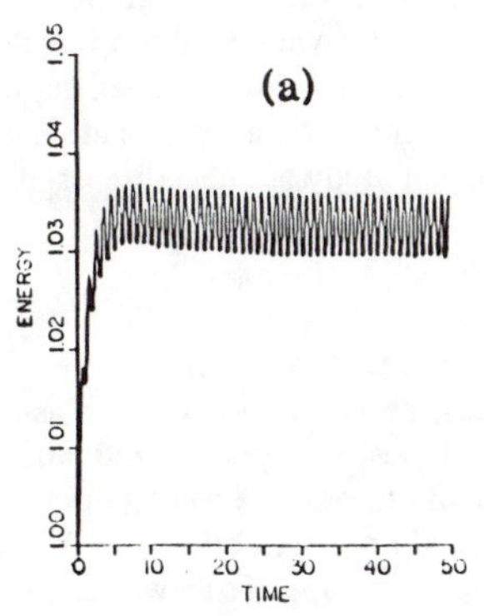

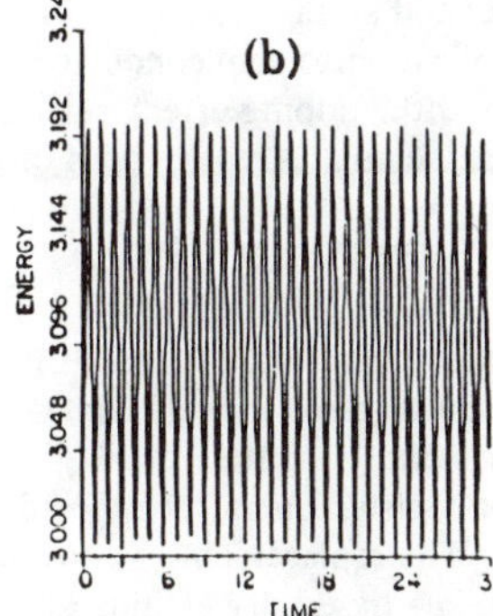

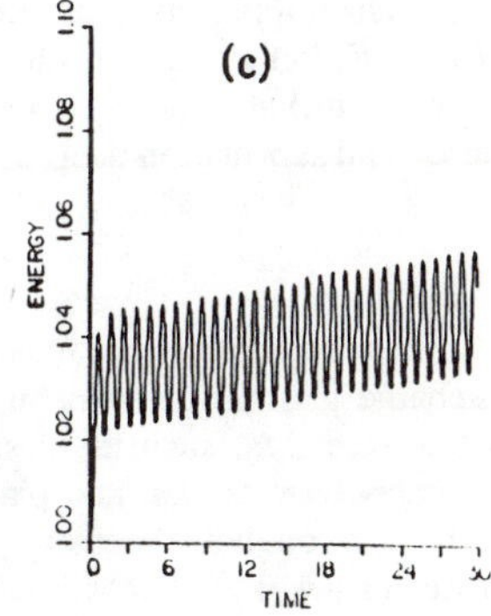

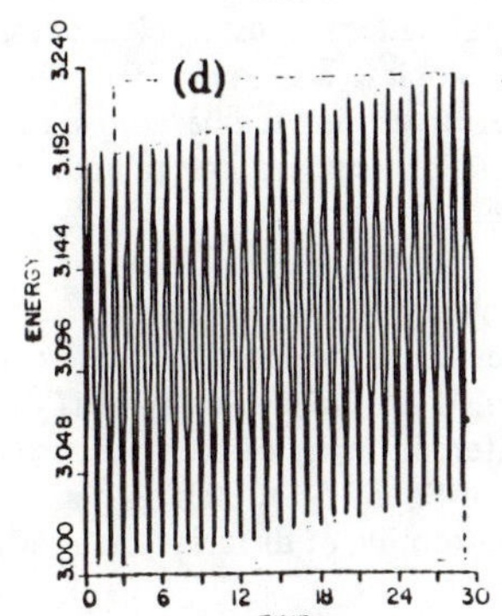

Fig. 3. (a) Parallel component of the energy vs time $\delta = 0.1$, $L = 0.5$, $v_c T = 0.5$.
(b) Total energy vs time $\delta = 0.1$, $L = 0.5$, $v_c T = 0.5$.
(c) Parallel component of the energy vs time $\delta = 0.1$, $L = 0.5$, $v_c T = 2$.
(d) Total energy vs time $\delta = 0.1$, $L = 0.5$, $v_c T = 2$.

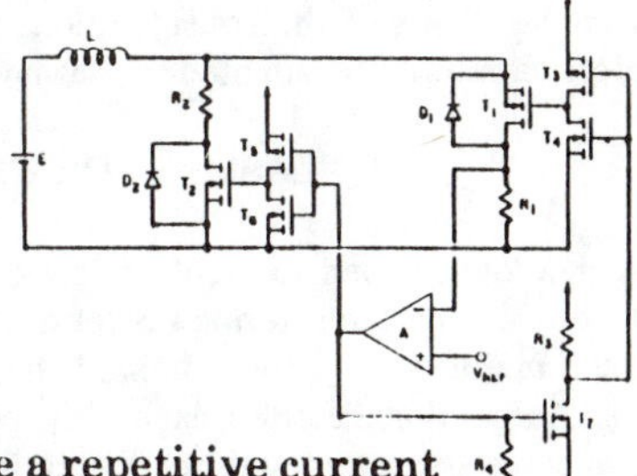

Fig. 4. Circuit used to generate a repetitive current ramp in the exciter coil.

APPLICATIONS OF RF PLASMAS TO ETCHING PROCESSES IN ADVANCED SEMICONDUCTOR TECHNOLOGY

Gottlieb S. Oehrlein
IBM Thomas J. Watson Research Center, Yorktown Heights, N.Y. 10598

ABSTRACT

Advanced semiconductor device manufacturing requires accurate transfer of submicron patterns into the layers of which an electronic device consists. Control of submicron dimensions necessitates anisotropic dry etching techniques, such as reactive ion etching (RIE), a RF glow discharge based etching method. The technological objectives of a dry etching process are control of anisotropy, etch selectivity of different materials, a high etch rate and minimization of detrimental plasma exposure related surface modifications. A brief review of the principles of reactive ion etching and its application to semiconductor processing is presented. Examples discussed include silicon etching and the doping effect, selective etching of silicon dioxide over silicon and sub-micron deep trench etching. Areas in need of further study are also discussed.

INTRODUCTION

The complexity of gas phase phenomena and plasma-surface interactions occuring in electric discharge plasmas of even pure stable gases is well-known. Despite these intricacies plasma based processes, such as plasma etching, sputter etching and deposition, plasma oxidation, plasma polymerization and plasma assisted chemical vapor deposition, have become pervasive in the semiconductor industry [1]. The replacement of conventional processes by plasma based processes has several reasons: Plasma based dry etching processes are superior to wet etching procedures because of etch directionality and capability of faithful pattern transfer, cleanliness, compatibility with automation and vacuum processing technologies, e.g. molecular beam epitaxy. Plasma based synthesis of materials often enables lower processing temperatures, thus providing processing compatibility with materials unstable at high temperature (aluminum, polymers) and minimizing diffusion effects.

The present article is a brief introduction to the use of rf plasmas to etching processes in silicon technology. The current status of several interesting problems will be described and areas in need of further study will be highlighted. Although the detailed plasma chemistry of rf discharges used for etching materials other than Si and SiO_2 is different, many of the basic processes are similar. Some of the considerations used in the study of Si and SiO_2 etching plasmas can therefore cautiously be adapted to plasmas utilized for the patterning of different materials.

RF PLASMAS FOR ETCHING APPLICATIONS

The technological need to etch materials directionally in the replication of micron or submicron scale device patterns during the fabrication of VLSI (very large scale integration) circuits has been the major driving force behind the emergence of plasma based anisotropic etching techniques. A directional etching capability, besides of enabling faithful pattern transfer, allows to solve problems associated with the increasing level of silicon device and circuit integration in novel ways. This can be illustrated taking the evolution of dynamic random access memories (DRAM) from 1-Megabit to 4-Megabit densities as an example. Higher levels of DRAM inte-

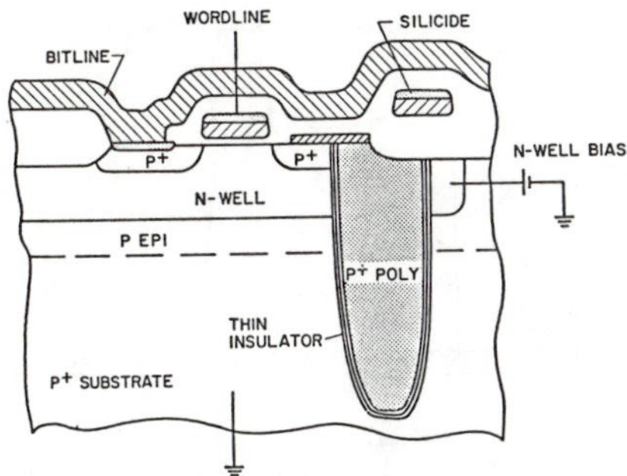

Fig. 1. Schematic of a substrate-plate trench-capacitor memory cell used for a 4-MBit dynamic random access memory (from Ref. 2).

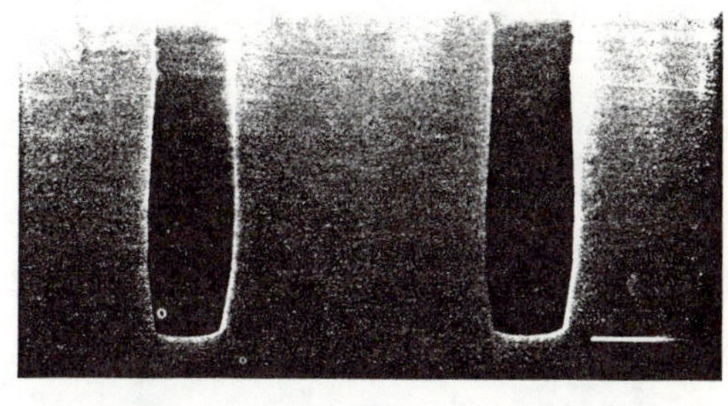

Fig. 2. Trenches in a silicon wafer formed by reactive ion etching. The white bar is 1 micron long. (Photo courtesy of D. Danner, IBM Research).

gration are achieved by reduction in cell size. In order to provide immunity to noise and leakage the stored charge can not be reduced as the lateral device dimensions are scaled down. In the past the thickness of the capacitor dielectric was reduced to compensate for the loss in capacitor area. This trend has resulted in the requirement of a silicon-dioxide thickness of $\simeq$10nm for a 1-Megabit DRAM. Further scaling of the oxide thickness at this point is no longer a desirable approach because of thin-oxide reliability concerns. The solution to this problem has been the replacement of the planar capacitor cell with a capacitor formed on the walls of a trench etched into the silicon substrate. Figure 1 is a schematic of a trench-capacitor memory cell used in a new 4-Megabit DRAM [2]. This structure, impossible to realize using wet etching technology, can be made utilizing a technique called reactive ion etching (RIE). Using RIE, 5- to 6-μm deep trenches can be formed through p-epi Si on a heavily doped Si substrate. Another example of the kind of structures which can be etched by RIE is shown in the scanning electron microscopy image of Fig. 2. Trenches $\simeq$4μm deep and 1μm wide dry etched into a Si wafer are depicted. In the particular case shown the trenches will be refilled with oxide and subsequently used for device isolation.

Perhaps the most widely utilized dry etching technique (in the accomplishment of directional pattern transfer) is reactive ion etching [1]. A reactive ion etch system is schematically illustrated in Fig. 3. In RIE a glow discharge plasma is used to generate, from a suitable feed gas (CF_4 in the case of Si and silicon dioxide), the gas phase etching environment which consists of positive and negative ions, electrons, radicals and neutrals. The material to be etched is placed on a high-frequency-driven (commonly 13.56MHz) capacitatively coupled electrode. After ignition of the plasma the electrode aquires a negative charge (due to the much greater electron mobility than ion mobility) and the material on the electrode will thus be exposed to energetic, positive ion bombardment. Chemical reactions between the radicals and neutrals (F, CF_3, CF_2, ..) and the material being etched (Si) occur at the surface and produce either volatile species (SiF_4) or their precursors (SiF, SiF_2, SiF_3). At the same time positive ions, such as CF_3^+, are accelerated across the plasma sheath and initiate and/or accelerate material removal due to sputtering. This combination of chemical activity of reactive species with sputtering can result in much greater vertical than lateral material erosion rates. Typical values of some operating parameters collected from the literature [1] are listed in Fig. 1.

The achievement of etch directionality in the case of RIE is due to energetic ion bombardment. An important clarifying experiment was performed by Coburn and Winters [3].

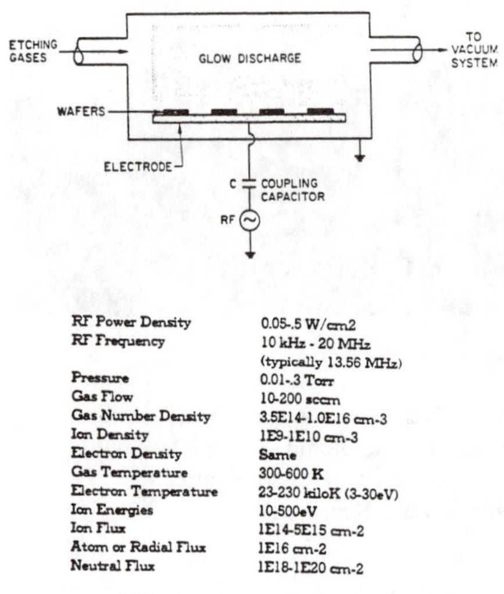

RF Power Density	0.05-.5 W/cm2
RF Frequency	10 kHz - 20 MHz (typically 13.56 MHz)
Pressure	0.01-.3 Torr
Gas Flow	10-200 sccm
Gas Number Density	3.5E14-1.0E16 cm-3
Ion Density	1E9-1E10 cm-3
Electron Density	Same
Gas Temperature	300-600 K
Electron Temperature	23-230 kiloK (3-30eV)
Ion Energies	10-500eV
Ion Flux	1E14-5E15 cm-2
Atom or Radial Flux	1E16 cm-2
Neutral Flux	1E18-1E20 cm-2

Fig. 3. Schematic diagram of apparatus used for reactive ion etching in semiconductor processing. Typical operating parameters are indicated.

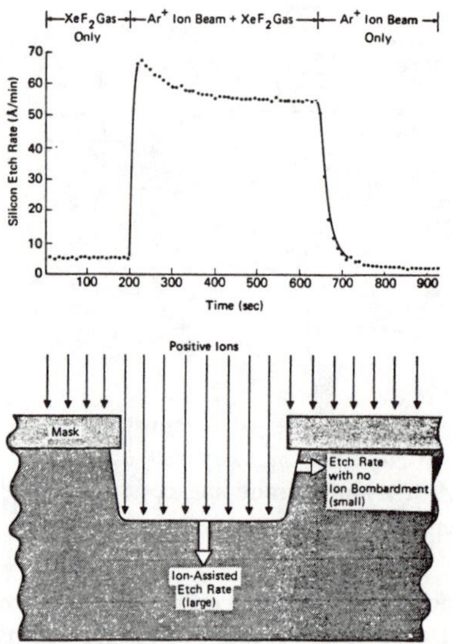

Fig. 4. Ion enhanced etching of Si by XeF_2 molecules due to 450eV Ar ions. Illustrative figure of ions incident on an etched feature (from Ref. 3).

They exposed a silicon surface to a well-defined dosage of chemical etchant, XeF_2, and simultaneous energetic Ar ion bombardment. A key result of their studies is displayed in Fig. 4: The silicon erosion rate obtained for a silicon surface simultaneously exposed to the XeF_2 chemical etchant and to the Ar ion beam is much greater than the sum of the etch rates for exposure to the ion beam and chemical etchant separately. As shown in Fig. 4, this synergism can explain very vividly the achievement of etch anisotropy under these conditions. Recent photoemission work by Yarmoff and McFeely [4] has shed some light on the mechanism of ion enhanced etching: They observed that the Ar ion beam tended to drive a disproportionation reaction in which SiF_3 molecules on the Si surface were converted into SiF_2 and volatile SiF_4 etch product. In general achievement of etch directionality is due to directed energy input into an etching reaction at a surface and can be accomplished by ion, electron or photon bombardment of a surface exposed to a chemical etchant.

ETCHING OF SILICON - DOPING EFFECT

Etching of Si can be accomplished using F-, Cl-, and Br-based chemistries and the etch products are volatile SiF_4, $SiCl_4$ and $SiBr_4$, respectively [1]. Most studies have focused on halocarbon chemistries, specifically CF_4. Mogab studied the etching of Si in a CF_4 plasma with negligible ion bombardment of the Si substrate [5]. He measured the intensity of the 704nm ($^2P_2$-$^2P^0_2$) atomic fluorine related optical emission. He observed a one-to-one correspondence of Si etch rate and atomic fluorine emission as the rf power was varied which showed that F atoms are directly involved and control the Si etching process.

A very interesting observation is that the Si etch rate depends on the electronic properties of the Si substrate. Figure 5 is from the work of Lee and Chen [6], who, using similar methodology as Mogab, determined the chemical (no ion bombardment) Si etch rate per fluorine atom

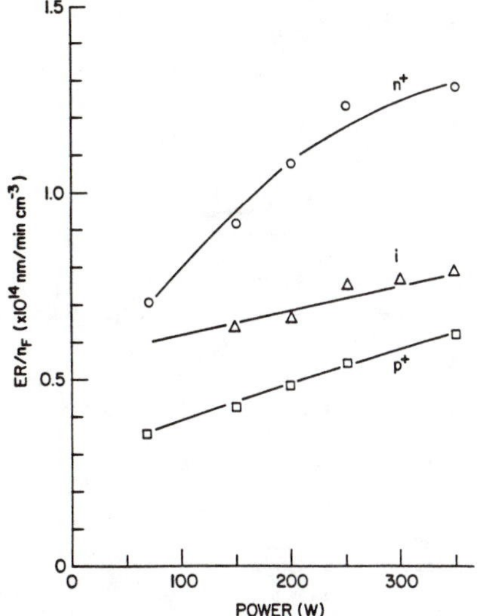

Fig. 5. Doping effect in plasma etching. Plotted is the chemical etch rate per fluorine gas phase atom for n-type, intrinsic; and p-type Si as a function of rf power (from Ref. 6).

Fig. 6. Space charge model of the doping effect observed in plasma etching (from Ref. 6).

as a function of rf power. Heavily As-doped Si (n-type) etches faster than intrinsic Si, which etches faster than heavily B-doped Si (p-type). [The Si etch rate per fluorine increases as a function of rf power due to heating of the Si substrate.] The effect is not chemical but electronic and has been explained by band bending effects at the semiconductor surface (see Fig. 6)[6]: Coulomb attraction between uncompensated donors (As^+) and chemisorbed halogens (F^-) enhances the Si etch rate for n-type Si, while Coulomb repulsion between uncompensated acceptors (B^-) and chemisorbed halogens (F^-) inhibits the Si etch rate for p-type Si. This doping effect makes the control of profile shapes in trench etching difficult, since typically a trench is formed through Si layers of different doping levels (see Fig. 1). Since the lateral etch rates (chemical etching only) of the differently doped Si layers are not the same, non-ideal trench profiles due to different, doping-level dependent amounts of mask undercutting would result. The solution to this problem has been the concept of side-wall passivation. The glow discharge chemistry is chosen so that etch inhibiting films can form as long as they are not exposed to ion bombardment. Ion bombardment leads to the dissolution of these films. Side-walls of trenches are not exposed to ion bombardment and will be covered by the etch-inhibiting films and prevent mask undercutting. The bottom of the trench which is exposed to ion bombardment is free of passivating film and etching reactions can proceed there. The detailed chemistry of side-wall passivants can not be discussed here because of space limitations.

SILICON DIOXIDE TO SILICON ETCH SELECTIVITY

A basic problem in semiconductor technology is selective etching of SiO_2 over Si. Silicon and SiO_2 etch at very similar rates in a CF_4 discharge. Addition of H_2 to CF_4 makes it possible to minimize the etching of Si as compared to the etching of SiO_2 [7]. Hydrogen addition to a CF_4 plasma causes the Si etch rate to decrease monotonically as the percentage of H_2 is raised and

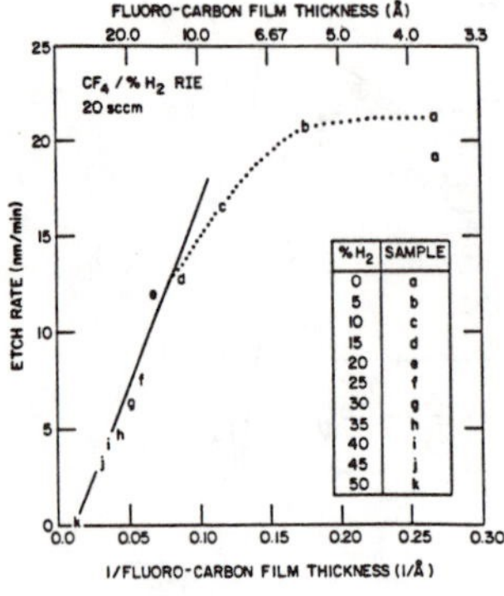

% H_2	SAMPLE
0	a
5	b
10	c
15	d
20	e
25	f
30	g
35	h
40	i
45	j
50	k

Fig. 8. Silicon etch rate depression by a fluorocarbon overlayer in a CF_4/H_2 plasma used for selective etching of SiO_2 over Si (from Ref. 8).

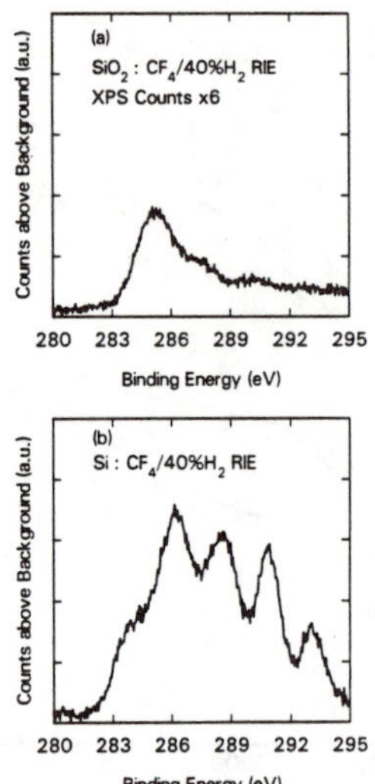

Fig. 8. Photoemission spectroscopy carbon 1s corelevel spectra of silicon dioxide (x6) and Si surfaces after reactive ion etching in CF_4/H_2 (from Ref. 8).

eventually stop, but results only in a small decrease of the silicon-dioxide etch rate ($\simeq$20% of maximum etch rate) [7]. The percentage of hydrogen where Si etching stops depends on the plasma etching process parameters, e.g. the total gas flow, pressure, etc., but is typically $\simeq$30-60%H_2 in CF_4/x%H_2. The role of hydrogen in controlling the Si etch rate is twofold: Atomic H scavenges atomic F in the gas phase to form HF molecules. The etch rate of Si, which depends on the F concentration, is consequently reduced. More important in slowing down the Si etch rate is the formation of a fluorocarbon film on the Si surface as a result of hydrogen addition to a CF_4 plasma. This is shown in Fig. 7, where the Si etch rate is plotted as a function of the inverse of the fluorocarbon film thickness measured on a Si surface under steady-state etching conditions [8]. [The percentage of H_2 addition to CF_4 necessary to cause a given fluorocarbon film thickness on Si is indicated in Fig. 7] Using x-ray photoelectron spectroscopy it was possible to observe and analyze silicon-fluorine bonding underneath the fluorocarbon film. From the dependence of the abundance of fluorosilyl species on the thickness of the fluorocarbon overlayer it was possible to elucidate the role of the fluorocarbon film in the slow-down of the Si etch rate. The fluorocarbon film "protects" the Si surface from attack of fluorine, rather than prevent the escape of SiF_4 etch product. Etch selectivity is a consequence of the presence of oxygen in SiO_2. During SiO_2 etching oxygen is continuously freed, reacts with the fluorocarbon film precursor to form volatile CO, CO_2 and COF_2 and prevents the formation of an etch inhibiting fluorocarbon film on the SiO_2 surface. This difference in the kind and degree of carbon contamination on CF_4/H_2 plasma etched Si and SiO_2 surfaces is demonstrated by carbon 1s corelevel spectra in Fig. 8. An etched Si surface (Fig. 8 (a)) exhibits a high intensity of C-CF, CF, CF_2 and CF_3 groups (in order of increasing binding energy), due to the existence of the fluorocarbon film. An XPS spectrum obtained with an etched SiO_2 surface exhibits only a very low intensity of fluorocarbon related groups and primarily C-C and Si-C type bonding.

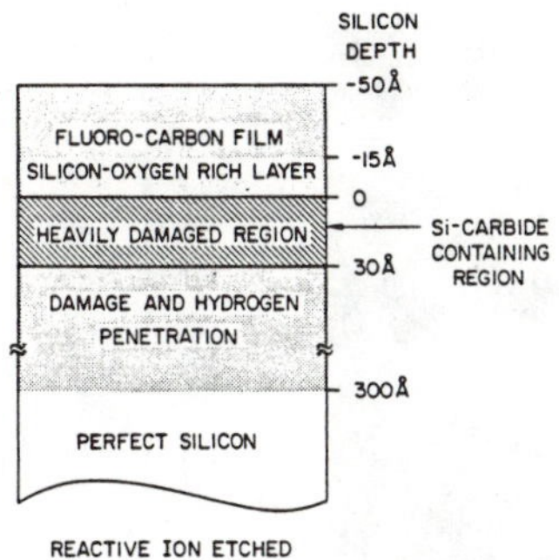

Fig. 9. Schematic view of changes in Si near-surface properties caused by reactive ion etching in CF_4/H_2 (from Ref. 9).

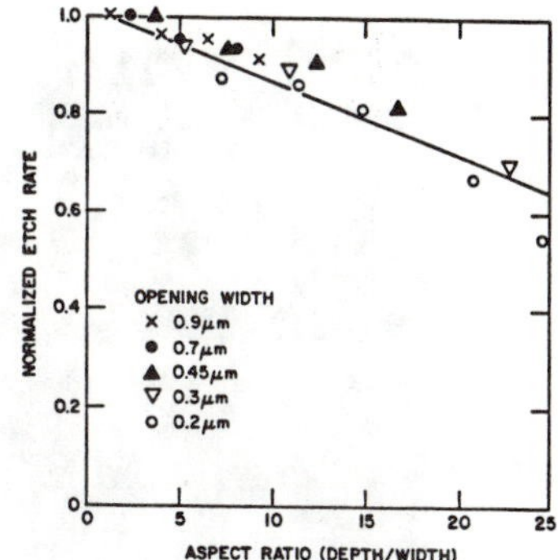

Fig. 10. Normalized silicon etch rate as a function of aspect ratio in sub-micron deep trench etching (from Ref. 10).

At this time RIE is the dry etching technique of choice because it offers the best combination of etch directionality, material-selectivity and process control. However, undesirable side-effects are associated with the achievement of etch directionality and material selectivity [9]. Etch-anisotropy is caused by energetic ion bombardment. Although the ion energies are typically below 500eV, the fluence is high ($\simeq 10^{15}$ions/cm^{2}sec) and bombardment damage can occur. Displacement damage will alter the near-surface region of the material which is being etched (or exposed to the plasma) and change its electrical properties. Etch-selectivity is primarily due to the deposition of etch inhibitors, which will remain on the exposed material after completion of the dry etching step (see above) and interfere with device processing following RIE. The changes in the silicon near-surface region incurred as a result of selective oxide removal by CF_4/H_2 RIE are schematically depicted in Fig. 9 [9]. These surface modifications necessitate surface cleaning/damage removal processes which have to be performed prior to further device processing following dry etching. Because of these complications improved etching tools/procedures which minimize contamination and damage effects are highly desirable.

SUB-MICRON TRENCH ETCHING

As lithographic capabilities improve feature sizes in semiconductor technology will continue to shrink. In trench etching the trench opening will therefore decrease while the etch depth of the trench will remain the same or become even greater. The aspect ratio (depth/width) will therefore increase. Reactive ion etching of trenches with submicron openings and high aspect ratio is more difficult than formation of trenches with smaller aspect ratio. Several effects contribute to this: For a given (sub-micron) opening width the etch rate decreases as a function of the aspect ratio (or depth of the trench). This is shown in Fig. 10, where results of work by Chin et al. [10] are displayed. The etch rate decreases almost linearly as the aspect ratio increases and is determined by the aspect ratio, regardless of the opening size (for opening widths less than 1 micron). This phenomenon has been attributed to a diverging electric field in the trench [10], diffusion effects on the supply of reactant to the bottom of the trench and consumption of reactant at the trench side-walls. Since the side-wall chemistry is poorly understood due to difficulties in examining it analytically, it is not possible to establish the dominant mechanism at this time. Figure 11 shows that the shape of the trench changes as the width decreases. A bottle-shaped profile due to undercutting of the mask starts to appear with trenches whose openings are smaller than one micron. This effect becomes more pronounced as the width of the trench decreases. It has been attributed to ions scattered from the mask and deflected to the opposite

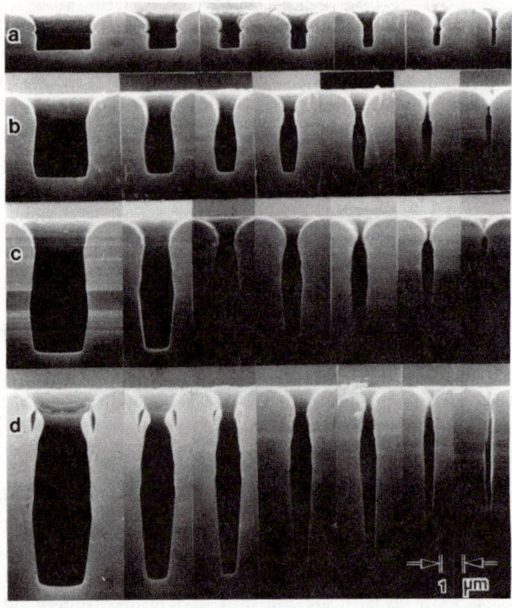

Fig. 11. Silicon trenches with various opening sizes. The etching times are (a) 5, (b) 20, (c) 40, and (d) 60 min, respectively (from Ref. 10).

trench wall, causing ion-enhanced etching and undercutting [10]. A prerequisite to controlling the profile shape of sub-micron trenches is clearly improved control of the mask profile shape.

RESEARCH NEEDS

Plasma etching processes depend on a large number of adjustable parameters, such as the applied rf power and frequency, the reactant gas composition, pressure and chamber residence time, the chamber configuration, the substrate temperature, the mask material, the electrode material, and so forth. The "output" of a plasma etch process depends in a non-linear way on these input parameters. Since at this time we do not have a valid model for the processes occuring in glow discharge etching plasmas, extensive experimentation is typically required in the development of suitable etching processes. An improved understanding of the science of rf discharges is an important prerequisite to optimize their use for electronic materials processing or other applications. At this time our understanding of even basic and widely used plasma based processes is at most qualitative. In the present author's view a three-fold approach is needed in order to improve our understanding. First, <u>measurements</u> need to be performed on <u>real rf plasma systems</u> in order to address the question as to what kind of phenomena are occuring. One would like to know electron densities and their energy distribution, the atom and radical concentrations, spatially and time resolved, the concentration of different ions and their energies, processes occuring selectively on certain surfaces and not on others, e.g. Si vs. SiO_2, bottom of trench vs. side-wall, and so forth. The measurements should ideally be performed non-intruisively and in-situ. In many cases new diagnostic techniques will have to be developped in order to measure the desired glow discharge parameters. A striking example is the case of the the determination of the atomic fluorine ground state concentration in a glow discharge. Presently this measurement can not be performed directly and one has to infer the fluorine concentration from an indirect, admittedly most likely reliable, procedure [1]. Since real glow discharges make well-controlled experiments difficult, e.g. due to the coupling of most parameters a controlled change in one quantity changes invariably other quantities, the question as to the relative importance of a specific quantity in causing a given result has to be addressed by model system studies. This is the second required component. Here one would like to study the interaction of fluxes of atoms and radicals with well-specified surfaces, alone or in combination with ion (mass/energy analyzed), electron and photon bombardment, measure the energy dependence of the cross sections for the production of important species found in glow discharges, perform controlled experiments to establish the rate and importance of plasma chemical reactions, and so forth. The third component is numerical modelling. The results of the model system approach on cross sections, sticking coefficients, reaction rates etc. and chosen values of the controllable operating parameters such as rf power and frequency, gas pressure etc. are used as inputs of a computer model of a glow discharge used for a given application. The output of the numerical model can be compared to the results of measurements performed on real systems. This combined approach should yield, at least for prototypical glow discharges, e.g. CF_4 RIE of Si, a valid model. This model could subsequently be used to optimize the utilization of rf glow discharges in certain applications.

Acknowledgements: I would like to thank Drs. J. W. Coburn, D. Danner, S. H. Dhong, Y. H. Lee, and N. C.-C. Lu for permission to reproduce figures from their published work.

REFERENCES

1. Books on this subject are:
B. Chapman, "Glow Discharge Processes", John Wiley & Sons, New York (1980);
J. W. Coburn, "Plasma Etching and Reactive Ion Etching", American Vacuum Society Monograph Series, New York (1982);
The Electrochemical Society (Pennington, NJ) and the Materials Research Society (Pittsburgh, PA) organize regularly symposia on Plasma Processing and publish proceedings. The most recent of these are: "Proceedings of the Fifth Symposium on Plasma Processing", edited by G. S. Mathad, G. C. Schwartz and G. Smolinsky (The Electrochem. Soc., Pennington, 1985) and "Plasma Processing", eds. J. W. Coburn, R. A. Gottscho, and D. W. Hess, (Materials Research Society, Vol. 68 of the Symposia Proceedings Series, Pittsburgh, 1986).
2. N.C.C. Lu, P. E. Cottrell, W. J. Craig, S. Dash, D. L. Critchlow, R. L. Mohler, B. J. Machesney, T. H. Ning, W. P. Noble, R. M Parent, R. E. Scheuerlein, E. J. Sprogis, and L. M. Terman, IEEE J. Solid-State Circuits SC-21, 627 (1986).
3. J. W. Coburn and H. F. Winters, J. Appl. Phys. 50, 3189 (1979).
4. J. A. Yarmoff and F. R. McFeely, Phys. Rev. B (in press).
5. C. J. Mogab, J. Electrochem. Soc. 124, 1262 (1977).
6. Y. H. Lee and M.-M. Chen, J. Vac. Sci. Technol. B4, 468 (1986).
7. R. A. H. Heinecke, Solid State Electronics 18, 1146 (1975); L. M. Ephrath, J. Electrochem. Soc. 126, 1419 (1979).
8. G. S. Oehrlein and H. L. Williams, J. Appl. Phys. (to be published July 1987); G. S. Oehrlein, S. W. Robey, and M. A. Jaso, in "Plasma Processing and Synthesis of Materials", eds. D. Apelian and J. Szekely, (Materials Research Society, Vol. 98 of the Symposia Proceedings Series, Pittsburgh, 1987).
9. G. S. Oehrlein, G. J. Coyle, J. G. Clabes, and Y. H. Lee Surface & Interface Analysis 9, 275 (1986); See also, G. S. Oehrlein, Physics Today, 39, 26 (1986).
10. D. Chin, S. H. Dhong, and G. J. Long, J. Electrochem. Soc. 132, 1705 (1985).

RF HEATING THE IONOSPHERE

G.J. Morales
Physics Dept., University of California at Los Angeles
Los Angeles, CA 90024-1547

ABSTRACT

A survey of recent developments in the modification of the earth's ionosphere by powerful radio waves is presented from a general perspective of RF heating of plasmas.

MOTIVATION

The study of the interaction of powerful radio waves with the earth's ionosphere enlarges the base for understanding the general problem of heating confined plasmas with RF waves. The earth's ionosphere provides a steady-state plasma in which the electron and ion temperatures are nearly in equilibrium ($T_e \sim 1 - 2T_i$). Due to the large size of the ionosphere and the relatively strong magnetic field of the earth, the ionospheric plasma exhibits good heat confinement. By appropriate selection of the RF frequency, the heated region can be located (say in the F region) far from boundaries. Also, the ionosphere exhibits large low frequency fluctuations analogous to those encountered in laboratory confinement devices, hence the generic problem of RF heating a turbulent plasma is present.

Currently ionosphere heating studies are providing a test ground for modern developments in plasma physics. Relevant examples to the RF community are: mode conversion, electron acceleration and current drive, and beat excitation.

Because the ionosphere is amenable to investigation under widely different plasma environments (e.g., at polar and mid-latitudes) it is possible to explore phenomena which are difficult to isolate in a laboratory device. For instance in the polar ionosphere the density gradient is almost parallel to the magnetic field, thus permitting the study of effects not explored yet in much detail by lab plasma physicists. Other examples of interest are: convective heat transport, memory effects, and transition between collisionless and collisional physics.

Finally, RF heating the ionosphere provides a valuable tool for understanding the fundamental processess that govern the dynamics of the natural ionospheric plasma.

IONOSPHERE PARAMETERS

The experiments of relevance to the RF heating community are those in which the reflection layer for the ground-launched RF wave is located in the F region, which is typically at a height of 250-300 km. In this region the complicated neutral atom chemistry, although important, is not overwhelming. The typical plasma frequency is $\omega_p/2\pi \sim 3$ -10 MHz and the electron cyclotron frequency is $\Omega_e/2\pi \sim 1.4$ MHz.

The electron temperature is $T_e \sim 0.1$ eV and is comparable to the ion temperature, $T_e \sim 1\text{-}2T_i$. At these heights the dominant ion species is 0^+ and results in an ion cyclotron frequency $\Omega_i/2\pi \sim 47$ Hz. The collision frequency is dominated by Coulomb processes with a typical ordering 1KHz $> \nu_{ei} \gtrsim \nu_{en}$, with the equal sign holding at the lower heights. The macroscopic density scale length is $L \equiv (d\ln N/dz)^{-1} \sim$ 20-50 km, but shorter scale length distortions are present in the density profile. The electron Larmor radius is ~ 1.5 cm, the ion radius is ~ 250 cm, and the Debye length is $\lambda_D \sim 1$ cm, which implies that $k_D L \sim 10^5$ for these plasmas. The electron distribution function consists of a background Maxwellian plus an energetic photoelectron tail extending to energies of several eV and with a fractional density on the order of 10^{-4}.

ACTIVE EXPERIMENTAL FACILITIES

At the present time there are 3 active RF heating ionospheric facilities in western countries. The Arecibo observatory in Puerto Rico, located at mid-latitude, has been used for RF heating studies for almost 15 years, and its present capabilities are: $P_{RF} \sim 600$ KW, 23 db antenna gain, and frequency range $\omega/2\pi \sim 3\text{-}10$ MHz. The angle between ∇n and $\underline{B}_o$ is on the order of 40°. At high latitudes the HEATER facility in Tromso, Norway has been operational for nearly 5 years, and its characteristics are: $P_{RF} \sim$ 1MW, 24 db antenna gain, $\omega_{RF}/2\pi \sim 2.5 - 8$ MHz. In Fairbanks, Alaska the HIPAS facility is in operation with $P_{RF} \sim 1$ MW, 17db antenna gain, and $\omega_{RF}/2\pi \sim 4.5$ MHz. The angle between ∇n and $\underline{B}_o$ for these polar facilities is on the order of 13°.

DIAGNOSTIC CAPABILITIES

The principal diagnostic tool used in ionospheric RF heating experiments is Thomson backscattering from electron plasma waves (plasma line) and from low frequency density fluctuations (ion line). By scattering from the ambient noise spectrum the zero order plasma parameters can be deduced. Scattering from waves enhanced by the RF then permits the study of various nonlinear processes. Recently, Hagfors and his collaborators[1] have developed an ingenious chirping variation of the Thomson scattering diagnostic that permits the identification of short scale density cavities. Other coding schemes used to obtain sharp height resolution are time compression[2] (Duncan and Sheerin) and phase inversion (Barker code) of the Thomson radar pulse. The Thomson radar at Arecibo operates at 430 MHz, and provides information about phenomena having 35 cm wavelength. The equivalent radar at Tromso operates at 933 MHz with bi-static capabilities. Radio-Star scintillations[3] are used to monitor the generation and dynamics of density distortions with scales on the order of 100 meters. Monitoring the enhanced airglow stimulated during RF heating provides information[4] about the integrated distortions produced on the tail of the electron distribution function. A highly desireable diagnostic is the in-situ measurement of wave and plasma parameters inside the heated volume, i.e., the equivalent of inserting a probe in the center of a tokamak plasma. In the ionosphere this can be accomplished by flying a well instrumented rocket through the RF reflection layer. Due to the high cost and involved logistics of such an operation, few experiments of this type have been performed. In a later section we discuss some interesting results recently obtained from a rocket fly-by in Tromso.

DIFFICULTIES AND UNCERTAINTIES

One of the major obstacles in interpreting the outcome of heating experiments is the highly variable nature of the ionospheric plasma. As a consequence, many experimentalists favor the usage of long time statistical sampling in order to improve the signal to noise ratio. While in some instances this is a worthwhile approach, the danger exists of averaging over completely different physics processes that are triggered under various conditions. For this reason it is useful, when possible, to investigate real-time phenomena with short RF pulses. Because the information about T_e and T_i is derived from Thomson scattering due to thermal fluctuations, it is difficult to obtain meaningful results when the plasma is driven far from equilibrium by the RF. Considerable uncertainty also exists about the actual radiation pattern associated with RF antenna arrays. Idealized calculations are combined with airplane fly-throughs to estimate the actual beam divergence and sidelobe patterns. The situation is somewhat analogous to that encountered in predicting the $k_\parallel$ spectrum of an ICRF antenna in a tokamak. In addition to the antenna uncertainty, the modelling of the RF electric field at the reflection layer is not yet satisfactory. The difficult problem of wave reflection and resonance when $\underline{k}$, ∇n, and $\underline{B}_0$ have arbitrary orientation remains to be solved. Because of the heavy reliance on the Thomson scattering diagnostic, most of the information obtained in heating experiments comes from a narrow height interval and within it from a restricted length scale. Much remains to be learned from the simultaneous usage of several radars having different operating frequencies (e.g., sampling waves at 1-10 meters and at 30 cm).

TOPICS OF PRESENT INTEREST

Research in this area is increasingly being directed toward understanding the connection between microscopic processes, such as generation of electrostatic waves and electron acceleration, and large scale transport modifications. Some of the topics being explored are: role of mode conversion[5], ponderomotive modifications by electrostatic[1,2] and electromagnetic[6] waves, generation of large scale density cavities[7], structure of self-focusing[8], fast electron production[4], stimulated generation of secondary waves[9,10]. A related area of potential interest to the development of communication systems is the controlled generation of low frequency signals[11] (VLF, ELF, ULF) resulting from the modulation of natural ionospheric currents.

RECENT DEVELOPMENTS

Direct Conversion. When a long wavelength electric field of the form $E_0 \exp[i(k_0 z - \omega_0 t)]$ interacts with a nearly static distortion of the density profile $\delta n = \tilde{n} \exp(ikz)$ having $k \gg k_0$, a beat current

$\tilde{j} = (ie^2E_o/m\omega_o)\tilde{n}\ \exp[i(kz-\omega_ot)]$ is generated which can excite plasma waves resonantly if Re $\varepsilon(k, \omega_o) \simeq 0$. In experiments aimed at examining the response of the ionosphere to short RF pulses [~ 10 msec] it has been found[12] that this process causes a rapid growth of plasma waves which serve as the seed from which nonlinear wave interactions develop. The unique signatures of this process are: no RF power threshold, linear dependence on RF amplitude, instantaneous growth upon arrival of RF, early time secular growth, and frequency centered around ω_o for nearly static distortions. All of these signatures have been observed experimentally at Arecibo using short RF heating pulses (~ 10 msec) separated by an off-time on the order of 40 msec, and by recording the signals on real time.

Memory Effects. In a follow-up experimental study of direct conversion at Arecibo a paradoxical result has been discovered[13]. It is found that the amplitude of the plasma wave excited within 1-2 msec depends on the length of the RF pulse, i.e., the present behavior depends on the future. Of course, causality is not violated, what actually occurs is that the experiments are performed by firing the RF source repetitively with an off-time on the order of 40 msec. Consequently, the apparent violation of causality is in fact a memory effect, i.e., prior RF pulses are capable of sustaining a high level of irregularities in the density profile that permit the direct conversion process to occur. It is found that a threshold pulse length of 5 msec is required in order for rapid conversion to be observed. For pulse lengths larger than 10 msec direct conversion can be observed for several hours. The sustainment of the process is independet of the average power for $\langle P_{RF}\rangle > 10$ KW. What is of general interest here to RF heating of plasmas is the fact that microscopic RF wave absorption processes are tightly coupled to the global transport properties of the plasma.

Relation to ICRF Heating. Motivated by the ionospheric experiments, an analysis has been made[14] of the direct conversion of ion Bernstein waves resulting from the beat between a fast wave and a drift wave, as may be encountered in a tokamak heating experiment. It is found that this process provides an enhanced damping of the fast wave, which for coherent fluctuations can be significant at a level $|\delta n|/n \sim 1\%$. The process can be a useful high-harmonic heater of large devices, but can also result in the parasitic launching of Bernstein waves by fast couplers, thus giving rise to edge heating. It is expected that some of these processes may be enhanced in plasmas with large edge fluctutations, as is characteristic of compact tokamaks.

Thermal Cavities. In order to assess the connection between microscopic processes and global plasma modifications an Ohmic heating transport code study[7] has been made of a day-time ionosphere. At $P_{RF} = 1$ MW, 20 db antenna gain, a steady-state $\delta T_e/T_e \sim 0.5$ is achieved within 15 sec after RF turn-on. The pressure gradient created at the RF reflection layer causes the growth of a density cavity well beyond the δT_e saturation and having several km in extent. After 10 min the depth of the cavity is $|\delta n|/n \sim 5\%$ and continues to grow as $\sqrt{t}$; an assessment of the relevant recombination chemistry predicts an eventual saturation below the 10% level. Recent experiments at Arecibo performed by Duncan and collaborators, have indeed observed such global density modifications, but the level of density depletion found can be 50% or larger. Although the details of the underlying physics creating such a large effect have not yet been resolved, it is known that the solar source of plasma must be turned-off, i.e., the observations are made in late-evening experiments. The study of this effect is presently an active topic of research.

Rocket Fly-Through. A rather interesting and technically challenging experiment has been performed by Rose,et al [15] in Tromso. A rocket instrumented with Langmuir probes, magnetic loops, and energy analyzers has been flown through the beam footprint of the ground-launched RF. This rare opportunity at an in-situ measurement has validated the existence of an electromagnetic Airy cut-off, with electron heating emanating from it. Secondary waves and enhanced fast electron tails (energy $> 100T_e$) are also detected. Density modifications of the thermal-cavity-type are reported to attain a level $|\delta n|/n \sim 2\%$, consistent with the transport code analysis.

Stimulated Electromagnetic Emissions. An interesting observation by Thide, originally made in Tromso[10], and subsequently repeated in Arecibo, is that when the ionosphere is RF heated, it is stimulated into emitting electromagnetic waves over a rather broad-band (~ 200KHz) centered around the RF heater frequency. Although some of the features in the stimulated emissions could be identified with parametric decay instabilities, the broad and universal character of the emissions has not been yet satisfactorily explained. It is interesting to note that ICRF experiments in JET also exhibit an analogous spontaneous emission. Since early laboratory experiments[16] of electron heating by Landau acceleration have consistently yielded the spontaneous generation of sideband signals, it is suggestive that some of the spontaneous emissions may be related to distortions in the electron distribution function.

Electron Acceleration. To assess the possible role of distortions in the electron distribution function, an analysis has been made[9] of the electron acceleration resulting from localized fields (driven-Airy pattern) excited by mode conversion. Below a threshold electric field the fast electron tail exhibits an unidirectional density enhancement $\delta n_T/n_T = (E_o/E_s)^2$ where $E_s = mv_T{}^2/(e\pi L)$, v_T is the effective tail velocity, and L the density gradient scale length. E_o is the pump field associated with the RF wave, which can be related to the incident power P_o through $E_o = (8\eta P_o/\omega_o L)^{1/2}$, with η the mode conversion efficiency. Above a threshold field (say $E_o/E_s \sim 6$ for certain ionospheric conditions) the tail distribution develops a region of positive slope. An estimate of the bandwidth of the unstable sideband waves that can be excited by the bump is $\Delta\omega/\omega_o \sim 5 \times 10^{-2}$, which is comparable to the frequency band over which stimulated electromagnetic emissions occur.

Current Modulation. Over the past few years several experimental and theoretical studies have been made of low frequency wave generation by modulating natural ionospheric currents with the RF heating wave, as reviewed in Ref. 11. An analysis[17] of relevance to the RF heating community pertains to the modulation of field aligned currents at frequencies $\omega < \Omega_i$. The formulation allows for the simultaneous excitation of shear and compressional modes by the in-situ antenna generated through thermal modulation of the conductivity tensor. It is found that 90% of the energy radiated is in the shear mode, and 10% in the compressional, with an efficiency less than 10^{-5}. The shear mode exhibits a collimated headlight-type pattern emanating from the heated region. An analogous behavior has been observed in a tokamak experiment[18] which uses a field aligned wire antenna.

Acknowledgments

My collaborators in theoretical studies reported here are Drs. J.E. Maggs and M.M. Shoucri, and the experimentalists are Dr. L.M. Duncan and Prof. A.Y. Wong. Valuable information was provided by Prof. T. Hagfors. This work has been sponsored by ONR and NSF.

References

1. T. Hagfors, W.k Birkmayer, and M. Sulzer, J. Geophys. Res. 89, 6841 (1984). W. Birkmayer, T. Hagfors, and W. Kofman, Phys. Rev. Lett. 57, 1008 (1986).
2. L.M. Duncan and J. P. Sheerin, J. Geophys. Res. 90, 8371 (1985).
3. A. Frey, P. Stubbe, and H. Kopka, Geophys. Res. Lett. 11, 523 (1984).
4. H.C. Carlson, V.B. Wickwar, and G. P. Mantas, J. Atmos. and Terr. Phys. 44, 1089 (1982)
5. E. Mjolhus and T. Fla, J. Geophys. Res. 89, 3921 (1984)
6. J.A. Fejer, H.M. Ierick, R.F. Woodman, J. Rottger, M. Sulzer, R.A. Behnke, and A. Veldhuis, J. Geophys. Res. 88, 2907 (1983).
7. M.M. Shoucri, G.J. Morales, and J.E. Maggs, J. Geophys. Res. 89, 2907 (1984).
8. A. Frey and L.M. Duncan, Geophys. Res. Lett. 11, 677 (1984)
9. M.M. Shoucri, G.J. Morales, and J.E. Maggs, J. Geophys. Res. 92, 246 (1987)
10. B. Thide, H. Derblum, A. Hedberg, H. Kopka, and P. Stubbe, Radio Science 18, 851 (1983)
11. M.T. Rietveld, J. Atmos and Terr. Phys. 47, 1283 (1985).
12. A.Y. Wong, G. J. Morales, D. Eggleston, J. Santoru, and R. Behnke, Phys. Rev. Lett. 47, 1340 (1981).
13. G.J. Morales, A.Y. Wong, J. Santoru, L. Wang, and L.M.Duncan, Radio Science 17, 1313 (1982).
14. G.J. Morales, S.N. Antani, and B.D. Fried, Phys. Fluids 28, 3302 (1985).
15. G. Rose, B. Grandal, E. Neske, W. Ott, K. Spenner, J. Holtet, K. Maseide, and J. Troim, J. Geophys. Res. 90, 2851 (1985).
16. T.P. Starke and J.H. Malmberg, Phys. Fluids 21, 22412 (1978); E. Mark, R. Hatakeyama, and N. Sato, Plasma Phys. 20, 415 (1978).
17. M.M. Shoucri, G.J. Morales, and J.E. Maggs, Phys. Fluids 28, 2458 (1985).
18. C. Borg, M. H. Brennan, R. C. Cross, and L. Giannone, Proc. Int. Conf. Plasma Phys., Lausanne, Switzerland, (1982), Vol. II, p. 230.

PROPAGATION OF LANGMUIR WAVES IN AN INHOMOGENEOUS PLASMA

Burton D. Fried and George J. Morales
UCLA, Los Angeles, CA 90024*

ABSTRACT

The integral equation for Langmuir wave propagation excited by both localized antennae and sinusoidal beat waves is solved for an inhomogeneous plasma with constant density gradient. Comparison of the kinetic and fluid solutions shows clearly the effects of spatial Landau damping in the kinetic case.

INTRODUCTION

The usual WKB treatment of wave propagation in an inhomogeneous plasma is based on a differential equation approximation to the governing integral equation, obtained by an expansion of the dielectric tensor $\underline{\varepsilon}$ in powers of k/k_D, kr_{ce}, etc. However for propagation parallel to ∇n, in the case of an unmagnetized plasma, or parallel to $\underline{B}_0$ and ∇n in the magnetized case, the non-analytic dependence of $\underline{\varepsilon}$ on $\underline{k}$ prevents such an expansion. To develop techniques for solving the integral equation itself in such cases, we have analyzed the case of Langmuir wave propagation in an inhomogeneous plasma, the waves being launched by either a localized antenna (grid structure) or by the interference of two electromagnetic waves, producing an electrostatic beat wave.

ANALYSIS

If the unperturbed electron distribution function is assumed Maxwellian, $f_0 = n(x)\exp(-v^2/a^2)/a\pi^{1/2} = n(x)f_M(v)$ then the solution of the linearized Vlasov equation,

$$f(x,v) = (-e/T)f_M(v)\int_{Lv/|v|}^{x} dx'\, n(x')E(x')\exp[i\omega(x-x')/v], \quad (1)$$

together with Ampere's Law, gives directly the integral equation

$$E(x) = E_s(x) + \int_{-\infty}^{\infty} dx'\, K(|x-x'|)g(x')E(x') \quad (2)$$

$$K(x) = -2i\pi^{-1/2}\int_0^{\infty} du\, u\exp(-u^2 + ix/u) \qquad g(x) = \omega_p^2(x)/\omega^2 \quad (3)$$

with x measured in units of $a/\omega = 2^{1/2}/k_D$. (k_D is the Debye wave number when $\omega_p = \omega$.) For the external source field $E_s(x)$ we use either a plane (beat) wave $\exp(ip_s x)$ or a Gaussian, $\exp[-(x-x_s)^2/w^2]/w\pi^{1/2}$.

The Fourier transform of (2)

*Work supported by U.S.D.O.E.; Institute of Theoretical Physics, UCSB; TRW; and Culler Scientific Systems.

$$E(p) = \int_{-\infty}^{\infty} dx\ E(x)\exp(-ipx) = E_s(p) + \chi(p)\tilde{E}(p) \qquad (4)$$

$$\chi(p) = Z'(1/|p|)/p^2 \qquad \tilde{E}(x) = g(x)E(x) \qquad (5)$$

becomes a differential equation for a linear density profile, $g(x) = 1 - x/L$,

$$\chi\ E'(p) - i(1 - \chi)L\ E(p) = -\ iLE_s(p) \qquad (6)$$

with solution

$$E(p) = -iL \int_{-\infty}^{p} dp'\ [E_s(p')/\chi(p')]\exp[i\psi(p) - i\psi(p')] \qquad (7)$$

$$\psi(p) = L\int_{-\infty}^{p} dp'\ (1 - \chi^{-1}(p')] \qquad (8)$$

In the fluid limit, $|p|<<1$, where $\chi(p) \to (1 + 3p^2/2)$ we have, for $L >> 1$, an inhomogeneous Airy equation in $z = (2/3L)^{1/3}x$

$$d^2E/dz^2 + zE = S(z) = (2L^2/3)^{1/3}\ E_s(z) \qquad (9)$$

with solution

$$E(z) = -\pi[\int_{-\infty}^{z} dz'\ E_L(z')S(z')E_R(z) + \int_{z}^{\infty} dz'\ E_R(z')S(z')E_L(z)] \qquad (10)$$

$$E_L(z) = Ai(-z) \qquad E_R(z) = Bi(-z) + iAi(-z) \qquad (11)$$

In Figs. 1a, 1b and 1c we compare the solutions $|E(x)|$ for the kinetic and fluid models for $L = 100$, $-20 \leq x \leq 80$ with a Gaussian antenna source, of width $w = 1$ and x_s values of 40, 30 and -10. Both models show the standing waves resulting from reflection at the cut-off when the antenna is in the propagating region and the Landau damping for the kinetic case is clearly evident. The dimple in $|E(x)|$ at the antenna in the evanescent region occurs also for constant density plasma, as shown in Fig. 1d.

In Figs. 2a, 2b and 2c we show $|E(x)|$ for beat wave excitation at $p_s = 0$ (condenser plate problem[1]); $p_s = -0.4$ (leftward propagating beat wave); and $p_s = 0.2$ (rightward propagating beat wave). Again, the Landau damping effects are clearly evident, as is also true in Fig. 1d, which combines 10 separate plots of $|E(x)|$ vs. x for $0.4 \geq p_s \geq -0.5$.

The method of solving (2) described here is numerically efficient. With 1024 intervals in p and 256 in x, each solution requires less than one minute of computing time on the array-processor-based CHI computer at UCLA.

REFERENCES

1. Calunga, Mora and Pellat Phys. Fluids 28, 854 (1985).

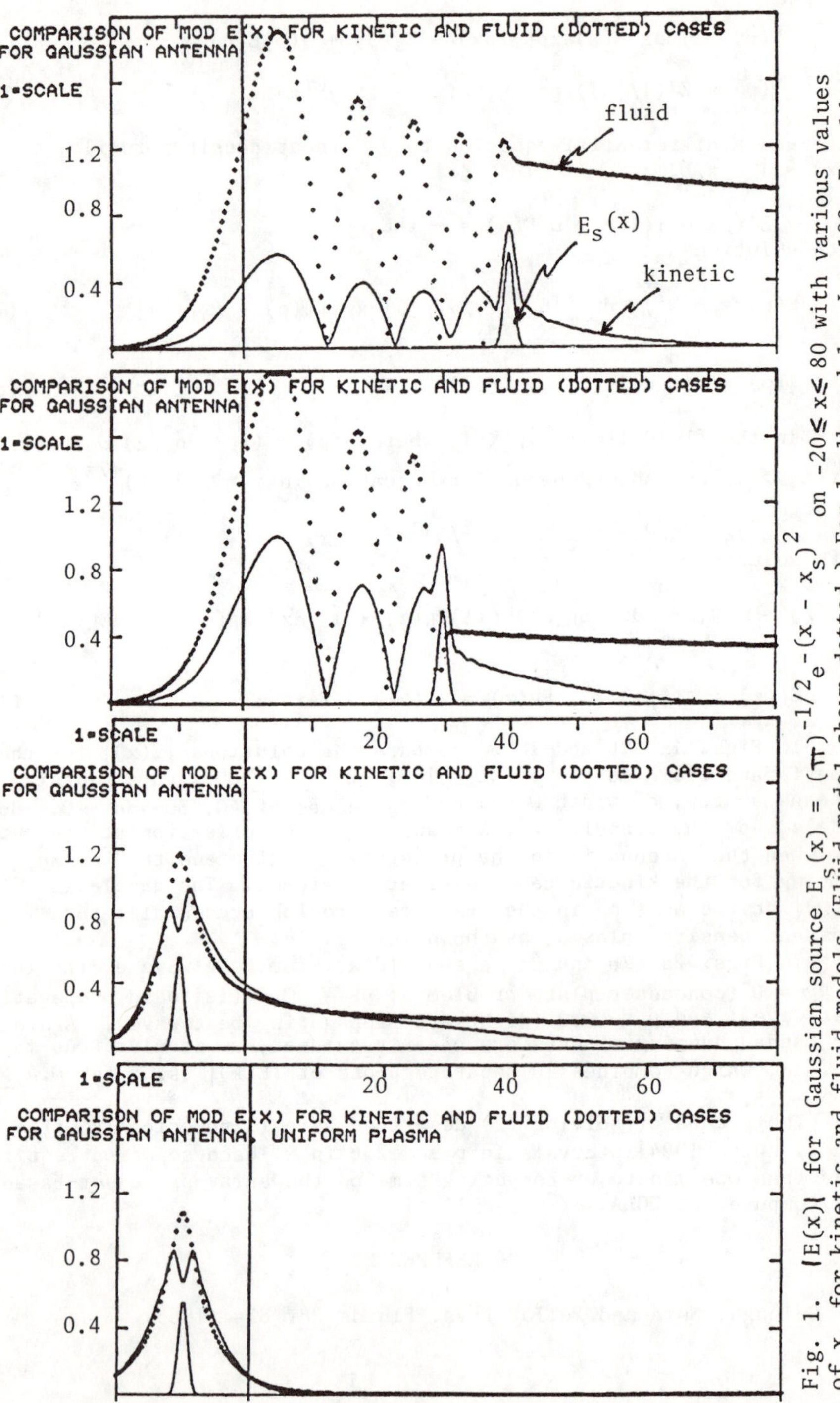

Fig. 1. $|E(x)|$ for Gaussian source $E_s(x) = (\pi)^{-1/2} e^{-(x - x_s)^2}$ on $-20 \leq x \leq 80$ with various values of x_s for kinetic and fluid models. (Fluid model shown dotted.) For 1a, 1b and 1c, L = 100. For 1d, L=0 (uniform plasma).

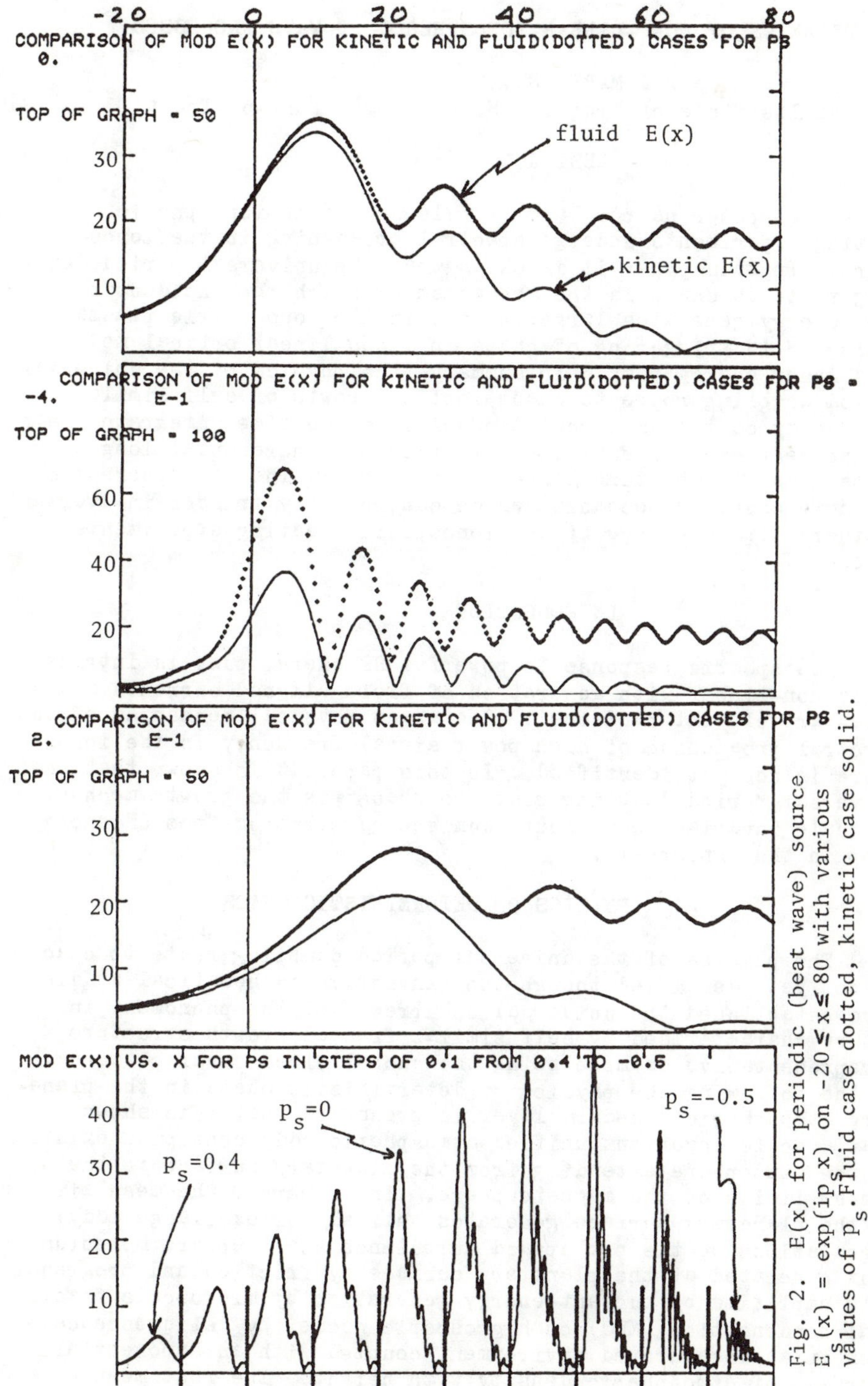

Fig. 2. E(x) for periodic (beat wave) source $E_s(x) = \exp(ip_s x)$ on $-20 \le x \le 80$ with various values of p_s. Fluid case dotted, kinetic case solid.

THE DYNAMICS OF NON -LINEAR IONOSPHERIC rf WAVE PROCESSES

A. MARY SELVAM
Indian Institute of Tropical Meteorology, Pune-5, India

ABSTRACT

High power narrow band rf pulses used in ionospheric heating experiments undergo spectral broadening in the ionosphere. In this paper it is shown that the universal period doubling route to chaos is the mechanism by which the incident rf pump energy generates larger eddies in the ionospheric plasma analogous to triggering of chaos in a non-linear optical medium by a laser energy pump. In summary, the physics of the universal period doubling route to chaos implies growth of self similar large eddy continuum circulations from space time integrated mean of inherent smaller scale perturbations at incremental length steps equal to the turbulence scale length. The ionosphere-troposphere coupling mechanism which can possibly trigger inadvertant weather/climate change by the ionospheric heating experiments is discussed.

INTRODUCTION

Ionospheric response to powerful HF energy pump is intrinsically non-linear with generation of field aligned currents and wave-particle interactions[1]. The exact physical mechanism of the spectral broadening of high power signal frequency in the ionosphere is not yet identified. In this paper it is shown that the universal period doubling route to chaos[2] is the growth mechanism of scale invariant eddy continuum energy spectrum from HF energy pump in the ionosphere.

PHYSICS OF DETERMINISTIC CHAOS

The physics of the universal period doubling route to chaos is not yet identified though the exhaustive mathematical studies have established the universal occurrence of the phenomena in nature characterised by self similar fractal growth structure, a representative example being the global cloud cover pattern[3]. In the following the physics of deterministic chaos in the planetary atmospheric boundary layer is presented and it is shown that a scale invariant unified atmospheric eddy continuum exists in the atmosphere extending from the planetary surface to the outermost limits of the magnetosphere. In summary[4], the mean airflow at the planetary surface generates helical vortex (large eddy) circulations by the net upward turbulence scale upward momentum flux generated at the planetary surface by friction and progressively amplified by buoyant energy generation by microscale fractional condensation (MFC) on hygroscopic nuclei by deliquescence even in an unsaturated environment coupled with an exponential decrease of atmospheric density with height. The root mean square (r.m.s.) circulation speed W of the large eddy of radius R is equal to the integrated mean w of the circulation speeds of turbulent eddies of representative radius r and is given by

$$W^2 = \frac{2}{\pi}\frac{r}{R} w^2 \qquad (1)$$

The turbulent eddies are carried upwards on the envelopes of the large eddies and assist in their (large eddies) further growth and also vertical mass exchange (mixing) of large eddy volume occurs by turbulent eddy fluctuations. The environment of turbulent eddies on the large eddy envelope is a region of buoyant energy generation by MFC and is identified by a microscale capping inversion (MCI) layer seen as the rising inversion of the daytime ABL in echosonde records and is associated with wind shear. The non dimensional steady state fractional volume dilution rate k of large eddy by turbulent eddy fluctuations is

$$k = \frac{w_*}{dW}.\frac{r}{R} \qquad (2)$$

where w_* is the turbulence scale vertical acceleration and dW the corresponding increase in large eddy circulation speed. The large eddy circulation speed therefore follows logarithmic law with respect to Z since

$$W = \frac{w_*}{k} \ln Z$$

as derived from (2). k is equal to 0.4 for Z=10 and is identified with the Von Kármán's constant. Therefore large eddy growth occurs from turbulence scale with inherent two way energy feedback (Eqn.1) at length step increments equal to the turbulence scale length and such a process may be identified with the universal period doubling route to chaos in the ABL, resulting in the formation of a scale invariant, self similar atmospheric eddy continuum energy structure with dominant eddies at decadic scale range intervals since the fractional volume dilution $k > 0.5$ for $Z < 10$, thereby erasing the signature of the large eddy for $Z < 10$. Global cloud cover pattern has fractal geometry because of the inherent self similarity of the three dimensional atmospheric eddy continuum circulations which form nested logarithmic spiral vortices spanning the complete spectrum of eddies from the microscopic to planetary scale eddies. Such a concept leads to the following conclusions as a natural consequence (1) the observed normal distribution characteristics of perturbation fields in the ABL are consistent (2) the eddy energy spectrum is the same as the cumulative normal probability density distribution and (3) the kinetic energy per unit mass of an eddy of frequency ν is equal to $H\nu$ where H is the spin angular momentum of the largest eddy in the continuum. Therefore the atmospheric eddy continuum energy structure follows quantum mechanical laws analogous to subatomic dynamics. The universal relation $2a^2=3d$ for the Feigenbaum's constants a and d is a statement of the law of conservation of energy since the threefold increase in spin angular momentum (3d) of large eddy generated by perturbation on either side of the primary turbulent eddy propogates outward as kinetic energy of large eddy ($2a^2$) into the environment.

The atmospheric eddy circulations result in vertical mass exchange in the ABL extending from the surface to the outermost limits of the magnetosphere and the resulting vertical space charge (aerosol borne) convection current is of the right sign and order of magnitude to account for the variation of the geomagnetic H component and also the fair weather atmospheric electric field. Any perturbation in the ionosphere would be transmitted to troposphere and vice versa. Numerous studies indicate significant correlation between geomagnetic field variations (ionospheric perturbation)and tropospheric weather activity. Observational evidence for the tropospheric eddy continuum extension into the ionosphere is seen in satellite observations which indicate that increased currents at ionospheric levels are accompanied by a simultaneous increase in wind speed at lower levels. Measurements with Poker Flat radar and instruments at Alaska and with NOAA radar at Fairbanks support this contention. The fine structure signal of the unified atmospheric eddy continuum is manifested as the luminescence phenomena of the auroral oval, where field aligned currents in discrete auroral arcs indicate bidirectional eddy energy/charge flow. Field aligned currents in discrete auroral arcs are visible manifestations of atmospheric vertical mixing with transport of negative charges downwards accompanied by simultaneous upward transport of terrestrial positive ions and may account for the observed stratospheric vertical electric current structures[5]. Therefore the aerosol currents originating from the planetary surface extend into the magnetosphere in the discrete auroral arcs and couple to solar wind energy pump accounting for the observed close association between the solar wind dynamic parameters, discrete aurorae and geomagnetic micropulsations. The signature of the clockwise ascent along logarithmic spiral curves of planetary scale eddy circulations in the northern hemisphere with maximum extent at noon time and anticlockwise descent on the local dawn and dusk side sectors on either side are seen in satellite observations[6] which show that at geomagnetic latitudes 70° the P_c 4-5 wave polarisation is found to be left handed in the local morning and right handed in the afternoon. Longitudinal (East-West) phase studies indicate that the waves propogate away from the noon meridian towards the dawn and dusk terminators.

CONCLUSION

Precipitation of energetic electrons from the radiation belts by the controlled injection from the ground of VLF radio waves[7] may possibly trigger cloud lightning discharges followed by massive electron precipitation[7]. Also ionospheric heating experiments and the numerous earth orbiting satellites may possibly create fine scale magnetospheric/ionospheric perturbations and lead to inadvertent modification of climate for e.g.

the African drought, anomalous El-Nino and abnormal hurricane activity.

REFERENCES

1. Spl. Issue. Active experiments in space plasmas. J. Atmos. Terr. Phys. 47, 1149 (1985)

2. R. G. Harrison and D. J. Biswas, Nature, 321, 394 (1986)

3. S.Lovejoy and D. Schertzer, Bull. Amer. Meteor. Soc., 67, 21 (1986).

4. A. Mary Selvam, Proc.IGARSS'87, University of Michigan, May (1987).

5. N.D'Angelo, I. B. Iverson and M. M. Madsen, J. Geophys. Res., 98 (D6), 9659 (1984).

6. I.A. Ansari and B. J. Fraser, Planet, Space Sci., 34, 519 (1986).

A DENSE, LOW-TEMPERATURE, PLASMA TARGET, DRIVEN BY AN IMMERSED, INDUCTIVE ANTENNA*

J. R. Trow and K. G. Moses
JAYCOR, Plasma Technology Division, Torrance, CA 90503

ABSTRACT

A rectangular plasma target (35x15x40 cm) with electron densities near 2×10^{13} cm^{-3} for hydrogen, near 6×10^{13} cm^{-3} for argon, and near 10^{14} cm^{-3} for xenon, has been produced by an rf inductive discharge. This plasma target is being developed as a neutralizer for a negative-hydrogen-ion-based neutral-beam heating system for magnetically confined fusion plasma. The results of operation with a 3-turn antenna at 1.4 MHz and up to 60 kW pulsed rf power are reported here.

INTRODUCTION

A high-density, low-temperature plasma, suitable for use as a neutralizer for negative ion beams, can be readily produced by a non-resonant rf field from a loosely-wound inductive antenna immersed in the discharge. The antenna is insulated to prevent high voltage arcing and to isolate it from the plasma. Energy from the rf field is initially transferred to electrons, which (after gaining sufficient energy) collisionally ionize the fill gas. The efficiency of the source can be greatly improved by retarding the loss of these hot electrons. Multicusp fields generated by arrays of permanent magnets are a proven method of reducing the loss rate of energetic electrons. The multicusp fields are intense only near the wall, so that the bulk of the plasma volume is virtually field free. Besides improving efficiency, this magnetic field configuration localizes the plasma gradients to the regions near the wall and produces a large volume of nearly uniform plasma in the center. The magnetic field in contact with the plasma can be strengthened and external stray fields eliminated by providing a low reluctance return path for the magnetic flux. This is accomplished by enclosing the plasma target in a mild steel box and mounting the magnets on the inside surfaces. The steel walls serve as a flux return path for the permanent magnets.

Neutral beam systems for heating magnetically confined fusion plasmas require beam energies of several hundred keV to penetrate to the center. A suitable plasma neutralizer[1] must have a total integrated line density (electrons+ions+neutrals) in the 10^{15} cm^{-2} range, a useful central volume of several cm^2 cross section to accommodate multiampere beams, and a high degree of transverse uniformity to assure optimum neutralization over the entire beam.

*Work performed under DOE Contract DE-AC03-84ER80153

A degree of ionization of 20-30% is needed to attain the higher neutral fraction yields afforded by a plasma target. Finally, the process of producing the plasma in the neutralizer must be electrically efficient to minimize the recirculating power requirements of the neutral beam system. An rf-generated, multicusp-confined plasma target of the type described in this paper in a dimensionally appropriate neutralizer cell would meet these requirements.

DESCRIPTION OF THE EXPERIMENT

The dimensions of the neutralizer test cell[2] are 35x15x40 cm. The transverse dimensions were chosen to accommodate a 25x3 cm slab, negative hydrogen ion beam. The multicusp field is produced by twelve rectangular rings of Nd-Fe-Bo permanent magnets. The magnets are separated from the discharge by a thin copper liner and a small vacuum gap which serves as a thermal barrier. This arrangement yields a field exceeding 4 kG at the inner edge of the liner, with the field intensity dropping to less than 20 gauss, 5 cm from the wall. The magnet array is completely enclosed by a steel box except at the entrance and exit apertures. However, these openings are adjacent to regions where the magnetic field is less than 20 gauss, and no detectable magnetic field can be found outside.

Rf power is supplied by a two-stage pulsed amplifier system with a 50 Ω output impedance. The inductive antenna is part of a parallel LC circuit, which is coupled to the rf amplifier via an autotransformer. The capacitor and the transformer are adjusted until the amplifier's output voltage and current are in phase and their ratio indicates a 50 Ω load. No attempt has been made to differentiate between power dissipated in the antenna circuit and power absorbed in the discharge.

The diagnostics for this work consisted of a fast ion gauge to determine the fill gas pressure and several Langmuir probes. The ion gauge is not operated when rf is present, but is used to calibrate a pulsed gas valve. The Langmuir probe is cylindrical, with a diameter of 0.51 mm and a length of 2.5 mm. The probe voltage is swept from -50 to +50 volts in approximately 50 μsec, timed to occur at peak density in the discharge. This voltage is also delivered (through a 10:1 resistive divider) to a transient recorder. The probe current signal (the voltage across a 1Ω series resistor) is impressed on the terminals of another transient recorder, and the ion current signal is measured by a third recorder via a gain of 10 clipping amplifier. All three recorder units are Biomation 610C's (6-bit 256 point) modified to utilize a common clock. The accumulated data is delivered via a CAMAC based system to a microcomputer. The I-V characteristic of the probe is reconstructed by plotting the current signals as a function of the voltage signal recorded at the same points in time.

RESULTS

Probe data was collected along the chamber axis at 5 cm intervals from the midplane to one edge. Data was obtained for hydrogen and argon at 2, 5, and 10 mTorr initial fill pressures. The results (Figure 1a) show that the hydrogen plasma was peaked in the center, whereas the argon plasma (Figure 1b) seems to have a fairly flat profile at the highest power used. The profiles for the hydrogen data are similar to those reported previously for a single-turn antenna,[2] but the 3-turn antenna yielded approximately 50% higher densities at 60 kW and 70% higher at 40 kW. No argon data was taken with the single-turn antenna. The plasma density at the midplane shows that the hydrogen (Figure 2a) plasma density varies linearly with power, with no apparent saturation; conversely, the argon data (Figure 2a) exhibits saturation. Limited data taken with xenon as the fill gas indicates it is similar in behavior to argon, but higher plasma densities were measured (peak densities at 10 mTorr were $\gtrsim 9\times10^{13}\ cm^{-3}$). In all cases 4-7 eV electron temperatures are measured.

The plasma densities obtained in the Ar and Xe studies are sufficient for a practical negative ion beam neutralizer; however, the degree of ionization in the H_2 discharge needs to be increased. The existing 40 cm test cell is sufficient as a plasma neutralizer when operated with argon with a fill pressure of 2 mTorr. The target thickness produced (not including the contribution from the plasma beyond the entrance and exit slots) is approximately $4\times10^{15}\ cm^{-2}$ and the degree of ionization over 35%. Since a plasma neutralizer yields approximately 20% more beam neutralized than would unionized gas (85% vs 65%), this cell, if used on a 1 megawatt negative ion beam, would deliver an additional 200 kW of neutral beam power far exceeding the 40 to 60 kW invested in powering the rf discharge.

REFERENCES

1. K. H. Berkner, R. V. Pyle, S. E. Savas, and K. R. Stalder, "Plasma Neutralizers for H^- and D^- Beams," Proc. 2nd Int. Symp. Prod. and Neut. Neg. Ions and Beams, Brookhaven National Lab., Oct 1980.
2. J. R. Trow and K. G. Moses, "Characteristics of an RF Plasma Neutralizer," Proc. 4th Int. Symp. Prod. and Neut. Neg. Ions and Beams, Brookhaven National Lab., Oct 1986.

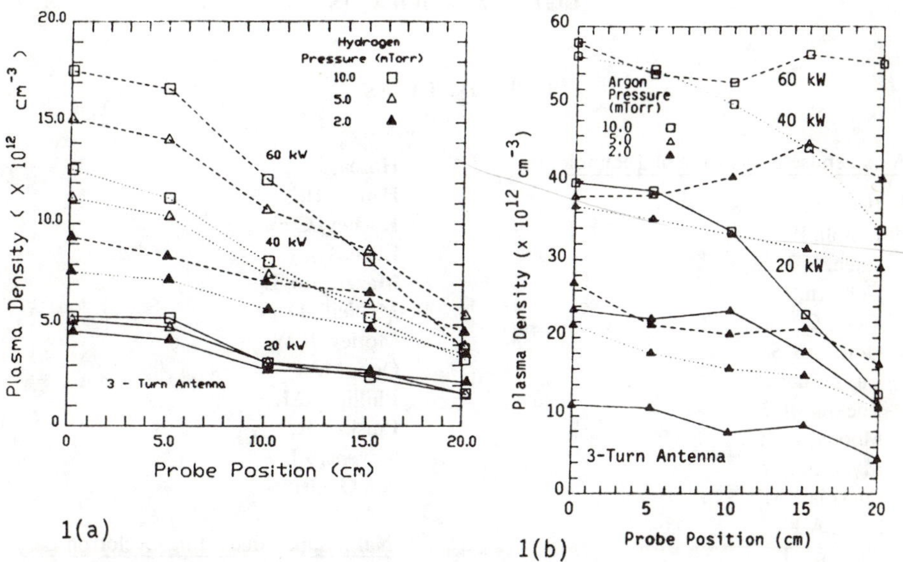

Fig. 1. Plasma density along the chamber axis as indicated by a Langmuir probe. The probe was moved from the midplane (Ø) to the edge (20) on successive shots. 1(a) hydrogen. 1(b) argon.

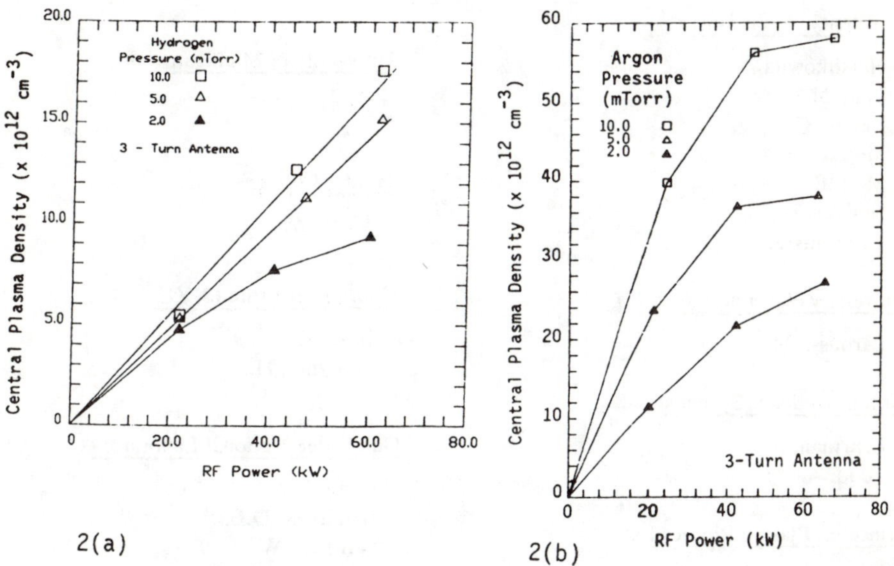

Fig. 2. Plasma density in the midplane (probe position Ø) as a function of applied rf power. 2(a) hydrogen. 2(b) argon.

List of Participants

UNITED STATES

Massachusetts Institute of Technology

Bers, A.
Bonoli, P.
Chen, K.I.
Colborn, J.
Fiore, C.
Golovato, S.
Lu, Zhihong
Luckhardt, S.
Moody, J.
Myer, R.
Porkolab, M.
Ram, A.K.
Shepard, T.
Takase, Y.

University of Southern Florida

Buckle, K.

University of Wisconsin

Hershkowitz, N.
Lam, N.T.
Litwin, C.
Majeski, R.
Mett, R.
Scharer, J.
Tataronis, J.

University of Tennessee

Laroussi, M.

Grumman Space Systems

Lehrman, I.
Todd, A.

Princeton Plasma Physics Lab.

Bell, R.
Bernabei, S.
Cavallo, A.
Chu, T.K.
Colestock, P.L.
Greene, G.J.
Hammett, G.
Hosea, J.
Hsuan, H.
Karney, C.F.F.
Luce, T.
Mazzucato, E.
McNeill, D.
Motley, R.W.
Ono, M.
Phillips, C.K.
Pinsker, R.
Stevens, J.E.
von Goeler, S.

Nat'l. Superconducting Cyclotron Lab.

Antaya, T.

North Carolina State University

Hankins, O.

University of Michigan

Smithe, D.

Hunter College

Kritz, A.

Courant Institute (NYU)

Imre, K.
Weitzner, H.

Oak Ridge National Laboratory

Baity, F.W.
Batchelor, D.B.
Gardner, W.
Goldfinger, R.C.
Hively, L.
Hoffman, D.
Jaeger, E.F.
Owens, T.L.
Ryan, P.
Swain, D.

GA Technologies

Chiu, S.C.
Harvey, R.
Hsu, J.Y.
Kwon, M.
Matsuda, K.
Mayberry, M.J.
Prater, R.

Science Application Intern'l. Corp.

D'Ippolito, D.
Myra, J.

Caltech

Bellan, P.

Georgia Institute of Technology

Thomas, C.E.

Lawrence Berkeley

Friedland, L.
Kaufman, A.

Auburn University

Cho, S.
Hartwell, G.
Swanson, D.G.
Wersinger, J.M.

Karlsruhe Nuclear Research Center

Schmidt, W.

IBM

Oehrlein, G.S.

University of Texas

Booth, W.
Oakes, M.E.
Richards, B.

University of Iowa

Goree, J.

UC–Irvine

McWilliams, R.
Wolf, N. (Dickinson)

Yale

Bernstein, I.

JAYCOR

Moses, K.

Argonne National Laboratory

Ehst, D.

Lawrence Livermore National Lab.

Dimonte, G. (TRW)
Kerbel, G.D.
McCoy, M.
Nevins, W.
Simonen, T.C.
Smith, G.
Stallard, B.W.

Department of Energy

Dagazian, R.
Sadowski, W.
Staten, S.

University of California, Los Angeles

Fried, B.
Lai, K.F.
Mau, T.K.
Morales, G.J.

EUROPEAN COMMUNITY

Frascati

DeMarco, F.
Tuccillo, A.A.

Culham Laboratory

Riviere, A.

CNR-EURATOM/Milano

Argenti, L.
Cima, G.

University of St. Andrews

Cairns, R.

Cadarache

Fidone, I.
Moreau, D.

Nancy University

Meyer, R.L.

JET

Bosia, G.
Cottrell, G.A.
Jacquinot, J.
Start, D.F.H.

Max-Planck Institute

Brambilla, M.
Puri, S.
Solder, F.X.
Steinmetz, K.

FOM

Verhoeven, A.

JAPAN

Hitachi Ltd.

Kinoshita, S.
Yoshioka, K.

JAERI

Imai, T.
Nagashima, T.
Uesugi, Y.

University of Tokyo

Toyama, H.

Kyoto University

Mutoh, T.

CANADA

MPB Technologies

Shkarofsky, I.P.

Institut de Recherche d'Hydro-Quebec

Fuchs, V.

Author Index

M

N

O

P

W

Y

Z

AIP Conference Proceedings

		L.C. Number	ISBN
No. 1	Feedback and Dynamic Control of Plasmas – 1970	70-141596	0-88318-100-2
No. 2	Particles and Fields – 1971 (Rochester)	71-184662	0-88318-101-0
No. 3	Thermal Expansion – 1971 (Corning)	72-76970	0-88318-102-9
No. 4	Superconductivity in *d*- and *f*-Band Metals (Rochester, 1971)	74-18879	0-88318-103-7
No. 5	Magnetism and Magnetic Materials – 1971 (2 parts) (Chicago)	59-2468	0-88318-104-5
No. 6	Particle Physics (Irvine, 1971)	72-81239	0-88318-105-3
No. 7	Exploring the History of Nuclear Physics – 1972	72-81883	0-88318-106-1
No. 8	Experimental Meson Spectroscopy –1972	72-88226	0-88318-107-X
No. 9	Cyclotrons – 1972 (Vancouver)	72-92798	0-88318-108-8
No. 10	Magnetism and Magnetic Materials – 1972	72-623469	0-88318-109-6
No. 11	Transport Phenomena – 1973 (Brown University Conference)	73-80682	0-88318-110-X
No. 12	Experiments on High Energy Particle Collisions – 1973 (Vanderbilt Conference)	73-81705	0-88318-111–8
No. 13	π-π Scattering – 1973 (Tallahassee Conference)	73-81704	0-88318-112-6
No. 14	Particles and Fields – 1973 (APS/DPF Berkeley)	73-91923	0-88318-113-4
No. 15	High Energy Collisions – 1973 (Stony Brook)	73-92324	0-88318-114-2
No. 16	Causality and Physical Theories (Wayne State University, 1973)	73-93420	0-88318-115-0
No. 17	Thermal Expansion – 1973 (Lake of the Ozarks)	73-94415	0-88318-116-9
No. 18	Magnetism and Magnetic Materials – 1973 (2 parts) (Boston)	59-2468	0-88318-117-7
No. 19	Physics and the Energy Problem – 1974 (APS Chicago)	73-94416	0-88318-118-5
No. 20	Tetrahedrally Bonded Amorphous Semiconductors (Yorktown Heights, 1974)	74-80145	0-88318-119-3
No. 21	Experimental Meson Spectroscopy – 1974 (Boston)	74-82628	0-88318-120-7
No. 22	Neutrinos – 1974 (Philadelphia)	74-82413	0-88318-121-5
No. 23	Particles and Fields – 1974 (APS/DPF Williamsburg)	74-27575	0-88318-122-3
No. 24	Magnetism and Magnetic Materials – 1974 (20th Annual Conference, San Francisco)	75-2647	0-88318-123-1
No. 25	Efficient Use of Energy (The APS Studies on the Technical Aspects of the More Efficient Use of Energy)	75-18227	0-88318-124-X

No. 26	High-Energy Physics and Nuclear Structure – 1975 (Santa Fe and Los Alamos)	75-26411	0-88318-125-8
No. 27	Topics in Statistical Mechanics and Biophysics: A Memorial to Julius L. Jackson (Wayne State University, 1975)	75-36309	0-88318-126-6
No. 28	Physics and Our World: A Symposium in Honor of Victor F. Weisskopf (M.I.T., 1974)	76-7207	0-88318-127-4
No. 29	Magnetism and Magnetic Materials – 1975 (21st Annual Conference, Philadelphia)	76-10931	0-88318-128-2
No. 30	Particle Searches and Discoveries – 1976 (Vanderbilt Conference)	76-19949	0-88318-129-0
No. 31	Structure and Excitations of Amorphous Solids (Williamsburg, VA, 1976)	76-22279	0-88318-130-4
No. 32	Materials Technology – 1976 (APS New York Meeting)	76-27967	0-88318-131-2
No. 33	Meson-Nuclear Physics – 1976 (Carnegie-Mellon Conference)	76-26811	0-88318-132-0
No. 34	Magnetism and Magnetic Materials – 1976 (Joint MMM-Intermag Conference, Pittsburgh)	76-47106	0-88318-133-9
No. 35	High Energy Physics with Polarized Beams and Targets (Argonne, 1976)	76-50181	0-88318-134-7
No. 36	Momentum Wave Functions – 1976 (Indiana University)	77-82145	0-88318-135-5
No. 37	Weak Interaction Physics – 1977 (Indiana University)	77-83344	0-88318-136-3
No. 38	Workshop on New Directions in Mossbauer Spectroscopy (Argonne, 1977)	77-90635	0-88318-137-1
No. 39	Physics Careers, Employment and Education (Penn State, 1977)	77-94053	0-88318-138-X
No. 40	Electrical Transport and Optical Properties of Inhomogeneous Media (Ohio State University, 1977)	78-54319	0-88318-139-8
No. 41	Nucleon-Nucleon Interactions – 1977 (Vancouver)	78-54249	0-88318-140-1
No. 42	Higher Energy Polarized Proton Beams (Ann Arbor, 1977)	78-55682	0-88318-141-X
No. 43	Particles and Fields – 1977 (APS/DPF, Argonne)	78-55683	0-88318-142-8
No. 44	Future Trends in Superconductive Electronics (Charlottesville, 1978)	77-9240	0-88318-143-6
No. 45	New Results in High Energy Physics – 1978 (Vanderbilt Conference)	78-67196	0-88318-144-4
No. 46	Topics in Nonlinear Dynamics (La Jolla Institute)	78-57870	0-88318-145-2
No. 47	Clustering Aspects of Nuclear Structure and Nuclear Reactions (Winnepeg, 1978)	78-64942	0-88318-146-0
No. 48	Current Trends in the Theory of Fields (Tallahassee, 1978)	78-72948	0-88318-147-9

No. 49	Cosmic Rays and Particle Physics – 1978 (Bartol Conference)	79-50489	0-88318-148-7
No. 50	Laser-Solid Interactions and Laser Processing – 1978 (Boston)	79-51564	0-88318-149-5
No. 51	High Energy Physics with Polarized Beams and Polarized Targets (Argonne, 1978)	79-64565	0-88318-150-9
No. 52	Long-Distance Neutrino Detection – 1978 (C.L. Cowan Memorial Symposium)	79-52078	0-88318-151-7
No. 53	Modulated Structures – 1979 (Kailua Kona, Hawaii)	79-53846	0-88318-152-5
No. 54	Meson-Nuclear Physics – 1979 (Houston)	79-53978	0-88318-153-3
No. 55	Quantum Chromodynamics (La Jolla, 1978)	79-54969	0-88318-154-1
No. 56	Particle Acceleration Mechanisms in Astrophysics (La Jolla, 1979)	79-55844	0-88318-155-X
No. 57	Nonlinear Dynamics and the Beam-Beam Interaction (Brookhaven, 1979)	79-57341	0-88318-156-8
No. 58	Inhomogeneous Superconductors – 1979 (Berkeley Springs, W.V.)	79-57620	0-88318-157-6
No. 59	Particles and Fields – 1979 (APS/DPF Montreal)	80-66631	0-88318-158-4
No. 60	History of the ZGS (Argonne, 1979)	80-67694	0-88318-159-2
No. 61	Aspects of the Kinetics and Dynamics of Surface Reactions (La Jolla Institute, 1979)	80-68004	0-88318-160-6
No. 62	High Energy e^+e^- Interactions (Vanderbilt, 1980)	80-53377	0-88318-161-4
No. 63	Supernovae Spectra (La Jolla, 1980)	80-70019	0-88318-162-2
No. 64	Laboratory EXAFS Facilities – 1980 (Univ. of Washington)	80-70579	0-88318-163-0
No. 65	Optics in Four Dimensions – 1980 (ICO, Ensenada)	80-70771	0-88318-164-9
No. 66	Physics in the Automotive Industry – 1980 (APS/AAPT Topical Conference)	80-70987	0-88318-165-7
No. 67	Experimental Meson Spectroscopy – 1980 (Sixth International Conference, Brookhaven)	80-71123	0-88318-166-5
No. 68	High Energy Physics – 1980 (XX International Conference, Madison)	81-65032	0-88318-167-3
No. 69	Polarization Phenomena in Nuclear Physics – 1980 (Fifth International Symposium, Santa Fe)	81-65107	0-88318-168-1
No. 70	Chemistry and Physics of Coal Utilization – 1980 (APS, Morgantown)	81-65106	0-88318-169-X
No. 71	Group Theory and its Applications in Physics – 1980 (Latin American School of Physics, Mexico City)	81-66132	0-88318-170-3
No. 72	Weak Interactions as a Probe of Unification (Virginia Polytechnic Institute – 1980)	81-67184	0-88318-171-1
No. 73	Tetrahedrally Bonded Amorphous Semiconductors (Carefree, Arizona, 1981)	81-67419	0-88318-172-X

No. 74	Perturbative Quantum Chromodynamics (Tallahassee, 1981)	81-70372	0-88318-173-8
No. 75	Low Energy X-Ray Diagnostics – 1981 (Monterey)	81-69841	0-88318-174-6
No. 76	Nonlinear Properties of Internal Waves (La Jolla Institute, 1981)	81-71062	0-88318-175-4
No. 77	Gamma Ray Transients and Related Astrophysical Phenomena (La Jolla Institute, 1981)	81-71543	0-88318-176-2
No. 78	Shock Waves in Condensed Matter – 1981 (Menlo Park)	82-70014	0-88318-177-0
No. 79	Pion Production and Absorption in Nuclei – 1981 (Indiana University Cyclotron Facility)	82-70678	0-88318-178-9
No. 80	Polarized Proton Ion Sources (Ann Arbor, 1981)	82-71025	0-88318-179-7
No. 81	Particles and Fields –1981: Testing the Standard Model (APS/DPF, Santa Cruz)	82-71156	0-88318-180-0
No. 82	Interpretation of Climate and Photochemical Models, Ozone and Temperature Measurements (La Jolla Institute, 1981)	82-71345	0-88318-181-9
No. 83	The Galactic Center (Cal. Inst. of Tech., 1982)	82-71635	0-88318-182-7
No. 84	Physics in the Steel Industry (APS/AISI, Lehigh University, 1981)	82-72033	0-88318-183-5
No. 85	Proton-Antiproton Collider Physics –1981 (Madison, Wisconsin)	82-72141	0-88318-184-3
No. 86	Momentum Wave Functions – 1982 (Adelaide, Australia)	82-72375	0-88318-185-1
No. 87	Physics of High Energy Particle Accelerators (Fermilab Summer School, 1981)	82-72421	0-88318-186-X
No. 88	Mathematical Methods in Hydrodynamics and Integrability in Dynamical Systems (La Jolla Institute, 1981)	82-72462	0-88318-187-8
No. 89	Neutron Scattering – 1981 (Argonne National Laboratory)	82-73094	0-88318-188-6
No. 90	Laser Techniques for Extreme Ultraviolt Spectroscopy (Boulder, 1982)	82-73205	0-88318-189-4
No. 91	Laser Acceleration of Particles (Los Alamos, 1982)	82-73361	0-88318-190-8
No. 92	The State of Particle Accelerators and High Energy Physics (Fermilab, 1981)	82-73861	0-88318-191-6
No. 93	Novel Results in Particle Physics (Vanderbilt, 1982)	82-73954	0-88318-192-4
No. 94	X-Ray and Atomic Inner-Shell Physics – 1982 (International Conference, U. of Oregon)	82-74075	0-88318-193-2
No. 95	High Energy Spin Physics – 1982 (Brookhaven National Laboratory)	83-70154	0-88318-194-0
No. 96	Science Underground (Los Alamos, 1982)	83-70377	0-88318-195-9

No. 97	The Interaction Between Medium Energy Nucleons in Nuclei – 1982 (Indiana University)	83-70649	0-88318-196-7
No. 98	Particles and Fields – 1982 (APS/DPF University of Maryland)	83-70807	0-88318-197-5
No. 99	Neutrino Mass and Gauge Structure of Weak Interactions (Telemark, 1982)	83-71072	0-88318-198-3
No. 100	Excimer Lasers – 1983 (OSA, Lake Tahoe, Nevada)	83-71437	0-88318-199-1
No. 101	Positron-Electron Pairs in Astrophysics (Goddard Space Flight Center, 1983)	83-71926	0-88318-200-9
No. 102	Intense Medium Energy Sources of Strangeness (UC-Sant Cruz, 1983)	83-72261	0-88318-201-7
No. 103	Quantum Fluids and Solids – 1983 (Sanibel Island, Florida)	83-72440	0-88318-202-5
No. 104	Physics, Technology and the Nuclear Arms Race (APS Baltimore –1983)	83-72533	0-88318-203-3
No. 105	Physics of High Energy Particle Accelerators (SLAC Summer School, 1982)	83-72986	0-88318-304-8
No. 106	Predictability of Fluid Motions (La Jolla Institute, 1983)	83-73641	0-88318-305-6
No. 107	Physics and Chemistry of Porous Media (Schlumberger-Doll Research, 1983)	83-73640	0-88318-306-4
No. 108	The Time Projection Chamber (TRIUMF, Vancouver, 1983)	83-83445	0-88318-307-2
No. 109	Random Walks and Their Applications in the Physical and Biological Sciences (NBS/La Jolla Institute, 1982)	84-70208	0-88318-308-0
No. 110	Hadron Substructure in Nuclear Physics (Indiana University, 1983)	84-70165	0-88318-309-9
No. 111	Production and Neutralization of Negative Ions and Beams (3rd Int'l Symposium, Brookhaven, 1983)	84-70379	0-88318-310-2
No. 112	Particles and Fields – 1983 (APS/DPF, Blacksburg, VA)	84-70378	0-88318-311-0
No. 113	Experimental Meson Spectroscopy – 1983 (Seventh International Conference, Brookhaven)	84-70910	0-88318-312-9
No. 114	Low Energy Tests of Conservation Laws in Particle Physics (Blacksburg, VA, 1983)	84-71157	0-88318-313-7
No. 115	High Energy Transients in Astrophysics (Santa Cruz, CA, 1983)	84-71205	0-88318-314-5
No. 116	Problems in Unification and Supergravity (La Jolla Institute, 1983)	84-71246	0-88318-315-3
No. 117	Polarized Proton Ion Sources (TRIUMF, Vancouver, 1983)	84-71235	0-88318-316-1

No. 118	Free Electron Generation of Extreme Ultraviolet Coherent Radiation (Brookhaven/OSA, 1983)	84-71539	0-88318-317-X
No. 119	Laser Techniques in the Extreme Ultraviolet (OSA, Boulder, Colorado, 1984)	84-72128	0-88318-318-8
No. 120	Optical Effects in Amorphous Semiconductors (Snowbird, Utah, 1984)	84-72419	0-88318-319-6
No. 121	High Energy e^+e^- Interactions (Vanderbilt, 1984)	84-72632	0-88318-320-X
No. 122	The Physics of VLSI (Xerox, Palo Alto, 1984)	84-72729	0-88318-321-8
No. 123	Intersections Between Particle and Nuclear Physics (Steamboat Springs, 1984)	84-72790	0-88318-322-6
No. 124	Neutron-Nucleus Collisions – A Probe of Nuclear Structure (Burr Oak State Park - 1984)	84-73216	0-88318-323-4
No. 125	Capture Gamma-Ray Spectroscopy and Related Topics – 1984 (Internat. Symposium, Knoxville)	84-73303	0-88318-324-2
No. 126	Solar Neutrinos and Neutrino Astronomy (Homestake, 1984)	84-63143	0-88318-325-0
No. 127	Physics of High Energy Particle Accelerators (BNL/SUNY Summer School, 1983)	85-70057	0-88318-326-9
No. 128	Nuclear Physics with Stored, Cooled Beams (McCormick's Creek State Park, Indiana, 1984)	85-71167	0-88318-327-7
√ No. 129	Radiofrequency Plasma Heating (Sixth Topical Conference, Callaway Gardens, GA, 1985)	85-48027	0-88318-328-5
No. 130	Laser Acceleration of Particles (Malibu, California, 1985)	85-48028	0-88318-329-3
No. 131	Workshop on Polarized ^{3}He Beams and Targets (Princeton, New Jersey, 1984)	85-48026	0-88318-330-7
No. 132	Hadron Spectroscopy–1985 (International Conference, Univ. of Maryland)	85-72537	0-88318-331-5
No. 133	Hadronic Probes and Nuclear Interactions (Arizona State University, 1985)	85-72638	0-88318-332-3
No. 134	The State of High Energy Physics (BNL/SUNY Summer School, 1983)	85-73170	0-88318-333-1
No. 135	Energy Sources: Conservation and Renewables (APS, Washington, DC, 1985)	85-73019	0-88318-334-X
No. 136	Atomic Theory Workshop on Relativistic and QED Effects in Heavy Atoms	85-73790	0-88318-335-8
No. 137	Polymer-Flow Interaction (La Jolla Institute, 1985)	85-73915	0-88318-336-6
No. 138	Frontiers in Electronic Materials and Processing (Houston, TX, 1985)	86-70108	0-88318-337-4
No. 139	High-Current, High-Brightness, and High-Duty Factor Ion Injectors (La Jolla Institute, 1985)	86-70245	0-88318-338-2

No. 140	Boron-Rich Solids (Albuquerque, NM, 1985)	86-70246	0-88318-339-0
No. 141	Gamma-Ray Bursts (Stanford, CA, 1984)	86-70761	0-88318-340-4
No. 142	Nuclear Structure at High Spin, Excitation, and Momentum Transfer (Indiana University, 1985)	86-70837	0-88318-341-2
No. 143	Mexican School of Particles and Fields (Oaxtepec, México, 1984)	86-81187	0-88318-342-0
No. 144	Magnetospheric Phenomena in Astrophysics (Los Alamos, 1984)	86-71149	0-88318-343-9
No. 145	Polarized Beams at SSC & Polarized Antiprotons (Ann Arbor, MI & Bodega Bay, CA, 1985)	86-71343	0-88318-344-7
No. 146	Advances in Laser Science–I (Dallas, TX, 1985)	86-71536	0-88318-345-5
No. 147	Short Wavelength Coherent Radiation: Generation and Applications (Monterey, CA, 1986)	86-71674	0-88318-346-3
No. 148	Space Colonization: Technology and The Liberal Arts (Geneva, NY, 1985)	86-71675	0-88318-347-1
No. 149	Physics and Chemistry of Protective Coatings (Universal City, CA, 1985)	86-72019	0-88318-348-X
No. 150	Intersections Between Particle and Nuclear Physics (Lake Louise, Canada, 1986)	86-72018	0-88318-349-8
No. 151	Neural Networks for Computing (Snowbird, UT, 1986)	86-72481	0-88318-351-X
No. 152	Heavy Ion Inertial Fusion (Washington, DC, 1986)	86-73185	0-88318-352-8
No. 153	Physics of Particle Accelerators (SLAC Summer School, 1985) (Fermilab Summer School, 1984)	87-70103	0-88318-353-6
No. 154	Physics and Chemistry of Porous Media—II (Ridge Field, CT, 1986)	83-73640	0-88318-354-4
No. 155	The Galactic Center: Proceedings of the Symposium Honoring C. H. Townes (Berkeley, CA, 1986)	86-73186	0-88318-355-2
No. 156	Advanced Accelerator Concepts (Madison, WI, 1986)	87-70635	0-88318-358-0
No. 157	Stability of Amorphous Silicon Alloy Materials and Devices (Palo Alto, CA, 1987)	87-70990	0-88318-359-9
No. 158	Production and Neutralization of Negative Ions and Beams (Brookhaven, NY, 1986)	87-71695	0-88318-358-7